CELLULAR AGING AND CELL DEATH

MODERN CELL BIOLOGY

SERIES EDITOR

Joe B. Harford
RiboGene, Inc.
Hayward, California

ADVISORY BOARD

RECENT VOLUMES PUBLISHED IN THE SERIES

CELLULAR AGING AND CELL DEATH

Editors

Nikki J. Holbrook

National Institute on Aging
National Institutes of Health
Baltimore, Maryland

George R. Martin
Fibrogen
Sunnyvale, California

Richard A. Lockshin

Department of Biological Sciences
St. John's University
Jamaica, New York

A JOHN WILEY & SONS, INC., PUBLICATION
NEW YORK • CHICHESTER • BRISBANE • TORONTO • SINGAPORE

Address All Inquiries to the Publisher
Wiley-Liss, Inc., 605 Third Avenue, New York, NY 10158-0012

Copyright © 1996 Wiley-Liss, Inc.

Printed in the United States of America

Library of Congress Cataloging-in-Publication Data

Cellular aging and cell death / editors, Nikki J. Holbrook, George R.
    Martin, Richard A. Lockshin.
        p.   cm. — (Modern cell biology ; v. 16)
    Includes bibliographical references and index.
    ISBN 0–471–12123–1 (alk. paper)
    1. Cells—Aging. 2. Aging—Molecular aspects. 3. Apoptosis.
I. Holbrook, Nikki J., 1953–   . II. Martin, George R., 1930–   .
III. Lockshin, R.A. (Richard A.) IV. Series.
    [DNLM: 1. Cell Aging. 2. Cell Death. W1 MO124T v. 16 1995 /QH
608 C394 1995]
QH573.M63 vol. 16
[QH608]
574.87 s—dc20
[574.87´65]
DNLM/DLC
for Library of Congress   95–39953
                                                             CIP

The text of this book is printed on acid-free paper.
10   9   8   7   6   5   4   3   2   1

Contents

Contributors

William H. Adler, Clinical Immunology Section, National Institute on Aging, National Institutes of Health, Baltimore, MD 21224 [51]

Cynthia A. Afshari, National Institute of Environmental Health Sciences, National Institutes of Health, Research Triangle Park, NC 27709 [109]

J. Carl Barrett, Laboratory of Molecular Carcinogenesis, Environmental Carcinogenesis Program, National Institute of Environmental Health Sciences, National Institutes of Health, Research Triangle Park, NC 27709 [109]

Marvin O. Boluyt, Laboratory of Cardiovascular Science, Gerontology Research Center, National Institute on Aging, National Institutes of Health, Baltimore, MD 21224 [81]

Eduardo Bonilla, Department of Neurology, Columbia University, New York, NY 10032 [19]

E. Anne Buckmaster, Department of Biochemistry, University of Cambridge, Cambridge CB2 1QW, United Kingdom; present address: University Laboratory of Physiology, University of Oxford, Oxford, United Kingdom [267]

Xi Chen, Department of Neurology, Columbia University, New York, NY 10032; present address: Department of Neurology, New York University, New York, NY 10016 [19]

F. Joseph Chrest, Clinical Immunology Section, National Institute on Aging, National Institutes of Health, Baltimore, MD 21224 [51]

Michael T. Crow, Laboratory of Cardiovascular Science, Gerontology Research Center, National Institute on Aging, National Institutes of Health, Baltimore, MD 21224 [81]

Angela J. DiBenedetto, Department of Pharmacology, University of Pennsylvania School of Medicine, Philadelphia, PA 19104 [255]

Monica Driscoll, Department of Molecular Biology and Biochemistry, Center for Advanced Biotechnology and Medicine, Rutgers University, Piscataway, NJ 08855 [235]

Stacey A. Duhon, Department of Psychology, University of Colorado, Boulder, CO 80309 [1]

Atanu Duttaroy, Bloomfield Centre for Research in Aging, Lady Davis Institute for Medical Research, Sir Mortimer B. Davis–Jewish General Hospital, McGill University, Montreal, Quebec H3T 1E2, Canada [139]

Susan N. Edwards, Department of Biochemistry, University of Cambridge, Cambridge CB2 1QW, United Kingdom; present address: Centre for Biotechnology, University of Turku, Turku, Finland [267]

The numbers in brackets are the opening page numbers of the contributors' articles.

Robert A. Floyd, Free Radical Biology and Aging Research Program, Oklahoma Medical Research Foundation, Oklahoma City, OK 73104 [35]

Carol W. Greider, Cold Spring Harbor Laboratory, Cold Spring Harbor, NY 11724 [123]

Calvin B. Harley, Geron Corporation, Menlo Park, CA 94025 [123]

Nikki J. Holbrook, Section on Gene Expression and Aging, National Institute on Aging, National Institutes of Health, Baltimore, MD 21224 [67]

Thomas E. Johnson, Institute of Behavioral Genetics, University of Colorado, Boulder, Colorado 80309 [1]

Edward G. Lakatta, Laboratory of Cardiovascular Science, Gerontology Research Center, National Institute on Aging, National Institutes of Health, Baltimore, MD 21224 [81]

Priti Lal, Clinical Immunology Section, National Institute on Aging, National Institutes of Health, Baltimore, MD 21224 [51]

Stephen Lee, Bloomfield Centre for Research in Aging, Lady Davis Institute for Medical Research, Sir Mortimer B. Davis–Jewish General Hospital, McGill University, Montreal, Quebec H3T 1E2, Canada; present address: Cell Biology and Metabolism Branch, National Institute of Child Health and Human Development, National Institutes of Health, Bethesda, MD [139]

Robyn Lints, Department of Molecular Biology and Biochemistry, Center for Advanced Biotechnology and Medicine, Rutgers University, Piscataway, NJ 08855 [235]

Gordon J. Lithgow, Institute for Behavioral Genetics, University of Colorado, Boulder, CO 80309; present address: University of Manchester, Manchester, United Kingdom [1]

Richard A. Lockshin, Department of Biological Sciences, St. John's University, Jamaica, NY 11439 [167]

Scott W. Lowe, Department of Biology, Massachusetts Institute of Technology, Cambridge, MA 02139; present address: Cold Spring Harbor Laboratory, Cold Spring Harbor, NY 11724 [209]

Yongquan Luo, Section of Geriatric Psychiatry, National Institute of Mental Health, National Institutes of Health, Bethesda, MD 20892 [283]

Carolanne E. Milligan, Department of Biology, University of Massachusetts, Amherst, MA 01003; present address: Department of Neurobiology and Anatomy, Bowman Gray School of Medicine, Wake Forest University, Winston-Salem, NC 27103 [181]

Shin Murakami, Institute for Behavioral Genetics, University of Colorado, Boulder, CO 80309 [1]

James E. Nagel, Clinical Immunology Section, National Institute on Aging, National Institutes of Health, Baltimore, MD 21224 [51]

Catherine D. Nobes, Department of Biochemistry, University of Cambridge, Cambridge CB2 1QW, United Kingdom; present address: MRC Laboratory of Molecular Cell Biology, University College London, London, United Kingdom [267]

Francesco Pallotti, Department of Neurology, Columbia University, New York, NY 10032 [19]

Randall N. Pittman, Department of Pharmacology, University of Pennsylvania School of Medicine, Philadelphia, PA 19104 [255]

Arlan Richardson, Aging Research and Education Center, Department of Physiology, University of Texas, San Antonio, TX 78284 [67]

H. Earl Ruley, Department of Microbiology and Immunology, Vanderbilt University School of Medicine, Nashville, TN 37232 [209]

Eric A. Schon, Department of Neurology, Columbia University, New York, NY 10032 [19]

Lawrence M. Schwartz, Department of Biology, University of Massachusetts, Amherst, MA 01003 [181]

Monica Sciacco, Department of Neurology, Columbia University, New York, NY 10032 [19]

Jerry W. Shay, Department of Cell Biology and Neuroscience, University of Texas Southwestern Medical Center, Dallas, TX 75235 [153]

David R. Shook, Institute for Behavioral Genetics, University of Colorado, Boulder, CO 80309 [1]

Tahereth Tabatabaie, Free Radical Biology and Aging Research Program, Oklahoma Medical Research Program, Oklahoma City, OK 73104 [35]

Aviva M. Tolkovsky, Department of Biochemistry, University of Cambridge, Cambridge CB2 1QW, United Kingdom [267]

Kanwar Virdee, Department of Biochemistry, University of Cambridge, Cambridge CB2 1QW, United Kingdom [267]

Eugenia Wang, Bloomfield Centre for Research in Aging, Lady Davis Institute for Medical Research, Sir Mortimer B. Davis–Jewish General Hospital, McGill University, Montreal, Quebec H3T 1E2, Canada [139]

Benjamin Wolozin, Section of Geriatric Psychiatry, National Institute of Mental Health, National Institutes of Health, Bethesda, MD 20892 [283]

Katherine Wood, Biochemistry Section, Surgical Neurology Branch, NINDS, National Institutes of Health, Bethesda, MD 20892 [283]

Woodring E. Wright, Department of Cell Biology and Neuroscience, University of Texas Southwestern Medical Center, Dallas, TX 75235 [153]

Zahra Zakeri, Department of Biology, Queens College, and Graduate Center of the City University of New York, Flushing, NY 11387 [167]

Preface

Numerous theories have been advanced to explain the still-mysterious process of aging, but it is increasingly clear that the great physiological compromises—cardiovascular problems, cancer, and deterioration in the immune and nervous systems—originate in the biology of cells. Whether cells are responding to variant levels of growth factors, deriving from circulatory or source limitations, or to random but cumulative damage from endogenous or environmentally derived stresses and toxins, the cells' responses influence the future physiology of the organism, often negatively. Factors under genetic control clearly contribute to the process, as evidenced by the striking variation in life span among different species. As we begin to understand these mechanisms, research in aging has come to explore the primary basis of cell proliferation and cell death.

The cumulative toxicity throughout an organism's life may lead to inappropriate growth responses, potentially clogging the circulatory system, corrupting other pathways, or facilitating malignancy. Alternatively, cells under stress can opt to die, depriving the organism of their function. Programmed cell death therefore plays at least two major roles in aging.

We address these issues of cell aging and cell death through three major themes. In the first part of the book, we address basic mechanisms responsible for damage and degeneration of cellular and molecular functions. The second section addresses the control of cellular proliferation and the switch to senescence. Finally, we consider the mechanisms and role of programmed cell death in development, aging, and disease.

Basic Mechanisms of Aging

In this section the contributing authors address basic mechanisms affecting the health of cells and the consequences of their effects on major physiological systems. Johnson et al. discuss longevity-conferring genes in invertebrates and the implications of these findings on human aging. Schon et al. review recent evidence confirming the theoretical possibility of oxidative damage to mitochondrial DNA. The resultant deletions in mitochondrial DNA molecules increase in incidence with age, and have been shown to underlie a number of genetic diseases that show age-related onset and increases in severity. Tabatabaie and Floyd address the role of oxygen radical-induced damage to DNA, proteins, and lipids as a factor in aging, as well as the potential for neutralizing these reactive species *in vivo*.

The characteristic responses of cells to challenge may be monitored and used to assess aging in two manners. Stress responses (genetic responses elicited by harmful environmental and endogenous factors) include activation of genes for antioxidant enzymes, heat shock proteins, and metal binding proteins. These responses may be used as indicators of stress, and the ability of cells to respond to specific stress may be assessed to reflect the ability of the organism or individual to cope with the demands of the environment. Richardson and Holbrook review evidence to suggest that such responses may be lost in senescent cells and with age.

The consequences of these cellular challenges and insults are decrements in the major physiological systems. A decline in immunological function is one hallmark of aging, and the cellular and molecular changes that underlie this decrement are reviewed by Nagel et al. Of the numerous changes that occur with age in the cardiovascular system, many are secondary to vascular changes brought about by normal cellular responses to physical stress, as well as by poorly regulated growth of the endothelium. The molecular and cellular changes observed in the aging cardiovascular system are reviewed by Crow et al.

Molecular Mechanisms Controlling Cellular Senescence

Other than stem cells, normal cells show an infinite life span *in vitro* and cease to proliferate after a certain number of passages. This phenomenon, referred to as cellular senescence, has been well studied. As first described by Hayflick and Moorehead, normal cells in culture undergo a defined number of population doublings before losing the ability to proliferate and entering into a post-replicative, senescent state. Since tumor cells do not show such a limit, the early authors proposed that senescence of an organism derived from the "clock" that kept tabs on cell proliferation. This hypothesis underwent numerous vicissitudes, as many laboratories failed to find that aged individuals ever "ran out of cells." Recently, however, molecular biology has relinked cell senescence with senescence of the organism. Various aspects of cellular senescence, including its relationship to quiescence, differentiation, transformation, cancer, and *in vivo* aging are presented in chapters by Afshari and Barrett, Lee et al., and Wright and Shay.

A theoretically necessary limitation to unlimited cellular proliferation has recently been confirmed and provides a provocative consideration. The manner in which DNA replicates guarantees that a small piece of one end of each chromosome will be lost at each turn of the cell cycle. Chromosomes normally have short stretches of structural DNA, named telomeres, to accommodate this loss. In normally limited somatic cells, either *in vitro* or isolated from

elderly individuals, telomere length reduces with age. A germ cell enzyme, telomerase, reconstructs these "throw-away" tabs, and this enzyme is reactivated in cancer cells. Greider and Harley describe the consideration that reduction in telomere length with age and with cell division may underlie cellular senescence.

Cell Death

Cell death, programmed cell death, or apoptosis, has major consequences for the process of aging. It is now recognized that most cell deaths are the result of controlled processes in which cells under stress actively self-destruct. Cell loss in the immune and central nervous systems leads to major physiologcial limitations, and lack of cell death leads to malignancy and autoimmune disorders. Thus there is a large and growing interest in apoptosis, particularly with respect to the molecular signals that induce and regulate programmed cell death, and the role of apoptosis in a variety of age-associated diseases and disabilities.

Lockshin and Zakeri introduce the section with an overview of programmed cell death. Milligan and Schwartz review the role of apoptosis in development and in mature tissues. The decision to enter cell division, to remain in a non-mitotic state (G_0), or to die appears to be a particularly delicate one. As Lowe and Ruley describe, the activation or nonactivation of cell cycle genes such as p53 appears to be either an essential aspect or a very good marker of cell death and potential for malignant growth. In post-mitotic tissues such as nerve and muscle, programmed cell death or apoptosis plays an important role in both pathological and non-pathological aspects of aging, including neurodegenerative diseases. The importance of this latter area is such that the subject is a major focus of developmental neurology, and we consequently have learned a great deal about the genetic, hormonal, and synaptic interactions necessary to keep neurons functioning. Lints and Driscoll review how cells can be killed by exhaustion, using a nematode mutation that may be a paradigm for excitotoxic or late-onset neuronal deaths. Pittman and DiBenedetto and Tolkovsky et al. discuss from different standpoints the recent recognition that nerves, and presumably most cells, require trophic support throughout their lives. Both chapters analyze the role of trophic support, in autonomic and sympathetic neurons, respectively, as well as other metabolic and hormonal supports. These exciting laboratory findings suggest mechanisms of neuronal loss in aging. This aspect is addressed by Wolozin, Luo, and Wood, who conclude the book and the section by putting the subject of neuronal cell death into the perspective of aging.

With our greater understanding and our consequent preventive intervention, we have driven back infirmities, such as frailty and heart disease, that were once considered to be synonymous with aging. Our understanding of the metabolism of cholesterol, injury responses in the vascular intima, and the cellular and molecular biology of hypoxic cardiac muscle have led to preventive and emergency procedures that have pushed back cardiovascular deaths almost twenty years. As we approach the 21st century, we know of some human genes that sensitize individuals to age-associated infirmi-

ties and other genes that segregate to centenarians, with rapid progress predicted as the genome is sequenced and decoded. These advances in cell and molecular biology both contribute to the rapid advance of our understanding of aging and suggest new means of reducing or slowing age-associated disease and disabilities. More significantly, this approach suggests preventive rather than reparative approaches. Our understanding of the life and death of cells suggests that global interventions to maintain the appropriate balance will be possible. Aging is clearly complex and multifactorial, but cells exist in a complex social structure. The activity, life, and death of each cell are dictated by its responses to a wide variety of stimuli derived from its matrix, diffusible chemical signals, and its interactions with its neighbors. Our increasing ability to understand and to adjust these stimuli could have a profound impact on our vitality in later life.

Nikki J. Holbrook
George R. Martin
Richard A. Lockshin
September 1995

Cellular Aging and Cell Death: 1–17
© 1996 Wiley-Liss, Inc.

Genetics of Aging and Longevity in Lower Organisms

Thomas E. Johnson, Gordon J. Lithgow, Shin Murakami,
Stacey A. Duhon, and David R. Shook

I. INTRODUCTION

Species with short life spans have been the primary targets for studies into the genetic basis of the aging process(es). Almost all genetic studies of aging and longevity have been performed on invertebrates. Invertebrates often have very short life spans, and this fact, together with the excellent genetic systems available for some yeasts, *Drosophila melanogaster*, and the nematode *Caenorhabditis elegans*, make them almost the only choice for studying the genetics of aging.

The mouse, with its 2-year life span, is an exception to this rule [Johnson, 1993]; several genetic studies, including the analysis of short-lived mice [Covelli et al., 1989; Takeda et al., 1981], have been performed and used in an attempt to infer causal relationships. The mouse is the focus of a recent NIA initiative on mammalian models of aging. Recently human "marker association" studies, where longevity is shown to be associated with defined regions of the human genome by characterizing molecular markers at certain candidate loci, have started to appear [Takata et al., 1987; Schächter et al., 1994; Rea and Middleton, 1994]. Genetic approaches have been used also to identify the processes causing replicative senescence in human tissue culture [Pereira-Smith and Smith, 1988; Wright and Shay, this volume].

Two fundamentally distinct but overlapping questions have been the focus of genetic studies. What are the molecular mechanisms limiting life span? How does the process of evolution lead to aging and senescence? Although the answer to the latter question has been obtained to the satisfaction of most in the field [Rose, 1991], studies into the molecular basis continue. The details of the mechanisms leading to senescence and aging have not been forthcoming. Genetic approaches promise to fill this gap.

II. IDENTIFYING THE GENES SPECIFYING LONGEVITY

II.A. Overview

Since many mutations result in life shortening (in the nematode, 50% of all mutants) [Johnson, 1984], the search for genes specifying longevity in invertebrates has focused on those genes (sometimes called *gerontogenes*) [Rattan, 1985] that can result in longer than wild-type life spans after appropriate manipulation. All mutations described in this review lead to increased longevity. Long-lived genotypes have been identified by using mendelian, quantitative, and molecular strategies. Martin and Turker [1988] have made the point that some life-shortening mutations, such as those causing Werner's syndrome, can also reveal basic aging processes, and a similar rationale has been put forward with regard to the senescence accelerated mouse (SAM) [Takeda et al., 1981]. Although this may be true, short-lived strains are more likely to involve alterations in processes other than those fundamental to aging.

II.B. Identification of Mutations Leading to Longer Life

Single-gene mutants affecting longevity have been identified in filamentous fungi [Osiewacz, 1992] and in screens for longevity mutants in *C. elegans* [Klass, 1983; Duhon, Lithgow, White, and Johnson, unpublished data]. Genetic analysis in *C. elegans*, as in other gentically well-studied species such as yeast or the fruit fly, typically involves the induction and identification of mutants that are qualitatively different from normal (wild type). This strategy has obvious historical precedence and practical value [Brenner, 1974] and is the method of choice in the dissection of biological processes [Botstein and Mauer, 1982].

Long-lived mutants have not been identified in any metazoan species other than *C. elegans* and perhaps in *Drosophila* [Maynard Smith, 1958]. The screens of Roberts and Iredale [1985] for sex-linked recessive or autosomal dominant mutations that slowed the senescent decline in motor function were unsuccessful. They argue that their failure to identify mutants in *Drosophila melanogaster* having increased longevity suggests that such mutations do not occur in *Drosophila*. However, their failure to detect mutations such as *grand-childless*, which has been reported to have a longer than wild-type life span [Maynard Smith, 1958], show that their arguments are specious and that their screen was not designed in such a way that even known mutants affecting life span could be detected. Unfortunately, the observations of Maynard Smith suggesting trade-offs between reduced fertility and increased life span in *D. subobscura* have never been replicated in *D. melanogaster*, where a variety of female-sterile and other mutants affecting fertility are available. Thus the question of whether mutations in single genes can result in an extended life span in species other than *C. elegans* and certain fungi remains open [review by Arking and Dudas, 1989].

There are several reasons that screens for long-lived mutants in *C. elegans* were successful. First, the self-fertilizing mode of reproduc-tion of *C. elegans* makes it easy to screen for recessive mutations. Second, the lack of inbreeding depression for life span or other life history traits [Johnson and Wood, 1982; Johnson and Hutchinson, 1993] makes it possible to assess life span in a straightforward manner unaffected by inbreeding depression. Third, the fact that strains of *C. elegans* can be cryogenically preserved and thus maintained indefinitely in a nongrowing state allows all laboratories using *C. elegans* to use the same wild-type reference strain (N2); this eliminated problems associated with the effects of genetic background that are very problematic in mice and fruit flies. Finally, there have been no serious attempts in other species to identify single-gene mutants that extend life span. It thus appears that only in the nematode will such genes be identified; this represents a tremendous advantage not shared by other metazoan species commonly used for genetic analyses.

II.B.1. Fungi. A brute-force approach to identify newly induced mutants that have longer than normal conidial life spans has been performed in *Neurospora* [Munkres and Furtek, 1984]. Senescent processes are under genetic control, and senescence can be slowed and even eliminated entirely in the fungus *Podospora anserina* by synergistic mutations in pairs of nuclear genes: *i viv* or *gr viv* [Tudzynski and Esser, 1979]. The mechanisms of action of these genes is still unknown [Osiewacz, 1992].

II.B.2. *C. elegans*. Klass [1983] induced forward mutations with ethyl methane sulfonate and then used a replica plate system to screen cohorts for those that survived longer than unmutagenized controls. Once the long-lived cohorts were identified, putative mutants could be rescued from genetically identical and pre-reproductive siblings that had been maintained on the master plate under conditions favoring the formation of dauer larvae. This strategy took advantage of several unique aspects of *C. elegans* biology: its high forward mutation rate, its ability to form offspring by self-fertilization thus facilitating the search for recessive mutations in the F_2, and its ability to form dauer larvae (an alternative, non-aging, third-larval stage) [Klass and Hirsh, 1976]. Subsequent analyses of these mutations were carried out in our laboratory [Friedman and Johnson, 1988a,b] by taking advantage of the lack of inbreeding depression that facilitates such genetic analyses [Johnson and Wood, 1982; Johnson and Hutchinson, 1993]. These experiments positioned *age-1* to linkage group 2 [Friedman and Johnson, 1988a] and associated the extended life span trait (Age) with a decrease in fertility or "brood size" (Brd).

Subsequent analysis of this region as a prelude to cloning *age-1* have confirmed this tight association but have separated the loci responsible for the two traits such that Brd maps to the *fer-15* interval and Age maps near *unc-4* [Johnson et al., 1993]. A series of transgenic lines that complement Brd and a fertility locus called *fer-15* but not Age confirm the separation of the Age and Brd traits [Johnson et al., 1993], as does the assignment of the Brd phenotype to the *fer-15* locus by deficiency mapping (Tedesco and Johnson, unpublished data). More recently, we have succeeded in mapping *age-1* to a small region of 150 kb (Lithgow and Johnson, unpublished data) and have identified two genes within this interval as *age-1* candidates by transgenic complementation (Murakami, Gehle, Lithgow, and Johnson, unpublished data). We anticipate the cloning of the *age-1* locus shortly.

There are three other published reports of mutations in *C. elegans* that result in longer life. *(1)* Mutations in *spe-26* dramatically reduce sperm formation and result in life extensions of about 80% for the hermaphrodite and the mated male [Van Voorhies, 1992]. *(2)* At 16°C and normal oxygen tension, *rad-8* mutant worms exhibit almost 50% longer mean life span than wild-type [Ishii et al., 1994]. *(3)* Temperature-sensitive, *daf-2* mutants, which cause dauer formation under conditions in which dauers are not usually seen, result in a more than twofold extension of mean life span at the permissive temperature [Kenyon et al., 1993], and this extension is blocked by the action of *daf-16*. (The dauer is an alternative

developmental path taken by *C. elegans* under "hard times" conditions [Riddle, 1988]; this stage functions as a "time out" from the normal aging pathway in that worms can be maintained as dauers for months and then induced to recover by feeding and still have normal adult life spans [Klass and Hirsh, 1976].) Several other dauer mutants also affect length of life: *daf-23* doubles the length of the normal adult life span, and *daf-12* mutants interact with both *daf-23* and *daf-2* to cause an enhancement of the life-prolongation phenotype to an almost fourfold increase in mean life span [Larsen et al., 1992]. This interaction is consistent with the model that *daf-12* is on one arm and *daf-2* and *daf-23* are on the other arm of a pathway downstream of sensory inputs and that these two arms interact with each other [Gottlieb and Ruvkun, 1994]. This model also suggests that the life-prolongation effects of *daf-23* should be blocked by *daf-16*, which has also been observed (P. Larsen, personal communication). The dominant, dauer-constitutive mutant *daf-28* [Malone and Thomas, 1994] can also prolong life slightly (J. Thomas, personal communication). These dauer-formation mutants define a signal-transduction pathway in which homologs to mammalian genes involved in signal transduction can be identified. *daf-1* codes for a translational product that has homology to a serine-threonine kinase in the *RAF* superfamily and that may be a cell-surface receptor [Georgi et al., 1990]. *daf-4* is the nematode homolog of human bone morphogenetic protein (BMP) receptor [Estevez et al., 1993], with the *daf-7* gene encoding a member of the TGF-β superfamily (P. Larsen et al., personal communication). Finally, *daf-12* encodes a transcription factor of the steroid-thyroid hormone receptor superfamily (P. Larsen et al., personal communication).

The relationship between *age-1* and the other Age genes is not completely clear. *age-1 daf-16* double mutants have wild-type or shorter life spans, similar to *daf-16* alone (Duhon and Johnson, unpublished data). The simple interpretation that *age-1* is on the dauer determination pathway may be specious because of nonspecific effects on life span resulting from *daf-16*. The long-life phenotype of *age-1* is similarly suppressed by morphological and behavioral mutations such as *dpy-10* and *unc-4* [Johnson, 1984; Johnson et al., 1990; Johnson, unpublished observations]. Thus, a more complete analysis of these results must be performed prior to a, perhaps overly simplistic, interpretation that *age-1* is on the dauer-formation pathway.

II.B.3. Drosophila. Only one screen [Roberts and Iredale, 1985] for long-lived mutants has been conducted, to our knowledge, in *Drosophila;* that screen was unsuccessful in finding any mutations. However, at least two reports describe anecdotal studies where mutants that were originally identified on some other basis have effects on life prolongation. First is the oft-cited observation of Maynard Smith [1958] that the *grandchildless* mutant in *D. subobscura* has an extended life span. He observed life spans of 67.6 days compared with 33.1 days for mated females and 58.7 days for virgin females. Unfortunately, these observations have not been replicated; with the excellent array of female-sterile mutants now available in *D. melanogaster,* this represents a serious oversight that one of the several *Drosophila* laboratories should correct. The second report is that of Shepherd et al. [1989] on flies transgenic for EF-1α. When the experiments were performed on more lines and in several genetic backgrounds the effect of the transgene was small relative to position of transgene insertion or genetic background [Stearns and Kaiser, 1993]. Many reports of shorter lived strains have been seen in the literature, but these cannot be interpreted in a straightforward manner with the exception of the excellent series of studies by Gould and Clark [1977] on X-ray-induced mutations on the X chromosome.

II.C. Breeding for Strains Leading to Longer Life

Quantitative genetic approaches to the identification of longevity genes include both selective breeding for increased age of reproduction [Luckinbill et al., 1984; Rose, 1984a]

and the generation of populations segregating polygenic determinants specifying length of life in the nematode [Johnson, 1987a; Brooks and Johnson, 1991; Ebert et al., 1993; Shook et al., submitted] and in the mouse [Yunis et al., 1984]. A major problem with quantitative approaches in general is the difficulty of identifying the effects of individual genes and the almost impossible task of determining the genes that are responsible for these relatively small effects [see Fleming et al., 1993, for an example of this difficulty].

II.C.1. Drosophila. There are several more detailed reviews of genetic approaches to studying aging in *Drosophila* [Mayer and Baker, 1985; Arking and Dudas, 1989; Rose, 1991]. Two groups selected for extended periods of survival and fertility in *D. melanogaster.* Luckinbill et al. [1984] succeeded in extending life expectancy by some 30 days using a newly established laboratory population obtained from recently collected wild flies. A similar, but less dramatic (10 days for females, 6 days for males), increase in life span was obtained by Rose [1984a; Clare and Luckinbill, 1985] in a "wild" population that had been adapted to laboratory growth for 130 generations. However, the Rose study employed five replicate lines as opposed to the duplicate lines used by Luckinbill and Arking. The long-lived lines had a small decrease in fertility early in life, suggesting a "trade off" between fertility in early life and fertility later in life (or longevity). One unexplained complication is the effect of controlled larval density; in low larval density (10 flies per vial) no life span differences between selected and control strains were detected [Luckinbill and Clare, 1986]; such studies have not been conducted by the Rose group, but maintenance at different adult densities does not affect the differential longevity [Graves and Mueller, 1993]. The stocks had been selected under uncontrolled density and points to the extreme care that must be used in performing such selections.

In addition to the reduction in early-life fertility, a number of physiological alterations have been implicated in the extension of life.

Rose et al. [1984] found that the long-lived stocks had lower ovary weight in early life. These strains also showed increased resistance to starvation, desiccation, and other environmental stresses [Service et al., 1985]. The proof that this resistance is causally connected to longevity was obtained by selective breeding for starvation or desiccation resistance, which also led to a 40% increase in longevity [Rose et al., 1992]. Metabolic changes were also detected in the lines selected in Detroit; Luckinbill et al. [1990] showed that glucose-6-phosphate dehydrogenase (G6PD) was differentially expressed in selected and control lines such that the long-lived lines had higher specific activities of G6PD than did controls. Resistance to reactive oxygen species has also been implicated in the extension of longevity. Arking et al. [1991] showed that one of the long-lived strains was more resistant to paraquat than the control; similar findings were made by Luckinbill on the other selected line. Fleming et al. [1993] have been applying two-dimensional gel technology to identify proteins specifically selected in long-lived strains; they find that a high-activity variant of superoxide dismutage (SOD) has been selected. Examination of this allelic variant in segregating populations has failed to show a significant correlation between enzyme activity and life span [Tyler et al., 1993]; so even in this best-case scenario it has not been possible to prove a causal relationship.

Of course, one might wish to use the power of *Drosophila* molecular genetics to dissect and understand the basis of the increased life span of these strains. All three groups have been stymied by their inability to separate and understand the effects of individual genes. Both Luckinbill et al. [1988] and Buck et al. [1993] showed that one of the selected lines had genes on chromosome 3 that were implicated in life extension. Arking et al. [1991] have suggested that this confirms that increased SOD and catalase are causally implicated in both increased paraquat resistance and life extension, but, since 60% of all *Drosophila* genes are on chromosome 3, this suggestion cannot be taken seriously. The differences in life span in the Rose lines result from the combined actions at as many as several hundred loci [Hutchinson and Rose, 1990; Fleming et al., 1993]. The highly polygenic nature of these lines and the fact that each line is still very outbred make it extremely difficult, if not impossible, to associate the differences in longevity with individual genes using today's technologies. Indeed, this inability, given the current strength of *Drosophila* molecular genetics, is a forceful argument [Johnson, 1988] against the recent decision by the NIA to commit significant resources for selective breeding for life span in mice where the ability to separate the effects of single genes in polygenic systems will be even more problematic and much more expensive than in *Drosophila*. Another problem in sexually reproducing species such as *Drosophila* and the mouse is inbreeding depression [Rose, 1984b; Rose, 1991]. Rose's selected lines are much more heterozygous than the controls [Fleming et al., 1993]. It is not clear to what extent life extension results from possible overdominance, and attempts to show a causal relationship between the one known gene (SOD) and life span have failed [Tyler et al., 1993]. This is no trivial problem, because most of biometric theory is based on the assumption that the alleles being analyzed show additive genetic variance. If the response to selection stems from increases in heterozygosity, the increased life span phenotype cannot be studied in inbred strains, and it seems almost certain that little progress can be made at the molecular genetic level in dissecting the processes responsible for increased longevity. Indeed, Rose and other laboratories have been careful to avoid problems of inbreeding in all selections. A systematic inbreeding of the selected and control lines would determine the extent to which the increases in life span result from overdominance and would reveal any underlying additive genetic variance that could be studied in inbred lines using the tools of molecular genetics [Johnson, 1988].

II.C.2. C. elegans. Selective breeding has not been fruitfully employed in the worm to obtain long-lived lines. Instead, long-lived strains have been identified among the inbred progeny resulting from the mating of two laboratory wild-type strains: N2 and Bergerac [Johnson and Wood, 1982; Johnson, 1987a]. Some of these "recombinant-inbred" (RI) strains have mean and maximum life spans that are as long as those seen in *age-1* mutants [Johnson, 1987a]. Variation in length of life results from new combinations of genes that differed between the parents.

In an attempt to circumvent the detailed and statistically complex methods of quantitative trait locus (QTL) mapping Ebert et al. [1993] used an analysis of a population segregating a variety of genotypes. As in earlier studies, they looked at offspring that originated from crosses between N2 and Bergerac. Rather than examining inbred populations derived from the F_1 of these crosses, however, they carefully maintained polymorphisms by crossing large populations and then examined multiple markers throughout the genomes of individual worms arising from these crosses. They found regions of the genome that were responsible for increased survival by examining worms that had outlived the majority of their comrades and comparing their marker distributions with the distribution of markers in the whole population obtained early in life. This depended critically on the development of markers that could be analyzed by the polymerase chain reaction (PCR). Williams et al. [1992] developed a series of 40 such markers that could be examined in multiplex reactions. Ebert et al. [1993] found several regions of the genome that were associated with increases in life span. The statistical aspects of these studies have been poorly explored, and, although a replication was carried out, it is not possible to ascertain the amount of genetic variance in longevity associated with any one of these markers. Moreover, using such a design, other aspects of life history that may have been responsible for the differential survival cannot be explored.

Brooks and Johnson [1991] and Shook et al. [1995] have explored the same crosses using a series of recombinant inbred strains (RIs) derived from the same two parents. Intermediate

levels of heritability for life span and for fertility were found again, confirming earlier studies [Johnson and Wood, 1982; Johnson, 1987a]. Brooks and Johnson [1991] found heritabilities ranging from 0.05 to 0.36, and these were confirmed by Shook et al. [1995] on a much larger population of RIs (79 strains). We found that regions on linkage groups 2, 3, and 4 were significantly associated with QTLs for hermaphrodite self-fertility and that QTLs for life span mapped to 2, 3, and the X. We found no evidence to suggest that reduced fertility was associated with increased life span. Thus, there was little evidence of antagonistic pleiotropy between life span and early fertility at any locus, although there was a considerable negative association between early and late reproduction [Brooks and Johnson, 1991].

II.C.3. Others. In mice, recombinant inbred strains [Yunis et al., 1984], congenic strains [Crew, 1993], and F_2s have been used to identify and map QTLs involved in the specification of length of life. Two long-term breeding studies are underway under a 1993 initiative of the NIA. These two studies involve breeding for increased immune competency or directly for life span itself. More details on the rodent work can be found in reviews targeted to studies of life span and aging using mice [Johnson, 1993; Yunis and Salazar, 1993].

II.D. Molecular Approaches to Identifying Genetic Alterations Leading to Longer Life

Molecular approaches have relied on the identification of candidate genes and subsequent manipulation of these candidates to see if life span is affected. The most extensive molecular searches for genes specifying the aging processes are those of Jazwinski using the baker's yeast *Saccharomyces cerevisiae*. The yeast life span is defined as the number of cell divisions a yeast cell can undergo. Jazwinski and his associates have searched for genes that are differentially expressed over the life span [Egilmez et al., 1989]. While several genes were down-regulated over the approximately 20 cell divisions of a yeast cell's life, one gene was identified (O30) that was up-regulated over that period. Even then, only a 50% increase in steady-state mRNA abundance was observed.

LAG1 and *LAG2* as well as *RAS1* and *RAS2* have been analyzed by disruption and by overexpression [D'Mello et al., 1994; Sun et al., 1994; Jazwinski, personal communication]. In *S. cerevisiae*, null mutants of *LAG1* show increased longevity as measured by the number of cell divisions that an individual cell can perform [D'Mello et al., 1994]. *LAG2* disruptions result in decreased numbers of cell divisions, while overexpression increases the number of divisions. Similarly, disruptions of *RAS1* and *RAS2* result in effects on the number of replicative divisions that are not straightforward to interpret but for the first time highlight differences between the two forms of *RAS* [Sun et al., 1994].

Direct approaches to altering life span have been used by several groups to test the "free radical theory" of aging. For example, Reveillaud et al. [1992] constructed transgenic strains of *Drosophila* in which bovine CuZn superoxide dismutase (SOD) was expressed in strains that previously had normal SOD levels. Modest effects on life span were found in some strains. Similar studies were conducted by Orr and Sohal [1993], who overexpressed *Drosophila* SOD and again found only modest effects on life span. Similarly, over-expression of catalase alone had only modest effects on life span [Orr and Sohal, 1992]. Both groups reasoned that since SOD and catalase are both needed to detoxify oxygen radicals, a coordinate up-regulation may be needed to lower the short-term toxicity of reactive intermediates. Orr and Sohal [1994] found that most transgenic strains coordinately expressing both of these enzymes had a 30% increase in mean and maximum life spans and a slowing of the rate of increase of mortality with chronological age. Although only "healthy" strains were chosen from all of the double transgenic strains that they obtained, this report stands as a very significant result that marks the state of the art in the analysis of aging using transgenic approaches.

III. AGING AND DIFFERENTIAL GENE ACTION

Alterations in gene expression during aging have been proposed to be an important factor causing senescence [Cutler, 1982; Danner and Holbrook, 1990]. Up-regulation or expression of genes only late in life might be the hallmark of genes showing "delayed age of gene action" [Rose, 1991]. Studies in *D. melanogaster* [Fleming et al., 1986a,b], *C. elegans* [Johnson and McCaffrey, 1985; Meheus et al., 1987], and rat brains, among others [Wilson et al., 1978; Cosgrove et al., 1987], indicate that there are few or no qualitative changes in the abundance of specific proteins during aging. However, quantitative changes in gene expression during aging have been reported in many different organisms [Reff, 1985; Richardson and Semsei, 1987; Danner and Holbrook, 1990; Rattan et al., 1992; Thakur et al., 1993]. Fabian and Johnson [1995] isolated cDNA clones representing transcripts that undergo changes in abundance over the adult life span of *C. elegans*. The clones were identified in a cDNA library made from old-worm mRNA by a differential screen using radiolabeled cDNA made from either young or old nematodes. We cloned nine distinct transcripts that decreased in abundance with age, two that increased slightly in abundance with age, and one that peaked in abundance during the middle part of the adult life span. Six of these were quantified using a dot blot assay of total RNA isolated at several ages from two strains that have wild-type life spans. All six mRNAs showed similar age-dependent abundance patterns in these two strains. Mutation of *age-1* did not appear to alter these patterns. Sequence analysis showed that three of the mRNAs that decreased in abundance with age corresponded to previously identified *C. elegans* vitellogenin genes. One transcript that showed a small increase in abundance with age was highly homologous with translation factor EF1-α. The other five clones represent novel nematode genes. Our analysis indicates that only a small fraction (~2%) of abundant transcripts are altered by more than approximately twofold during aging in *C. elegans*. A prediction of the developmentally programmed theory of aging [Russell, 1987; Johnson, 1987b], i.e., that a variety of senescence-specific transcripts should be turned on in old animals, was not supported by these studies. In contrast to development, the molecular etiology of aging and senescence is unknown in almost every instance, which makes aging a fascinating area for study.

Is aging and senescence programmed? Rose [1991] points out that it is a mistake to view aging as a process similar, for example, to development or to the cell cycle. In contrast to development, aging is almost always nonadaptive; i.e., aging or senescence occurs because of a lack of selection against traits that are deleterious only late in life, after reproduction has ceased [Charlesworth, 1980]. Instead, genes specifying aging are thought to be primarily involved in other biological processes. Accepted theories suggest that aging, senescence, and a limited life span have evolved not in response to direct selection, but rather as the inevitable consequence of natural selection in age-structured populations [Charlesworth, 1980]. Two mechanisms have been proposed to account for senescence: "antagonistic pleiotropy" and age specific gene action [Rose, 1991], also called mutation accumulation. The long-lived lines of *Drosophila* selected for late reproduction showed minor reductions in early fertility [Rose, 1984a; Luckinbill et al., 1984] and display "antagonistic pleiotropy between life span and fertility." This may be somewhat of a misnomer because *Drosophila* females reproduce almost until they die, so a direct selection for later age of reproduction is directly selecting for increased life span as well. Thus, the trade off is likely to be entirely at the level of reproduction. Quite unlike development, where many genes are differentially regulated, in senescence large batteries of genes are not regulated differentially [Johnson and McCaffrey, 1985; Danner and Holbrook, 1990].

IV. GENETIC ANALYSIS OF DEMOGRAPHICS IN INVERTEBRATES

An exponential acceleration of age-specific mortality rate with chronological age has become almost one of the hallmarks of organismic aging and senescent processes. Age-specific mortality rates in populations of *C. elegans* increases with near exponential kinetics throughout life [Johnson, 1987a; 1990; Brooks et al., 1994]. Some of the mortality rate change is not explained by the exponential model [Brooks et al., 1994], and a better fit can be obtained by using a two-stage Gompertz model to fit late-life mortality [Vaupel et al., 1994]. A mutation in the *age-1* gene results in a 65% increase in mean life span and a 110% increase in maximum life span at 25°C [Friedman and Johnson, 1988a]. *age-1* mutant hermaphrodites show a 50% slower rate of acceleration of mortality with chronological age [Johnson, 1990], demonstrating that this exponential acceleration is also under genetic control. Mutant males also show a lengthening of life and a slowing of the rate of mortality acceleration, although *age-1* mutant males still have significantly shorter life spans than do hermaphrodites of the same genotype. The slower rates of mortality acceleration are recessive characteristics of the *age-1* mutant allele examined.

Long-lived recombinant inbred lines also show alterations in rate of acceleration of mortality rate [Johnson, 1987a]. Longer lived strains have a slowing of the characteristic exponential increase in mortality rate. The length of developmental periods and the length of the reproductive period are unrelated to increased life span. Lengthened life is due entirely to an increase in postreproductive life span. Development, reproduction, and life span are each under independent genetic control. General motor activity decays linearly with chronological age in all genotypes. The decay in general motor activity is both correlated with and a predictor of life span, suggesting that both share at least one common rate-determining component.

Carey et al. [1992] and Curtsinger et al. [1992] showed that in medflies and in *Drosophila*, respectively, mortality rate increased exponentially for a few weeks and then leveled off or even decreased later in life. We have been able to mimic such a response using genetically heterogeneous populations of 79 RI genotypes [Brooks et al., 1994]. Age-specific mortality increased exponentially until 17 days and then remained constant until the last death at about 60 days. The period of constant age-specific mortality results from genetic heterogeneity. Subpopulations differ in mean life span, but each exhibits near exponential, albeit different, rates of increase in age-specific mortality. Thus, much of the heterogeneity in mortality rates later in life could result from genetic heterogeneity and not from an inherent effect of aging. These studies have elicited much comment [Vaupel et al., 1994; Wang et al., 1994; Curtsinger et al., 1994].

Researchers using *Drosophila* and *C. elegans* have used movement as a general index of vitality or health to ask whether longer lived populations are at increased risk for extended periods of late-life morbidity. Mutant strains of *age-1* actually appear to move better at all ages and to keep moving for longer, again suggesting that mutations that increase longevity increase overall vitality and "health" [Duhon and Johnson, in press]. There was a clear association between postponed senescence and flight duration in the long-lived *Drosophila* stocks as well. Graves et al. [1988] also compared other long-lived and a short-lived population of *D. melanogaster* for tethered flight duration and found almost a threefold difference favoring the long-lived lines when measured at 30–50 days of age. Graves and Rose [1990] showed that four long-lived lines averaged a 42 minute longer flight duration than their control short-lived lines.

V. STRESS RESPONSE SYSTEMS AND THE ACCUMULATION OF CONFORMATIONALLY ALTERED MACROMOLECULES

V.A. Accumulation of Conformationally Altered Protein

The free radical theory of aging [Harman, 1956] has gained significant ground with the demonstration that life span can be increased by the overexpression of enzyme activities that convert the superoxide radical to oxygen and H_2O [Orr and Sohal, 1994]. It has been postulated that reactive oxygen species (ROS) are responsible for aging due to the modification of macromolecular components of the cell, including proteins.

Various mechanistic models of aging have centered on changes in the protein complement of the cell: as a result of changes in the DNA template (somatic mutation), transcriptional or translational problems (error catastrophe), and protein turnover or conformational alteration [for a review, see Danner and Holbrook, 1990]. In support of these theories are data showing the accumulation of both covalently and noncovalently modified protein in a number of tissues and species at later ages. Increasing oxidation of proteins with age may be due to the increasing oxidation potential of the cell [Noy et al., 1985] or a slowing down in the rate of protein turnover [Sharma et al., 1979] or both. "Noncovalent" conformational change may result from specific oxidation, which causes conformational isomerization, and then subsequent reduction [Gafni, 1985]. In the nematode, there is extensive evidence for the accumulation of conformationally modified proteins with age [Zeelon et al., 1973; Bolla and Brot, 1975; Sharma and Rothstein, 1980] and a decrease in specific activity [Sharma and Rothstein, 1980]. The steady-state level of conformationally altered protein is a function of the rate of native to non-native conformation change, the rate of refolding to the correct conformation, and the rate of degradation of each species. In aging nematodes, the protein half-life is changed dramatically [Sharma et al., 1979], which could lead to a large alteration in the fraction of proteins in incorrect conformations. To date there are no estimates of the levels of these modified proteins in any long-lived mutants.

V.B. Loss of Stress Response in Older Age

There are numerous cellular systems that either prevent free radical damage or respond to the presence of denatured or modified protein. In response to a variety of environmental stresses, cellular systems initiate the preferential synthesis of heat shock proteins, many of which exhibit chaperone activity *in vitro* [reviewed by Morimoto et al., 1994]. Environmental stresses that induce hsp gene expression include heat shock, amino acid analogs, transition heavy metals, antineoplastic chemicals, and inhibitors of metabolism. Physiological states such as inflammation, oxidant injury, ischemia, infection, and aging also induce expression [reviewed by Morimoto et al., 1994]. Even nonstress conditions, such as cell cycle, growth factors, and differentiation, induce the expression of some hsp genes. Molecular chaperones prevent the aggregation of unfolded polypeptides, both nascent polypeptides and denatured proteins. Although individual chaperones exhibit chaperone activity *in vitro*, specific sets of chaperones form a chaperone machinery that prevents misfolding and aggregation [Frydman et al., 1994].

Heat shock genes are poorly induced in aged animals [Fargnoli et al., 1990; Blake et al., 1991; Udelsman et al., 1993; Wu et al., 1993] or cell cultures undergoing replicative senescence [Choi et al., 1990; Liu et al., 1989; Effros et al., 1994]. Both are deficient in hsp accumulation after stress. Hsp genes are regulated by the heat shock transcription factor (HSF); thus reduced HSF DNA binding activity in older animals [Heydari et al., 1993; Blake, 1991] could explain the reduction in levels of hsps. *C. elegans* shows increased sensitivity to thermal stress as it ages [Lithgow et al., 1994]. In short, aged animals may be compromised during environmental stress and accumulate malfolded or aggregated protein at a higher rate than young animals.

V.C. Involvement of Stress Response in Determining Life Span in *C. elegans*

Larsen [1992] and Vanfleteren [1993] showed that *age-1* mutants are resistant to oxidative stress and exhibit elevated levels of superoxide dismutase (SOD) and catalase in late life. Since hsps and genes modulating response to oxidative stress can be coordinately regulated and since cross-tolerance between these stresses have been reported [reviewed by Donati et al., 1990; Fleming et al., 1992], we hypothesized that *age-1* mutant strains may exhibit extended life span as a consequence of an enhanced stress response.

Thermotolerance was assessed by measuring fertility or survival. The fertility of *age-1* animals was somewhat less affected than wild type after mild thermal stress. However, survival of *age-1* mutants during thermal stress is consistently increased (Table I; Fig. 1)

TABLE I. Comparisons of Age and Non-Age Strains for Intrinsic Thermotolerance†

N2 (wild type)	TJ1052 (*age-1*)	BA713 (*fer-15*)	TJ401 (*fer-15 age-1*)
392.5 ± 101.9 (58)	528.0 ± 62.2 (55)**	557.9 ± 145.8 (56)	646.9 ± 130.1 (46)*
524.3 ± 111.1 (60)	724.7 ± 118.2 (48)**	509.2 ± 141.3 (57)	809.7 ± 152.9 (58)**
615.0 ± 56.5 (48)	933.2 ± 132.8 (50)**	536.1 ± 59.1 (55)	796.3 ± 100.5 (59)**
493.7 ± 91.7 (58)	736.2 ± 117.0 (59)**	504.3 ± 136.0 (58)	733.5 ± 88.3 (59)**
447.0 ± 80.2 (49)	666.3 ± 89.4 (60)**		
421.9 ± 87.1 (58)	696.9 ± 112.6 (55)**		
427.6 ± 63.4 (57)	629.5 ± 63.9 (54)**		
509.8 ± 72.4 (60)	781.3 ± 105.9 (59)**		

†Mean (minutes) ± SD (N).
*$p < 0.01$.
**$p < 0.0001$.

All Hermaphrodites

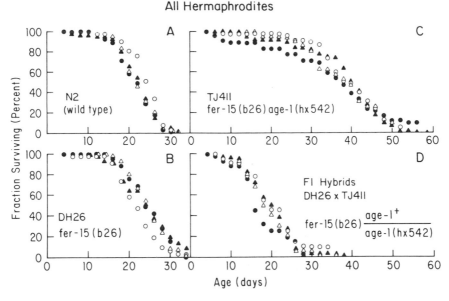

Fig 1. *Survival curves of two wild-type Age strains, an Age mutant strain, and a heterozygous strain of C. elegans. In each case four identical populations of hermaphrodites (shown by different symbols) were separated at 3 days of age and monitored daily in separate mass cultures of 50 worms each until all worms had died. Survival analyses and statistical tests were performed as previously described except that survivors were counted every 24 hours throughout life and Bonferroni corrections were used where appropriate. (A) Survival curves for the wild-type, N2 (Bristol) hermaphrodites; life expectancies of the four identical subpopulations were 20.5, 23.0, 20.9, and 21.0 days; culture 2 was found to be significantly longer lived than the other three cultures (P < 0.05)* and was not used in the further analyses described in Table 1; inclusion of this sample results in only minor modifications. (B) Survival curves for hermaphrodites of DH26, the parental strain in which the mutants were derived; life expectancies were 23.8, 20.7, 23.2, and 23.8 and were not significantly different from each other. (C) Survival curves for age-1 hermaphrodites; life expectancies were 34.3, 38.0, 37.6, and 36.6 days and were not significantly different from each other. (D) Survival curves for F_1 hermaphrodites from a cross of TJ411 hermaphrodites by DH26 males; life expectancies were 15.0, 18.7, 19.1, and 17.8 days and were not significantly different from each other. [Reproduced from Johnson, 1990, with permission of the publisher.]*

[Lithgow et al., 1994, 1995]. This new phenotype has been called Itt for Increased Thermo-Tolerance. Itt was mapped to the *age-1* interval on chromosome 2 [Lithgow et al., 1995] and was shown to be associated with *age-1* in both sterile and nonsterile worms [Lithgow et al., 1994]. Other Age strains, including constitutive dauer mutations *daf-2, daf-28, daf-4,* and *daf-7* are also Itt (Fig. 2) [Lithgow et al., 1995]. These genes define signal transduction in a developmental pathway leading to the differentiation of dauers.

Thermotolerance is associated with the expression of hsp genes in many systems but not in all [Parsell and Lindquist, 1994]. Proof of hsp involvement has come more recently from the construction of transgenic strains overexpressing Hsp70 in *Drosophila* [Welte et al., 1993], where embryos carrying extra Hsp70 genes are more thermotolerant. The small heat shock proteins have been implicated in thermotolerance in mammalian cell lines and Hsp104 in yeast thermotolerance [reviewed by Parsell and Lindquist, 1994].

Further evidence for the hsp's involvement in the determination of life span is that exposure to a nonlethal thermal stress induces both thermotolerance and increased life span [Maynard Smith, 1958; Lithgow et al., 1995] and that long-lived caloric-restricted animals have

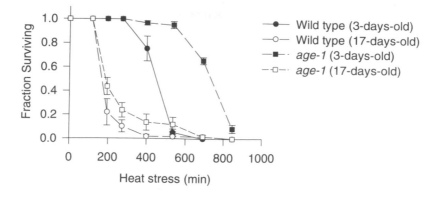

Fig. 2. *Intrinsic thermotolerance of* C. elegans *at 3 and 17 days of age. Fraction surviving was scored during a 35°C thermal stress of three plates containing 20 animals of each strain and each age. Each data point represents the mean fraction alive over three plates (SEM) at each time point. The mean survivals ± SD were as follows: 3-day-old N2 (wild type), 510 ± 72 minutes; 3-day-old TJ1052 (age-* 1[hx546]), 781 ± 106 minutes. Three-day-old N2 and TJ1052 are significantly different (p <0.0001).17-day-old N2 (wild type), 233 ± 102 minutes; 17-day-old TJ1052 (age-1[hx546]), 297 ± 171 minutes; 3-day-old and 17-day-old mean survivals are significantly different (p <0.0001) for both strains. [Reproduced from Lithgow et al., 1995, with permission of the publisher.]*

improved induction of hsp genes. However, a causal relationship between hsp abundance and life span has not yet been established.

A model has been proposed [Lithgow et al., 1995] in which the level of conformationally altered proteins is proportional to age-specific mortality rate. Molecular chaperones enhance the rate of removal of malfolded polypeptides and may also prevent aggregation of malfolded polypeptides. Animals that have elevated levels of chaperones, due either to a genetic mutation or to environmental induction of chaperone synthesis, will better cope with the age-related accumulation of conformationally altered proteins [Lithgow et al., 1995].

VI. PROGNOSIS FOR THE FUTURE

It is striking that the first increased longevity mutant (*age-1*) has increased levels of SOD and catalase [Larsen, 1992; Vanfleteren, 1993] as well as increased rates of induction of Hsp16 [Lithgow and Johnson, unpublished data]. *age-1* mutant strains are also more tolerant of ROS as monitored by resistance to paraquat or H_2O_2, and this resistance maps to the *age-1* site [Jensen, Melov, and Johnson, unpublished

data]. *age-1* and other Age strains are also more resistant to elevated temperatures [Lithgow et al., 1994; 1995] and UV [Murakami and Johnson, unpublished data] than is wild type. Finally, *age-1* mutations are associated with lower levels of mitochondrial deletions at later ages than is the wild type [Melov et al., 1994, 1995].

The race to identify mechanisms that limit life span is underway. Aging research in lower organisms has already provided a system for the examination of basic aging processes and will continue to provide candidate genes and physiological processes that could influence the onset of human disease and senescence. The construction of long-lived strains of invertebrates has demonstrated the role of genes in defining life span in these species. Some of the same strains have been instructive in the identification of physiological processes associated with longevity and have almost established causality. As these contributions continue to emerge we anticipate that targeted experiments in vertebrates will be undertaken with some success, and we look forward to the construction of transgenic strains in the mouse carrying multiple transgenes similar to those constructed in fruit flies [Orr and Sohal, 1994] and in *C. elegans* [Murakami and Johnson, unpublished data]. Basic aging research in invertebrates could lead to insights finally leading to direct intervention in mammalian and human aging processes.

REFERENCES

Arking R, Buck S, Berrios A, Dwyer S, Baker III GT (1991): Elevated paraquat resistance can be used as a bioassay for longevity in a genetically based long-lived strain of *Drosophila*. Dev Genet 12:362–370.

Arking R, Dudas SP (1989): Review of genetic investigations into the aging processes of *Drosophila*. J Am Geriatr Soc 37:757–773.

Blake MJ, Udelsman R, Feulner GJ, Norton DD, Holbrook NJ (1991): Stress-induced heat shock protein 70 expression in adrenal cortex: An adrenocorticotropic hormone-sensitive, age-dependent response. Proc Natl Acad Sci USA 88:9873–9877.

Bolla R, Brot N (1975): Age dependent changes in enzymes involved in macromolecular synthesis in *Turbatrix aceti*. Arch Biochem Biophys 169:227–236.

Botstein D, Mauer R (1982): Genetic approaches to the analysis of microbial development. Annu Rev Genet 16:61–83.

Brenner S (1974): The genetics of *Caenorhabditis elegans*. Genetics 77:71–94.

Brooks A, Johnson TE (1991): Genetic specification of life span and self-fertility in recombinant-inbred strains of *Caenorhabditis elegans*. Heredity 67:19–28.

Brooks A, Lithgow GJ, Johnson TE (1994): Mortality rates in a genetically heterogenous population of *Caenorhabditis elegans*. Science 263:668–671.

Buck S, Wells RA, Dudas SP, Baker III GT, Arking R (1993): Chromosomal localization and regulation of the longevity determinant genes in a selected strain of *Drosophila melanogaster*. Heredity 71:11–22.

Carey JR, Liedo P, Orozco D, Vaupel JW (1992): Slowing of mortality rates at older ages in large medfly cohorts. Science 258:457–461.

Charlesworth B (1980): Evolution in Age-Structured Populations. London: Cambridge University Press.

Choi HS, Lin Z, Li B, Liu AY-C (1990): Age-dependent decrease in the heat-inducible DNA sequence-specific binding activity in human diploid fibroblasts. J Biol Chem 265:18005–18011.

Clare MJ, Luckinbill LS (1985): Selection for life span in *Drosophila melanogaster*. Heredity 55:9–18.

Cosgrove JW, Atack JR, Rapoport SI (1987): Regional analysis of rat proteins during senescence. Exp Gerontol 22:187–198.

Covelli V, Mouton D, Di Majo V, Bouthillier Y, Bangrazi C, Mevel J-C, Rebessi S, Doria G, Biozzi G (1989): Inheritance of immune responsiveness, life span, and disease incidence in interline crosses of mice selected for high or low multispecific antibody production. J Immunol 142:1224–1234.

Crew MD (1993): Genes of the major histocompatibility complex and the evolutionary genetics of lifespan. Genetica 91:199–209.

Curtsinger JW, Fukui HH, Townsend DR, Vaupel JW (1992): Demography of genotypes: Failure of the limited life-span paradigm in *Drosophila melanogaster*. Science 258:461–463.

Curtsinger JW, Fukui HH, Xui L, Khazaeli A, Pletcher S (1994): Letters. Science 266:826.

Cutler RG (1982): Longevity is determined by specific genes: Testing the hypothesis. In Adelman RC, Roth GS (eds): Testing the Theories of Aging. Boca Raton, FL: CRC Press, pp 25–114.

Danner DB, Holbrook NJ (1990): Alterations in gene expression with aging. In Schneider EL, Rowe JW (eds): Handbook of the Biology of Aging, 3rd ed. San Diego: Academic Press, Inc., pp 97–115.

D'mello NP, Childress AM, Franklin DS, Kale SP, Pinswasdi C, Jazwinski SM (1994): Cloning and characterization of *LAG1*, a longevity-assurance gene in yeast. J Biol Chem 269:15451–15459.

Donati YRA, Slosman DO, Polla BS (1990): Oxidative injury and the heat shock response. Biochem Pharmacol 40:2571–2577.

Duhon SA, Johnson TE: Movement as an index of vitality: Comparing wild type and the *age-1* mutant of *Caenorhabditis elegans*. J Gerontol Biol Sci (in press).

Ebert RH, Cherkasova VA, Dennis RA, Wu JH, Ruggles S, Perrin TE, Shmookler Reis RJ (1993): Longevity-determining genes in *Caenorhabditis elegans*: Chromosomal mapping of multiple noninteractive loci. Genetics 135:1003–1010.

Effros RB, Zhu X, Walford RL (1994): Stress response of senescent T lymphocytes: Reduced hsp70 is independent of the proliferative block. J Gerontol Biol Sci 49:B65–B70.

Egilmez NJ, Chen JB, Jaswinski SM (1989): Specific alterations of transcript prevalence during the yeast life span. J Biol Chem 254:14312–14317.

Estevez M, Arrisano L, Wrana JL, Albert PS, Massague J, Riddle DL (1993): The *daf-4* gene encodes a bone morphogenetic protein receptor controlling *C. elegans* dauer development. Nature 365:644–649.

Fabian TJ, Johnson TE (1995): Identification of genes that are differentially expressed during aging in *Caenorhabditis elegans*. J Gerontol Biol Sci (in press).

Fargnoli J, Kunisada T, Fornace AJ Jr, Scheider EL, Holbrook NJ (1990): Decreased expression of heat shock protein 70 mRNA and protein after heat treatment in cells of aged rats. Proc Natl Acad Sci USA 87:846–850.

Fleming JE, Melnikoff PS, Latter GI, Chandra D, Bensch KG (1986a) Age-dependent changes in the expression of *Drosophila* mitochondrial proteins. Mech Ageing Dev 34:63–72.

Fleming JE, Quattrocki E, Latter G, Miquel J, Marcuson R, Zuckerkandl E, Bensch KG (1986b) Age-dependent changes in proteins of *Drosophila melanogaster*. Science 231:1157–1159.

Fleming JE, Reveillaud I, Niedzwiecki A (1992): Role of oxidative stress in *Drosophila* aging. Mutat Res 275:267–279.

Fleming JE, Spicer GS, Garrison RC, Rose MR (1993): Two-dimensional protein electrophoretic analysis of postponed aging in *Drosophila*. Genetics 91:183–198.

Friedman DB, Johnson TE (1988a): A mutation in the *age-1* gene in *Caenorhabditis elegans* lengthens life and reduces hermaphrodite fertility. Genetics 118:75–86.

Friedman DB, Johnson TE (1988b): Three mutants that extend both mean and maximum life span of the nematode, *Caenorhabditis elegans*, define the *age-1* gene. J Gerontol Biol Sci 43:B102–B109.

Frydman J, Nimmesgern E, Ohtsuka K, Hartl U (1994): Folding of nascent polypeptide chains in a high molecular mass assembly with molecular chaperones. Nature 370:111–117.

Gafni A (1985): Age-related modifications in a muscle enzyme. In Adelman RC, Dekker EE (eds): Modifications of Proteins During Aging. New York: Alan R Liss, pp 19–39.

Georgi LL, Albert PS, Riddle DL (1990): *daf-1*, a *C. elegans* gene controlling dauer larva development, encodes a novel receptor protein kinase. Cell 61:635–645.

Gottlieb S, Ruvkun G (1994): *daf-2, daf-16* and *daf-23*: Genetically interacting genes controlling dauer formation in *Caenorhabditis elegans*. Genetics 137:107–120.

Gould AB, Clark AM (1977): X-ray induced mutations causing adult life-shortening in *Drosophila melanogaster*. Exp Gerontol 12:107–112.

Graves JL, Luckinbill LS, Nichols A (1988): Flight duration and wing beat frequency in long- and short-lived *Drosophila melanogaster*. J Insect Physiol 34:1021–1026.

Graves JL, Mueller LD (1993): Population density effects on longevity. Genetica 91:99–109.

Graves JL, Rose MR (1990): Flight duration in *Drosophila melanogaster* selected for postponed senescence. In Harrison DE (ed): Genetic Effects on Aging II. Caldwell, NJ: Telford Press, pp 59–65.

Harman D (1956): Aging: A theory based on free radical and radiation chemistry. J Gerontol 11:298–300.

Heydari AR, Wu B, Takahashi R, Strong R, Richardson A (1993): Expression of heat shock protein 70 is altered by age and diet at the level of transcription. Mol Cell Biol 13:2909–2918.

Hutchinson EW, Rose MR (1990): Quantitative genetic analysis of postponed aging in *Drosophila melanogaster*. In Harrison DE (ed): Genetic Effects on Aging II. Caldwell, NJ: Telford Press, pp 65–85.

Ishii N, Suzuki N, Hartman PS, Suzuki K. (1994): The effects of temperature on the longevity of a radiation-sensitive mutant *rad-8* of the nematode *Caenorhabditis elegans*. J Gerontol Biol Sci 49:B117–B120.

Johnson TE (1984): Analysis of the biological basis of aging in the nematode, with special emphasis on *Caenorhabditis elegans*. In Johnson TE, Mitchell DH (eds): Invertebrate Models in Aging Research. Boca Raton, FL: CRC Press, pp 59–93.

Johnson TE (1987a): Aging can be genetically dissected into component processes using long-lived lines of *Caenorhabditis elegans*. Proc Natl Acad Sci USA 84:3777–3781.

Johnson TE (1987b): Developmentally programmed aging: Future directions. In Warner HR (ed): Modern Biological Theories of Aging. New York: Raven Press, pp 63–76.

Johnson TE (1988): Thoughts on selection of long-lived rodents. Growth Dev Aging 52:207–209.

Johnson TE (1990): The increased life span of *age-1* mutants in *Caenorhabditis elegans* results from lowering the Gompertz rate of aging. Science 249:908–912.

Johnson TE (1993): Genetic influences on aging in mammals and invertebrates. Aging Clin Exp Res 5:299–307.

Johnson TE, Friedman DB, Foltz N, Fitzpatrick PA, Shoemaker JE (1990): Genetic variants and mutations of *Caenorhabditis elegans* provide tools for dissecting the aging processes. In Harrison DE (ed): Genetic Effects on Aging, Vol II, Caldwell, NJ: Telford Press, pp 101–126.

Johnson TE, Hutchinson EW (1993): Absence of strong heterosis for life span and other life history traits in *Caenorhabditis elegans*. Genetics 134:463–474.

Johnson TE, Lithgow GJ (1992): The search for the genetic basis of aging: The identification of gerontogenes in the nematode *Caenorhabditis elegans*. J Am Geriatr Soc 40:936–945.

Johnson TE, McCaffrey G (1985): Programmed aging or error catastrophe? An examination by two-dimensional polyacrylamide gel electrophoresis. Mech Ageing Dev 30:285–297.

Johnson TE, Tedesco PM, Lithgow GJ (1993): Comparing mutants, selective breeding, and transgenics in the dissection of aging processes of *Caenorhabditis elegans*. Genetica 91:65–77.

Johnson TE, Wood WB (1982): Genetic analysis of lifespan in *Caenorhabditis elegans*. Proc Natl Acad Sci USA 79:6603–6607.

Kenyon C, Chang J, Gensch E, Rudner A, Tabtiang R (1993) A *C. elegans* mutant that lives twice as long as wild type. Nature 366:461–464.

Klass MR (1983): A method for the isolation of longevity mutants in the nematode *Caenorhabditis elegans* and initial results. Mech Ageing Dev 22:679–286.

Klass M, Hirsh D (1976): Non-ageing developmental variant of *Caenorhabditis elegans*. Nature 260:523–525.

Larsen PL (1992): Aging and resistance to oxidative damage in *Caenorhabditis elegans*. Proc Natl Acad Sci USA 90:8905–8909.

Larsen PL, Albert PS, Riddle DL (1995): Genes that regulate both development and longevity in *Caenorhabditis elegans*. Genetics 139:1567–1583.

Lithgow GJ, White TM, Hinerfeld DA, Johnson TE (1994): Thermotolerance of a long-lived mutant of *Caenorhabditis elegans*. J Gerontol Biol Sci 49:B270–B276.

Lithgow GJ, White TM, Melov S, Johnson TE (1995): Longevity mutants of *Caenorhabditis elegans* exhibit increased intrinsic thermotolerance. Proc Natl Acad Sci USA (in press).

Liu AY-C, Lin Z, Choi HS, Sorhage F, Li B (1989): Attenuated induction of heat shock gene expression in aging diploid fibroblasts. J Biol Chem 164:12037–12045.

Luckinbill LS, Arking R, Clare MJ, Cirocco WC, Muck SA (1984): Selection for delayed senescence in *Drosophila melanogaster*. Evolution 38:996–1003.

Luckinbill LS, Clare R (1986): A density threshold for the expression of longevity in *Drosophila melanogaster*. Heredity 56:329–335.

Luckinbill LS, Graves JL, Reed AH, Koetsawang S (1988) Localizing genes that defer senescence in *Drosophila melanogaster*. Heredity 60:367–374.

Luckinbill LS, Riha V, Rhine S, Grudzien TA (1990): The role of glucose-6-phosphate dehydrogenase in the evolution of longevity in *Drosophila melanogaster*. Heredity 65:29–38.

Malone EA, Thomas JH (1994): A screen for nonconditional dauer-constitutive mutations in *Caenorhabditis elegans*. Genetics 136:879–886.

Martin GM, Turker MS (1988): Model systems for the genetic analysis of mechanisms of aging. J Gerontol Biol Sci 43:B33–B39.

Mayer PJ, Baker III GT (1985): Genetic aspects of *Drosophila* as a model system of eukaryotic aging. Int Rev Cytol 95:61–102.

Maynard Smith J (1958): The effects of temperature and of egg-laying on the longevity of *Drosophila subobscura*. J Exp Biol 35:832–842.

Meheus LA, Van Beeumen JJ, Coomans AV, Vanfleteren JR (1987): Age-specific nuclear proteins in the nematode worm *Caenorhabditis elegans*. Biochem J 245:257–261.

Melov S, Hertz GZ, Stormo GD, Johnson TE (1994): Detection of deletions in the mitochondrial genome of *Caenorhabditis elegans*. Nucleic Acids Res 22:1075–1078.

Melov S, Lithgow GJ, Fischer DR, Tedesco PM, Johnson TE (1995): Increased deletions in the mitochondrial genome with increased age of *Caenorhabditis elegans*. Submitted.

Morimoto RI, Tissieres A, Georgopoulos C (1994): The Biology of Heat Shock Proteins and Molecular Chaperones. Cold Spring Harbor, NY: Cold Spring Harbor Laboratory Press.

Munkres KD, Furtek CA (1984): Selection of conidial longevity mutants of *Neurospora crassa*. Mech Ageing Dev 25:47.

Noy N, Schwartz H, Gafni A (1985): Age-related changes in the redox status of rat muscle cells and their role in enzyme-aging. Mech Ageing Dev 29:63–69.

Orr WC, Sohal RS (1992): The effects of catalase gene overexpression on life span and resistance to oxidative stress in transgenic *Drosophila melanogaster*. Arch Biochem Biophys 297:35–41.

Orr WC, Sohal RS (1993): Effects of Cu-Zn superoxide dismutase overexpression on life span and resistance to oxidative stress in transgenic *Drosophila melanogaster*. Arch Biochem Biophys 301:34–40.

Orr WC, Sohal RS (1994): Extension of life-span by overexpression of superoxide dismutase and catalase in *Drosophila melanogaster*. Science 263:1128–1130.

Osiewacz HD (1992): The genetic control of aging in the ascomycete *Podospora anserina*. In Zwilling R, Balduini C (eds): Biology of Aging. Berlin: Springer Verlag, pp 153–164.

Parsell DA, Lindquist S (1994): Heat shock proteins and stress tolerance. In Morimoto RI, Tissieres A, Georgopoulos C (eds): The Biology of Heat Shock Proteins and Molecular Chaperones. Cold Spring Harbor, NY: Cold Spring Harbor Laboratory Press, pp 457–494.

Pereira-Smith OM, Smith JR (1988): Genetic analysis of indefinite division in human cells: Identification of four complementation groups. Proc Natl Acad Sci USA 85:6042–6046.

Rattan SIS (1985): Beyond the present crisis in gerontology. Bioessays 2:226–228.

Rattan SIS, Derventzi A, Clark B (1992): Protein synthesis, posttranslational modifications, and aging. Ann NY Acad Sci 663:48–62.

Rea IM, Middleton D (1994): Is the phenotypic combination A1 B8Cw7DR3 a marker for male longevity? J Am Geriatr Soc 42:978–983.

Reff ME (1985): RNA and protein metabolism. In Finch CE, Schneider EL (eds): Handbook of the Biology of Aging, 2nd ed. New York: Van Nostrand-Reinhold, pp 225–254.

Reveillaud IA, Kongpachith R, Park, Fleming JE (1992): Stress resistance of *Drosophila* transgenic for bovine CuZn superoxide dismutase. Free Rad Res Commun 17:73–85.

Reznik AZ, Gershon D (1979): The effect of age on the protein degradation system in the nematode *Turbatrix aceti*. Mech Ageing Dev 11:403–415.

Richardson A, Semsei I (1987): Effect of aging on transcription and translation. Rev Biol Res Aging 3:467–483.

Riddle DL (1988): The dauer larva. In Wood WB (ed): The Nematode *Caenorhabditis elegans*. Cold Spring Harbor, NY: Cold Spring Harbor Laboratory Press, pp 393–412.

Roberts PA, Iredale RB (1985): Can mutagenesis reveal major genes affecting senescence? Exp Gerontol 20:119–121.

Rose MR (1984a): Laboratory evolution of postponed senescence in Drosophila melanogaster. Evolution 38:1004–1010.

Rose MR (1984b): Genetic covariation in Drosophila life history: Untangling the data. Am Nat 123:565.

Rose MR (1991): Evolutionary Biology of Aging. New York: Oxford University Press.

Rose MR, Dorey AM, Coyle AM, Service PM (1984): The morphology of postponed senescence in Drosophila melanogaster. Can J Zool 62:1576–1580.

Rose MR, Vu LN, Park SU, Graves JL Jr (1992): Selection on stress resistance increases longevity in Drosophila melanogaster. Exp Gerontol 27:241–250.

Russell RL (1987): Evidence for and against the theory of developmentally programmed aging. In Warner HR, Butler RM, Sprott RL, Schneider EL (eds): Modern Biological Theories of Aging. New York: Raven, pp 35–61.

Schächter F, Faure-Delanef L, Guénot F, Rouger H, Froguel P, Lesueur-Ginot L, Cohen D (1994): Genetic associations with human longevity at the APOE and ACE loci. Nat Genet 6:29–32.

Service PM, Hutchinson EW, MacKinley MD, Rose MR (1985): Resistance to environmental stress in Drosophila melanogaster selected for postponed senescence. Physiol Zool 58:380–389.

Sharma HK, Prasanna HR, Lane RS, Rothstein M (1979): The effect of age on enolase turnover in the free-living nematode, Turbatrix aceti. Arch Biochem Biophys 194:275–282.

Sharma HK, Rothstein M (1980): Altered enolase in aged Turbatrix aceti results from conformational changes in the enzyme. Proc Natl Acad Sci USA 77:5865–5868.

Shepherd JCW, Walldorf U, Hug P, Gehring WJ (1989) Fruit flies with additional expression of the elongation factor EF-1α live longer. Proc Natl Acad Sci USA 86:7520–7521.

Shook DR, Brooks A, Johnson TE (1996): Mapping quantitative trait loci specifying hermaphrodite life span or self fertility in the nematode Caenorhabditis elegans. Genetics (in press).

Stearns, SC, Kaiser, M (1993): The effects of enhanced expression of EF-1α on lifespan in Drosophila melanogaster. Genetica 91:167–182.

Sun J, Kale SP, Childress AM, Pinswasdi C, Jazwinski SM (1994): Divergent roles of RAS1 and RAS2 in yeast longevity. J Biol Chem 269:18638–18645.

Takata H, Susuki M, Ishii T, Sekiguchi S, Iri H (1987): Influence of major histocompatability complex region genes on human longevity among Okinawan-Japanese centenarians and nonagenarians. Lancet II: 824–826.

Takeda T, Hosokawa M, Takeshita S, Irino M, Higuchi K, Matsushita T, Tomita Y, Yauhira K, Hamamoto H, Shimizu K, Ishii M, Yamamuro T (1981): A new murine model of accelerated senescence. Mech Ageing Dev 17:183–194.

Thakur MK, Oka T, Natori Y (1993): Gene expression and aging. Mech Aging Dev 66:283–298.

Tudzynski P, Esser K (1979): Chromosomal and extrachromosomal control of senescence in the ascomycete Podospora anserina. Mol Gen Genet 173:71–84.

Tyler RH, Brar H, Singh M, Latorre A, Graves JL, Mueller LD, Rose MD, Ayala FJ (1993): The effect of superoxide dismutase alleles on aging in Drosophila. Genetica 91:143–149.

Udelsman R, Blake MJ, Stagg CA, Li D, Putney DJ, Holbrook NJ (1993): Vascular heat shock protein expression in response to stress. J Clin Invest 91:465–473.

Vanfleteren JR (1993): Oxidative stress and ageing in Caenorhabditis elegans. Biochem J 292:605–608.

Vaupel JW, Johnson TE, Lithgow GJ (1994): Rates of mortality in populations of Caenorhabditis elegans (Technical Comment). Science 266:826.

VanVoorhies WA (1992): Production of sperm reduces nematode lifespan. Nature 360:456–458.

Wang J-L, Müller H-G, Capra WB, Carey JR (1994): Letters. Science 266:827.

Welte MA, Tetrault JM, Dellavalle RP, Lindquist SL (1993): A new method for manipulating transgenes: Engineering heat tolerance in a complex, multicellular organism. Curr Biol 3:842–853.

Williams BD, Schrank B, Huynh C, Shownkeen R, Waterston RH (1992): A genetic mapping system in Caenorhabditis elegans based on polymorphic sequence-tagged-sites. Genetics 131:609–624.

Wilson DL, Hall ME, Stone GC (1978): Test of some aging hypotheses using two-dimensional protein mapping. Gerontology 24:426–433.

Wu B, Gu MJ, Heydari AR, Richardson A (1993): The effect of age on the synthesis of two heat shock proteins in the hsp70 family. J Gerontol Biol Sci 48:B50–B56.

Yunis EJ, Salazar M (1993): Genetics of life span in mice. Genetica 91:211–223.

Yunis EJ, Watson ALM, Gelman RS, Sylvia SJ, Bronson R, Dorf ME (1984): Traits that influence longevity in mice. Genetics 108:999–1011.

Zeelon P, Gershon H, Gershon D (1973): Inactive enzyme molecules in aging organisms. Nematode fructose-1, 6-diphosphate aldolase. Biochemistry 12:1743–1750.

ABOUT THE AUTHORS

THOMAS E. JOHNSON is Associate Professor for Behavioral Genetics at the University of Colorado at Boulder, where he teaches undergraduate and graduate courses including such topics as behavioral genetics, molecular genetics, and gerontology. After receiving a B.Sc. in 1970 from the Massachusetts Institute of Technology, he pursued a Ph.D. (1975) at the University of Washington in Seattle under the supervision of Dr. Benjamin Hall for work on *Neurospora crassa*. This was followed by postdoctoral research at Cornell University in Ithaca, NY with Dr. Adrian Srb and at the University of Colorado with Dr. Wood, where he began the dissection of the aging processes in *Caenorhabditis elegans* using genetic approaches. Among numerous other awards, he is the 1993 recipient of the Busse Research Award for Biomedical Gerontology, presented at the International Association for Gerontology meeting in Budapest, Hungary, and the 1995 Nathan Shock Award for the National Institute on Aging. Dr. Johnson is on the Board of Managing Editors for *Mutation Research DNAging* and is Associate Editor for both *Experimental Cerontology* and *Journals of Gerontology, Biological Sciences*. He is also Chair for the Gordon Conference on the Biology of Aging in 1997. His work continues to focus on the genetic basis of the aging processes, primarily in *C. elegans*.

GORDON J. LITHGOW was a Research Associate at the Institute for Behavioral Genetics at the University of Colorado at Boulder; he is currently an Instructor at the University of Manchester. After receiving a B.Sc. (Honors) in Applied Microbiology in 1985 from the University of Strathclyde in Glasgow, Scotland, he obtained a Ph.D. (1989) in Genetics at the University of Glasgow under the supervision of Dr. A.J.P. Brown for work on the yeast *Sarccharomyces cerevisiae*. This was followed by two years of postdoctoral research at Ciba Geigy AG in Basel, Switzerland, and nearly four years at the University of Colorado with Dr. Thomas E. Johnson in molecular genetics of aging in the nematode. Dr. Lithgow has had several papers and reviews published and has a book chapter in press. His research focuses on life-span extension in a series of long-lived mutations in *C. elegans*, specifically the role of molecular chaperones in maintaining protein conformation in aging worms.

SHIN MURAKAMI is a Research Associate at the Institute for Behavioral Genetics at the University of Colorado at Boulder, where he does research on molecular biology of *C. elegans*. He received his B.Sc. in 1988 and his Ph.D. in 1993 from Kyoto University in Japan. His doctoral and one year of postdoctoral research was done on the molecular genetics/biology of the cell cycle and chromosome segregation in yeast in the laboratory of Drs. Mitsuhiro Yanagida and Osami Niwa. He continues his postdoctoral research in molecular biology of aging in the nematode *C. elegans* in the laboratory of Dr. Thomas E. Johnson. Dr. Murakami served as a Fellow of the Japan Society of Promotion of Science (JSPS) from 1992 to 1994 and is currently a Fellow of the JSPS for Research Abroad. His current research involves exploration of the genetic and molecular controls underlying longer life mutations in *C. elegans*.

STACEY A. DUHON is a graduate student at the University of Colorado at Boulder, where she is pursuing her doctoral degree in Psychology. After receiving a B.S. in 1989 from Grambling State University in Louisiana, she obtained her master's degree at the University of Colorado in the Laboratory of Tom Johnson, for work on Aging in the nematode *C. elegans*. In 1993, she became the graduate student representative for the Gerontological Society of America, Biological Section. Her doctoral research focuses specifically on identifying and characterizing genes that extend life span.

DAVID R. SHOOK is a graduate student at the University of Colorado at Boulder, where he is pursuing a doctoral degree in Environmental, Population and Organismic Biology. After receiving an A.B. in 1986 from the University of California at Berkeley, he obtained his master's degree at the University of Colorado in the laboratory of Dr. Mike Klymkowsky for work on the role of intermediate filaments during gastrulation in the frog *Xenopus*. His current doctoral thesis work is on the quantitative genetics of aging in the nematode *C. elegans* in the laboratory of Dr. Thomas Johnson.

Cellular Aging and Cell Death: 19–34
© 1996 Wiley-Liss, Inc.

Mitochondrial DNA Mutations and Aging

Eric A. Schon, Monica Sciacco, Francesco Pallotti, Xi Chen, and Eduardo Bonilla

I. OVERVIEW

Mitochondria are the main sources of energy in the cell. They contain their own DNA (mtDNA), whose genes encode components of the respiratory chain/oxidative phosphorylation system. Mitochondria, which are maternally inherited, are essential for the normal functioning of all cells in the body and are absolutely critical for the function of those tissues that are highly dependent on aerobic metabolism, such as muscle and brain.

Pathological defects in mitochondrial function result in a heterogeneous group of disorders known collectively as the *mitochondrial encephalomyopathies*, but until a few years ago the molecular bases of these defects were completely unknown. Beginning in 1988, a number of laboratories found that unique giant deletions of mtDNA (Δ-mtDNAs) were associated with Kearns-Sayre syndrome (KSS), a spontaneous and often-fatal disorder characterized by progressive external ophthalmoplegia (PEO), respiratory chain deficiency, and massive mitochondrial proliferation ("ragged-red fibers" [RRF]) in muscle. A related disorder—familial PEO—was found to be an autosomal dominantly inherited disease that was characterized by large amounts of *multiple* species of Δ-mtDNAs in muscle.

In the last few years, the field of mitochondrial genetics has taken a major turn: Deletions in mitochondrial DNA that had previously been thought to occur only in patients with these often-fatal mitochondrial encephalomyopathies have now also been found to occur at extremely low levels (observable only by the polymerase chain reaction [PCR]) in normal individuals, with the amount varying among tissues. Moreover, these mutations appear to accumulate with age. While any single species of Δ-mtDNA may be present in some age tissues at levels in the neighborhood of only 0.1% of total mtDNA, it could well be that up to 5%–10% of total mtDNA in aged muscle may be mutated, resulting in a level of dysfunctional mitochondria likely to result in clinical consequences. Importantly, similar accumulations of Δ-mtDNAs have been found in elderly brain and also appear to be distributed heterogeneously in different brain regions. These findings may have relevance to age-related cognitive impairment, including neurodegenerative diseases such as Alzheimer's, Huntington's, and Parkinson's diseases.

These unexpected and surprising findings are the first pieces of evidence to validate, at the molecular genetic level, the "mitochondrial theory of aging." Moreover, because the mitochondrial respiratory chain is the single greatest source of free radicals in the cell, these data may also be the beginning of the unification of this hypothesis with that of the "free-radical theory of aging."

II. MITOCHONDRIAL GENETICS

The human mitochondrial genome (Fig. 1) is a 16,569 bp circle of double-stranded DNA [Anderson et al., 1981]. It contains genes encoding two ribosomal RNAs, 22 transfer RNAs, and 13 polypeptides, all of which are subunits of components of the respiratory chain complexes. Of the 13 structural genes, 7 encode subunits of complex I (NADH–CoQ oxidoreductase), 1 encodes the cytochrome *b*

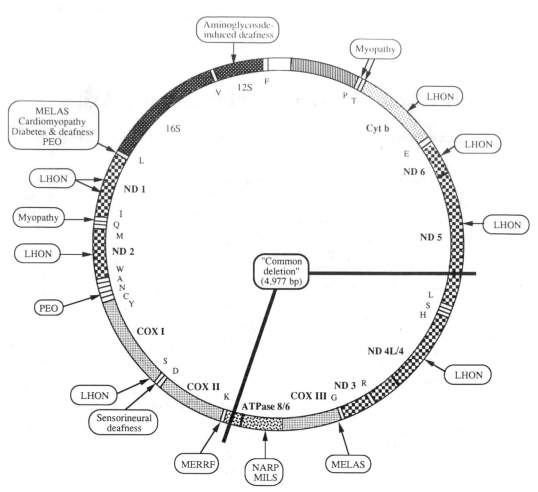

Fig. 1. *Morbidity map of the human mitochondrial genome. The structural genes for the mtDNA-encoded 12S and 16S ribosomal RNAs, the subunits of NADH–coenzyme Q oxidoreductase (ND), cytochrome c oxidase (COX), cytochrome b (Cyt b), and ATP synthetase (ATPase) and 22 tRNAs (1-letter amino acid nomenclature) are shown. A typical mtDNA deletion (the "common deletion"), as well as the various known pathogenic point mutations, are indicated (see text).*

subunit of complex III (CoQ–cytochrome *c* oxidoreductase), 3 encode subunits of complex IV (cytochrome *c* oxidase [COX]), and 2 encode subunits of complex V (ATP synthetase). Each of these complexes also contains subunits encoded by nuclear genes, which are imported from the cytoplasm and assembled, together with the mtDNA-encoded subunits, into the respective holoenzymes, each of which is located in the mitochondrial inner membrane. Complex II (succinate dehydrogenase–CoQ oxidoreductase), of which succinate dehydro-

genase (SDH) is a component, is encoded entirely by nuclear genes; SDH thus serves as an independent marker for mitochondrial number and activity.

Mitochondria (and mtDNAs) are maternally inherited [Giles et al., 1980]. Thus, pathological defects in mtDNA-encoded genes should result in pedigrees exhibiting *maternal inheritance*, i.e., the disease should pass only through females, and essentially all children (both boys and girls) should inherit the error. Moreover, because there are hundreds or even thousands of mitochondria in each cell, with an average of 5 mtDNAs per organelle [Satoh and Kuroiwa, 1991], mutations in mtDNA may result in two populations of mtDNAs (wild type and mutated), a condition known as *heteroplasmy*.

From the foregoing, a crucial fact emerges that impinges on the role of mtDNA in normal and diseased states: Mitochondrial genetics is *population* genetics [Wallace, 1992]. Both mtDNA replication and mitochondrial division are stochastic processes unrelated to the cell cycle or to the timing of nuclear DNA replication. Thus, a dividing cell may potentially donate a different complement of organelles and genomes to its progeny (i.e., *mitotic segregation*). This process becomes clinically important if an individual contains a heteroplasmic population of both wild-type and mutated mtDNAs causing a mitochondrial disease. The phenotypic expression of a mutation may vary in both space (among cells or tissues) and time (during development or during the course of a life span), based merely on the random processes of mitotic segregation. Of course, there may also be active selection processes going on as well, in which certain cells may either eliminate or concentrate a population of mutant mtDNAs. These effects will combine to generate, for example, a respiratory chain deficiency in some tissues but not others, but only if the number of mutant mtDNAs exceeds a certain *threshold*. This threshold varies from tissue to tissue and is related to the requirements for aerobic respiration and energy production, with brain, retina, and muscle exhibiting the highest energy requirements. Thus, many mitochondrial disorders are *encephalomyopathies* [DiMauro et al., 1990], although other organ systems—notably kidney, blood, and pancreas—are often affected as well.

III. MITOCHONDRIAL DISEASES

The two main biochemical features in many (but not all) mitochondrial diseases are *lactic acidosis* and *respiratory chain deficiency*: Lactate is often elevated in both blood and cere-brospinal fluid, and the enzymatic activities of specific respiratory chain complexes are often decreased. The morphological hallmark of mitochondrial diseases is the presence of RRF in the muscle biopsy. RRF are observed with the modified Gomori trichrome stain, which visualizes the massive accumulation of mito-chondria as purple-red blotches against a tur-quoise background [Engel and Cunningham, 1963]. Alternatively, RRF may be visualized using cytochemistry to detect the increased activity of SDH in a subset of muscle fibers [Bonilla et al., 1992]. Not all mitochondrial diseases display RRF, but the presence of RRF is a nearly invariant sign that a mitochondrial disorder is present.

III.A. Pathogenic Point Mutations in mtDNA

Numerous mtDNA point mutations have been described, located in all regions of the genome [for review, see Schon et al., 1994]. Among the more notable are described.

III.A.1. Leber's hereditary optic neuropathy (LHON). LHON was the first mitochondrial disease to be defined at the molecular level. Wallace and coworkers [1988] found maternal inheritance in a number of LHON pedigrees of a G→A transition at mtDNA position 11778, in codon 340 of the ND4 gene of complex I. This mutation was nonconservative and converted the codon from Arg→His. The mutation is homoplasmic in some families but heteroplasmic in others. Numerous other mutations have now been found that are also associated with LHON. These include mutations in the ND1, ND2, ND4, ND5, ND6, CO I, and cyt *b* genes [reviewed by Wallace, 1992].

III.A.2. Myoclonus epilepsy with ragged-red fibers (MERRF). MERRF is characterized by myoclonus, ataxia, weakness, generalized seizures, mental retardation, and hearing loss [Fukuhara et al., 1980]. Muscle biopsies show RRF. MERRF has been found to be associated with an A→G transition at nt-8344 in the tRNA^{Lys} gene [Shoffner et al., 1990; Yoneda et al., 1990]. A second mtDNA mutation has also been found in MERRF patients, and interestingly it, too, is in the tRNA^{Lys} gene, at nt-8356 [Silvestri et al., 1992; Zeviani et al., 1993]. Both mutations are heteroplasmic.

III.A.3. Mitochondrial encephalomyopathy with lactic acidosis and stroke-like episodes (MELAS).
MELAS is characterized by seizures, migraine-like headaches, lactic acidosis, episodic vomiting, short stature, mental retardation, and recurrent cerebral stroke-like episodes, causing hemiparesis, hemianopia, or cortical blindness [Hirano and Pavlakis, 1994]. MELAS is associated mainly with two different point mutations. The first mutation, found in about 80% of cases, is an A→G transition at nt-3243 in the tRNA$^{Leu(UUR)}$ gene [Goto et al., 1990]; the second mutation, found in about 10% of cases, is an A→G transition at nt-3271, also in the tRNA$^{Leu(UUR)}$ gene [Goto et al., 1991]. All mutations in MELAS are heteroplasmic, with the proportion of mutated mtDNAs usually exceeding 80% in muscle [Ciafaloni et al., 1992].

III.A.4. Maternally inherited PEO.
The tRNA$^{Leu(UUR)}$ mutation at nt-3243 is not confined exclusively to MELAS. PEO is characterized by paralysis of the extraocular muscles and is often accompanied by some degree of limb weakness. Importantly, PEO has also been associated with sporadic mtDNA deletions (see below). Within the group of "deletion-minus" PEO patients, approximately one-third had a positive family history; and most of these patients have now been found to harbor the MELAS-3243 mutation [Moraes et al., 1993]. There are qualitative differences between the two disorders. We have now found significant differences in the localized concentration of the nt-3243 mutation at the single muscle fiber level in PEO-3243 compared with those in MELAS-3243, implying that the pattern of the spatial distribution of mutant mtDNAs at the cellular level leads to clinically distinct phenotypes [Petruzzella et al., 1994].

III.A.5. NARP and MILS.
Neuropathy, ataxia, and retinitis pigmentosa (NARP) is characterized by developmental delay, dementia, retinitis pigmentosa, seizures, ataxia, proximal neurogenic muscle weakness, and sensory neuropathy. It is associated with a T→G transversion at nt-8893 of the ATPase 6 gene (i.e., subunit 6 of complex V), which converts amino acid 156 from Leu→Arg [Holt et al., 1990]. The mutation has always been found to be heteroplasmic, and the clinical severity appeared to be correlated with the amount of mutant mitochondrial genomes present in heteroplasmic family members. Another mutation at nt-8993 in the ATPase 6 gene is also associated with NARP: a T→C transition converting Leu→Pro [de Vries et al., 1993; Santorelli et al., 1994]. When the proportion of the NARP-8993 mutation is extremely high (>90%), it produces a clinically different phenotype—maternally inherited Leigh syndrome (MILS) [Tatuch et al., 1992]—which is a devastating and fatal encephalopathy. Because the mutation is in the ATPase 6 gene, it has been postulated that the fundamental defect in the NARP/MILS mutation lies in the proton channel portion of complex V (the F_0 portion of ATPase), with the positively charged arginine residue interfering with the mobility of protons as they traverse the channel in the F_0 "stalk" of the ATPase complex [Tatuch et al., 1992].

III.A.6. Other point mutations.
There are mutations in a number of other tRNA genes [Schon et al., 1994], and a point mutation in 12S rRNA associated with maternally inherited cochlear deafness [Prezant et al., 1993].

III.B. Pathogenic mtDNA Rearrangements: KSS and Sporadic Ocular Myopathy (OM)

OM is due to a paralysis of the extraocular muscles (PEO) that causes ptosis (droopy eyelids) and an inability to move the eyes. PEO can also be part of a much more severe, and ultimately fatal, multisystem disorder called Kearns-Sayre syndrome (KSS). KSS is defined by the triad of PEO, pigmentary retinopathy, and onset before age 20, plus at least one of the following: CSF protein content above 100 mg/dl, heart block, or cerebellar ataxia. Common but nonspecific features include dementia, neurosensory hearing loss, and endocrine abnormalities (short stature, diabetes, hypoparathyroidism). Muscle biopsy shows RRF.

OM shares with KSS ocular myopathy and RRF in muscle, but, unlike KSS, there is no systemic involvement, and the disease is usually compatible with a normal lifespan. In both KSS and OM, biochemical studies of muscle often show multiple respiratory chain enzyme defects, especially COX deficiency.

Historically, PEO and KSS were difficult to classify, because most patients were *sporadic*, with no apparent genetic component. However, in 1988, giant deletions of mtDNA (Δ-mtDNA)—up to 9 kb of the mitochondrial genome—were documented in most KSS patients and in many sporadic PEO patients [Holt et al., 1988; Zeviani et al, 1988; Moraes et al., 1989]. In these disorders, the Δ-mtDNAs can be observed easily by Southern blot hybridization analysis as a large population (up to 80% of total mtDNA) of a *single* mtDNA species migrating more rapidly than full-length mtDNAs in electrophoretic gels. All patients with deletions have been heteroplasmic. The size and location of the deletion and the proportion of Δ-mtDNA differs among patients and does not appear to be correlated with the presentation or the severity of the disease phenotype. One particular Δ-mtDNA has been found in about one-third of all patients and has therefore been called the *common deletion* [Schon et al., 1989; Mita et al., 1990]. It is 4,977 bp long and removes DNA between the ATPase 8 and the ND5 genes (see Fig. 1). Deletions of mtDNAs have also been identified in Pearson's pancreas/marrow syndrome, a hematopoietic disease of infancy characterized by refractory sideroblastic anemia, vacuolization of bone marrow precursors, and exocrine pancreas dysfunction, in which the population of Δ-mtDNA is most pronounced in blood [Rötig et al., 1989]. A few patients with Pearson's syndrome survive their blood disorder and live long enough to show the symptoms of KSS, as the Δ-mtDNA population declines in blood and begins to accumulate in muscle and brain. It appears that Δ-mtDNAs are transcribed into RNA but are not translated, probably because the deletions remove essential tRNAs that are required for protein synthesis [Nakase et al., 1990]. For this reason, even genes not encompassed by the deletion are not translated.

Only a single species of Δ-mtDNA is found in patients with sporadic PEO, KSS, or Pearson's syndrome, implying that the population of Δ-mtDNAs is a *clonal expansion of an initial mutation event* occurring early in oogenesis or embryogenesis [Chen et al., 1995b]. This would also explain why these diseases arise spontaneously. It is currently not clear whether a woman with KSS can transmit the disease to her children.

Besides sporadic Δ-mtDNAs in KSS and PEO, *autosomal dominant* inheritance of Δ-mtDNAs has also been demonstrated in a disorder characterized by PEO and mitochondrial myopathy [Zeviani et al., 1989]. In autosomal dominant PEO (AD-PEO), different deletions coexist within the same muscle in different affected family members. Thus, as opposed to the "single" deletions found in sporadic PEO, AD-PEO is associated with *multiple deletions* that are apparently generated over the life span of the individual. The genetic defect in AD-PEO most likely resides in a nuclear gene product affecting the proclivity of the mtDNA to suffer deletions.

In an important new finding, adult-onset noninsulin-dependent type II diabetes mellitus (NIDDM), often accompanied by deafness, has been found in maternally related members of families. In many cases, the muscle biopsy showed few or no RRF, but biochemical analyses showed combined defects of complexes I, III, and IV. Some of these families harbored large-scale mtDNA duplications and/or deletions [Ballinger et al., 1992, 1994; Dunbar et al., 1993], whereas others harbored mtDNA point mutations [Reardon et al., 1992; van den Ouweland et al., 1992, 1994; Remes et al., 1993]. Approximately 1%–2% of NIDDM patients examined to date have harbored mtDNA mutations [Kadowaki et al., 1994].

There are two recurrent themes that run through these descriptions of mitochondrial diseases. First, the symptoms are remarkably heterogeneous and affect numerous organ systems. Second, and perhaps even more relevant to aging, the symptoms often encountered in these disorders are those that are more commonly associated with aging. These include endocrine abnormalities (e.g., diabetes), cognitive abnormalities (e.g., progressive mental deterioration, dementia), motor abnormalities (e.g., myopathy, ataxia), cardiovascular abnormalities (e.g., cardiomyopathy, conduction block), and sensory abnormalities (e.g., visual impairment, hearing loss). Thus, in a very real sense, many of the symptoms of mitochondrial disorders, which as a rule affect primarily the young, appear to be "aging writ large." It is with this conceptual background that investigators began to probe for a wider role for mtDNA mutations in normal individuals.

IV. MUTATIONS OF mtDNA IN AGING
IV.A. Genetic Analyses

In the last few years it has become clear that tissues from *normal individuals*, and especially terminally differentiated tissues with high oxidative requirements, such as muscle and brain, contain low amounts of deletions of mitochondrial DNA [Ikebe et al., 1990; Ozawa et al., 1990a,b; Linnane et al., 1990; Yen et al., 1991; Cortopassi and Arnheim, 1991, Cortopassi et al., 1992; Corral-Debrinski et al., 1991, 1992; Hattori et al., 1991a; Sugiyama et al., 1991; Zhang et al., 1992; Torii et al., 1992; Simonetti et al., 1992; Soong et al., 1992; Chen et al., 1993]. These Δ-mtDNAs, which are observable only after amplification by PCR, are qualitatively similar to the highly abundant populations of Δ-mtDNA that have been described in patients with sporadic PEO and KSS, sporadic Pearson's syndrome, and AD-PEO.

Ikebe et al. [1990] searched for the presence of Δ-mtDNAs in brain tissue of normal subjects and of patients with Parkinson's disease (PD). Focusing on the "common deletion," they found that this Δ-mtDNA species was present at extremely low levels in brain from all five PD patients analyzed (aged 51–77 years) and in brain from two normal age-

matched controls (aged 64 and 73 years), but not in four younger subjects (aged 38–57 years). Using a semiquantitative approach, they estimated that the amount of the "common deletion" was about 0.3% in control striatum and about 5% in PD striatum [Ozawa et al., 1990a]. Yen et al. [1991] found the "common deletion" in liver DNA from 5 of 8 subjects aged 31–50 years, in 9 of 11 subjects aged 41–50, and in all 34 subjects over age 50. Cortopassi and Arnheim [1990] found that normal adult heart and brain tissue, but not fetal heart of brain tissue, contained the "common deletion" and estimated that there was 0.1% of this Δ-mtDNA in the heart muscle of middle-aged adults. Finally, Δ-mtDNAs were also observed in heart tissue from patients with ischemic heart disease and other cardiac abnormalities [Ozawa et al., 1990b; Corral-Debrinski et al., 1991; Hattori et al., 1991a].

Because of the potential significance of accumulations of Δ-mtDNAs in aging, we felt that it was important to quantitate more accurately the amount of Δ-mtDNA in tissues. Accordingly, we developed a quantitative "competitive PCR" method [Siebert and Larrick, 1992] to measure more accurately the accumulation of Δ-mtDNA species, focusing on the "common deletion" as a marker Δ-mtDNA species [Simonetti et al., 1992]. Using this procedure, we calculated the amount of "common deletion" in total muscle DNA from 8 normal subjects, ranging in age from 0.5 to 84 years. There was a dramatic positive correlation between the amount of "common deletion" and the age of the subject, with an exponential 10,000-fold increase in this Δ-mtDNA over the course of the normal human life span. It must be stressed, however, that this single Δ-mtDNA species was present in aged muscle at a level of only 0.1% of total mtDNA.

We have since found other species of Δ-mtDNAs in aging [Chen et al., 1993], including a recently sequenced Δ-mtDNA species (Fig. 2), corroborating the work of others [Zhang et al., 1992]. Using qualitative PCR, we were usually able to observe PCR fragments corresponding to other unique deletions

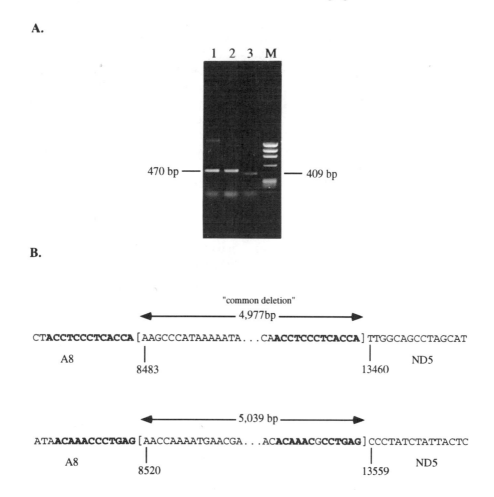

A.

B.

"common deletion"

— 4,977bp —

CT**ACCTCCCTCACCA** [AAGCCCATAAAAATA...CA**ACCTCCCTCACCA**] TTGGCAGCCTAGCAT

A8 | | ND5
                    8483   13460

— 5,039 bp —

ATA**ACAAACCCTGAG** [AACCAAAATGAACGA...AC**ACAAACGCCTGAG**] CCCTATCTATTACTC

A8 | | ND5
                    8520   13559

Fig. 2. *Example of a Δ-mtDNA found in normal human aging. A: PCR analysis showing amplification of mtDNA from three aged subjects by primers at nt-8273–8305 ("forward") and nt-13,720–13,692 ("backward"), showing amplified fragments (and sizes) of PCR products spanning the "common deletion" (lanes 1 and 2) and a previously unidentified deletion (lane 3). M, markers of HaeIII-digested* ΦX174 *DNA. B: Sequence of the PCR fragments in A, showing the region flanking the Δ-mtDNAs. The deletion size is indicated above the sequences (top, common deletion; bottom, newly discovered deletion). Note the direct repeats flanking the deletion (in bold). The genes flanking the deletion breakpoints (brackets) are indicated.*

found in KSS patients [Mita et al., 1990; Moraes et al., 1991]. These results imply that the common deletion, while frequently found in KSS patients, is certainly not the only Δ-mtDNA present in aged muscle and that it is quite likely that aged muscle harbors numerous (hundreds or even thousands of) species of Δ-mtDNA. Thus, even though any single species of Δ-mtDNA is present at extremely low levels, the *total* amount of Δ-mtDNAs may

reach levels that could be physiologically significant in terms of the decline in oxidative metabolism in aging. This is currently an active area of investigation.

In humans, some tissues, such as skeletal and heart muscle, accumulate Δ-mtDNAs to relatively high levels as compared with other tissues, such as liver, kidney, and lung [Simonetti et al., 1992]. One likely reason for this quantitative difference could be that Δ-mtDNAs accumulate in long-lived postmitotic tissues but are eliminated in tissues that turn over [Byrne et al., 1991]. In fact, this explanation is probably insufficient. First, as opposed to what has been observed in humans [Simonetti et al., 1992; Cortopassi et al., 1992]. Δ-mtDNAs were found to accumulate quite readily in liver from aging rodents [Piko et al., 1988; Gadaleta et al., 1992; Chen et al., 1993; Edris et al., 1994]. Second, there were significant differences in Δ-mtDNA levels among different regions of human brain, with cerebellum containing far fewer Δ-mtDNAs than frontal cortex [Corral-Debrinski et al., 1992; Soong et al., 1992]. Clearly, factors other than cellular turnover must also influence the rate of accumulation of Δ-mtDNAs in aging.

At present, we do not know if the accumulation of Δ-mtDNAs is an epiphenomenon or is related to the age-related decline in cellular energy levels in a more fundamental way. We believe the latter is more likely, based on our observation that the accumulation of Δ-mtDNAs is species specific. Mice have mitochondrial genomes whose structure is highly similar to that of humans [Bibb et al., 1981]. We therefore synthesized PCR primers flanking regions in mouse mtDNAs containing a long direct repeat and which we thought might correspond to the version of the human "common deletion." We were able to amplify and sequence a PCR product from mouse tissue corresponding exactly to the predicted Δ-mtDNA. Importantly, we were able to observe the predicted Δ-mtDNA PCR product only in old mice (i.e., almost 2 years old) but not in young animals (i.e.., 3 weeks old) [Chen et al., 1993; see also Brossas et al., 1994; Chung et al., 1994]. We note that, by contrast, a 2-year-old human does not contain a significant amount of Δ-mtDNA. These results agree with electron microscopic observations of rearranged mtDNAs in aged rat and mouse tissues [Piko et al., 1988]. Moreover, we have found that mtDNA deletions accumulate in other organisms with life spans significantly shorter than that of humans, including mice and flies [Chen et al., 1993]. Other investigators have obtained similar results, not only in mice [Brossas et al., 1994; Chung et al., 1994], but in other short-lived organisms, such as rats [Gadaleta et al., 1992; Edris et al., 1994] and worms [Melov et al., 1994]. These results are consistent with earlier findings that cytoplasmic factors may affect life span in organisms such as flies [Yonemura et al., 1991] and fungi [Vierny et al., 1982; Egilmez and Jazwinski, 1989; Jazwinski et al., 1989].

Taken together, these results imply that mutations of mtDNA accumulate in a species-specific manner and that this phenomenon may be directly related to the aging process. Cause and effect, of course, have not yet been established and are not inferred here.

IV.B. Point Mutations in mtDNA in Aging

Most of the effort in the aging field has been focused on mtDNA deletions. Since mitochondria are thought to have poor repair systems for DNA [Clayton et al., 1974; LeDoux et al., 1992], point mutations in mtDNAs (pm-mtDNAs) would be expected to accumulate at least as much as do Δ-mtDNAs. There are only a few publications addressing this issue, and on the surface they are contradictory. Using PCR/RFLP analysis, Münscher et al. [1993b] found an A→G transition at nt-8344 (a pathogenic mutation associated with MERRF) in extraocular muscle in 11 of 14 normal older adults; up to 2.4% pm-mtDNA was calculated in the oldest samples. Focusing on a different pm-mtDNA—the pathogenic A→G transition at nt-3243 in the tRNA$^{Leu(UUR)}$ gene that is associated with MELAS—Münscher et al. [1993a] and Zhang et al. [1993] obtained qualitatively similar results (i.e., older individuals had "PCR-observable" mutations, whereas younger subjects did not). On the other hand, Münscher et al. [1993a] did not find two other pathogenic tRNA mutations in their aged individuals.

Furthermore, Monnat's group identified only 1 mutation in 49 kb (i.e., <0.002%) of cloned normal blood cell DNA [Monnat and Loeb, 1985], and only 3 mutations in 81.7 kb (i.e., <0.004%) of cloned leukemic cell DNA [Monnat et al., 1985]. Similar results were obtained by Bodenteich et al. [1991], who sequenced about 32 kb of cloned mtDNA isolated from human retina from a 71-year-old subject (and which presumably had been exposed to a lifetime of ultraviolet radiation). They found only one heteroplasmic mutation, implying that somatic point mutations occur extremely rarely in aging (i.e., <1/32,000 or <0.003%), if at all. The discrepancy among these reports may be due to a number of factors, including methodology (e.g., PCR/RFLP vs. cloning/sequencing), the specific mutation involved (e.g., a known pathogenic site vs. a neutral polymorphism), technique (e.g., sensitivity of the respective assays, and also the issue of whether retinal DNA containing UV-induced thymine dimers can be amplified by PCR or can be cloned and detected by dideoxy sequencing), the tissue (e.g., muscle vs. blood vs. retina), and the mode of quantitation. The bottom line is that the question of the accumulation of pm-mtDNAs in aging remains open and, like the quantitation of total Δ-mtDNAs, has become an area of active investigation.

IV. C. Morphological Analyses of Muscle

In 1989, Müller-Höcker showed that there is an age-dependent increase in the proportion of COX-negative fibers in normal human hearts. These data have now been extended to other tissues, including human skeletal muscle [Müller-Höcker, 1990], extraocular muscle (a tissue that is severely affected in KSS and PEO) [Müller-Höcker et al., 1992], and the oxyphilic cells of the parathyroid gland [Müller-Höcker, 1992]. Similar findings have been observed in rat heart and kidney [Nagley et al., 1992]. We feel it is significant that the COX-negative fibers observed in these studies were often RRF, because, as noted above, such fibers are a morphological hallmark of many mitochondrial disorders in which the RRF contain massive amounts of mutated mtDNAs [Bonilla et al., 1992].

We have looked for the accumulation of COX-negative fibers and for the presence of RRF in normal human muscle at various ages, using histochemistry to detect COX and SDH activities. We found COX-negative fibers (about 1%) in muscle, some of which were RRF, from two individuals, aged 65 and 67 years; COX-negative RRF are typically found in mitochondrial disorders associated with mtDNA mutations (both deletions and point mutations). Interestingly, however, we also found COX-negative fibers that were *not* RRF and that corresponded to type I and type II fibers (Fig. 3A,B). In addition, one of the samples also showed the converse, that is, RRFs that were COX *positive* (Fig. 3C,D): These are reminiscent of the COX-positive RRF that are found in MELAS muscle and that contain high amounts of mutated mtDNAs [Moraes et al., 1992; Petruzzella et al., 1994]. No COX-negative fibers or RRFs were observed in samples from four younger individuals between the ages of 30 and 52 years.

Our preliminary observations on normal muscle at various ages are in agreement with those of Müller-Höcker [1990] and suggest that COX deficiency and mitochondrial proliferation (i.e., RRFs) at the individual fiber level may begin around the fifth or sixth decade of life. We suspect that one cause of this phenomenon is an accumulation of mtDNA mutations in specific fibers above a certain (presently unknown) minimum threshold. In other words, there may be an accumulation of Δ-mtDNAs and pm-mtDNAs in *all* muscle fibers, whether COX positive or COX negative, but the total amount of mutated mtDNAs in COX-negative fibers may be higher than that in COX-positive fibers. The relationship between the two, however, appears complex, as one can have COX deficiency without visible evidence of RRF. Specifically, when we studied muscle biopsies from four aged subjects (62, 63, 65, and 67 years old), we found that they were negative for the presence of Δ-mtDNAs by Southern blot analysis (as expected), but by PCR we found the "common deletion" in two of these subjects. One of these had COX-negative RRF in the muscle biopsy, but we were surprised to find no observable morphological abnormalities in the other. Conversely, of the two subjects without the "common deletion,"

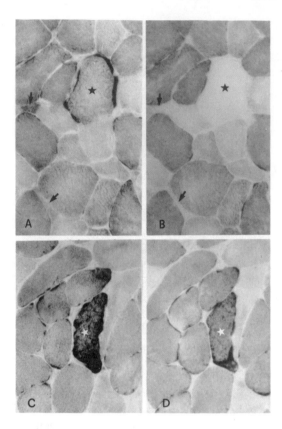

Fig. 3. *Unusual histochemistry in aged muscle. Serial muscle sections stained for SDH (A,C) and COX (B,D) activities from two aged individuals. Note COX-negative RRF (black stars) and COX-negative non-RRF (arrows) in a 65-year-old subject (top) and COX-positive RRF (white stars) in a 62-year-old subject (bottom).*

one had no observable morphological abnormalities but the other had both COX-positive RRF and COX-negative non-RRF (see Fig. 3). Assays for the presence of the MELAS-3243, MELAS-3271, MERRF-8344, and MERRF-8356 mutations were negative in all four samples. These results imply that, if mtDNA mutations are associated with respiratory chain deficiency, simply focusing on RRF is insufficient and will require analysis of a range of fiber types in aged muscle.

V. CONCLUDING REMARKS

The recent, and surprising, finding that small amounts of Δ-mtDNA molecules accumulate

in the muscle and brain of aged normal individuals opens up an entirely new area of investigation in the field of mitochondrial genetics. It implies that both normal and diseased states of aging, including cardiomyopathies and neurodegenerative disorders, as well as the idiopathic decline in muscle strength and cognition with age, may in fact be related to mitochondrial dysfunction arising from accumulations of mutated mtDNAs. In addition, this observation now parallels the documented findings of age-related mutations in nuclear DNA (including mtDNA sequences located in the nucleus, some of which have been implicated in both aging and tumorigenesis [Shay et al., 1991; Shay and Werbin, 1992]) and adds a new layer of complexity to the developing analysis of nuclear–mitochondrial interactions in normal and pathological cellular function. It also demands that we look at known mitochondrial encephalomyopathies in a new light.

The observation of mtDNA mutations in aging also seems to support work by others on changes in the biochemistry and morphology of muscle in aging. It has long been known that there are morphological changes in aged muscle, including single-fiber necrosis, group atrophy, and fiber loss [Walton, 1988]. Declines in muscle respiratory chain function have been documented at both the enzymatic level [Trounce et al., 1989; Byrne et al., 1991] and the transcriptional level [Gadaleta et al., 1990; Fernandez-Silva et al., 1991]. In addition, there are defects in mitochondrial calcium homeostasis in cardiac myocytes [Darley-Usmar et al., 1990] and an increase in the numbers of respiratory-deficient heart muscle fibers with age [Müller-Höcker, 1989].

The discovery of the age-dependent accumulation of Δ-mtDNAs in normal aging implies that not only nuclear, but also mitochondrial mutations, may play a role in the aging process. While any single individual mutated mtDNA may be present at a level of <0.1% of total mtDNA, there might be hundreds or even thousands of individual species of mutated mtDNAs. In other words, while any one mtDNA mutation may be present in relatively small amounts, it may well be that in the aggregate up to 5–10% of total mtDNA in aged muscle may be mutated, resulting in a level of dysfunctional mitochondria likely to result in clinical consequences. Since mitochondrial function is absolutely essential for normal respiration in cells, cumulative defects in these organelles would likely have serious deleterious consequences in tissues that are highly dependent on aerobic respiration, such as muscle and brain. However, as with many new findings, the clinical relevance of the presence of low levels of Δ-mtDNAs in aging is unclear at present, the cumulative level of all species of Δ-mtDNAs is not yet known, nor is it immediately obvious whether (or why) small amounts of Δ-mtDNAs can have clinically observable phenotypic consequences.

With respect to mtDNA point mutations, the jury is still out. Although mitochondria cannot repair thymine dimers [Clayton et al., 1974], it is clear that there are robust repair systems for other types of DNA damage [LeDoux et al., 1992; Driggers et al., 1994]. Thus, pm-mtDNAs might also accumulate in aging, but the literature on the role of pm-mtDNAs in aging is both sparse and conflicting. This is an area that deserves particular attention in the immediate future.

The accumulation of mutated mtDNAs may also be relevant to neurodegenerative diseases associated with aging, such as Alzheimer's, Huntington's, and Parkinson's diseases, particularly in light of the fact that all three disorders have been associated with impairment of mitochondrial respiratory chain function [Parker et al., 1989, 1990a,b; Hattori et al, 1991b]. The accumulation of mutated mtDNAs in specific sets of neurons, in a manner analogous to what seems to be occurring in RRF in muscle (which, like neurons, are highly oxidative postmitotic cells), may explain the morphological and functional "drop out" of neurons in aging and in age-related cognitive impairment. However, as with other issues in this newly emerging area, conflicting data abound [Mann et al., 1992; DiDonato et al., 1993; Chen et al., 1995a], and no consensus has yet emerged.

We note that the mitochondrial respiratory chain is the single greatest source of free radicals in the body and that cumulative defects in mtDNA integrity may be intimately associated with damage by reactive oxygen species (ROS). Thus, the unexpected and surprising findings discussed in this chapter are the first pieces of evidence of validate, at the molecular genetic level, the "mitochondrial theory of aging" [Miquel et al., 1980; Linnane et al., 1989]. Moreover, because the mitochondrial respiratory chain is the single greatest source of free radicals in the cell, these data may also be the beginning of the unification of this hypothesis with the "free-radical theory of aging" [Harman, 1981; Richter et al., 1988; Ames, 1993, Mecocci et al., 1993]. Finally, the study of mtDNA mutations may point the way to potential amelioration of ROS-associated mtDNA damage during aging by biochemical intervention.

ACKNOWLEDGMENTS

This work was supported by grants from the National Institutes of Health, the Muscular Dystrophy Association, the Myoclonus Research Foundation, the Dana Foundation, and the Procter & Gamble Company.

REFERENCES

Ames BN, Shigenaga MK, Hagen TM (1993): Oxidants, antioxidants, and the degenerative diseases of aging. Proc Natl Acad Sci USA 90:7915–7922.

Anderson S, Bankier AT, Barrell BG, de Bruijn MHL, Couslon AR, Drouin J, Eperon IC, Nierlich DP, Roe BA, Sanger F, Schreier PH, Smith AJH, Staden R, Young IG (1981): Sequence and organization of the human mitochondrial genome. Nature 290:457–465.

Ballinger SW, Shoffner JM, Gebhart S, Koontz DA, Wallace DC (1994): Mitochondrial diabetes revisited. Nature Genet 7:458–459.

Ballinger SW, Shoffner JM, Hedaya EV, Trounce I, Polak MA, Koontz DA, Wallace DC (1992): Maternally transmitted diabetes and deafness associated with a 10.4 kb mitochondrial DNA deletion. Nature Genet 1:11–15.

Bibb MJ, Van Etten RA, Wright CT, Walberg MW, Clayton DA (1981): Sequence and gene organization of mouse mitochondrial DNA. Cell 26:167–180.

Bodenteich A, Mitchell LG, Merril CR (1991): A lifetime of retinal light exposure does not appear to increase mitochondrial mutations. Gene 108:305–310.

Bonilla E, Sciacco M, Tanji K, Sparaco M, Petruzzella V, Moraes CT (1992): New morphological approaches to the study of mitochondrial encaphalomyopathies. Brain Pathol 2:113–119.

Brossas J-Y, Barreau E, Courtois Y, Treton J (1994): Multiple deletions in mitochondrial DNA are present in senescent mouse brain. Biochem Biophys Res Commun 202:654–659.

Byrne E, Dennett X, Trounce I (1991): Oxidative energy failure in post-mitotic cells: A major factor in senescence. Rev Neurol (Paris) 147:532–535.

Chen X, Bonilla E, Sciacco M, Schon EA (1995a): Paucity of deleted mitochondrial DNAs in brain regions of Huntington's disease patients. Biochim Biophys Acta 1271:229–233.

Chen X, Prosser R, Simonetti S, Sadlock J, Jagiello G, Schon EA (1995b): Rearranged mitochondrial genomes are present in human oocytes. Am J Hum Genet 57:239–247.

Chen X, Simonetti S, DiMauro S, Schon EA (1993): Accumulation of mitochondrial DNA deletions in organisms with various lifespans. Bull Mol Biol Med 18:57–66.

Chung SS, Weindruch R, Schwarze SR, McKenzie DI, Aiken JM (1994): Multiple age-associated mitochondrial DNA deletions in skeletal muscle of mice. Aging Clin Exp Res 6:193–200.

Ciafaloni E, Ricci E, Shanske S, Moraes CT, Silvestri G, Hirano M, Simonetti S, Angelini C, Donati A, Garcia C, Martinuzzi A, Mosewich R, Servidei S, Zammarchi E, Bonilla E, DeVivo DC, Rowland LP, Schon EA, DiMauro S (1992): MELAS: Clinical features, biochemistry, and molecular genetics. Ann Neurol 31:391–398.

Clayton DA, Doda JN, Friedberg EC (1974): The absence of a pyrimidine dimer repair mechanism in mammalian mitochondria. Proc Natl Acad Sci USA 71:2777–2781.

Corral-Debrinski M, Horton T, Lott MT, Shoffner JM, Beal MF, Wallace DC (1992): Mitochondrial DNA deletions in human brain: Regional variability and increase with advanced age. Nature Genet 2:324–329.

Corral-Debrinski M, Stepien G, Shoffner JM, Lott MT, Kanter K, Wallace DC (1991): Hypoxemia is associated with mitochondrial DNA damage and gene induction. JAMA 26:1812–1816.

Cortopassi GA, Arnheim N (1990): Detection of a specific mitochondrial DNA deletion in tissues of older humans. Nucleic Acids Res 18:6927–6933.

Cortopassi GA, Shibata D, Soong N-W, Arnheim N (1992): A pattern of accumulation of a somatic deletion of mitochondrial DNA in aging human tissues. Proc Natl Acad Sci USA 89:7370–7374.

Darley-Usmar VM, Smith DR, O'Leary VJ, Stone D, Hardy DL, Clark JB (1990): Hypoxia-reoxygenation induced damage in the myocardium: The role of mitochondria. Biochem Soc Transm 18:526–528.

de Vries dD, van Engelen BGM, Gabreëls FJM, Ruitenbeek W, van Oost BA (1993): A second missense mutation in the mitochondrial ATPase6 gene in Leigh's syndrome. Ann Neurol 34:410–412.

DiDonato S, Zeviani M, Giovannini P, Savarese N, Rimoldi M, Mariotti C, Girotti F, Caraceni T (1993): Respiratory chain and mitochondrial DNA in muscle and brain in Parkinson's disease patients. Neurology 43:2262–2268.

DiMauro S, Bonilla E, Lombes A, Shanske S, Minetti C, Moraes CT (1990): Mitochondrial encephalomyopathies. Neurol Clin 8:483–506.

Driggers WJ, LeDoux SP, Wilson GL (1993): Repair of oxidative damage within the mitochondrial DNA of RINr 38 cells. J Biol Chem 268:22042–22045.

Dunbar DR, Moonie PA, Swingler RJ, Davidson D, Roberts R, Holt IJ (1993): Maternally transmitted partial direct tandem duplication of mitochondrial DNA associated with diabetes mellitus. Hum Mol Genet 2:1619–1624.

Edris W, Burgett B, Stine OC, Filburn CR (1994): Detection and quantitation by competitive PCR of an age-associated increase in a 4.8-kb deletion in rat mitochondrial DNA. Mutat Res 316:69–78.

Egilmez NK Jazwinski SM (1989): Evidence for the involvement of a cytoplasmic factor in the aging of the yeast Saccharomyces cerevisiae. J Bacteriol 171:37–42.

Engel WK, Cunningham CG (1963): Rapid examination of muscle tissue: An improved trichrome stain method for fresh-frozen biopsy sections. Neurology 13:919–923.

Fernandez-Silva P, Petruzzella V, Fracasso F, Gadaleta MN, Cantatore P (1991): Reduced synthesis of mtRNA in isolated mitochondria of senescent rat brain. Biochem Biophys Res Commun 176:645–653.

Fukuhara N, Tokigushi S, Shirakawa K, Tsubaki T (1980): Myoclonus epilepsy associated with ragged-red fibers (mitochondrial abnormalities): Disease entity or syndrome? Light and electron microscopic studies of two cases and review of the literature. J Neurol Sci 47:117–133.

Gadaleta MN, Petruzzella V, Renis M, Fracasso F, Cantatore P (1990): Reduced transcription of mitochondrial DNA in the senescent rat. Eur J Biochem 187:501–506.

Gadaleta MN, Rainaldi G, Lezza AMS, Milella F, Fracasso F, Cantatore P (1992): Mitochondrial DNA copy number and mitochondrial DNA deletion in adult and senescent rats. Mutat Res 275:181–193.

Giles RE, Blanc H, Cann HM, Wallace DC (1980): Maternal inheritance of human mitochondrial DNA. Proc Natl Acad Sci USA 77:6715–6719.

Goto Y-i, Nonaka I, Horai S (1990): A mutation in the tRNA$^{Leu(UUR)}$ gene associated with the MELAS subgroup of mitochondrial encephalomyopathies. Nature 348:651–653.

Goto Y-i, Nonaka I, Horai S (1991): A new mtDNA mutation associated with mitochondrial myopathy, encephalopathy, lactic acidosis and stroke-like episodes (MELAS). Biochim Biophys Acta 1097:238–240.

Harman D (1981): The aging process. Proc Natl Acad Sci USA 78:7124–7128.

Hattori K, Tanaka M, Sugiyama S, Oyabashi T, Ito T, Satake T, Hanaki Y, Asai J, Nagano M, Ozawa T (1991a): Age-dependent increase in deleted mitochondrial DNA in the human heart: Possible contributory factor to presbycardia. Am Heart J 121:1735–1742.

Hattori N, Tanaka M, Ozawa T, Mizuno Y (1991b): Immunohistochemical studies on complexes I, II, III, and IV of mitochondria in Parkinson's disease. Ann Neurol 30:563–571.

Hirano M, Pavlakis SG (1994): Mitochondrial myopathy, encephalopathy, lactic acidosis, and strokelike episodes (MELAS): Current concepts. J Child Neurol 9:4–13.

Holt IJ, Harding AE, Morgan-Hughes JA (1988): Deletions of mitochondrial DNA in patients with mitochondrial myopathies. Nature 331:717–719.

Holt IJ, Harding AE, Petty RKH, Morgan-Hughes JA (1990): A new mitochondrial disease associated with mitochondrial DNA heteroplasmy. Am J Hum Genet 46:428–433.

Ikebe S-i, Tanaka M, Ohno K, Sato W, Hattori K, Kondo T, Mizuno Y, Ozawa T (1990): Increase of deleted mitochondrial DNA in the striatum in Parkinson's disease and senescence. Biochem Biophys Res Commun 170:1044–1048.

Jazwinski SM, Egilmez NK, Chen JB (1989): Replication control and cellular life span. Exp Gerontol 24:432–436.

Kadowaki T, Kadowaki H, Mori Y, Tobe K, Sakuta R, Suzuki Y, Tanabe Y, Sakura H, Awata T, Goto Y-i, Hayakawa T, Matsuoka K, Kawamori R, Kamada T, Horai S, Nonaka I, Hagura R, Akanuma Y, Yazaki Y (1994): A subtype of diabetes mellitus associated with a mutation of mitochondrial DNA. N Engl J Med 330:962–968.

LeDoux SP, Wilson GL, Beecham EJ, Stevnsner T, Wassermann K, Bohr VA (1992): Repair of mitochondrial DNA after various types of DNA damage in Chinese hamster ovary cells. Carcinogenesis 13:1967–1973.

Linnane AW, Baumer A, Maxwell RJ, Preston H, Zhang C, Marzuki S (1990): Mitochondrial gene mutation: The aging process and degenerative diseases. Biochem Int 22:1067–1076.

Linnane AW, Marzuki S, Ozawa T, Tanaka M (1989): Mitochondrial DNA mutations as an important contributor to aging and degenerative diseases. Lancet i:642–645.

Mann VM, Cooper JM, Schapira AHV (1992): Quantitation of a mitochondrial DNA deletion in Parkinson's disease. FEBS Lett 299:218–222.

Mecocci P, MacGarvey U, Kaufman AE, Koontz D, Shoffner JM, Wallace DC, Beal MF (1993): Oxidative damage to mitochondrial DNA shows marked age-dependent increases in human brain. Ann Neurol 34:609–616.

Melov S, Hertz GZ, Stormo DG, Johnson TE (1994): Detection of deletions in the mitochondrial genome of Caenorhabditis elegans. Nucleic Acids Res 22:1075–1078.

Miquel J, Economos AC, Fleming J, Johnson JE (1980): Mitochondrial role in cell aging. Exp Gerontol 15:575–591.

Mita S, Rizzuto R, Moraes CT, Shanske S, Arnaudo E, Fabrizi G, Koga Y, DiMauro S, Schon EA (1990): Recombination via flanking direct repeats is a major cause of large-scale deletions of human mitochondrial DNA. Nucleic Acids Res 18:561–567.

Monnat RJ Jr, Loeb LA (1985): Nucleotide sequence preservation of human mitochondrial DNA. Proc Natl Acad Sci USA 82:2895–2899.

Monnat RJ Jr, Maxwell CL, Loeb LA (1985): Nucleotide sequence preservation of human leukemic mitochondrial DNA. Cancer Res 45:1809–1814.

Moraes CT, Andreetta F, Bonilla E, Shanske S, DiMauro S, Schon EA (1991): Replication-competent human mitochondrial DNA lacking the heavy-strand promoter region. Mol Cell Biol 11:1631–1637.

Moraes CT, Ciacci F, Silvestri G, Shanske S, Sciacco M, Hirano M, Schon EA, Bonilla E, DiMauro S (1993): Atypical clinical presentations associated with the MELAS mutation at position 3243 of human mitochondrial DNA. Neuromusc Disord 3:43–49.

Moraes CT, DiMauro S, Zeviani M, Lombes A, Shanske S, Miranda AF, Nakase H, Bonilla E, Wernec LC, Servidei S, Nonaki I, Koga Y, Spiro A, Brownell KW, Schmidt B, Shotland DL, Zupanc MD, DeVivo DC, Schon EA, Rowland LP (1989): Mitochondrial DNA deletions in progressive external ophthalmoplegia and Kearns-Sayre syndrome. N Engl J Med 320:1293–1299.

Moraes CT, Ricci E, Bonilla E, DeMauro S, Schon EA (1992): The mitochondrial tRNA$^{Leu(UUR)}$ mutation in MELAS: Genetic, biochemical, and morphological correlations in skeletal muscle. Am J Hum Genet 50:934–949.

Müller-Höcker J (1989): Cytochrome-c-oxidase deficient cardiomyocytes in the human heart—an age-related phenomenon. A histochemical ultracytochemical study. Am J Pathol 134:1167–1173.

Müller-Höcker J (1990): Cytochrome-c-oxidase deficient fibres in the limb muscle and diaphragm of men without muscular disease: An age related alteration. J Neurol Sci 106:14–21.

Müller-Höcker J (1992): Random cytochrome-c-oxidase deficiency of oxyphil cell nodules in the parathyroid gland. A mitochondrial cytopathy related to cell aging? Pathol Res Pract 188:701–706.

Müller-Höcker J, Schneiderbanger K, Stefani FH, Kadenbach B (1992): Progressive loss of cytochrome-c-oxidase in the extraocular muscles during aging. A cytochemical–immunocytochemical study. Mutat Res 275:115–124.

Münscher C, Müller-Höcker J, Kadenbach B (1993a):

Human aging is associated with various point mutations in tRNA genes of mitochondrial DNA. Biol Chem Hoppe Seyler 374:1099–1104.

Münscher C, Rieger T, Müller-Höcker J, Kadenbach B (1993b): The point mutation of mitochondrial DNA characteristic for MERRF disease is found also in healthy people of different ages. FEBS Lett 317:27–30.

Nagley P, Mackay IR, Baumer A, Maxwell RJ, Vaillant F, Wang Z-X, Zhang C, Linnane AW (1992): Mitochondrial DNA mutation associated with aging and degenerative disease. Ann NY Acad Sci 673:92–102.

Nakase H, Moraes CT, Rizzuto R, Lombes A, DiMauro S, Schon EA (1990): Transcription and translation of deleted mitochondrial genomes in Kearns-Sayre syndrome: Implications for pathogenesis. Am J Hum Genet 46:418–427.

Ozawa T, Tanaka M, Ikebe S-i, Ohno K, Kondo T, Mizuno Y (1990a): Quantitative determination of deleted mitochondrial DNA relative to normal DNA in parkinsonian striatum by a kinetic PCR analysis. Biochem Biophys Res Commun 172:483–489.

Ozawa T, Tanaka M, Sugiyama S, Hattori K, Ito T, Ohno K, Takahashi A, Sato W, Takada G, Mayumi B, Yamamoto K, Adachi K, Koga Y, Toshima H (1990b): Multiple mitochondrial DNA deletions exist in cardiomyocytes of patients with hypertrophic or dilated cardiomyopathy. Biochem Biophys Res Commun 170:830–836.

Parker WD, Boyson SJ, Parks JK (1989): Abnormalities of the electron transport chain in idiopathic Parkinson's disease. Ann Neurol 26:719–723.

Parker WD, Boyson SJ, Luder AS, Parks JK (1990a): Evidence for a defect in NADH ubiquinone oxidoreductase (complex I) in Huntington's disease. Neurology 40:1231–1234.

Parker WD, Filley CM, Parks JK (1990b): Cytochrome oxidase deficiency in Alzheimer's disease. Neurology 40:1302–1303.

Petruzzella V, Moraes CT, Sano MC, Bonilla E, DiMauro S, Schon EA (1994): Extremely high levels of mutant mtDNAs co-localize with cytochrome c oxidase–negative ragged-red fibers in patients harboring a point mutation at nt-3243. Hum Mol Genet 3:449–454.

Piko L, Hougham AJ, Bulpitt KJ (1988): Studies of sequence heterogeneity of mitochondrial DNA from rat and mouse tissues; Evidence for an increased frequency of deletions/additions with aging. Mech Ageing Dev 43:279–293.

Prezant TR, Agapian JV, Bohlman MC, Bu X, Oztas S, Qui W-Q, Arnos KS, Cortopassi GA, Jaber L, Rotter JI, Shohat M, Fischel-Ghodsian N (1993): Mitochondrial ribosomal RNA mutation associated with both antibiotic-induced and non-syndromic deafness. Nature Genet 4:289–294.

Reardon W, Ross RJ, Sweeney MG, Luxon LM, Harding AE, Trembath RC (1992): Diabetes mellitus associated with a pathogenic point mutation in mitochondrial DNA. Lancet 340:1376–1379.

Remes AM, Majamaa K, Herva R, Hassinen IE (1993): Adult-onset diabetes mellitus and neurosensory hearing loss in maternal relatives of MELAS patients in a family with the tRNA(Leu(UUR)) mutation. Neurology 43:1015–1020.

Richter C, Park J-W, Ames BN (1988): Normal oxidative damage to mitochondrial and nuclear DNA is extensive. Proc Natl Acad Sci USA 85:6465–6467.

Rötig A, Colonna M, Bonnefont JP, Blanche S, Fisher A, Saudubray JM, Munnich A (1989): Mitochondrial DNA deletion in Pearson's marrows/pancreas syndrome. Lancet i:902–903.

Santorelli FM, Shanske S, Jain KD, Tick D, Schon EA, DiMauro S (1994): A T→C mutation at nt 8993 of mitochondrial DNA in a child with Leigh syndrome. Neurology 44:972–974.

Satoh M, Kuroiwa T (1991): Organization of multiple nucleoids and DNA molecules in mitochondria of a human cell. Exp Cell Res 196:137–140.

Schon EA, Hirano M, DiMauro S (1994): Mitochondrial encephalomyopathies: Clinical and molecular analysis. J Bioenerg Biomembr 26:291–299.

Schon EA, Rizzuto R, Moraes CT, Nakase H, Zeviani M, DiMauro S (1989): A direct repeat is a hotspot for large-scale deletions of human mitochondrial DNA. Science 244:346–349.

Shay JW, Baba T, Zhan QM, Kamimura N, Cuthbert JA (1991): HeLaTG cells have mitochondrial DNA inserted into the c-myc oncogene. Oncogene 6:1869–1874.

Shay JW, Werbin H (1992): New evidence for the insertion of mitochondrial DNA into the human genome: Significance for cancer and aging. Mutat Res 275:227–235.

Shoffner JM, Lott MT, Lezza AMS, Seibel P, Ballinger SW, Wallace DC (1990): Myoclonic epilepsy and ragged-red fiber disease (MERRF) is associated with a mitochondrial DNA tRNA(Lys) mutation. Cell 61:931–937.

Siebert PD, Larrick JW (1992): Competitive PCR. Nature 359:557–558.

Silvestri G, Moraes CT, Shanske S, Oh SJ, DiMauro S (1992): A new mtDNA mutation in the tRNA(Lys) gene associated with myoclonic epilepsy and ragged-red fibers (MERRF). Am J Hum Genet 51:1213–1217.

Simonetti S, Chen X, DiMauro S, Schon EA (1992): Accumulation of deletions in human mitochondrial DNA during normal aging: Analysis by quantitative PCR. Biochem Biophys Acta 1180:113–122.

Soong NW, Hinton DR, Cortopassi G, Arnheim N (1992): Mosaicism for a specific somatic mitochondrial DNA mutation in adult human brain. Nature Genet 2:318–323.

Sugiyama S, Hattori K, Ozawa T (1991): Quantitative analysis of age-associated accumulation of mitochondrial DNA with deletion in human hearts. Biochem Biophys Res Commun 180:894–899.

Tatuch Y, Christodoulou J, Feigenbaum A, Clarke JTR, Wherret J, Smith C, Rudd N, Petrova-Benedict R,

Robinson BH (1992): Heteroplasmic mtDNA mutation (T→G) at 8993 can cause Leigh's disease when the percentage of abnormal mtDNA is high. Am J Hum Genet 50:852–858.

Torii K, Sugiyama S, Tanaka M, Takagi K, Hanaki Y, Iida K-i, Matsuyama M, Hirabayashi N, Uno Y, Ozawa T (1992): Aging-associated deletions of human diaphragmatic mitochondrial DNA. Am J Respir Cell Mol Biol 6:543–549.

Trounce I, Byrne E, Marzuki S (1989): Decline in skeletal muscle mitochondrial respiratory chain function: Possible factor in aging. Lanced i:637–639.

van den Ouweland JMW, Lemkes HHPJ, Ruitenbeek W, Sandkuijl LA, de Vijlder MF, Struyvenberg PAA, van de Kamp JJP, Maassen JA (1992): Mutation in mitochondrial tRNA$^{Leu(UUR)}$ gene in a large pedigree with maternally transmitted type II diabetes mellitus and deafness. Nature Genet 1:368–371.

van den Ouweland JMW, Lemkes HHPJ, Trembath RC, Ross R, Velho G, Cohen D, Froguel P, Maassen JA (1994): Maternally inherited diabetes and deafness is a distinct subtype of diabetes and associates with a single point mutation in the mitochondrial tRNA$^{Leu(UUR)}$ gene. Diabetes 43:743–751.

Vierny C, Keller A-M, Begel O, Belcour L (1982): A sequence of mitochondrial DNA is associated with the onset of senescence in a fungus. Nature 297:157–159.

Wallace DC (1992): Diseases of the mitochondrial DNA. Annu Rev Biochem 61:1175–1212.

Wallace DC, Singh G, Lott MT, Hodge JA, Schurr TG, Lezza AMS, Elsas II LJ, Nikoskelainen EK (1988): Mitochondrial DNA mutation associated with Leber's hereditary optic neuropathy. Science 242:1427–1430.

Walton J (1988): Disorders of Voluntary Muscle. London: Churchill Livingstone.

Yen T-C, Su J-H, King K-L, Wei Y-H (1991): Aging-associated 5 kb deletion in human liver mitochondrial DNA. Biochem Biophys Res Commun 178:124–131.

Yoneda M, Tanno Y, Horai S, Ozawa T, Miyatake T, Tsuji S (1990): A common mitochondrial DNA mutation in the t-RNA^{Lys} of patients with myoclonus epilepsy associated with ragged-red fibers. Biochem Int 21:789–796.

Yonemura I, Motoyama T, Hasekura H, Boettcher B (1991): Cytoplasmic influence on the expression of nuclear genes affecting life span in *Drosophila melanogaster*. Heredity 66:259–264.

Zeviani M, Moraes CT, DiMauro S, Nakase H, Bonilla E, Schon EA, Rowland LP (1988): Deletions of mitochondrial DNA in Kearns-Sayre syndrome. Neurology 38:1339–1346.

Zeviani M, Muntoni F, Savarese N, Serra G, Tiranti V, Carrara F, Mariotti C, DiDonato S (1993): A MERRF/MELAS overlap syndrome associated with a new point mutation in the mitochondrial tRNA^{Lys} gene. Eur J Hum Genet 1:80–87.

Zeviani M, Servidei S, Gellera C, Bertini E, DeMauro S, DiDonato S (1989): An autosomal dominant disorder with multiple deletions of mitochondrial DNA starting at the D-loop region. Nature 339:309–311.

Zhang C, Baumer A, Maxwell RJ, Linnane AW, Nagley P (1992): Multiple mitochondrial DNA deletions in an elderly human individual. FEBS Lett 297:34–38.

Zhang C, Linnane AW, Nagley P (1993): Occurrence of a particular base substitution (3243 A to G) in mitochondrial DNA of tissues of aging humans. Biochem Biophys Res Commun 195:1104–1110.

ABOUT THE AUTHORS

ERIC A. SCHON is Associate Professor of Genetics and Development (in Neurology) at Columbia University. He received his Ph.D. from the University of Cincinnati, studying goat globin gene structure and evolution, and did his postdoctoral work at Harvard and Columbia, focusing on DNA topology in mammalian gene promoter regions and on the evolution of retroposons. His current research is focused on various aspects of human mitochondrial genetics, with particular emphasis on mitochondrial diseases, on mitochondrial biogenesis and communication, and on cytochrome *c* oxidase structure and function.

MONICA SCIACCO is a postdoctoral fellow at Columbia University. She received her M.D. *summa cum laude* from the University of Milan. Her research focuses on genetic and morphological approaches to the study of mitochondrial encephalomyopathies, with emphasis on quantitative PCR analysis of mtDNA mutations in single muscle fibers.

FRANCESCO PALLOTTI received his M.D. from the Universitá degli Studi in Bologna, Italy. He is currently a postdoctoral fellow at Columbia University, studying mitochondrial DNA defects in aging and in neurodegenerative diseases.

XI CHEN is a Research Scientist in the Department of Neurology at New York University. He received his M.D. from Fujian Medical College (P.R.C.), an M.S. from Tongji Medical University (P.R.C.), and his Ph.D. from Shanghai Second Medical University (P.R.C.). From 1991–1994, he was a postdoctoral fellow at Columbia University, studying mitochondrial DNA mutations in normal human aging.

EDUARDO BONILLA is Professor of Neurology and Pathology at Columbia University. He received his M.D. *cum laude* from the Javeriana University in Bogota, Colombia, and pursued further medical training at Case Western Reserve University (Cleveland), New York Medical College, and the University of Pennsylvania. His current research focuses on clinical, morphological, and molecular approaches to the study of mitochondrial and other neuromuscular diseases, with emphasis on the use of histochemistry, immunohistochemistry, and *in situ* hybridization in muscle and brain.

Cellular Aging and Cell Death: 35–49
© 1996 Wiley-Liss, Inc.

Protein Damage and Oxidative Stress

Tahereh Tabatabaie and Robert A. Floyd

I. INTRODUCTION

It is becoming widely recognized that oxidative damage to proteins increases with age. Oxidative damage to proteins is considered a result of the continuous oxidative stress to which aerobic systems are subjected. It is highly likely that the buildup of oxidized protein plays a crucial role in the aging process and/or the way cells respond to stress, especially oxidative stress, with age. The buildup of oxidized protein reflects alterations in equilibrium processes influencing the rate of formation of oxidized proteins versus their catabolism. The buildup of oxidized protein may ultimately have as its root cause age-dependent differential alterations in the readout of various genes. This chapter presents some of the current major ideas and selected pertinent supportive data in a general fashion and thus as such should not be considered an exhaustive review.

II. OXIDATIVE STRESS

Oxidative stress is the stress imposed on a biological system that requires oxygen to sustain life. Oxidative damage is a result of oxidative stress. The extent of oxidative damage depends upon many factors, including the rate of production of semireduced oxygen species during aerobic metabolism as well as the ability of the biological system to withstand oxidative stress. Thus, aerobic systems have continuous oxidative damage potential, $\vec{P}_O$, imposed on them at all times simply because oxygen is utilized and in so doing semireduced toxic oxygen byproducts are formed. The an-

tioxidant defense capacity, $\vec{A}_c$, composed of the many natural antioxidants and antioxidant enzymes, oppose and in most cases maintain the imposed oxidative damage potential in check. This concept is presented in Figure 1. It should be pointed out that under most circumstances the oxidative damage potential is greater than the antioxidant defense capacity. Therefore the net result is that a small amount of oxidative damage potential exists at all times, leading to a sustained and normally constant amount of oxidative damage.

Oxygen metabolism produces a small percentage of semireduced oxygen species at all times, and these are the source of the imposed oxidative damage potential. Figure 2 presents the major points of this concept. It is considered that approximately 5% or less of the total oxygen consumed may lead to the formation of toxic semireduced oxygen byproducts. The fact that 5% or less of the total oxygen consumed forms toxic oxygen byproducts seems to be inconsequential, especially when it is also considered that nature has evolved a large array of systems to protect against the toxic oxygen byproducts. Even though a small percentage of toxic oxygen byproducts are formed per unit time, this can amount to as much as 3–5 tons produced in a life time for a human being.

Table I states that for every 100 molecules of oxygen consumed by the average human cell when at rest, i.e., not exercising, one molecule of oxygen is used to oxidize proteins. We have based our calculations on the assumption that an average human cell is cuboidal in shape with 10μ in one dimension. We have also assumed that the average cell is 15% protein and that

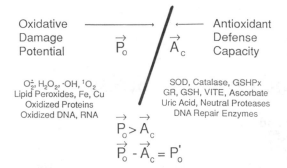

Fig. 1. *An overview representing the oxidative stress concept where the oxidative damage potential is maintained in a dynamic equilibrium with the antioxidant defense capacity. The oxidative damage potential is not completely quenched, so a small amount of oxidative damage is occurring at all times. SOD, superoxide dismutase; GSHPx, glutathione peroxidase; GR, glutathione reductase; VITE, vitamin E. [Reproduced from Floyd, 1994, with permission of the publisher.]*

the average protein has a molecular weight of 50,000 Daltons. We have utilized a value of 4 nmol of oxidized protein measured as carbonyl content/mg protein [Oliver et al., 1990] as an average measure and have assumed that there is only one oxidized amino acid residue per protein when the protein is oxidized and that the oxidized protein catabolism half life is 1 hour. Utilizing these assumptions, we calculate that 0.9% of the oxygen molecules consumed by the average human cell (10^{12} O_2/cell-day) is diverted to cytosolic protein oxidation. Clearly there are many assumptions, and the calculations were done simply to help grasp the relative magnitude of the oxygen flux. The calculations do clearly show that protein oxidation is a fact that must be considered in oxidative stress.

Oxygen Metabolism

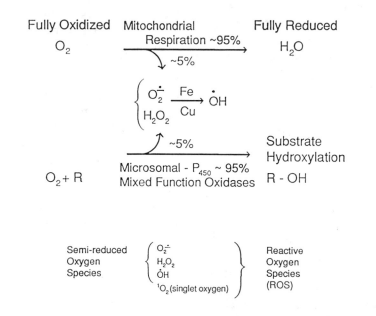

Fig. 2. *An overview illustrating that, even though a majority of the oxygen consumed is utilized either in mitochondrial respiration or in microsomal hydroxy-lation processes, a small percentage is diverted to semireduced oxygen species. [Reproduced from Floyd, 1994, with permission of the publisher.]*

TABLE I. Normal Oxygen Metabolism

Human cellular consumption
  20×10^6 O_2/μ^2-cell surface per second
  95% Used in metabolic processes
Semireduced oxygen species produced
  2–5% (H_2O_2, $O_2^{\cdot-}$, $\cdot OH$)
Normal steady-state levels of semireduced
oxygen species
  $H_2O_2 \sim 10$ nM
  $O_2^{\cdot-} \sim 0.1$ nM
  $\cdot OH \sim 1$ pM to 1 fM
Oxidized protein
  ~ 1 oxidized protein produced for each 100
    molecules of oxygen consumed
Oxidized DNA and RNA
  ~ 1 oxidized nucleic acid produced for each 200
    molecules of oxygen

III. OXYGEN METABOLISM AND FENTON CHEMISTRY

The one-electron reduction of molecular oxygen generates superoxide radical or $O_2^{\cdot-}$. Even though $O_2^{\cdot-}$ is not a remarkably reactive species, its spontaneous or superoxide dismutase–catalyzed dismutation leads to formation of hydrogen peroxide (H_2O_2) and the interaction between the two latter species leads to formation of hydroxyl radicals ($\dot{O}H$), as shown in equations 1 and 2. The hydroxyl radical is an extremely reactive species. It reacts with most substances with diffusion-limited rate constants (10^9–10^{10} M^{-1} S^{-1}). The high reactivity of $\dot{O}H$ implies that this species will have a very short half life and will react in the immediate surroundings of its site of formation.

$$O_2^{\cdot-} + O_2^{\cdot-} + 2H^+ \Leftrightarrow H_2O_2 + O_2 \quad (1)$$

$$O_2^{\cdot-} + H_2O_2 \rightarrow \dot{O}H + \overline{O}H + O_2 \quad (2)$$

Equation 2 is the so called Haber-Weiss reaction. The Haber-Weiss reaction is known to be too slow to account for formation of any significant level of $\dot{O}H$ under physiological conditions. In the presence of trace concentrations of transition metal ions, however, the rate of this reaction is greatly enhanced. Under these conditions, the so-called Fenton reaction or the metal ion-catalyzed Haber-Weiss reaction take place, as depicted in the equations 3–5:

$$Fe(II) + H_2O_2 \rightarrow Fe(III) + \dot{O}H + \overline{O}H \quad (3)$$
$$\text{(Fenton reaction)}$$

$$Fe(III) + O_2^{\cdot-} \rightarrow Fe(II) + O_2 \quad (4)$$

$$H_2O_2 + O_2^{\cdot-} \rightarrow \dot{O}H + \overline{O}H + O_2 \quad (5)$$
(metal ion–catalyzed Haber-Weiss reaction)

IV. AGE-DEPENDENT INCREASE IN OXIDIZED PROTEIN

It is becoming clear that the amount of oxidized protein increases with age. This has been found true for a wide range of subjects, including insects, laboratory rodents, and humans [Stadtman, 1992]. Figure 3 demonstrates the general accumulation of oxidized protein with age. Oxidized proteins increase in an exponential fashion with age. Oxidized protein is much higher than the chronological age of patients suffering from Werner's syndrome or from progeria, both diseases associated with accelerated aging. It is interesting to note that skin fibroblasts from normal individuals show increased oxidized protein in a general exponential fashion as a function of the age of the donor. Protein oxidation increases in normal human brain as a function of age, as the data in Figure 3 show.

V. SYSTEMS MEDIATING PROTEIN OXIDATION

Most systems that generate reactive oxygen species (ROS) are capable of bringing about *in vitro* and *in vivo* protein oxidation. Ionizing radiation, ozone, singlet oxygen-producing systems, and the Udenfriend system are among them. However, among the various oxidant systems that may be involved in the *in vivo* oxidation of proteins, two systems have been recently studied in greater detail. These include the so-called mixed-function oxidase (MFO) and the metal ion–catalyzed oxidant (MCO) systems [Stadtman, 1990].

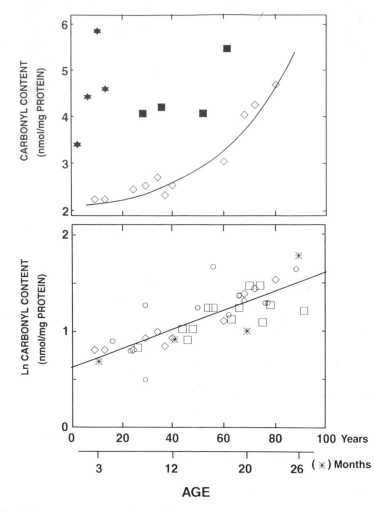

Fig. 3. *Age-related changes in the amount of oxidized protein.* **Top:** *carbonyl content of cultured human dermal fibroblasts from normal humans* (◇), *patients with progeria* (★), *and patients with Werner's syndrome* (■). **Bottom:** *semilog plot of carbonyl content of proteins versus age for cultured dermal fibroblasts from normal humans* (◇), *the occipital lobe of human brain tissue* (O), *human eye lens cortex* (□), *and from rat liver hepatocytes* (*). *The months scale refers to rat liver; the years scale to the human data.* [*Reproduced from Stadtman, 1992, with permission of the publisher.*]

Metal ion–catalyzed oxidations are basically Fenton reactions and require a transition metal ion, namely, iron or copper, and H_2O_2 or a substrate such as ascorbate that forms H_2O_2 upon autoxidation. The MFO system is comprised of an enzyme (usually a flavoenzyme) that serves as an intermediate carrier in the transfer of electrons from an electron donor (e.g., NAD[P]H) to the ultimate electron acceptor (Fe[III] or Cu[II]), either directly or through a stepwise reduction of other electron transfer intermediates (usually a heme protein or a protein containing an iron–sulfur center). A variety of enzymes are known to take part in MFO-type oxidations, including microsomal cytochrome P450 reductase, xanthine oxidase, NADH oxidase, and redoxin reductase. Cytochrome P450(LM$_2$), ferredoxin, and redoxin are among the known electron transfer intermediates.

Inactivation of a variety of metabolic enzymes by MFO systems has been shown [Stadtman, 1990; Fucci et al., 1983]. Among them are alcohol dehydrogenase, aspartokinase A, creatine kinase, enolase, glutamine synthetase, lactate dehydrogenase, pyruvate kinase, phosphoglycerate kinase, and glyceraldehyde 3-phosphate dehydrogenase. However, several other enzymes have been shown to be resistant to inactivation by MFO systems.

The detailed studies of Stadtman and his coworkers [Fucci et al., 1983; Stadtman & Wittenberger, 1985; Nakamura et al., 1985] on the oxidative modification and inactivation of the enzyme glutamine synthetase (GS) have provided more mechanistic insights into these oxidative processes. Inactivation of GS by all MFO systems has been shown to require oxygen. All of these systems are stimulated by Fe(III) and inhibited by catalase, Mn(II), ethylenediaminetetraacetic acid (EDTA), o-phenanthroline, and histidine [Levine et al., 1981; Oliver et al., 1982].

The inhibitory effect of catalase points to the role that H_2O_2 plays in this type of oxidation. Similarly, the inhibition brought about by metal ion chelators (EDTA, o-phenanthroline) and Mn(II), which inhibits the reduction of Fe(III) to Fe(II), indicates a role for Fe(II) (or transition metal ions in general) in these processes. Therefore, even though the MFO systems differ from MCO systems with respect to the components involved in the (initial) electron-transfer processes, the ultimate oxidizing agents are similar in both cases. In other words, the components of the MFO system serve only to provide Fe(II) and H_2O_2 through redox reactions occurring between the enzymes and the corresponding substrates. For this reason, more recently both systems have been referred to as MCO systems by Stadtman and his coworkers [Stadtman, 1990].

A mechanism for oxidative modification of proteins by these systems is depicted in Figure 4. As illustrated, H_2O_2 is formed by direct reduction of O_2 by the MFO system (through oxidation of a substrate) or by a one-electron transfer to O_2 to form $O_2^{\cdot-}$, which generates H_2O_2 upon dismutation (spontaneous or superoxide dismutase–mediated). Fe(III) is in turn reduced to Fe(II) either by $O_2^{\cdot-}$ or via the oxidation of the cosubstrate (NAD[P]H, etc.) by the enzyme system. Fe(II) generated as such is believed to bind the protein at a metal-binding site. The consequent reaction of the protein–Fe(II) complex with H_2O_2 thus results in formation of ROS (OH$^{\cdot}$, $Fe[O]^{2+}$, $[Fe[OH]_2]^{2+}$, $Fe[OH]^{3+}$, etc.). The formed ROS preferentially attack the side chains of the amino acids located at or in the vicinity of the metal-binding site, and this gives rise to the observed site specificity of the MCO or MFO-catalyzed oxidations. The metal ion–mediated site-specific nature of these reactions has been inferred from several obser-

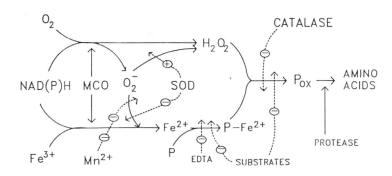

Fig. 4.   *The proposed mechanism for oxidation of proteins by MCO systems. P, protein; PO$_{ox}$, oxidized protein. The symbols + and – indicate stimulation and inactivation, respectively. [Reproduced from Stadtman and Oliver, 1991, with permission of the publisher.]*

vations: (1) these reactions are relatively insensitive to inhibition by free radical scavengers such as mannitol, dimethyl sulfoxide, urate, and formate; (2) in most cases one or only a few amino acids have been found to be modified by such systems (in contrast to oxidation by, for example, ionizing radiation, which oxidizes a large number of amino acids in the protein); (3) enzymes most sensitive to MFO- or MCO-catalyzed oxidations require metal ions for their activity and therefore possess metal ion-binding sites; and (4) these oxidations are in most cases inhibited by metal ion chelators that complex the metal and prevent its binding to the protein [Stadtman, 1990; Stadtman and Oliver, 1991].

EDTA has been shown to have a paradoxical effect on MCO- and MFO-catalyzed oxidations. It has been found to have an inhibitory effect on the inactivation of GS [Stadtman and Wittenberger, 1985; Nakamura et al., 1985] and several other enzymes by MFO and MCO systems [Fucci et al., 1983] but to enhance the inactivation of several other enzymes by the same systems. As shown in Figure 4, EDTA may interfere with the binding of the metal ion to the protein and thus inhibit the oxidative events. In other cases, however, it has been suggested that the protein may lack a high-affinity binding site for the metal ion per se, but may bind the metal–EDTA complex quite effectively. In such cases, EDTA may serve as a carrier for the metal ion, and therefore its presence stimulates the oxidation.

The site-specific metal ion–catalyzed oxidation/inactivation mechanism was originally proposed by Samuni et al. [1981] for copper-induced sensitization of the oxidative damage to penicillinase caused by O_2^{-} and later by Shinar et al. [1983] for ascorbate- and copper-induced inactivation of acetylcholine esterase.

VI. CARBONYL FORMATION AS A MARKER OF PROTEIN OXIDATION

A variety of amino acids are targets for modification by oxidants. However, depending on the nature of the oxidant, different amino acid residues may undergo oxidative modification. Table II shows the amino acid residues that are modified by several different oxidant systems, along with final products of such oxidations. For example, exposure to ozone or to hydroxyl radicals produced by ionizing radiation results in preferential oxidation of tyrosine, tryptophan, histidine, methionine, and cysteine residues. Depending on the conditions, these modifications may result in either cross-linking or fragmentation of the protein [Davies, 1987].

Oxidation of proteins by MFO or MCO systems, however, results in modification of proline, arginine, histidine, methionine, tyrosine,

TABLE II. Oxidative Modification of Amino Acids

Oxidation system	Residue modified	Products	References
MCO*	Proline	Glutamyl semialdehyde	Amici et al., 1989
	Arginine	Glutamyl semialdehyde	Climent et al., 1989
			Amici et al., 1989
	Lysine	α-Aminoadipyl semialdehyde	Amici et al., 1989
MCO, x-rays, ozone, etc.	Histidine	Asparagine–aspartate	Farber and Levine, 1986
		2-oxo-histidine	Ochida and Kawakishi, 1993
			Floyd and Wong, 1994 (unpublished data)
MCO, x-rays, ozone, etc.	Methionine	Methionine sulfoxide	Swallow, 1960
MCO, x-rays, etc.	Tyrosine	Tyr–tyr cross links	Giulivi and Davies, 1994
MCO, ozone, etc.	Cysteine	-S-S-cross links, mixed disulfides	Swallow, 1960

*MCO denotes both metal ion-catalyzed oxidant and mixed function oxidase systems.

lysine, and cysteine [Stadtman et al., 1992]. As indicated in Table II, oxidation of three of these amino acids, proline, arginine, and lysine, by MFO and MCO systems, and possibly by other oxidants, gives rise to derivatives that contain free carbonyl groups. Assessment of the carbonyl content of a protein thus provides a measure of the oxidative damage to the protein. Figure 5 illustrates a plausible mechanism by which the interaction between a protein-bound Fe(II) and H_2O_2 results in formation of the carbonyl derivative of, in this case, a lysine residue [Stadtman, 1990]. In this mechanism it has been assumed that Fe(II) binds the ε-amino group of the lysine residue ($-CH_2-NH_2$) and forms a Fe(II)–protein coordination complex. Interaction between this complex and H_2O_2 results in the formation of protein-bound $Fe(II), \overline{O}H$, and $\dot{O}H$ (Fenton reaction). The $\dot{O}H$ thus formed abstracts a hydrogen atom from the carbon atom bearing the ε-amino group and forms a carbon-centered radical. Subsequent transfer of an electron from the alkyl radical

to Fe(III) leads to the formation of the imine derivative ($-CH=NH$) of the lysine residue and regeneration of Fe(II). Spontaneous hydrolysis of the imine group yields the aldehyde derivative of the lysine residue. Fe(II) dissociates from the protein at this stage because of the destruction of the protein metal-binding site (ε-amino group). This is an example of a caged reaction, since the formed $\dot{O}H$, by virtue of its extremely high reactivity, reacts at a site in the vicinity of its site of generation, namely, with functional groups of the amino acids at the metal-binding site of the protein. This accounts for the observed lack of effectiveness of radical scavengers on MFO or MCO oxidations.

Carbonyl derivatives of the proteins can be formed by mechanisms other than MCO-type reactions. We have found that oxidation of proteins by oxidants such as ozone and singlet oxygen (produced by irradiation of the sensitizer dye methylene blue with visible light) are potent inducers of carbonyl groups [Tabatabaie and Floyd, 1994].

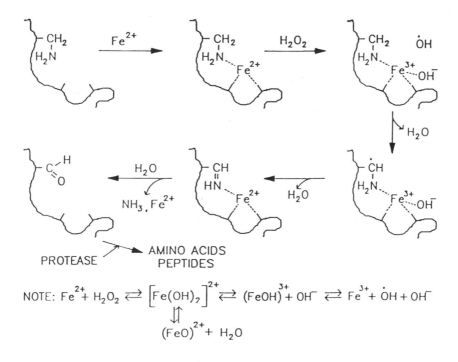

Fig. 5. *A possible mechanism for the site-specific metal ion–catalyzed oxidation of a lysine residue of a protein to carbonyl derivative. [Reproduced from Stadtman, 1990, with permission of the publisher.]*

Several sensitive methods for determination of protein carbonyl content have been developed by Levine and his coworkers [1990; 1994]. All of these methods take advantage of the characteristic reactivity (electrophilicity) of the carbonyl group to add a label (radioactive, fluorescence, etc.) to the oxidized protein. Addition of the label provides a means for detection and quantification of the original carbonyls. These methods include (1) reduction of the carbonyls to corresponding alcohols with tritiated sodium borohydride and subsequent measurement of the incorporated radioactivity [Lenz et al., 1989]; (2) reaction of the carbonyl groups with the classic carbonyl reagent 2,4-dinitrophenylhydrazine (2,4-DNP) to form a protein-bound hydrazone derivative that can be detected at its absorbance maximum at ca. 360–370 nm [Levine, 1983]; (3) reaction of the carbonyl groups with fluorescein thiosemicarbazide to form fluorescent thiosemicarbazones [Ahn et al., 1987]; (4) reaction of the carbonyl groups with fluorescine amine to form a Schiff base followed by reduction to the secondary amine by cyanoborohydride and the subsequent measurement of the fluorescence [Climent et al., 1989].

The 2,4-DNP method is so far the most widely used method for determination of oxidized proteins. Improvement of the sensitivity and specificity of this method by linking it to high performance liquid chromatography (HPLC) has been recently reported [Levine et al., 1994].

It has to be pointed out, however, that determination of carbonyl content of proteins as a measure of oxidative damage to proteins may result in evaluations that may be erroneously high or low. Low erroneous results stem from the fact that, as shown in Table II, modifications of some amino acids (e.g., cysteine and histidine) do not result in formation of carbonyl derivatives. This shortcoming may be overcome by supplementing this method with other assays that measure other oxidative modifications of the proteins such as dityrosin determination and measurement of sulfhydryl groups. Erroneous high results, on the other hand, may arise from introduction of carbonyl groups into proteins by mechanisms other than oxidation. As illustrated in Scheme 1, enzymatic and nonenzymatic glycation of proteins that leads to formation of a Schiff base (1) that forms a ketoamine (2) by undergoing Amadori rearrangement is one possibility.

In addition, α–β-unsaturated alkenals produced during lipid peroxidation can react with and bind the protein sulfhydryl groups to form thioether adducts carrying carbonyl moieties, as shown in Scheme 2 [Lenz et al., 1989; Esterbauer and Zollner, 1989].

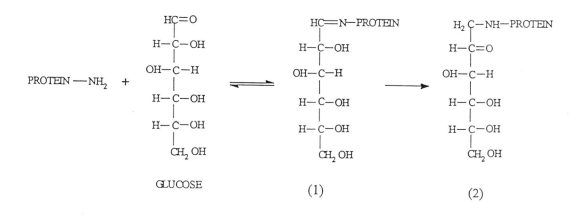

Scheme 1.

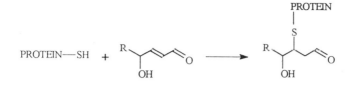

Scheme 2.

4-Hydroxy-2-nonenal has also been found to add to a lysine moiety of glucose 6-phosphate dehydrogenase (G-6-PDH) and inactivate it in the process [Szweda et al., 1993]. This process leads to introduction of 2,4-DNP detectable carbonyls into the enzyme (Scheme 3).

If the protein possesses enzymatic activity, amino acid modification at or near the active site can result in the loss of activity. In such cases the carbonyl content usually parallels the extent of inactivation. This has been shown to be the case with G-6-PDH [Amici et al., 1989]. However, there exists a possibility that amino acid modification and carbonyl formation occur at a site not critical for enzymatic activity. In such cases, increased carbonyl content is observed while activity remains unchanged. For example, we have found that treatment of glutathione reductase with a system comprised of Fe(II) and O_2, while causing a significant increase in the carbonyl content of this enzyme, has very little effect on its activity [Tabatabaie and Floyd, 1994].

VII. INCREASED SUSCEPTIBILITY OF OXIDATIVELY MODIFIED PROTEINS TO PROTEOLYTIC DEGRADATION

Oxidative modification of proteins renders them susceptible to proteolysis. Studies of Davies and his coworkers [Davies, 1987; Davies et al., 1987a,b; Davies and Delsignore, 1987; Davies and Goldberg, 1987; Salo et al., 1990] have demonstrated that cells are equipped with proteolytic pathways that preferentially degrade oxidized proteins.

By detailed studies of the oxidative modification of bovine serum albumin (as a model protein), these investigators have shown that modification of primary structure of the protein as a result of oxidative insult leads to alteration of protein secondary and tertiary structures. Such conformational changes (denaturation) may expose the previously shielded peptide bonds to protease attack [Salo et al., 1990]. A 670 kD neutral/alkaline protease termed *macroxy proteinase* (MOP) has been isolated from human and rabbit erythrocytes and rabbit reticulocytes and has been shown to be responsible for 70%–80% of the proteolytic activity toward oxidatively modified proteins in the former and 60%–70% of such activity in the latter [Pacifici et al., 1989]. This protease system, MOP, was found to possess a preference for hydrophobic or aromatic amino acids. In addition, these studies have demonstrated that under mild oxidative conditions the increase in proteolytic susceptibility of the proteins is accompanied by an increase in hydrophobicity. In normal undamaged protein,

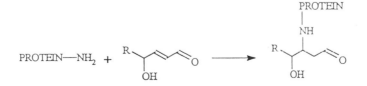

Scheme 3.

hydrophobic residues are usually shielded from proteolysis by being buried within the proteins' tertiary and quaternary structures. These proteins are thus generally resistant to proteolysis by proteases such as MOP. Oxidative modification of protein structure, however, can lead to exposure of such cleavage sites to the protease.

Similarly, isolation of a 75,000 kD soluble alkaline protease with high specificity for oxidized GS from *Escherichia coli* has been reported by Roseman and Levine [1987]. Furthermore, four different proteases with distinct preferential activity toward the oxidized form of GS have been isolated from rat liver [Rivett, 1985]. These include a lysosomal protease (cathepsin D), two Ca^{2+}-dependent cytosolic proteases (calpains), and a Ca^{2+}-independent, alkaline 300 kD cytosolic enzyme.

The existence of proteases with specificity for oxidized protein may be considered as a detoxification pathway to prevent the accumulation of these non-functional, potentially cytotoxic species [Davies and Goldberg, 1987]. At the same time, oxidative modification of proteins may serve as a signal for protein degradation.

VIII. PROTEIN OXIDATION OCCURS *IN VIVO*

The discovery that oxidation of proteins leads to carbonyl formation and the development of sensitive assays for detection and quantification of the carbonyls have provided researchers with a strong tool for investigation of the phenomenon of protein oxidation *in vivo*. Thus, in recent years, several interesting studies have been reported that point to a possible linkage between protein oxidation and different pathophysiological conditions and aging.

Accumulation of "altered" or "abnormal" forms of many enzymes in different aging models has been observed by a number of investigators [Salo et al., 1990; Pacifici et al., 1989]. These investigators have shown that these "altered" enzymes are almost invariably less active than the normal or "young" enzymes. A number of physical changes have been found to accompany the loss of activity.

These include changes in heat lability, the number of sulfhydryl groups, ultraviolet and circular dichroism spectra, and K_ms for substrates [Stadtman, 1988]. A variety of mechanisms have been proposed to account for the array of the changes that have been observed in aging proteins. Interestingly, however, many enzymes that are known to accumulate in a less active or inactive form during aging have also been found to be readily inactivated by MCO or MFO systems [Fucci et al., 1983; Oliver et al., 1987b].

Investigation of the relationship between aging and the level of oxidized proteins in several different systems has established that aging is accompanied by accumulation of oxidized proteins in the cells, among other changes. By measuring the protein carbonyl content, Oliver et al. [1987a] have shown that the levels of oxidatively modified proteins increase with age in circulating human erythrocytes. The increase in carbonyl content was found to be correlated with a decrease in the activities of several marker enzymes, including phosphoglycerate kinase, glyceraldehyde 3-phosphate dehydrogenase, and G-6-PDH.

In cultured human fibroblasts from donors of different ages, relatively little change in the level of oxidized proteins was observed until the age of 60. Fibroblasts from donors over 60 years of age, however, were shown to contain significantly higher levels of oxidatively modified proteins. Interestingly, cultured cells from individuals with progeria or Werner's syndrome, two genetic disorders characterized by premature or accelerated aging, were shown to contain significantly higher carbonyl content than the age-matched controls (Fig. 3).

Starke-Reed and Oliver [1989] have shown that protein carbonyl content is higher in hepatocytes from old (26-month-old) rats than young (3-month-old) rats, with the largest incremental increase occurring between 20 and 26 months. Specific activities of GS and G-6-PDH, two enzymes highly susceptible to oxidative inactivation, were shown to decline progressively with age even though their immunological cross-reactivity remained relatively unchanged (i.e., total enzyme content remained unchanged).

As mentioned earlier, increased protein oxidation in aging human brain (compared with young controls) has been shown by Smith et al. [1991]. The increase was found to be more pronounced in the frontal pole than the occipital pole. Age-matched Alzheimer's disease (AD) patients' brains were shown to contain relatively comparable levels of oxidized proteins. Glutamine synthetase and creatine kinase (another MCO-sensitive enzyme) activities were found to be lowered in both the normal aged group and the AD group. A distinction between AD patients and the age-matched controls, however, was found with respect to GS activity. Frontal cortical GS activity was demonstrated to be significantly lower in AD subjects, while immunodiffusion studies showed that the total content of GS was identical between the two groups. Therefore, in AD, a portion of the GS content of the frontal cortex exists in an "altered," less active (or inactive) form. Considering the marked vulnerability of GS to oxidative inactivation, this regional difference in GS activity can be attributed to a regional increase in free radical processes.

Ischemia/reperfusion-induced global cortical injury to Mongolian gerbil brain has been reported to lead to significantly increased levels of oxidized proteins [Oliver et al., 1990]. This increase, which was shown to occur during the reperfusion phase, was found to be accompanied by a decline in GS activity. Similarly, formation of protein carbonyls in rat lung undergoing an ischemic insult has been reported by Ayene et al. [1992]. Reperfusion was shown to increase further the carbonyl content of the ischemic lung.

IX. ACCUMULATION OF OXIDIZED PROTEINS IN AGING CELLS

Accumulation of oxidized proteins in the cells is the result of an imbalance between the rate of their generation and the rate by which they are removed from the cells by specific proteolytic enzymes. To this end several scenarios can be envisioned: occurrence of an age-dependent increase in the rate of protein oxidation without an accompanying increase in the rate of degradation; an unchanged rate of oxidation but an age-dependent decrease in the degradation rate; or an age-dependent increase in the oxidation with an accompanying decrease in the proteolysis rate (the same factors that accelerate protein oxidation may lead to a decreased proteolytic efficiency). All these pathways result in a gradual, age-related accumulation of oxidized proteins in the cells. Acceleration of oxidative events may be brought about by a variety of factors, including, among others, the following: increased availability of transition metal ions due to an age-related decrease in the efficacy of metal-binding proteins; depletion of cellular antioxidant defenses such as glutathione (which subsequently limits the ability of glutathione peroxidase (GSHPx) to remove H_2O_2 and organic hydroperoxides), vitamin E and carotenoids; or decrease in the activity of protective enzymes such as catalase, superoxide dismutase, and GSHPx.

The harmful consequences of such accelerated oxidative events may be exacerbated by an age-dependent decrease in the ability of the cells to degrade and remove the oxidized proteins [Stadtman, 1992]. An age-related decrease in the activity or the level of the proteases responsible for degradation of oxidized proteins has been observed. Starke-Reed and Oliver [1989] have shown the level of alkaline proteases that remove oxidized proteins in hepatocytes of 26-month-old rats to be only 20% of that of 3-month-old rats. Studies of Carney et al. [1991] showed a similar trend in gerbil brain. While the level of oxidized proteins was found to be 180% of that of young (3-month-old) animals, activities of GS and neutral proteases in old gerbil brains were only 65% and 33%, respectively, of those of young controls.

X. IS REVERSAL OF AGE-ASSOCIATED PHENOMENA POSSIBLE?

The same study [Carney et al., 1991] has revealed a more interesting aspect of age-related phenomena: the possibility of reversal. Chronic treatment (14 days) of old gerbils with the spin trapping reagent phenyl-*N-tert*-butyl nitrone (PBN) decreased the level of oxidized proteins and increased the levels of GS and neutral proteases, bringing all these parameters up to the corresponding levels observed in young animals (Fig. 6). Spin traps are compounds that react with reactive, unstable free radicals and form more stable radical species. Most interestingly, chronic administration of PBN was found to restore the old gerbils' short-term spatial memory (determined by means of a radial arm maze), which was determined to be significantly lower, to that of the young ger-

bils. This finding points to the possible connection between the accumulation of oxidized form of proteins with the malfunctioning of the central nervous system (e.g., memory loss) so that reduction of the level of the oxidized proteins may result in improvement of the cognitive function.

The complete effects of PBN, however, persist only as long as the compound is administered (Fig. 6). As illustrated in Figure 6, termination of PBN treatment led to a gradual reversal of all the biochemical parameters measured (level of oxidized protein, GS activity) to those characteristic of old animals. The

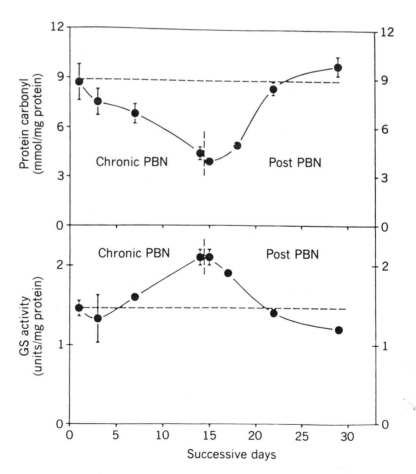

Fig. 6. *The protein carbonyl (top) and GS activity (bottom) of cerebral cortex of old male gerbils administered PBN chronically (32 mg/kg twice daily) for 14 days and then after PBN dosing was stopped. [Reproduced from Carney and Floyd, 1991, with permission of the publisher.]*

reversibility of the action of PBN indicates that age-related modifications are genetically determined and that PBN therapy, while capable of reversing some of the manifestations of these genetic changes, is unable to influence events on a genetic level. This point, however, does not lessen the importance of this and similar compounds in the possible therapeutic interventions with age-dependent brain modifications.

XI. CONCLUSION

Development of methods for determination of the carbonyl groups as a measure of protein oxidation has opened a new and fascinating avenue of research into the relationship existing between protein oxidation and aging, as well as different pathological conditions in that oxidative stress plays a role. There are a few existing studies which clearly point to such relationships, but whether the observed protein oxidation is the cause or the effect of the pathology is yet to be determined. Increased protein oxidation, however, is a certain marker of an increased oxidative insult on the system under study. A deeper understanding of this phenomenon will undoubtedly help researchers design strategies to prevent and/or intervene more effectively in the age-related disorders.

ACKNOWLEDGMENTS

The research cited from the authors' laboratory was funded in part by NIH grants AG 09690 and NS 23307.

REFERENCES

Ahn B, Rhee SG, Stadtman ER (1987): Use of fluorescein hydrazide and fluorescein thiosemicarbazide reagents for the fluorometric determination of protein carbonyl groups and for the detection of oxidized protein on polyacrylamide gels. Anal Biochem 161:245–257.

Amici A, Levine RL, Tsai L, Stadtman ER (1989): Conversion of amino acid residues in proteins and amino acid homopolymers to carbonyl derivatives by metal-catalyzed oxidation reactions. J Biol Chem 264:3341–3346.

Ayene IS, Dodia C, Fisher AB (1992): Role of oxygen in oxidation of lipid and protein during ischemia/reperfusion in isolated perfused rat lung. Arch Biochem Biophys 296:183–189.

Carney JM, Floyd RA (1991): Protection against oxidative damage to CNS by α-phenyl-*tert*-butyl nitrone (PBN) and other spin-trapping agents: A novel series of nonlipid free radical scavengers. J Mol Neurosci 3:47–57.

Carney JM, Starke-Reed PE, Oliver CN, Landrum RW, Chen MS, Wu JF, Floyd RA (1991): Reversal of age-related increase in brain protein oxidation, decrease in enzyme activity, and loss in temporal and spacial memory by chronic administration of the spin-trapping compound N-*tert*-butyl-α-phenylnitrone. Proc Natl Acad Sci USA 88:3633–3636.

Climent I, Tsai L, Levine RL (1989): Derivatization of γ-glutamyl semialdehyde residues in oxidized proteins by fluoresceinamine. Anal Biochem 182:226–232.

Davies KJA (1987): Protein damage and degradation by oxygen radicals. J Biol Chem 262:9895–9901.

Davies KJA, Delsignore ME (1987): Protein damage and degradation by oxygen radicals: III. Modification of secondary and tertiary structure. J Biol Chem 262(20):9908–9913.

Davies KJA, Delsignore ME, Lin SW (1987a): Protein damage and degradation by oxygen radicals. II. Modification of amino acids. J Biol Chem 262:9902–9907.

Davies KJA, Goldberg AL (1987): Oxygen radicals stimulate intracellular proteolysis and lipid peroxidation by independent mechanisms in erythrocytes. J Biol Chem 262(17):8220–8226.

Davies KJA, Lin SW, Pacifici RE (1987b): Protein damage and degradation by oxygen radicals. IV. Degradation of denatured protein. J Biol Chem 262:9914–9920.

Esterbauer H, Zollner H (1989): Methods for determination of aldehydic lipid peroxidation products. Free Rad Biol Med 7:197–203.

Farber JM, Levine RL (1986): Sequence of a peptide susceptible to mixed-function oxidation. J Biol Chem 261:4574–4578.

Floyd RA (1994): Measurement of oxidative stress in vivo. In Davies KJA(ed): The Oxygen Paradox in Biology and Medicine. CLEUP Press (in press).

Fucci L, Oliver CN, Coon MJ, Stadtman ER (1983): Inactivation of key metabolic enzymes by mixed-function oxidation reactions: Possible implication in protein turnover and ageing. Proc Natl Acad Sci USA 80:1521–1525.

Giulivi C, Davies KJA (1994): Dityrosine: A marker for oxidatively modified proteins and selective proteolysis. Methods Enzymol 233:363–371.

Lenz AG, Costabel U, Shaltiel S, Levine RL (1989): Determination of carbonyl groups in oxidatively modified proteins by reduction with tritiated sodium borohydride. Anal Biochem 177:419–425.

Levine RL (1983): Oxidative modification of glutamine

synthetase I. Inactivation is due to loss of one histidine residue. J Biol Chem 258:11823–11827.

Levine RL, Garland D, Oliver CN, Amici A, Climent I, Lenz A-G, Ahn B-W, Shaltiel S, Stadtman ER (1990): Determination of carbonyl content in oxidatively modified proteins. Methods Enzymol 186:464–478.

Levine RL, Oliver CN, Fulks RM, Stadtman ER (1981): Turnover of bacterial glutamine synthetase: Oxidative inactivation precedes proteolysis. Proc Natl Acad Sci USA 78:2120–2124.

Levine RL, Williams JA, Stadtman ER, Shacter E (1994): Carbonyl assays for determination of oxidatively modified proteins. Methods Enzymol 233:346–357.

Nakamura K, Oliver CN, Stadtman ER (1985): Inactivation of glutamine synthetase by a purified rabbit liver microsomal cytochrome P-450 system. Arch Biochem Biophys 240:319–329.

Oliver CN, Ahn B-W, Moerman EJ, Goldstein S, Stadtman ER (1987a): Age-related changes in oxidized proteins. J Biol Chem 262:5488–5491.

Oliver CN, Levine RL, Stadtman ER (1982): Regulation of glutamine synthetase degradation. In Orston LN, Sligar SG (eds): Experiences in Biochemical Perception. New York: Academic Press, Inc., pp 233–249.

Oliver CN, Levine RL, Stadtman ER (1987b): A role of mixed-function oxidation reactions in the accumulation of altered enzyme forms during aging. J Am Geriatr Soc 35:947–956.

Oliver CN, Starke-Reed PE, Stadtman ER, Liu GJ, Carney JM, Floyd RA (1990): Oxidative damage to brain proteins, loss of glutamine synthetase activity, and production of free radicals during ischemia/reperfusion-induced injury to gerbil brain. Proc Natl Acad Sci USA 87:5144–5147.

Pacifici RE, Salo DC, Davies KJA (1989): Macroxyproteinase (M.O.P.): A 670 kDa proteinase complex that degrades oxidatively denatured proteins in red blood cells. Free Rad Biol Med 7:521–536.

Rivett AJ (1985): Preferential degradation of the oxidatively modified form of glutamine synthetase by intracellular mammalian proteases. J Biol Chem 260:300–305.

Roseman JE, Levine RL (1987): Purification of a protease from Escherichia coli with specificity for oxidized glutamine synthetase. J Biol Chem 262:2101–2110.

Rothstein M (1977): Recent developments in the age-related alteration of enzymes: A review. Mech Ageing Dev 6:241–257.

Rothstein M (1984): Changes in enzymatic proteins during aging. In Roy AK, Chaterjee B (eds): Molecular Basis of Aging. New York: Academic Press, pp 209–232.

Sato DC, Pacifici RE, Lin SW, Giulivi C, Davies KJA (1990): Superoxide dismutase undergoes proteolysis and fragmentation following oxidative modification and inactivation. J Biol Chem 265:11919–11927.

Samuni A, Chevion M, Czapski G (1981): Unusual copper-induced sensitization of the biological damage due to superoxide radicals. J Biol Chem 256:12632–12635.

Shinar E, Navok T, Chevion M (1983): The analogous mechanisms of enzymatic inactivation induced by ascorbate and superoxide in the presence of copper. J Biol Chem 258:14778–14783.

Smith CD, Carney JM, Starke-Reed PE, Oliver CN, Stadtman ER, Floyd RA, Markesbery WR (1991): Excess brain protein oxidation and enzyme dysfunction in normal aging and in Alzheimer's disease. Proc Natl Acad Sci USA 88:10540–10543.

Stadtman ER (1988): Minireview: protein modification in aging. J Gerontol 43:112–120.

Stadtman ER (1990): Metal ion catalyzed oxidation of proteins: Biochemical mechanism and biological consequences. Free Rad Biol Med 9:315–325.

Stadtman ER (1992): Protein oxidation and aging. Science 257:1220–1224.

Stadtman ER, Oliver CN (1991): Metal-catalyzed oxidation of proteins. J Biol Chem 266:2005–2008.

Stadtman ER, Starke-Reed PE, Oliver CN, Carney JM, Floyd RA (1992): Protein modification in aging. In Emerit I, Chance B (eds). Free Radicals and Aging. Basel, Switzerland: Birkhauser Verlag, pp 64–72.

Stadtman ER, Wittenberger ME (1985): Inactivation of Escherichia coli glutamine synthetase by xanthine oxidase, nicotinate hydroxylase, horseradish peroxidase, or glucose oxidase: Effects of ferredoxin, putidaredoxin, and menadione. Arch Biochem Biophys 239:379–387.

Starke-Reed PE, Oliver CN (1989): Protein oxidation and proteolysis during aging and oxidative stress. Arch Biochem Biophys 275:559–567.

Swallow AJ (1960): Effect of ionizing radiation on proteins, RCO groups, peptide bond cleavage, inactivation–SH oxidation. In Swallow AJ (ed): Radiation Chemistry of Organic Compounds. New York: Pergamon Press, pp 211–224.

Szweda LI, Uchida K, Tsai L, Stadtman ER (1993): Inactivation of glucose-6-phosphate dehydrogenase by 4-hydroxy-2-nonenal. J Biol Chem 268:3342–3347.

Tabatabaie T, Floyd RA (1994): Susceptibility of glutathione peroxidase and glutathione reductase to oxidative damage and the protective effect of spin trapping agents. Arch Biochem Biophys 314:112–119.

Uchida K, Kawakishi S (1993): 2-Oxo-histidine as a novel biological marker for oxidatively modified proteins. FEBS Lett 332:208–210.

ABOUT THE AUTHORS

TAHEREH TABATABAIE is a Senior Research Scientist at the Free Radical Biology and Aging Research Program at the Oklahoma Medical Research Foundation. She received a B.S. degree in Chemistry from National University of Iran in Tehran, Iran. She then pursued a doctorate degree in the Department of Chemistry and Biochemistry, University of Oklahoma, where she earned a PhD in 1992 for studying the oxidation chemistry and biochemistry of the serotonergic neurotoxin 5,7-dihydroxytryptamine in the laboratory of Professor Glenn Dryhurst. She has received several awards from the University of Oklahoma for her achievements in graduate school. These include Cleo Cross Outstanding International Student Scholarship and Graduate Research Excellence Award. She has been a member of Robert A. Floyd's research group since 1992 and is presently performing research in the area of protein oxidation in relation to pathologic conditions where oxidative stress may play a role.

ROBERT A. FLOYD is currently Head of the Free Radical Biology and Aging Research Program at the Oklahoma Medical Research Foundation and Chairman of the Scientific Advisory Board of Centaur Pharmaceuticals, Inc., a company he cofounded. He has been actively conducting research on the role of free radicals in biomedical problems since 1971 and began research in the general area of oxygen and injury to biological systems as a postdoctorate in Britton Chance's laboratory at the University of Pennsylvania. Longstanding areas of active research interests include oxidative damage to DNA as this relates to the mechanisms of mutagenesis and carcinogenesis as well as oxidative damage to brain tissue during stroke, Parkinson's disease development, and aging. He has pioneered in the development of ultrasensitive methods to assess free radical formation rates in tissue and their mediation of damage to biological molecules. His collaborative effort with Dr. John Carney in these areas have lead to novel observations regarding the protective role of spin trapping compounds in the aging brain. For this effort they received The Glenn Foundation for Medical Research award for aging research in 1992. He co-discovered that the singlet oxygen generator, methylene blue plus light, was very effective in killing the HIV virus as well as other RNA viruses. He has an active research program in this area. He has published over 130 refereed scientific papers in the area of free radicals and biomedical problems and has served as an NIH peer reviewer on many study sections as well as served as a member of the Basic Biological Research Panel of the National Task Force on NIH Strategic Planning. He serves on the editorial boards of *Cancer Research, Free Radical Biology and Medicine, Archives of Biochemistry,* and *Biophysics and Archives of Gerontology and Geriatrics.*

Cellular Aging and Cell Death: 51–65
© 1996 Wiley-Liss, Inc.

Immune Function, Cell Death, and Aging

James E. Nagel, F. Joseph Chrest, Priti Lal, and William H. Adler

I. INTRODUCTION

Lymphoid cells have proved particularly useful in the study of programmed cell death (PCD). This is, in part, because PCD is an important regulator of immune system homeostasis and plays a major role in T- and B-lymphocyte maturation and selection. Furthermore, lymphocytes can be readily separated into clonal populations with defined phenotypes, which facilitates the study of apoptosis in the immune system.

II. APOPTOSIS AND T-LYMPHOCYTES

II.A. Apoptosis in the Thymus

During ontogeny, PCD is the normal response of immature thymocytes to activation [Smith et al., 1989]. Furthermore, apoptosis plays a role in the negative selection of T-cell precursors in the thymus [Blackman et al., 1990].

Whereas post-thymic T cells remain in interphase for prolonged periods of time, immature thymocytes turn over at a very rapid rate. Of the approximately 5×10^{10} immature ($CD4^+$, $CD8^+$) thymocytes, only about 3% develop into mature T cells and exit to the peripheral circulation. Double positive (i.e., $CD4^+$, $CD8^+$) thymocytes undergo one of three fates. Those whose T-cell receptor (TcR) has a high avidity for major histocompatibility complex (MHC) ligands will be negatively selected in the thymus and undergo PCD [MacDonald and Lees, 1990]. Thymocytes with low avidity for MHC ligands will be positively selected into $CD4^+$ or $CD8^+$ T-cell lineages [Hogquist et al., 1994]. Of these cells, those whose TcR is specific for class I MHC will become $CD8^+$, whereas those whose TcR is specific for class II MHC will become $CD4^+$ cells [Kisielow et al., 1988]. Thymocytes whose TcR is apparently unable to recognize any MHC ligand do not exit from the thymus and undergo apoptosis *in situ*.

II.B. Apoptosis in Mature T Cells

Transmembrane signaling initiated by anti-CD3 through its interaction with the TcR is a widely used model for studying T-cell activation [Weiss, 1989]. Typically, signal transduction through the TcR–CD3 complex produces a series of biochemical and molecular events that leads to T-cell proliferation [Hadden, 1988]. In addition to inducing the proliferation of mature T cells, anti-CD3 antibodies directed against the TcR–CD3 complex induce immature thymocytes, transformed T cells, and mature T cells to undergo PCD [Cohen et al., 1992].

Studies involving anti-CD3–induced apoptosis of $CD4^+/CD8^+$ immature thymocytes have led to an understanding of the negative selection and clonal deletion process associated with the development of the mature T-cell repertoire [Finkel et al., 1989; MacDonald and Lees, 1990; McConkey et al., 1989, Smith et al., 1989]. Although the role of apoptosis in regulating the response of mature T cells has been studied to a lesser extent, a number of recent studies have begun to address this issue and how it relates to peripheral T-cell tolerance. The results of these studies are discussed in a recent review of antigen-induced death of mature T cells [Kabelitz et al., 1994].

II.C. Murine T Cells

Our study of anti-CD3–induced apoptosis in mature murine T cells [Chrest et al., 1995] was performed using purified G_0 spleen cells from young and old C57Bl/6 mice. Details on isolation of G_0 T cells and activation with anti-CD3 have been previously described [Proust et al., 1991]. Flow cytometry was used to identify and quantify the percentages of cells undergoing apoptosis following 20 hours activation with anti-CD3 [Chrest et al., 1993]. Our initial findings (Fig. l) using flow cytometry indicated that anti-CD3 activation produced well-defined light scatter populations that, when sorted, could be identified as apoptotic or viable (nonapoptotic) cells.

In paired experiments where cells from young and old animals were examined under identical conditions, we were able to demonstrate, as shown in Fig. 2 that activated T cells from older animals consistently yielded higher levels of apoptotic cells and fewer viable cells than did T cells from the young. Cell cycle data (Fig. 3) indicate that the onset of apoptosis, as indicated by the sub-G_1 peak measured at 20 hours, precedes entry into the S phase of the cell cycle, suggesting that DNA synthesis is not a prerequisite for induction of apoptotic T-

cell death. As cells progress through the cell cycle, the occurrence of apoptotic cell death continues to increase, with higher levels recorded in older mice (Table I).

Since others have demonstrated an age-related increase in G_1 arrest [Staiano-Coico et al., 1984], we wished to determine if the higher levels of apoptosis seen at 40 hours activation were related to G_1 arrest or were the result of cell cycle progression and subsequent DNA synthesis. To address this issue we utilized the nick translation assay previously described [Gorczyca et al., 1993] to detect DNA strand breaks associated with apoptosis and simultaneously measured DNA content (cell cycle). The data in Figure 4 demonstrate that apoptosis is primarily restricted to the G_1 phase of the cell cycle. While these data suggest a relationship between the onset of apoptosis and G_1 arrest, it does not rule out the possibility that some apoptotic events may be the result of cell division and re-entry into G_1. These results provide convincing evidence linking apoptosis to the decreased proliferative capacity of T cells from older animals.

LIGHT SCATTER PROPERTIES OF 20 HR ANTI-CD3 ACTIVATED G_0 T-CELLS FROM YOUNG AND OLD MICE

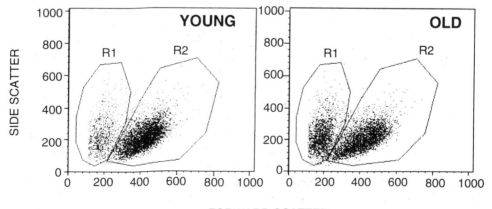

Fig. 1. *Typical forward and side scatter dot plots of 20 hour anti-CD3 activated T cells from young (**left**) and old (**right**) mice. Region Rl represents apoptotic cells; region R2 represents viable cells. Percentages of cells in each region were calculated using the LYSYS II software (Becton-Dickinson).*

PERCENTAGES OF APOPTOTIC AND VIABLE T-CELL FOLLOWING 20 HR ACTIVATION WITH IMMOBILIZED ANTI-CD3

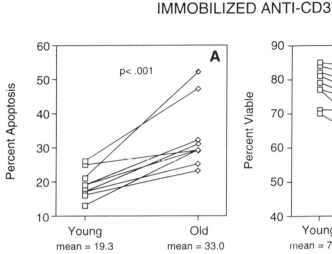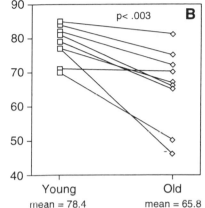

Fig. 2. *Results from nine experiments comparing percentages of apoptotic and viable cells 20 hours after immobilized anti-CD3 activation of G_0T cells from young and old mice. The lines connecting data points between young and old indicate results obtained in individual experiments comparing percentages of apoptotic (A) and viable (B) cells. The mean percentage of apoptotic cells was significantly greater ($p < 0.001$; paired t-test) in the old ($33.0 + 9.8\%$; Mean $\pm$ 1 SD) than in the young ($19.3 \pm 4.2\%$) and the percentage of viable cells was significantly decreased ($p < 0.003$) in the old ($65.8 \pm 11.3\%$) compared with the young ($78.4 \pm 5.3\%$).*

II.D. Human T Cells

Unlike murine splenic T cells, we have observed, as have others [Kabelitz et al., 1994], that anti-CD3 activation of normal, resting human peripheral blood T cells does not lead to any measurable level of apoptosis. However, the observation that activated T cells become more sensitive to apoptotic T-cell death [Kabelitz et al., 1994; Radvanyi et al., 1993] is explained in part by the controlling mechanisms associated with bcl-2, a known regulator of apoptosis [Korsmeyer, 1992]. Resting T cells have relatively high levels of bcl-2, which likely accounts for the resistance to apoptosis. In activated T cells, there is a down-regulation of bdl-2 [Akbar et al.,1993; Korsmeyer, 1992], leading to a corresponding increased sensitivity to apoptosis. Resting T cells that express Fas antigen are resistant to the effects of anti-Fas–induced apoptosis, whereas activated T cells, which do not express Fas, are respon-

sive to the effects of Fas-mediated apoptosis [Kabelitz et al., 1994]. It has been suggested that the down-regulation of bcl-2 renders activated T cells sensitive to Fas-driven apoptotic T-cell death.

The expression of bcl-2 is higher in CD45RB (naive) T cells, whereas highly differentiated CD45RO (memory) T-cells express higher levels of Fas antigen [Salmon et al., 1994]. The regulatory role of these two molecules may explain the age-related increase in memory cells and decrease in naive cells seen in peripheral blood [Sanders et al., 1988].

One of the most widely reported and consistent findings related to aging and the immune system is the increased presence of memory T cells in both older animals and humans. While the implications associated with an increase in memory T cells are many, studies that suggest memory T cells have a greater propensity to undergo apoptosis may provide

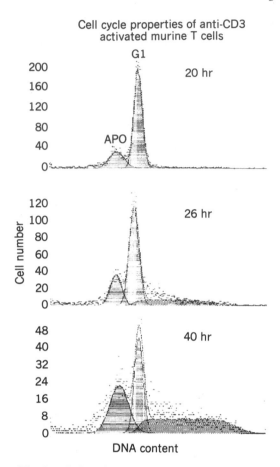

Cell cycle properties of anti-CD3
activated murine T cells

Fig. 3. *Cell cycle properties of anti-CD3 activated T cells demonstrating the presence (22%) of apoptotic cells (APO) as early as 20 hours following activation. Increased levels of apoptosis are seen at 26 hours (26%), when cells have entered S phase, and at 40 hours (32%).*

TABLE I. Percentages of Apoptotic and Cell Cycle Events in 40 Hour Anti-CD3 Activated T Cells

Experiments	Apoptotic (sub-G_1)	Viable (cell cycle)
1		
Young	29	66
Old	57	35
2		
Young	30	65
Old	46	50

ing of two light and two heavy chains. Both the light and heavy chains have variable (V) regions that are responsible for antigen binding and nonvariant or constant (C) regions that are responsible for most of the biological properties of immunoglobulins. In humans, approximately 10^{10} B-lymphocyte precursors are produced daily by hematopoietic stem cells located primarily in the bone marrow.

III.A. B-Cell Development

Early B-lineage cells, through a series of complex, stepwise events, assemble under genetic control first the heavy, then the light immunoglobulin chains. The heavy chain V region is assembled in a two-step process involving three distinct genes, V_H, D, and J_H. In the first recombination, one D family gene segment is translocated next to a J_H family gene segment. This translocation is followed by another, bringing a V_H gene segment next to the DJ_H segments to form a $V_H DJ_H$ complex that codes for the entire V region of the immunoglobulin heavy chain [Schatz et al., 1992]. $V_H DJ_H$ recombination occurs only in those cells committed to becoming B-lymphocytes and represents a critical control point in their development. In the human, there are four J_H segments, approximately 500 V_H segments, and 10 or more D segments available in the germline. Through recombination, the number of unique V_H-region molecules then becomes approximately 20,000.

Following assembly of the $V_H DJ_H$ complex, V region gene transcription is initiated from a promotor that lies upstream of each variable region gene. The primary transcript extends

additional information related to the age-related decline of immune function. It would be of interest to know the levels of bcl-2 and Fas in activated T cells from young and old donors and determine if apoptosis accounts for age defects associated with T-cell proliferation.

III. APOPTOSIS AND B LYMPHOCYTES

B lymphocytes are characterized by their unique ability to synthesize immunoglobulins and therefore to produce antibodies. Immunoglobulins are heterodimeric molecules consist-

MEASUREMENT OF DNA STRAND BREAKS
ASSOCIATED WITH APOPTOSIS
40 HR ANTI-CD3 ACTIVATED T-CELLS

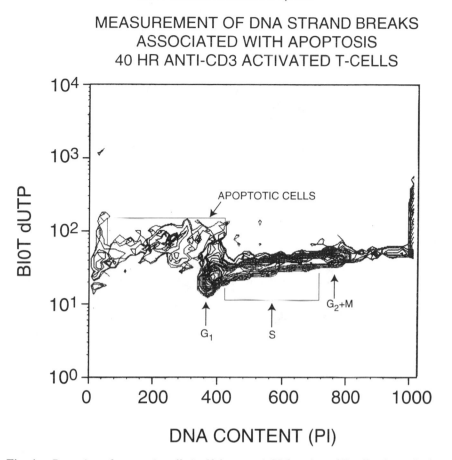

Fig. 4. *Detection of apoptotic cells in 40 hour anti-CD3 activated T-cell cultures by* in situ *nick translation. Incorporation of biotinylated dUTP is restricted to cells in the G_1 phase of the cell cycle.*

downstream through the exons that encode the first constant region (Cμ) after which it is ended. Multiple adenine nucleotides are added to one of several polyadenylation sites at the 3′ end of the Cμ mRNA, thereby supplying a poly-A tail to the final message. This generates mRNAs for both the secreted and membrane-bound forms of the Cμ (IgM) protein.

Beyond the large number of unique V_H regions created through recombination of the $V_H DJ_H$ gene segments, additional heavy chain diversity is generated during $V_H DJ_H$ joining. This so-called junctional diversity occurs because the ends of the V_H, D, and J_H segments may join at any of several nucleotides in the germline sequence and in the process add or delete one or several bases. About two-thirds

of the VDJ joining events result in nonfunctional recombinations that are out of frame and thus produce missense or nonsense sequences downstream from the $V_H DJ_H$ region. One-third of the transcripts have a productive rearrangement that has the initiation codon of the V_H segment in the same translational reading frame as the J_H segment and Cμ exons and produce full length VDJ-Cμ transcripts. Since there are two alleles, a pre-B cell that initially fails to produce a productive rearrangement gets a second chance by using its other DNA strand. Allelic exclusion allows only one successful variable region rearrangement per cell, and therefore only one of the alleles on both the heavy and light chains is expressed. The consequence of allelic exclusion is that a single

B cell produces an immunoglobulin with identical heavy and light chain variable regions.

The assembly of the immunoglobulin light chain is similar to that of the heavy chain, although the light chain V region has no D gene segments and is encoded by only V and J segments. As in the heavy chain, the primary mRNA transcript begins upstream of the VJ complex. Following the splicing out of an intron, the C gene joins the VJ transcript to produce an mRNA that produces either kappa (κ) or lambda (λ) protein. The light chain combines in the endoplasmic reticulum with the previously produced heavy chain to form a complete IgM molecule expressed on the cell surface. The diversity generated through the construction of the heavy and light chain V regions, as well as the pairings of the light and heavy chains themselves, creates many unique antibody molecules. These antibodies, however, are of the IgM class and generally have low affinity for antigen. As individual B cells continue to mature, they express the same heavy chain V region but may undergo isotype switching where they change or switch their C region [Snapper and Finkelman, 1993]. This means that a cell that initially expresses receptors for IgM may switch its constant region gene and produce IgA, IgG, or IgE class antibodies. Immunoglobulin class switching is initiated by the action of several different cytokines, including IL-4, IFN-γ, and TGF-β, on specialized switch (S) regions located upstream from the C region genes. Additional antibody diversity is also generated through a process called *somatic hypermutation*. Stimulated B cells, particularly in germinal centers, have regions in their heavy and light chain V regions that spontaneously mutate at a high frequency (approximately 1 mutation per 1,000 bp per cell division). These mutations "fine-tune" the antibody response. When antigen concentration falls, the B cells with the more avid antibody are preferentially selected, leaving an individual with high affinity antibodies against the particular antigen.

Apoptosis is also instrumental in generating diversity in the B cell repertoire. Overall,

during maturation from pro-B to mature B-lymphocyte, over 90% of the precursor cells die. During the assembly and rearrangement of the immunoglobulin V region chains, PCD determines which precursor cells survive to become mature B lymphocytes [Rolink and Melchers, 1991]. Approximately 75% of the precursor cells in the bone marrow undergo apoptosis at the transition between the large pro-B cells and the small pre-B stage [Ehlich et al., 1993]. A second selection process occurs at the immature IgM$^+$IgD$^-$ B cell stage. The selection process at this stage is based on the specificity of the IgM on the cell's surface. Immature B cells expressing IgM for self-antigens undergo clonal deletion or anergy [Schwartz, 1993]. Most naïve B lymphocytes die, generally within a few days of reaching maturity. However, if a naïve B cell encounters an antigen, it up-regulates the transcription of a gene called *bcl-2*, thus avoiding apoptosis, and becomes a memory B cell [Nuñez et al., 1991].

III.B. Bcl-2 Gene

bcl-2 is an M_r 25,000 integral membrane protein. It was initially described by Tsujimoto et al. [1984] as an oncogene associated with human follicular B-cell lymphomas (B-cell lymphoma/leukemia-2). In at least 90% of these lymphomas, the bcl-2 gene is translocated from its normal location at 18q21 to a position adjacent to the powerful enhancer elements of the immunoglobulin heavy chain (IgH) locus at 14q32. The heavy chain enhancer elements cause a deregulation of the translocated bcl-2 gene that results in the overproduction of both bcl-2 mRNA and bcl-2 protein. This t(14:18) is not unique to follicular lymphomas and occurs in up to 50% of cases of benign follicular hyperplasia in localized lymphoid tissue during an infection. Overexpression of bcl-2 also occurs in many other human malignancies, including non-Hodgkin's lymphoma, chronic lymphocytic leukemia (CLL), neuroblastomas, and adenocarcinomas of the prostate, lung, and nasopharynx [Miyashita and Reed, 1992; Hanada et al., 1993; McDonnell et al., 1993; Hague et al., 1994; Reed et al., 1994].

While not initiating neoplastic transformation, bcl-2 was first noted to prolong the survival without inducing proliferation, of IL-3–dependent immature pre-B lymphocytes and myeloid cells [Vaux et al., 1988]. Antisense inhibition of bcl-2 expedites cell death [Reed et al., 1990] and, normally, levels of bcl-2 decrease during cell maturation. Clinically important is that bcl-2 protein levels correlate with a cell's relative sensitivity or resistance to x-irradiation and to a variety of chemotherapeutic drugs [Miyashita and Reed, 1992; Miyashita et al. 1993]. Since many anticancer drugs act by inducing apoptosis, the overexpression of bcl-2 in a tumor often signals a poor clinical prognosis (see Lowe and Ruley, this volume). It appears that the inactivation of the tumor suppressor gene p53, which occurs in over half of all solid tumors, regulates bcl-2. While p53 normally promotes PCD, it also appears to up-regulate bcl-2 expression [Miyashita et al., 1994], thus preventing cell death.

Several cDNAs have recently been cloned that demonstrate an emerging family of bcl-2–related proteins. Mammalian members of this family include bcl-x (bcl-x_L and bcl-x_S) [Boise et al., 1993], *bax* [Oltvai et al., 1993], mcl-1 [Kozopas et al., 1993,] and A1 [Lin et al., 1993]. In addition, the Epstein-Barr virus gene product BHRF1 shares a high degree of similarity with the carboxyl portion of bcl-2 protein [Henderson et al., 1993; Hickish et al., 1994], as does the LMW5-HL open reading frame of the African swine fever virus [Neilan et al.,1993]. The core regions of the BHRF1, LMW5-HL, and bcl-2 proteins are highly conserved [Williams and Smith, 1993]. The *Caenorhabditis elegans* cell survival gene *ced-9* encodes a functional homolog of bcl-2 [Hengartner and Horvitz, 1994] (see Lints and Driscoll, this volume). In addition to direct p53 regulation of bcl-2 expression, bcl-2 appears regulated at the protein level by another bcl-2 family member, *bax* protein [Oltvai et al., 1993], which is a p53 immediate early response gene. Bcl-2 and *bax* proteins appear to compete with one another to control the relative susceptibility of cells to p53-mediated apoptosis [Selvakumaran et al., 1994].

Despite considerable interest and study, the mechanism(s) by which bcl-2 prevents or retards apoptotic cell death remain unestablished. Because few of the morphologic changes characteristic of apoptosis (e.g., chromatin condensation, membrane bleeding, cell shrinkage, nuclear fragmentation, DNA degradation) occur in cells overexpressing bcl-2, the gene is thought to act on an early event in the apoptotic pathway. bcl-2 has a unique subcellular distribution, being localized to the inner mitochondrial membrane, nuclear envelope, and endoplasmic reticulum. Some research suggests that bcl-2 may function by altering intracellular calcium pools or by decreasing antioxidant synthesis [Baffy et al., 1993; Kane et al., 1993]. Three classes of soluble mediators are recognized to prevent PCD generally through the up-regulation of bcl-2 expression. Cytokines such as IL-4, TNF-α, and IFN-γ prevent apoptosis in certain cell types or cell lines by increasing bcl-2. CD40 is expressed on mature B cells, and activation through this molecule prevents apoptosis. In murine, and perhaps human, B cells CD2 molecules induce either proliferation or apoptosis through up-regulation of bcl-2 [Genaro et al., 1994].

III.C. Fas/APO-1 (CD95)

Fas/APO-1 is an M_r 48,000 surface protein that when bound by a receptor-specific monoclonal antibody induces apoptosis in many cell types. It was independently described by separate groups of investigators each of which gave it a unique name [Trauth et al., 1989; Yonehara et al., 1989], but both proteins were subsequently revealed to have complete sequence identity.

Fas/APO-1 is a member of the tumor necrosis factor (TNF)/nerve growth factor (NGF) receptor superfamily that also includes TNF p55, TNF p75, the low-affinity NGF receptor, the B-cell antigen CD40, CD27, and CD30 [Barclay et al., 1993]. Recently, the natural ligand of the Fas/APO-1 receptor [Takahashi et al., 1994] and the human Fas gene were cloned [Cheng et al., 1995]. The Fas gene spans more than 26 kb of DNA and has nine exons. Fas/APO-1 is expressed on normal activated

T and B lymphocytes, thymocytes, and several nonlymphoid tissues, including the liver, lung, heart, ovary, and testis. Additionally, it is expressed in many human tumor cell lines of both hematopoietic and nonhematopoietic origin [Owen-Schaub et al., 1994].

The mechanism(s) through which Fas/APO-1 induces PCD is unknown, and its expression is not uniformly associated with the induction of PCD. Cells can apparently initiate independent programs leading to either necrosis or apoptosis depending upon the signal they receive. Despite both being members of the TNF/NGF receptor superfamily and sharing a 65 bp "death domain," Fas/APO-1 and TNF p55 clearly function through independent signaling pathways [Schulze-Ostoff et al., 1994]. Bcl-2 appears capable, in part, of interfering with the apoptotic process initiated by Fas/APO-1 and TNF [Itoh et al., 1993]. Among other differences, the triggering of Fas/APO-1 does not induce transcription of NF-κB which is readily activated by TNF and accounts for many of its biological activities. Primer extension analysis reveals that human Fas has multiple transcription initiation sites, one of which appears to be *c-myb*, which is important in G_1/S transition in normal activated T cells [Gewirtz et al., 1989].

Lymphoproliferation (*lpr*) mice carry a point mutation of, or due to a retroviral insert are unable to express the Fas/APO-1 gene. These animals accumulate double-negative T cells in their spleen and lymph nodes and autoimmune B cells, inferring that Fas/APO-1 killing is important in the normal elimination of self-reactive T and B cells and the induction of immune tolerance [Suda et al., 1993]. Fas/APO-1 has also been suggested to play a role in the induction of peripheral T-cell tolerance and in the antigen-stimulated elimination of mature T cells [Russell et al., 1993].

IV. PROGRAMMED CELL DEATH IN AIDS

While the loss of CD4$^+$ cells is arguably the most characteristic feature of AIDS, the mechanism(s) that account for the preferential disappearance of this cell population is not known [Fauci, 1988]. *In vitro* infection of cultured T-cell lines with HIV-1 usually is not associated with large-scale cell death or even marked cytopathic changes. In fact, in many cases it is difficult to demonstrate that HIV-1–infected cells contain virus. This is especially true for peripheral blood cells from asymptomatic HIV-1–infected individuals, in whom very few cells appear to carry the virus, and it is difficult to demonstrate HIV-1 in the blood without PCR amplification [Brinchmann et al., 1991]. Furthermore, the loss of the CD4$^+$ cells seems to proceed in the absence of direct HIV-1 infection of these cells [Everall et al., 1991; Genis et al., 1992]. Because of this curious phenomenon, there is a great deal of speculation on and investigation of the nature of the destruction of the CD4$^+$ cells [Clerici et al., 1989], as well as other cells such as neurons, in HIV-1–infected people. One important consideration in these studies is the role that PCD or apoptosis plays in the destruction of non-HIV-1–infected cells [Laurent-Crawford et al., 1991; Ameisen et al., 1994].

As previously outlined, PCD has many important functions in the normal immune system. In the thymus it serves to eliminate self-reactive T-cell clones and provides a control mechanism over the genesis of an autoimmune response that could occur during a normal immune response to a pathogen [Brinchmann et al., 1991; Cameron et al., 1994]. A normal immune response to a pathogen involves interactions between T, B, and accessory cells along with antigen. Recognition of the antigen is also linked to recognition of self (MHC) in order for the reaction to proceed to the generation of antibody or cell-mediated immunity. Therefore, in any immune process there is recognition of self along with the recognition of non-self. Normally, the self reacting cells are eliminated by PCD.

There are two types of cell surface receptors for HIV-1 on the CD4$^+$ helper cell. One is the T cell antigen receptor in that the helper T-cell recognizes the virus as an antigen, and the other is the CD4 epitope that can bind the envelope proteins of the virus directly in a nonimmune mechanism [Habeshaw et al., 1990]. Binding of the CD4 epitope has been found to cause the helper cell activation process leading to apoptosis in noninfected T cells [Banda et al., 1992].

There is an increasing amount of evidence indicating that there is an apototic process in AIDS that causes the loss of the helper CD4$^+$ T cells, as well as other cells in the body that are not themselves infected with HIV-1 [Bishop et al., 1993; Saha et al., 1994]. However, there is a report that shows that an apoptotic process in an HIV-1 infection does not correlate with progression of the disease [Meyaard et al., 1994]. If infection cannot explain the loss of cells, then cellular necrosis due to infection or inflammation would seem to be less probable in accounting for the cell death. Crosslinking of the CD4 epitope on helper T cells has been demonstrated to lead to the apoptosis of the uninfected cells [Banda et al., 1992]. This process can be inhibited by drugs that prevent cellular activation (cyclosporin A), which provides further evidence that the crosslinking activates the cells but that the activation does not proceed to an immune response since it does not involve the TcR or perhaps the accessory cells and their soluble cytokines [Laurent-Crawford et al., 1993]. Furthermore, there is evidence that cytokines secreted by accessory cells can regulate apoptosis [Coffman et al., 1991; D'Andrea et al., 1993; Groux et al,. 1993; Chehimi et al., 1994]. If HIV-1 disrupts accessory cell function it could alter the ability of the T cell to avoid apoptosis by depriving the T cell of a cytokine that it needs to progress through an activation cycle to an immune response [Liu and Janeway, 1990], or by providing a cytokine that actually promotes entry into an apoptotic cell cycle [Shi et al., 1989]. There is controversy as to the mechanism for the loss of accessory cell function. It may be that they are HIV-1 infected or that they themselves undergo apoptosis in the central lymphoid organs due to a lack of T-cell regulatory activity or perhaps are destroyed by CD8 lymphocytes that recognize the HIV-1 antigen on the surface of infected accessory cells. Furthermore, CD8$^+$ cells themselves may undergo an editing process through apoptosis in their response to viral infection [Selin and Welsh, 1994].

There is evidence that an HIV-1 infection can induce an autoimmune reaction [Fujinami and Oldstone, 1985; Golding et al., 1988, 1989; Kion and Hoffmann, 1991; Morrow et al., 1991, Schattner and Bentwich, 1993]. On the one hand, it has been reported that some of the HIV-1 antigens share homology with HLA antigens to a variable degree. It is known that viral and bacterial antigens can show homology with certain HLA antigens, and this may provide a mechanism for a chronic low-grade stimulus to the immune system to augment clonal expansion and maintenance. However, to avoid a pathologic autoimmune response it would be necessary to eliminate certain of the alloreactive cells. Another possible mechanism for the induction of an autoimmune response during a viral infection would rely not on any homology between HLA and viral antigens. In this case, the recognition of the viral antigens in concert with the recognition of self and perhaps recognition of damaged or altered HLA antigens would result in the initiation of an immune response to self-antigens. This theory has been used to explain many autoimmune disorders. Furthermore, there is evidence that the HIV-1 proteins can bind to certain of the histocompatibility antigens and set up a condition for the development of either a superantigen or an autoimmune response [Sette et al., 1989; Arthur et al., 1992; Chicz et al., 1992, 1993].

Another theory used to explain lymphocyte apoptosis in an HIV-1 infection has been the proposal that HIV-1 in some ways acts as a superantigen that could induce T cells having certain variable region shifts in the β chain of the TcR to proceed to an apoptotic event [Odaka et al., 1990; Imberti et al., 1991; Abe et al., 1992; Dalgleish et al., 1992; Foo-Philips et al., 1992; Lafon et al., 1992]. This particular theory has not been well documented at this time and would involve only those T cells that recognize the virus as an antigen. This mechanism would not explain the loss of cells in the infected person that do not bind the antigen or those that are not infected.

One explanation to account for apoptosis in uninfected cells is based on the observation that monocytes carry a high viral load [Koenig et al., 1986; Mosier and Sieburg, 1994], which alters their pattern of lymphokine secretion as well as their ability to provide second signals in an immune response [Wang et al., 1993]. Lymphocytes that need certain lymphokines in order to remain viable may not receive them [Clerici et al., 1993]. Second signals that may be necessary for progression through a cell cycle may not be provided and would also lead to apoptosis. Furthermore, the infected cells would provide the necessary viral proteins that could crosslink the CD4 epitope on the helper T cell.

Still another mechanism in an HIV-1 infection could lead to cellular activation and apoptosis. This relates to the presence of opportunistic infections in the compromised host. Such infections are generally thought to affect cellular function by contributing to an overall increased activation of the immune system due simply to inflammation and pathogen load.

In addition to serving a regulatory role in the development and maintenance of the immune system, apoptosis contributes to the depletion of infected T cells of the immune system [Howie et al., 1994]. In the case of infectious mononucleosis, infected CD45RO (memory) $CD4^+$ and $CD8^+$ T cells selectively undergo apoptosis in response to primary infection by Epstein-Barr virus [Uehara et al., 1992]. Additionally, there are proposals that the general process of apoptosis plays a role in many other human diseases. These include malignancy, neurologic disorders, heart disease, intestinal disease, kidney disease and aging, in addition to its role in autoimmune disorders [Barr and Tomei, 1994].

A recognized feature of an HIV-1 infection is the relationship between the age of the patient, the time to develop AIDS, and the rate of progression to death [Goedert et al., 1989]. As demonstrated in patients with hemophilia who had been infected with contaminated Factor VIII, the time from the infection with HIV-1 to AIDS condition was strongly age related. A feature of the immune system in old persons [Nagel et al., 1989] and old mice [Chrest et al., 1993] is an increase in apoptosis in stimu-

lated cultures of peripheral blood lymphocytes. It could be hypothesized that the additive effects of age-related apoptosis of lymphocytes and the HIV-1–induced apoptosis contribute to the shortening of time needed to develop AIDS in the older infected patient. However, this phenomenon could also account for the decline in the ability of the older patient to repopulate the lymphocyte population destroyed by HIV-1. It has been reported that even in young people there is a delay in T-cell reconstitution following chemotherapy [Mackall et al., 1995]. A combination of these processes might be the basis of the age-related changes.

There are other features of an aging immune system that may promote HIV-1–induced apoptosis of $CD4^+$ and other cells in the infected patient. One is the marked increase in memory T cells in the elderly [Lerner et al., 1989; Chrest et al., 1995], and the other is the increased levels of certain inflammatory cytokines found in the sera of the elderly. Of those cytokines IFN-γ is known to be expressed at higher levels by cells from elderly donors, and this particular cytokine has been implicated in the apoptotic process. As a person or experimental animal ages there is a shift in the phenotype of the T cell to the memory $CD45RO^+$ phenotype, with the concurrent loss of the naive T-cell phenotype. If memory T cells are more prone to apoptosis [Akbar et al., 1993; Chrest et al., 1995] then the elderly infected patient would have a faster loss of T cells from the lymphoid tissue and a faster progression of the disease. Therefore, there may be several reasons for the more rapid progression of HIV-1 infection to AIDS in the elderly.

REFERENCES

Abe R, Foo-Philips M, Granger LG, Kanagawa O (1992): Characterization of the Mls^f system I. A novel "polymorphism" of endogenous superantigens. J Immunol 149:3429–3439.

Akbar AN, Borthick N, Salmon M, Gombert W, Bofill M, Shamsadeen N, Pilling D, Pett S, Grundy JE, Janossy G (1993): The significance of low bcl-2 expression by $CD45RO^+$ T-cells in normal individuals and patients with acute viral infections. The role of apoptosis in T-cell memory. J Exp Med 178:427–438.

Ameisen JC, Estaquier J, Idziorek (1994): From AIDS to parasite infection: Pathogen-mediated subversion of programmed cell death as a mechanism for immune dysregulation. Immunol Rev 142:9–51.

Arthur LO, Bess JW, Sowder RC 2nd, Benveniste RE, Mann DL, Chermann JC, Henderson LE (1992): Cellular proteins bound to immunodeficiency viruses: Implications for pathogenesis and vaccines. Science 258:1935–1938.

Baffy G, Miyashita T, Williamson JR, Reed JC (1993): Apoptosis induced by withdrawal of interleukin-3 (IL-3): From an IL-3–dependent hematopoietic cell line is associated with repartitioning of intracellular calcium and is blocked by enforced Bcl-2 oncoprotein production. J Biol Chem 268:6511 6519.

Banda NK, Bernier J, Kurahara DK, Kurrle R, Haigwood N, Sekaly RP, Finkel TH (1992): Crosslinking CD4 by HIV gp 120 primes T-cells for activation-induced apoptosis. J Exp Med 176:1099–1106.

Barclay AN, Birkeland ML, Brown MH, Beyers AD, Davis SJ, Somoza C, Williams AF (1993): Protein superfamilies and cell surface molecules. In Barclay AN, Birkeland ML, Brown MH, Beyers AD, Davis SJ, Somoza C, Williams AF (eds): The Leukocyte Antigen Facts Book. London: Academic Press, pp 38–87.

Barr PJ, Tomei LD (1994): Apoptosis and its role in human disease. Biotechnology 12:487–493.

Bishop SA, Gruffydd-Jones TJ, Harbour DA, Stokes CR (1993): Programmed cell death (apoptosis): As a mechanism of cell death in peripheral blood mononuclear cells from cats infected with feline immunodeficiency virus (FIV). Clin Exp Immunol 93:65–71.

Blackman M, Kappler J, Marrack P (1990): The role of T-cell receptor in positive and negative selection of developing T-cells. Science 248:1335–1341.

Boise LH, Gonzalez-Garcia M, Postema CE, Ding L, Lindsten T, Turka LA, Mao X, Nuñez G, Thompson CB (1993): bcl-x, a bcl-2-related gene that functions as a dominant regulator of apoptotic cell death. Cell 74:597–608.

Brinchmann JE, Albert J, Vartdal F (1991): Few infected CD4+ T-cells but a high proportion of replication-competent provirus copies in asymptomatic human immunodeficiency virus type 1 infection. J Virol 65:2019–2023.

Cameron PU, Pope M, Gezelter S, Steinman RM (1994): Infection and apoptotic cell death of CD4+ T-cells during an immune response to HIV-1-pulsed dendritic cells. AIDS Res Hum Retroviruses 10:61–71.

Chchimi, J, Starr SE, Frank I, D'Andrea A, Ma X, MacGregor RR, Sennelier J, Trinchieri G (1994): Impaired IL-12 production in HIV-infected patients. J Exp Med 179:1361–1366.

Cheng J, Liu C, Koopman WJ, Mountz JD (1995): Characterization of human Fas gene. Exon/intron organization and promotor region. J Immunol 154:1239–1245.

Chicz RM, Urban RG, Lane WS, Gorga JC, Stern LJ, Vignali DA, Strominger JL (1992): Predominant naturally processed peptides bound to HLA-DRl are derived from MHC-related molecules and are heterogeneous in size. Nature 358:764–768.

Chicz RM, Urban RG, Gorga JC, Vignali DA, Lane WS, Strominger JL (1993): Specificity and promiscuity among naturally processed peptides bound to HLA-DR alleles. J Exp Med 178:27–47.

Chrest FJ, Buchholz MA, Kim YH, Kwon TK, Nordin AA (1993): Identification and Quantitation of apoptotic cells following anti-CD3 activation of murine G_0 T-cells. Cytometry 14:883–890.

Chrest FJ, Buchholz MA, Kim YH, Kwon TK, Nordin AA (1995): Anti-CD3-induced apoptosis in T-cells from young and old mice. Cytometry 20:33–42.

Clerici M, Lucey DR, Berzofsky JA, Pinto LA, Wynn TA, Blatt SP, Dolan MJ, Hendrix CW, Wolf SF, Shearer GM (1993): Restoration of HIV-specific cell-mediated immune response by IL-12 in vitro. Science 262:1721–1724.

Clerici M, Stocks NI, Zajac RA, Boswell RN, Luccy DR, Via CS, Shearer, GM (1989): Detection of three distinct patterns of T helper cell dysfunction in asymptomatic, human immunodeficiency virus seropositive patients. J Clin Invest 84:1892–1899.

Coffman RL, Varkila K, Scott P, Chatelain R (1991): Role of cytokines in the differentiation of CD4 T-cell subsets in vivo. Immunol Rev 123:189–207.

Cohen JJ, Duke RC, Fadok VA, Sellins KS (1992): Apoptosis and programmed cell death in immunity. Annu Rev Immunol 10:267–293.

Dalgleish AG, Wilson S, Gompels M, Ludlam C, Gazzard B, Coates AM, Habershaw J (1992): T-cell receptor variable gene products and early HIV-1 infection. Lancet 339:824–828.

D'Andrea A, Aste-Amezaga M, Valiante NM, Ma XJ, Kubin M, Trinchieri G (1993): Interleukin 10 (IL-10): inhibits human lymphocyte interferon gamma-production by suppressing natural killer cell stimulatory factor/IL-12 synthesis in accessory cells. J Exp Med 9:1041–1048.

Ehlich A, Schaal S, Gu H, Kitamura D, Muller W, Rajewsky K (1993): Immunoglobulin heavy and light chain genes rearrange independently at early stages of B cell development. Cell 72:695–704.

Everall IP, Luthert, PJ, Lantos, PL (1991): Neuronal loss in the frontal cortex in HIV-infection. Lancet 337:1119–1121.

Fauci AS (1988): The human immunodeficiency virus: Infectivity and mechanisms of pathogenesis. Science 239:617–622.

Finkel TH, Cambier JC, Kubo RT, Born WK, Marrack P, Kappler JW (1989): The thymus has two functionally distinct populations of immature $\alpha\beta$ T-cells: One population is deleted by ligation of $\alpha\beta$ TCR. Cell 58:1047–1054.

Foo-Philips M, Kozak CA, Principato MA, Abe R (1992): Characterization of the Mls^f system II. Identification of mouse mammary tumor virus proviruses involved in the clonal deletion of self-Mls^f-reactive T-cells. J Immunol 149:3440–3447.

Fujinami RS, Oldstone MBA (1985): Amino acid homology between the encephalitogenic site of myelin basic protein and virus: Mechanism for autoimmunity. Science 230:1043–1045.

Genaro AM, Gonzálo JA, Boscá L, Martinez C (1994): CD2-CD48 interactions prevent apoptosis in murine B lymphocytes by upregulating bcl-2 expression. Eur J Immunol 10:2515–2521.

Genis, P, Jett M, Bernton EW, Boyle T, Gelbard HA, Dzenko K, Keane RW, Resnick L, Mizrachi Y, Volsky DJ, Epstein LG, Gendelman HE (1992): Cytokines and arachidonic metabolites produced during human immunodeficiency virus (HIV): Infected macrophage-astroglia interactions: implications for the neuropathogenesis of HIV disease. J Exp Med 176:1703–1718.

Gewirtz AM, Anfossi G, Venturelli D, Valpreda S, Sims R, Calabretta B (1989): G_1/S transition in normal human T-lymphocytes requires a nuclear protein encoded by c-myb. Science 245:180–183.

Goedert JJ, Kessler CM, Aledort LM, Bigger RJ, Andes WA, White GC 2nd, Drummond JE, Vaidya K, Mann DL, Eyster ME, Ragni MV, Lederman MM, Cohen AR, Bray GL, Rosenberg PS, Friedman RM, Hilgartner MW, Blattner WA, Kroner B, Gail MH (1989): A prospective study of human immunodeficiency virus type 1 infection and the development of AIDS in subjects with hemophilia. N Engl J Med 321:1141–1148.

Golding H, Robey FA, Gates FT 3rd, Linder W, Beining PR, Hoffman T, Golding B (1988): Identification of homologous regions in human immunodeficiency virus I gp41 and human MHC class II β 1 domain. I. Monoclonal antibodies against the gp41-derived peptide and patients' sera react with native HLA class II antigens, suggesting a role for autoimmunity in the pathogenesis of acquired immune deficiency syndrome. J Exp Med 167:914–923.

Golding H, Shearer GM, Hillman K, Lucas P, Manischewitz J, Zajac RA, Clerici M, Gress RE, Boswell RN, Golding B (1989): Common epitope in human immunodeficiency virus (HIV): I-gp41 and HLA class II elicits immunosuppressive autoantibodies capable of contributing to immune dysfunction in HIV-1–infected individuals. J Clin Invest 83:1430–1435.

Gorczyca W, Melamed MR, Darzynkiewicz Z (1993): Apoptosis of S-phase HL-60 cells induced by DNA topoisomerase inhibitors: Detection of DNA strand breaks by flow cytometry using the in situ nick translation assay. Toxicol Lett 67:249–258.

Gougeon ML, Montagnier L (1993): Apoptosis in AIDS. Science 260:1269–1270.

Groux H, Monté D, Plouvier B, Capron A, Ameisen JC (1993): CD3-mediated apoptosis of human medullary thymocytes and activated peripheral T-cells: Respective roles of interleukin-1, interleukin-2, interferon gamma and accessory cells. Eur J Immunol 23:1623–1629.

Habeshaw JA, Dalgleish AG, Bountiff L, Newell AL, Wilks D, Walker LC, Manca F (1990): AIDS pathogenesis: HIV envelope and its interaction with cell proteins. Immunol Today 11:418–425.

Hadden JW (1988): Transmembrane signals in the activation of T lymphocytes by mitogenic antigens. Immunol Today 9:235–239.

Hague A, Moorghen M, Hicks D, Chapman M, Paraskeva C (1994): bcl-2 expression in human colorectal adenomas and carcinomas. Oncogene 9:3367–3370.

Hanada M, Krajewski S, Tanaka S, Cazals-Hatem D, Spengler BA, Ross RA, Biedler JL, Reed JC (1993): Regulation of Bcl-2 oncoprotein levels with differentiation of human neuroblastoma cells. Cancer Res 53:4978–4986.

Henderson S, Huen D, Rowe M, Dawson C, Johnson G, Rickinson A (1993): Epstein-Barr virus-coded BHRF1 protein, a viral homologue of Bcl-2, protects human B cells from programmed cell death. Proc Natl Acad Sci USA 90:8479–8483.

Hengartner MO, Horvitz HR (1994): $C.$ $elegans$ cell survival gene ced-9 encodes a functional homologue of the mammalian proto-oncogene bcl-2. Cell 76:665–676.

Hickish T, Robertson D, Clarke P, Hill M, di Stefano F, Clarke C, Cunningham D (1994): Ultrastructural localization of BHRF1: An Epstein-Barr virus gene product which has homology with bcl-2. Cancer Res 54:2808–2811.

Hogquist KA, Jameson SC, Heath WR, Howard JL, Bevan MJ, Carbone FR (1994): T-cell receptor antagonist peptides induce positive selection. Cell 76:17–27.

Howie SEM, Harrison DJ, Wyllie AH (1994): Lymphocyte apoptosis-Mechanisms and implications in disease. Immunol Rev 142:141–156.

Imberti L, Sottini A, Bettinardi A, Puoti M, Primi D (1991): Selective depletion in HIV infection of T-cells that bear specific T-cell receptor V_β sequences. Science 254:860–862.

Itoh N, Tsujimoto Y, Nagata S (1993): Effect of bcl-2 on Fas-mediated cell death. J Immunol 151:621 627.

Kabelitz D, Oberg HH, Pohl T, Pechhold K (1994): Antigen induced death of mature T lymphocytes: Analysis by flow cytometry. Immunol Rev 142:158–174.

Kane DJ, Sarafian TA, Anton R, Hahn H, Gralla EB, Valentine JS, Ord T, Bredesen DE (1993): Bcl-2 inhibition of neural death: Decreased generation of reactive oxygen species. Science 262:1274–1277.

Kion TA, Hoffmann GW (1991): Anti-HIV and anti-anti-MHC antibodies in alloimmune and autoimmune mice. Science 253:1138–1140.

Kisielow P, Teh HS, Bluthmann H, von Boehmer H (1988): Positive selection of antigen-specific T-cells in thymus by restricting MHC molecules. Nature 335:730–733.

Koenig S, Gendelman HE, Orenstein JM, Dal Canto MC, Perzeshkpour GH, Yungbluth M, Janotta F, Aksamit A, Martin, MA, Fauci AS (1986): Detection of AIDS virus in macrophages in brain tissue from AIDS patients with encephalopathy. Science 233:1089–1093.

Korsmeyer SJ (1992): Bcl-2: A repressor of lymphocyte death. Immunol Today 13:285–288.

Kozopas KM, Yang T, Buchan HL, Zhou P, Craig RW (1993): MCL1, a gene expressed in programmed myeloid cell differentiation, has sequence similarity to BCL2. Proc Natl Acad Sci USA 90: 3516–3520.

Lafon M, Lafage M, Martinez-Arends A, Ramirez R, Vuillier F, Charron D, Lotteau V, Scott-Algara D (1992): Evidence for a viral superantigen in humans. Nature 358:507–510.

Laurent-Crawford AG, Krust B, Muller S, Rivière Y, Rey-Cuillé MA, Béchet JM, Montagnier L, Hovanessian AG (1991): The cytopathic effect of HIV is associated with apoptosis. Virology 185:829–839.

Laurent-Crawford AG, Krust B, Rivière Y, Desgranges C, Muller S, Kieny MP, Dauguet C, Hovanessian AG (1993): Membrane expression of HIV envelope glycoproteins triggers apoptosis in CD4 cells. AIDS Res Hum Retroviruses 9:761–773.

Lemer A, Yamada T, Miller RA (1989): Pgp-1 hi T Lymphocytes accumulate with age in mice and respond poorly to concanavalin A. Eur J Immunol 19:977–982.

Lin EY, Orlofsky A, Berger MS, Prystowsky MB (1993): Characterization of A1, a novel hemopoietic-specific early-response gene with sequence similarity to bcl-2. J Immunol 151:1979–1988.

Liu Y, Janeway CA Jr (1990): IFNγ plays a crucial role in induced cell death of effector T-cell: A possible third mechanism of self tolerance. J Exp Med 172:1735–1739.

McConkey DJ, Hartzell P, Amador-Perez JF, Orrenius S, Jondal M (1989): Calcium-dependent killing of immature thymocytes by stimulation via the CD3/T-cell receptor complex. J Immunol 143:1801–1806.

MacDonald HR, Lees RK (1990): Programmed death of autoreactive thymocytes. Nature 343:642–644.

McDonnell TJ, Marin MC, Hsu B, Brisbay SM, McConnell K, Tu SM, Campbell ML, Rodriguez-Villanueva J (1993) The bcl-2 oncogene: apoptosis and neoplasia. Radiat Res 136:307–312.

Mackall CL, Fleisher TA, Brown MR, Andrich MP, Chen CC, Fuerstein IM, Horowitz MC, Magrath IT, Shad AT, Steinberg SM, Wexler LH, Gress RE (1995): Age, thymopoiesis and CD4+ T lymphocyte regeneration after intensive chemotherapy. N Engl J Med 332:143–149.

Meyaard L, Otto SA, Keet IP, Roos MT, Miedema F (1994): Programmed death of cells in human immunodeficiency virus infection. No correlation with progression to disease. J Clin Invest 93:982–988.

Miyashita T, Reed JC (1992): bcl-2 gene transfer increases relative resistance of S49.1 and WEHI7.2 lymphoid cells to cell death and DNA fragmentation induced by glucocorticoids and multiple chemotherapeutic drugs. Cancer Res 52:5407–5411.

Miyashita T, Reed JC (1993): Bcl-2 oncoprotein blocks chemotherapy-induced apoptosis in a human leukemia cell line. Blood 81:151–157.

Miyashihta T, Krajewski S, Krajewska M, Wang HG, Lin HK, Liebermann DS, Hoffman B, Reed JC (1994): Tumor suppressor p53 is a regulator of bcl-2 and bax gene expression *in vitro* and *in vivo*. Oncogene 9:1799–1805.

Morrow WJ, Isenberg DA, Sobol RE, Stricker RB, Kieber-Emmons T (1991): AIDS virus infection and autoimmunity: A perspective of the clinical, immunological, and molecular origins of the autoallergic pathologies associated with HIV disease. Clin Immunol Immunopathol 58:163–180.

Mosier, D, Sieburg, H (1994): Macrophage-tropic HIV: Critical for AIDS pathogenesis? Immunol Today 15:332–339.

Nagel JE, Chopra RK, Powers DC, Adler WH (1989): Effect of age on the high affinity interleukin 2 receptor of phytohaemagglutinin stimulated peripheral blood lymphocytes. Clin Exp Immunol 75:286–291.

Neilan JG, Lu Z, Afonso CL, Kutish GF, Sussman MD, Rock DL (1993): An African swine fever virus gene with similarity to the protooncogene bcl-2 and the Epstein-Barr virus gene BHRF1. J Virol 67:4391–4394.

Nuñez G, Hockenbery D, McDonnell TJ, Sorensen CM, Korsmeyer SJ (1991): Bcl-2 maintains B cell memory. Nature 353:71–73.

Odaka C, Kizaki H, Tadakuma T (1990): T-cell receptor-mediated DNA fragmentation and cell death in T-cell hybridoma. J Immunol 144:2096–2101.

Oltvai ZN, Milliman CL, Korsmeyer SJ (1993): bcl-2 heterodimerizes in vivo with a conserved homolog, Bax, that accelerates programmed cell death. Cell 74:609–619.

Owen-Schaub LB, Radinsky R, Kuzel E, Berry K, Yonehara S (1994): Anti-Fas on nonhematopoietic tumors: Levels of Fas/APO-1 and bcl-2 are not predictive of biological responsiveness. Cancer Res 54:1580–1586.

Proust JJ, Shaper NL, Buchholz MA, Nordin AA (1991): T-cell activation in the absence of interleukin 2 (IL2): Results in the induction of high affinity IL2 receptor unable to transmit a proliferative signal. Eur J Immunol 21:335–341.

Radvanyi LG, Mills GB, Miller RG (1993): Religation of the T-cell receptor after primary activation of mature T-cells inhibits proliferation and induces apoptotic cell death. J Immunol 150:5704–5715.

Reed JC, Haldar S, Cuddy M, Croce C, Makover D (1990): bcl-2–Mediated tumorigenicity in a T-lymphoid cell line: Synergy with c-myc and inhibition by bcl-2 antisense. Proc Natl Acad Sci USA 87:3660–3664.

Reed JC, Kitada S, Takayama S, Miyashita T (1994): Regulation of chemoresistance by the bcl-2 oncoprotein in non-Hodgkin's lymphoma and lymphocytic leukemia cell lines. Ann Oncol 5 (Suppl 1):S61–S65.

Rolink A, Melchers F (1991): Molecular and cellular origins of B lymphocyte diversity. Cell 66:1081–1094.

Russell JH, Rush B, Weaver C, Wang R (1993): Mature T-cells of autoimmune lpr/lpr mice have a defect in antigen-stimulated suicide. Proc Natl Acad Sci USA 90:4409–4413.

Saha K, Yuen PH, Wong PK (1994): Murine retrovirus-induced depletion of T-cells is mediated through activation-induced death by apoptosis. J Virol 68:2735–2740.

Salmon M, Pilling D, Borthwick NJ, Viner N, Janossy G, Bacon PA, Akbar AN (1994): The progressive differentiation of primed T-cells is associated with an increasing susceptibility to apoptosis. Eur J Immunol 24:892–899.

Sanders, ME, Makgoba MW, Shaw S (1988): Human naive and memory T-cells. Immunol Today 9:195–198.

Schattner A, Bentwich Z (1993): Autoimmunity in human immunodeficiency virus infection. Clin Aspects Autoimmunity 5:19–27.

Schatz DG, Oettinger MA, Schlissel MS (1992): V(D)J recombination: molecular biology and regulation. Annu Rev Immunol 10:359–383.

Schulze-Osthoff K, Krammer PH, Dröge W (1994): Divergent signaling via APO-1/Fas and the TNF receptor, two homologous molecules involved in physiologic cell death. EMBO J 13:4587–4596.

Schwartz RH (1993): Immunological Tolerance. In Paul WE (ed): Fundamental Immunology, 3rd ed. New York: Raven Press, pp 678–731.

Selin LK, Welsh RM (1994): Specificity and editing by apoptosis of virus-induced cytotoxic T lymphocytes. Curr Opin Immunol 6:553–559.

Selvakumaran M, Lin HK, Miyashita T, Wang HG, Krajewski S, Reed JC, Hoffman B, Liebermann D (1994): Immediate early up-regulation of bax expression by p53 but not TGFβ1: A paradigm for distinct apoptotic pathways. Oncogene 9:1791–1798.

Sette A, Buus S, Appella E, Smith JA, Chesnut R, Miles C, Colon SM, Grey HM (1989): Prediction of major histocompatibility complex binding regions of protein antigens by sequence pattern analysis. Proc Natl Acad Sci USA 86:3296–3300.

Shi YF, Sahai BM, Green DR (1989): Cyclosporin A inhibits activation-induced cell death in T-cell hybridomas and thymocytes. Nature 339:625–626.

Smith CA, Williams GT, Kingston R, Jenkinson EJ, Owen JJ (1989): Antibodies to CD3/T-cell receptor complex induce death by apoptosis in immature T-cells in thymic cultures. Nature 337:181–183.

Snapper CM, Finkelman FD (1993): Immunoglobulin Class Switching. In Paul WE (ed): Fundamental Immunology, 3rd ed. New York: Raven Press, pp 837–863.

Staiano-Coico L, Darzynkiewicz Z, Melamed MR, Weksler ME (1984): Impaired proliferation of T lymphocytes detected in elderly humans by flow cytometry. J Immunol 132:1788–1792.

Suda S, Takahashi T, Goldstein P, Nagata S (1993): Molecular cloning and expression of the Fas ligand, a member of the tumor necrosis family. Cell 75:1169–1178.

Takahashi T, Tanaka M, Inazawa J, Abe T, Suda T, Nagata S (1994): Human Fas ligand: Gene structure, chromosomal location and species specificity. Int Immunol 6:1567–1574.

Trauth BC, Klas C, Peters AMJ, Matzku S, Moller P, Falk W, Debatin KM, Krammer P (1989): Monoclonal antibody-mediated tumor regression by induction of apoptosis. Science 245:301–305.

Tsujimoto Y, Finger LR, Yunis J, Nowell PC, Croce CM (1984): Cloning of the chromosome breakpoint of neoplastic B cells with the t(14;18): Chromosome translocation. Science 226:1097–1099.

Uehara T, Miyawaki T, Ohta K, Tamaru Y, Yokoi T, Nakamura S, Taniguchi N (1992): Apoptotic cell death of primed CD45RO⁺ T lymphocytes in Epstein-Barr virus-induced infectious mononucleosis. Blood 80:452–458.

Vaux DL, Cory S, Adams JM (1988): Bcl-2 gene promotes haemopoietic cell survival and cooperates with c-myc to immortalize pre-B cells. Nature 335:440–442.

Wang R, Murphy K, Loh DY, Weaver C, Russell JH (1993): Differential activation of antigen-stimulated suicide and cytokine production pathways in CD4⁺ T-cells is regulated by the antigen-presenting cell. J Immunol 150:3832–3842.

Weiss A (1989): T Lymphocyte Activation. In Paul WE (ed): Fundamental Immunology, 2nd ed. New York: Raven Press, pp 359–384.

Williams GT, Smith CA (1993): Molecular regulation of apoptosis: genetic controls on cell death. Cell 74:777–779.

Yonehara S, Ishii A, Yonehara M (1989): A cell-killing monoclonal antibody (anti-Fas) to a cell-surface antigen co-downregulated with the receptor of tumor necrosis factor. J Exp Med 169:1747–1756.

ABOUT THE AUTHORS

JAMES E. NAGEL is a Senior Investigator in the Clinical Immunology Section of the National Institute on Aging, NIH. He is on the faculty of the Johns Hopkins University School of Medicine. Dr. Nagel received a B.A. from Southern Illinois University and an M.D. from the University of Illinois at Chicago. Speciality training in Pediatrics was obtained at the University of Illinois Hospital and the Children's Hospital of Pittsburgh. This was followed by Fellowship training in Allergy and Clinical Immunology at the University of Pittsburgh. Dr. Nagel joined the NIH in 1977 as a Clinical Associate. His current research is concerned with lymphokine production by T cells from elderly individuals.

F. JOSEPH CHREST is a Biologist working in the Clinical Immunology Section of the National Institute on Aging, NIH. Mr. Chrest received his B.S. in Biology from Towson State University. He has been involved in the use of flow cytometry and cell sorting for studying age-associated cellular defects of the immune system. His current research is related to the role of apoptosis and the use of multiparameter flow cytometry to study apoptotic T-cell death in both murine and human T-cell activation systems.

PRITI LAL is a Fogarty Center Visiting Fellow in the Clinical Immunology Section of the National Institute on Aging, NIH. She received an B.S. from The University of Delhi and an M.B.B.S. from Maulana Azad Medical College in New Delhi India. Housestaff training in Pathology was obtained at the University of Delhi and the Indian Institute of Pathology. Current research is concerned with apoptosis in lymphocytes from elderly people as it relates to the age related loss of immune function.

WILLIAM H. ADLER is Chief of the Clinical Immunology Section of the National Institute on Aging, NIH. Research in the Section is directed to the studies of the age associated immune deficiency. He is on the faculty of Johns Hopkins University School of Medicine and participates in teaching courses on Immunology for the Medical School and the School of Hygiene and Public Health. He received an A.B. from Harvard College and an M.D. from the State University of New York at Buffalo. Housestaff and fellowship training was obtained at the University of Florida in the Departments of Pediatrics and Pathology. Postdoctoral research in Immunology was in the laboratory of Dr. Richard T. Smith at Florida and concerned investigations of cell mediated immunity and T-cell function. The work produced the first descriptions of the basis for the immune deficiency of aging as a T cell deficit and the effects of endotoxin and enterotoxin mitogenic effects on lymphoid cells. Current research is concerned with the activation, signal transduction, and lymphokine production of cells from elderly individuals that represent the features of age-related loss of immune function.

Cellular Aging and Cell Death: 67–80

Aging and the Cellular Response to Stress: Reduction in the Heat Shock Response

Arlan Richardson and Nikki J. Holbrook

I. INTRODUCTION

Advanced age is associated with a reduction in most physiological functions and in particular with a decreased ability to maintain homeostasis during episodes of acute stress [Shock, 1977; Shock et al., 1984]. It is generally believed that this inability to maintain homeostasis is at least partially responsible for the increase in morbidity/mortality that is observed with advancing age. Current concepts suggest that aging, though multifactorial, results at least in part from damage to molecules, cells, and tissues that exceeds the capacity of the organism to adapt to and/or repair the damage [Martin et al., 1993]. Cells have evolved complex genetic systems to detect specific forms of damage and to activate the expression of genes whose products are presumed to increase the resistance of the cell to the damage and/or aid in its repair. The types of genes induced are dependent on the nature of the stress. Heat stress induces a particular set of genes, oxidative stress another, and DNA damage still another, although there is clearly overlap in the responses. The effectiveness of these genetic responses to stress is likely to be a major factor in resistance to disease and aging. In this chapter we focus on research into the effect of age on the expression of heat shock proteins, because the transcriptional regulation of this system has been well studied and the current data show that this system is indeed altered with aging.

II. HEAT SHOCK PROTEINS: GENERAL FEATURES

The induction of heat shock proteins (Hsp) is the most highly conserved and best understood cellular response to stress occurring in every organism examined from bacteria to man [Lindquist, 1986; Heikkila, 1993; Lindquist and Craig, 1988; Ang et al., 1991]. Hsp encompass a group of related proteins that were originally characterized based on their induction following heat stress. However, the so-called heat shock response is likewise evoked by a variety of other adverse conditions, including metabolic stress, traumatic injury, and even psychological stress [Morimoto et al., 1990, 1992; Morimoto et al., 1993]. Several classes of Hsp genes have been identified and are grouped according to the molecular weights of the proteins they encode. The major families include Hsp110, Hsp90, Hsp70, Hsp60, and the low molecular weight (22 to 27kD) Hsp, with Hsp70 being the most abundant and highly stress-inducible of the Hsp. Certain Hsp are present constitutively in the cell and appear to serve vital cell functions as "chaperones" to assist in the assembly, disassembly, and transport of other intracellular proteins [Welch, 1993; Craig et al., 1994; Gething and Sambrook, 1992]. Thus, the Hsp facilitate the removal as well as the replacement of damaged proteins. Increased expression of Hsp during conditions of stress is presumed to be related to a higher requirement for such chaperone functions. Whatever the precise roles for the various Hsp, their beneficial effects during stress have been established in a variety of studies that have linked Hsp expression with enhanced thermotolerance and to resistance to cytotoxic effects of other stresses [Riabowol et al., 1988; Johnston and Kucey, 1988; Li et al., 1991; Angelidis et al., 1991].

III. DECLINE IN STRESS-INDUCED HsP70 EXPRESSION DURING AGING

III.A. Hsp70 Expression in Response to Heat and Related Stresses

A number of laboratories have demonstrated that mammalian aging is associated with an alteration in the ability of cells to express Hsp70 in response to the stress of heat, and these studies are listed in Table I. It is apparent from the data in Table I that a number of studies have been conducted with a variety of cell types/tissues and animals. Hsp70 expression has been shown to decline with aging in heat-stressed cells, including lung and skin fibroblasts [Fargnoli et al., 1990], hepatocytes [Heydari et al., 1993], splenocytes [Pahlavani et al. submitted], and mononuclear cells [Deguchi et al., 1988]. In addition, heat-induced Hsp70 expression has been measured in whole lung tissue heat stressed *in vitro* [Fargnoli et al., 1990] and in the hippocampus of intact animals subjected to heat stress [Pardue et al., 1992], where it has likewise been shown to decline with age. Thus, an age-related attenuation in the induction of Hsp70 expression in reponse to hyperthermia appears to be a common phenomenon. Although the magnitude of the age-related decline varies somewhat, the response appears roughly to be reduced by half.

The induction of Hsp70 has also been studied in cultured senescent cells that have reached the end of their replicative life span and are unable to proliferate. Liu et al. [1989] first showed that induction of Hsp70 expression by heat shock was significantly reduced in late compared with early passage human IMR90 lung fibroblasts. More recently Luce and Cristofalo [1992] showed that the induction of Hsp expression following heat stress was likewise reduced in late passage WI38 fibroblasts

compared with early passage cells. This change in the induction of Hsp70 expression with *in vitro* aging or senescence is not restricted to fibroblasts as Hsp70 expression also decreases with passage number in human T cells [Effros et al., 1994].

Age-related decreases occur in Hsp70 expression following treatment of cells with various inducers of the response other than heat. For example, induction of Hsp70 in response to arsenite and the amino acid analog canavanine declines with cellular senescence in human diploid fibroblasts [Luce and Cristofalo, 1992; Liu et al., 1989]. Likewise, induction of Hsp70 synthesis by mitogens decreases with age in lymphocytes isolated from young and elderly human subjects [Faassen et al., 1989].

III.B. Hsp70 Expression in Response to Restraint

Recent studies have indicated that even behavioral or psychological stress can evoke the heat shock response in rodent models. For example, surgical stress [Udelsman et al., 1991], as well as restraint or immobilization stress [Blake et al., 1991b; Udelsman et al., 1993], leads to the expression of Hsp70 mRNA and protein selectively in the adrenal gland and vasculature of rats. Surprisingly, both the adrenal and the vascular responses were shown to be linked to acute neuroendocrine stress responses; however, Hsp70 induction is differentially regulated in the two tissues. The adrenal response, localized within the cortical region of the gland, was found to be dependent on the hypothalamic–pituitary–adrenal (HPA) axis and specifically required adrenocorticotropic hormone (ACTH) [Udelsman et al., 1993, 1994]. The vascular response, on the other hand, was shown to be restricted to the smooth muscle cell layer of vessels and appeared to be mediated through the α_1-adrenergic receptor. Figure 1 shows the effect of restraint on the induction of Hsp70 mRNA in adrenals isolated from rats ranging in age from 2–4 to 24 months. The values shown are for stress-inducible members of the Hsp70 family and do not include the constitutively expressed

TABLE I. List of Studies on the Effect of Aging on the Induction of Hsp70 by Heat Shock in Mammalian Cells

Strain (Tissue)	Assay for Hsp70 expression	Ages studied	Percent decrease with age	References
Human subjects (male and female)				
Mononuclear cells	Transcription	23–29 & 75–89 years	30	Deguchi et al. [1988]
Nonhuman primates (male and female):				
Rhesus (Splenocytes)	Protein level	4–6 & 17–31 years	75	Pahlavani et al. [1995]
Rats (male):				
Wistar (lung, skin)	Synthesis mRNA level	5 & 24 months	Decrease 55%	Fargnoli et al. [1990]
Wistar (brain, lung, skin)	mRNA level	5–6 & 24–25 months	50–75	Blake et al. [1991a]
F344 (hippocampus)	mRNA level	4 & 30 months	85	Pardue et al. [1992]
F344 (hepatocytes)	Synthesis	5–7 & 25–27 months	37	Wu et al. [1993]
F344 (hepatocytes)	Synthesis mRNA level Transcription	4–6 & 26–28 months	45 40 31	Heydari et al. [1993]
F344 (splenocytes)	Protein level Synthesis mRNA level Transcription	4–6 & 26–28 months	75 37 52 85	Pahlavani et al. [1995]
Cultured cells (cell senescence)				
Human IMR-90 Lung diploid Fibroblasts	Synthesis mRNA level Transcription	15–51 population doubling levels	65 80 50	Liu et al. [1989]
Human IMR-90 Lung diploid Fibroblasts	mRNA level	21 & 45 population doubling levels	70 80	Liu et al. [1991]
Human WI-38 Diploid Fibroblasts	Synthesis mRNA level	50 & 90% of *in vitro* life span	Decrease Decrease	Luce and Cristofalo [1992]
Human T-Lymphocytes	Synthesis mRNA level	11–80% of *in vitro* life span	50 50	Effros et al. [1994]

Hsc70 transcripts, which do not vary significantly with age. A twofold decline in restraint-induced Hsp70 expression is seen comparing 2–4-month- and 6-month-old rats. Little change in the magnitude of induction occurs over the next 6–8 months but by 18 months of age the level of induction again falls and by 24 months of age little induction is observed. Thus, the attenuation in restraint-induced expression appears to occur more or less continuously over the life span.

In summary, the current data indicate that the induction of Hsp70 expression by a variety of stresses generally declines with cellular and organismic aging both *in vivo* and *in vitro*. In fact, the only example in the literature in which no age-related decline in the induction of Hsp70 expression has been observed was reported in *Drosophila melanogaster*, where an age-related increase in the expression of Hsp70 in response to hyperthermia was reported [Fleming et al., 1988; Niedzwiecki et al., 1991].

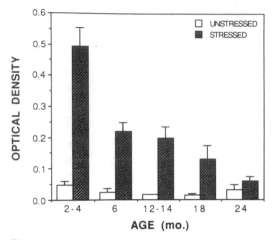

Fig. 1. *Decline in restraint-induced adrenal Hsp70 mRNA expression. Total RNA was isolated from rats of various ages following 90 minutes of restraint and analyzed for Hsp70 expression by Northern analysis. The optical density corresponding to induced (3.0) Hsp70 transcripts was obtained from individual samples, and the data are presented as the mean ± SEM for 8–12 animals in the restrained groups and 5–8 animals in the control, unrestrained group.*

III.C. Expression of Other HSPs During Aging

Although most of the research in the area of aging has focused on Hsp70, there have been a few reports in which the expression of heat shock genes other than Hsp70 has been studied. Liu et al. [1989] compared the synthesis of other heat-inducible Hsp gene families in early and late passage cultures of human fibroblasts after heat shock. They found that, in addition to the Hsp70 family, Hsp with molecular weights of 90, 89, 64, 50, and 25 also showed a reduced expression in late passage cells following heat shock or the addition of canavanine. This is not surprising since the heat-induced activation of all of these Hsp are presumed to be mediated through a common or similar element in the promoters of the genes (discussed in greater detail below). Surprisingly, however, in the restraint model discussed above, examination of a variety of Hsp genes at the mRNA level revealed that only Hsp70 and Hsp27 were induced by restraint in the vasculature and only Hsp70 was induced in the

adrenal gland. These observations suggest an additional level of control for this *in vivo* response that is not apparent in most *in vitro* models studied. Like Hsp70, expression of Hsp27 following restraint was markedly attenuated with age.

Wu et al. [1993] have studied the effect of age on the synthesis of Hsc70 in rat hepatocytes. Hsc70 is a constitutively expressed member of the Hsp70 gene family, which in rodents is only mildly induced or is not induced by heat and classic inducers of the heat shock response in most cell types [O'Malley et al., 1985]. In contrast to Hsp70, aging had no effect on the synthesis of Hsc70, either basally, or following heat stress. Heydari et al. [1994] also showed that the basal and heat-induced expression (mRNA levels and transcription) of Hsc70 did not change with age in rat hepatocytes. However, Pahlavani et al. [1995] recently found that the basal and heat-induced levels of Hsc70 were significantly reduced in splenocytes isolated from old rats compared with those from young rats. Thus, the effect of age on the expression of Hsc70 appears to vary from tissue to tissue. Heydari et al. [1994] and Spindler et al. [1990] have examined the effect of age on the expression of GRP78 in the livers of rats and mice, respectively. GRP78 is another member of the Hsp70 family that is not inducible by heat. Both groups reported a decline in the basal levels of the GRP78 mRNA transcript with aging.

IV. REDUCED Hsp70 EXPRESSION DURING AGING ARISES AT THE LEVEL OF TRANSCRIPTION

Expression of Hsp is regulated primarily at the level of transcription, and the current model for the mechanism of regulation is shown schematically in Figure 2. The transcription of Hsp70 is mediated by one or more of a family of transcription factors (HSF) that interact with a specific regulatory element, the heat shock element (HSE) present in the promoter of Hsp70 as well as other Hsp genes [Morimoto, 1993; Sorger, 1991]. The Hse is composed of contiguous alternating repeats of the sequence NGAAN, three of which are required for high-affinity binding of the HSF [Amin et al., 1988;

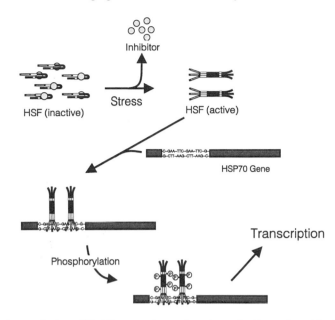

Fig. 2. *Model for the regulation of Hsp70 transcription. The model shown is based on models proposed by Rabindran et al. [1993] and Morimoto [1993]. The Hsf molecule is shown with the DNA-binding domain at the N terminus (shaded area) and the adjacent cluster of hydrophobic amino acids organized into heptad repeats (leucine zippers), and a heptad repeat at the C terminus (solid area).*

Fernandes et al., 1994; Perisic et al., 1989]. In most mammalian cells, HSF are present constitutively, but in an inactive (i.e., non-DNA–binding) form in unstressed cells. Stress results in the activation of HSF to a form that binds the HSE. Although the activation process is still poorly understood, it requires the oligomerization of HSF monomers to a trimeric state and is associated with hyperphosphorylation of the transcription factor [Baler et al., 1993; Sistonen et al., 1994; Sarge et al., 1993]. Binding of HSF trimers to the HSE is necessary for transactivation of heat shock promoters, but is apparently not sufficient to ensure transcription as a number of conditions have been described in which binding of mammalian HSF to DNA is uncoupled from Hsp transcription [Hensold et al., 1990; Jurivich et al., 1992; Takahashi et al., 1994].

IV.A. Age-Related Decline in the DNA Binding Activity of HSF

A number of the studies listed in Table I have provided evidence from nuclear run-on assays that the age-related attenuation in Hsp70 ex-

pression following heat stress is associated with reduced transcription of the gene. Furthermore, four different model systems have provided evidence that the age-related decline involves reduced HSF binding to DNA, and data from these studies are summarized in Figure 3. First, Liu et al. [1989] found that when early and late passage human fibroblasts were transfected with a plasmid containing the human Hsp70 promoter fused to the bacterial chloramphenicol acetyltransferase gene, late passage cells expressed less of the bacterial enzyme in response to heat shock than early passage cells, indicating that the promoter was less active in the late passage cells. Using gel mobility shift assays to measure the levels of HSF capable of binding to the HSE in extracts from early and late passage cells, they demonstrated a >90% decline in heat stress–induced HSE binding activity in aged fibroblasts [Choi et al., 1990; Liu et al., 1991]. (Fig. 3, right). This decrease in HSE binding activity was comparable with the reduced expression of the Hsp70 promoter–CAT reporter gene in aged cells. Such alteration during *in vitro* aging ap-

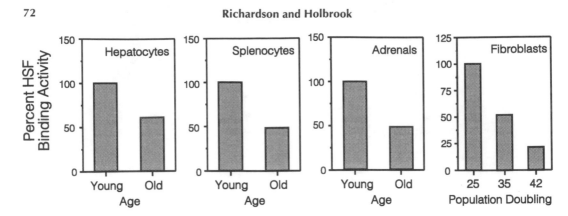

Fig. 3. *The effect of age on the DNA-binding activity of HSF. The DNA binding activity of HSF was determined in cell extracts by the gel mobility shift assay using a consensus HSE and is expressed as a percentage of the youngest animals or cells. The data shown are from heat-shocked hepatocytes [Heydari et al., 1993] and splenocytes [Pahlavani et al.,* *1995] isolated from young (4–6 month) and old (24–26 month) male F344 rats, adrenal extracts isolated from young (5–6 month) and old (25–26 month) male F344 rats exposed to restraint stress [Fawcett et al., 1995], and heat shocked human IMR-90 diploid fibroblasts at various population doublings [Liu et al., 1991].*

pears to reflect changes that occur *in vivo* as evidenced by independent studies performed in each of our laboratories.

Heydari et al. [1993] also used gel mobility shift assays to show a reduction in DNA-binding activity in hepatocytes from aged Fischer 344 (SF344) rats subjected to heat stress *in vitro*. Although the magnitude of the age-related decline in HSF binding activity was less in the *in vivo* aged hepatocytes (40% decline in DNA-binding activity) than in the *in vitro* aged fibroblasts, it correlated well with the age-related decline in transcription determined in nuclear run-on assays (Fig. 4). Recent studies of Fawcett et al. [1995] have demonstrated that restraint-induced Hsp70 expression in adrenal tissue is also mediated via HSF. Examining two different strains of rats, Wistar and F344, they have observed strain-related differences in both the biochemical and DNA-binding properties of HSF. Adrenal extracts from unstressed Wistar rats exhibit low levels of HSE-binding activity that are markedly increased by restraint or administration of ACTH. In contrast, these investigators noted that adrenal extracts from F344 rats show appreciable DNA-binding activity even in the absence of stress. The actual amount of DNA-binding activity does not increase significantly in response to restraint, but the mobility of HSE-binding complexes is al-

tered by restraint, suggesting that though capable of binding to DNA in the absence of stress, the transcription factor undergoes a conformational change in response to stress that is correlated with increased transcription of Hsp70. Comparing the level of HSF-binding activity in extracts prepared from old and young restrained animals, they have observed less DNA-binding activity in aged extracts from both the F344 (60% decline) and Wistar (40% decline) strains. Finally, Pahlavani et al. [1995] have observed a 55% decrease with age in HSE-binding activity in splenocytes isolated from F344 rats. Thus, it appears that the level of HSF that binds to the HSE after a stress is significantly reduced with age in a variety of mammalian cells. Because the binding of HSF to the HSE in the promoter of Hsp70 appears to play a critical role in the induction of Hsp70 transcription (see Fig. 2), the current data suggest that the reduced ability of cells from old animals to express Hsp70 in response to a stress occurs because less HSF is binding to the Hsp70 promoter.

IV.B. HSF Protein Levels Are Similar in Young and Aged Animals

Based on current models (see Fig. 2), the age-related decrease in the levels of HSF capable of binding to the HSE could occur by at least two distinct mechanisms. First, decreased Hsf-binding activity could arise from a decrease in the expression of HSF. In other words, less HSF-binding activity is observed in tissues of old rats because less of the HSF protein is present in the old tissues. Second, decreased HSF-binding activity could occur as a result of a decrease in the post-translational activation of HSF by stress. That is, similar levels of HSF are present in nonstressed hepatocytes isolated from young and aged rats; but, in response to stress, the conversion of HSF from its inactive, non-DNA–binding form to its active HSE-binding form is reduced in the hepatocytes isolated from old rats.

Two distinct HSF, designated HSF1 and HSF2, have been shown to exist in mammalian species [Sarge et al., 1991; Rabindran et al., 1991; Schuetz et al., 1991]. HSF1 has been shown to be involved in the regulation of HSF expression following heat stress [Sarge et al., 1993] and recent studies have shown that HSF1 is likewise responsible for restraint-induced Hsp70 expression in the adrenal gland [Fawcett et al., 1994]. Both of our laboratories have addressed the question of whether HSF1 protein levels decline with aging using polyclonal antisera raised against mouse HSF1 to measure HSF1 levels on Western blots. HSF1 levels have been determined in both unstressed and in vitro heat-stressed rat hepatocytes and in adrenal tissue derived from unstressed and restrained rats. The results indicate that the levels of HSF1 protein are similar in both adrenal tissue and hepatocytes from old and young rats [Fawcett et al., 1995; Heydari et al., 1994]. Thus, it appears that the age-related decrease in HSF-binding activity is not due to a decrease in the amount of HSF1 present in tissues isolated from old rats, but rather due to its activation to a DNA-binding form.

Why is HSF1 activation reduced in aged animals? One possible explanation for the age-related decrease in the activation of HSF1 by stress is that cells from old animals perceive restraint to be less stressful than do young animals. This is an especially important consideration in the restraint model, because Hsp70

expression is mediated, at least in part, by hormonal action. Therefore, it is possible that the lower induction of HSP70 could occur secondary to less hormone or reduced receptor activity. Both the HPA and sympathetic axes have been reported to undergo alterations with age [Roberts and Tumer, 1987; Nielson et al., 1992; Sapolsky, 1990]. However, no differences were observed in the HPA response (as determined by levels of plasma ACTH and corticosterone) in young and aged rats exposed to restraint stress, although they showed different levels of Hsp70 expression [Udelsman et al., 1993]. In addition, we have shown that the induction of Hsp70 expression remains low in aged animals injected with a dose of ACTH sufficient to produce maximum glucocorticoid secretion [Holbrook et al., unpublished findings]. Thus, the studies with restraint stress strongly suggest that the age-related decrease in induction of Hsp70 occurs because of a defect in the transcriptional regulation of Hsp70.

As Figure 2 shows, the oligomerization of HSF1 monomers to trimers is a key event in its stress-induced activation to a DNA-binding form. Current evidence indicates that the transcription of Hsp70 is under negative regulation [Lis and Wu, 1993; Morimoto, 1993; Clos et al., 1990]. A regulatory factor, presumably a protein, has been proposed to stabilize HSF1 in its monomeric form and prevent oligomerization. Hsp70 has itself been proposed to serve such a function, but this issue is highly controversial [Abravaya et al., 1992; Mosser et al., 1993; Morimoto, 1993; Rabindran et al., 1994]. Heat shock (as well as other stresses) is believed to disrupt the binding of the regulatory molecule and allow oligomerization. Based on this model of HSF activation, it is logical to propose that the mechanism responsible for the decrease in HSF oligomerization with age involves an increase in the level or activity of the putative negative regulatory protein. Three studies have addressed this issue. Choi et al. [1990], performed mixing experiments with various proportions of extracts prepared from low passage and high passage IMR90 fibroblasts subjected to heat stress.

They reported that Hse-binding activity was not an additive function of the activities present in old and young cell extracts, but rather the binding activity decreased more rapidly than predicted when increasing amounts of late passage cell extracts were mixed with young cell extracts. These findings suggest the presence of an inhibitor of HSF DNA-binding activity in the senescent cells.

Our laboratories have performed similar mixing experiments with extracts from either adrenal or liver tissue of rats. In contrast with what was observed in human diploid fibroblasts, no evidence for an inhibitor of DNA-binding activity was observed in adrenal extracts from old rats. Rather, the level of DNA-binding activity was proportional to the amounts of young and old extract added to the DNA-binding reaction [Fawcett et al., 1995]. An identical observation was made with extracts from hepatocytes isolated from old rats [Heydari et al., unpublished findings]. Thus, studies with extracts from animal tissues suggest that the age-related decline in HSF-binding activity is not due to an age-related accumulation of an inhibitory molecule.

Another mechanism that would explain the age-related decline in HSF-binding activity is an alteration in the HSF molecule in tissues of old rats. For example, an accumulation of abnormal HSF molecules, with decreased ability to oligomerize, could result in an age-related decline in the activation of HSF and, therefore, a decrease in the expression of Hsp70. This mechanism is plausible because research in the field of aging over the past two decades has shown that a variety of proteins become altered with age [Rothstein, 1979, 1981, 1983]. These alterations have been shown to arise from conformational changes rather than from errors in translation, and the conformational changes in proteins of old organisms occur through postsynthetic modifications, e.g., protein oxidation, amino acid racemization, deamination, glycation, conformational drift, and so forth [Rothstein, 1979, 1981, 1983; Stadtman, 1992]. Rothstein proposed in 1979 that the accumulation of abnormal proteins in old cells oc-

curred because of an age-related decrease in the rate of protein turnover. A decrease in protein turnover would result in a longer dwell time for protein molecules within a cell. Rothstein rationalized that the longer dwell time in the cell increased the probability that a protein would be altered by various types of postsynthetic modifications, e.g., conformational drift, oxidation, deamidation, and glycation. More recently, Stadtman's laboratory has shown that protein oxidation increases markedly with age, especially in liver, and that the increase in protein oxidation is associated with decreased protein degradation [Stadtman, 1992; Smith et al., 1991; Starke-Reed and Oliver, 1989]. As one would predict from the model proposed by Rothstein, the likelihood that a protein will become altered with age appears to be dependent on the rate of turnover of the protein. Proteins that are rapidly turned over have less potential for becoming modified with age than proteins that have slower rates of turnover [Rothstein, 1979]. In this context, it is important to note that HSF has a relatively slow turnover rate. Thus, HSF might be potentially vulnerable to post-translational modifications, and these might alter the biological activity of HSF, e.g., its ability to oligomerize and bind to the Hsp70 promoter.

V. REVERSAL OF AGE-RELATED DECLINES IN Hsp70 EXPRESSION
V.A. Effect of Caloric Restriction on Hsp70 Expression

Research over the past two decades has established that dietary restriction (restriction of energy) increases the longevity of rodents by reducing the rate of aging [Richardson, 1985; Heydari and Richardson, 1992; Masoro, 1985, 1988]. This provides investigators with a powerful tool for determining which are important for aging. Thus, it is of interest to determine whether dietary restriction can indeed prevent or reverse the decline in Hsp70 expression that occurs with aging. Hedari et al. [1993] have studied the effect of dietary restriction on the ability of hepatocytes to express Hsp70. Fig-

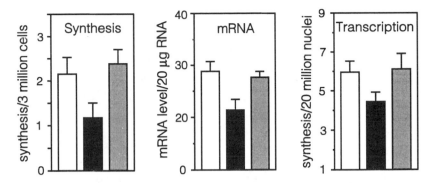

Fig. 4. *Effect of dietary restriction on the induction of Hsp70 expression. Hepatocytes isolated from young (4–6 month, open bars) and old (26–28 month, closed bars) rats fed* ad libitum *and old (26–28 month, shaded bars) rats fed a caloric-restricted diet. The synthesis, mRNA levels, and nuclear transcription of Hsc70 are shown. The data are taken from Heydari et al. (1993).*

ure 4 shows data from this study in which the induction of Hsp70 expression by a brief heat shock was measured in hepatocytes isolated from young or old rats fed *ad libitum* and old rats fed a caloric-restricted diet. The levels of Hsp70 synthesis and mRNA and the nuclear transcription of Hsp70 were approximately 50–60% less in hepatocytes isolated from old rats fed *ad libitum* compared with hepatocytes isolated from old rats fed a caloric-restricted diet. The data in Figure 4 show that the level of Hsp70 expression induced by heat shock in the hepatocytes isolated from old rats fed the restricted diet was similar to the level of Hsp70 expression observed in hepatocytes isolated from young rats fed *ad libitum*. In addition, Heydari et al. [1993] correlated the increase in Hsp70 transcription in the hepatocytes from the caloric-restricted rats to an increase in HSF-binding activity. Thus, it appears that dietary restriction reversed the effect of aging on the induction of Hsp70 expression by altering the transcription of Hsp70 through changes in HSF.

More recently, Heydari et al. [1995] compared the effect of dietary restriction on the expression of Hsc70 and GRP78 in hepatocytes isolated from rats. Dietary restriction had no effect on the basal levels of Hsc70 mRNA or the nuclear transcription of Hsc70. However, when hepatocytes were exposed to a brief heat shock, the induction of Hsc70 expression was almost 70% higher in hepatocytes isolated from

restricted rats compared with rats fed *ad libitum*. In contrast to Hsc70, the basal levels of GRP78 were reduced by dietary restriction, which is consistent with the observation by Spindler et al. [1990] that dietary restriction reduced the level of GRP78 in the livers of mice. Thus, the effect of dietary restriction on the expression of heat shock genes appears to vary considerably from heat shock gene to heat shock gene.

V.B. Modulation of the Vascular Response to Restraint by Transplantation

To study the mechanisms responsible for the age-related decline in vascular heat shock protein expression, Holbrook and colleagues have developed a cross-transplantation model in which thoracic aortas harvested from young rats are transplanted into aged rats in an infrarenal position, and *vice versa* [Udelsman et al., 1995]. Using this model, they have addressed the question: Is the age-related decline in response inherent to the aortic tissue, or is it due to the environment in which the aorta resides? Results of these studies indicate that the heat shock protein response in aged aortas is partially rejuvenated by transplanting them into young animals. In contrast, young aortas behave more like old vessels and show less Hsp70 induction in response to restraint when they are transplanted into old hosts. These findings are consistent with the observation noted below that the magnitude of the vascular Hsp70 response is directly correlated with elevations in systemic blood pressure that are controlled primarily via peripheral vasoconstriction and subject to hormonal influence.

VI. CONCLUSIONS AND FUTURE DIRECTIONS

Hsp70 expression is an excellent example of a cellular defense mechanism that has evolved to protect cells and organisms from heat and other types of stress. Therefore, changes in this system could seriously compromise the capacity of an organism to respond to changes in its environment. Because aging and senescence are characterized by a reduced ability to maintain homeostasis in response to stress, a number of investigators have compared the abilities of young and old cells to express Hsp70. At the present time it appears that aging, either at the organismic or at the cellular level, is associated with a decreased ability to express Hsp70 in response to hyperthermia as well as to a variety of other stresses. In addition, studies from several laboratories also indicate that aging is associated with a general defect that occurs at the transcriptional level and appears to involve the transcription factor HSF1. Cells from old animals or cells that have senescenced in culture have decreased active HSF after exposure to stress. Furthermore, dietary restriction, which retards aging and enhances survival, appears to reverse the age-related decline in the transcription of Hsp70 and the deficit in HSF-binding activity. It remains to be determined why HSF activity is reduced in aged cells. Is it due to a defect in the signal-transduction processes that leads to transcription factor activation? Or is it due to a defect or alteration in the HSF protein itself or accessory proteins involved in transforming the inactive HSF to an active DNA-binding form in which oligomerization appears to be a key event. While a biochemical understanding of the process that underlies the shift in equilibrium from monomer to tri-

mer undoubtedly involves the leucine zipper domain of the protein [Rabindran et al., 1993; Nakai and Morimoto, 1993], an understanding of the signals and molecules that influence the monomer/trimer equilibrium remains elusive. Since the observed differences in DNA binding in old and young animals are related to these events it is unlikely that these points will be resolved soon.

One would predict that an age-related defect in the ability of cells to express Hsp70 in response to stress would make senescent organisms more vulnerable. This is particularly true with respect to hyperthermia, as it has been shown that cells become more thermosensitive when the expression of Hsp70 is inhibited [Johnston and Kucey, 1988; Riabowol et al., 1988]. In contrast, enhanced Hsp70 expression is associated with enhanced resistance to heat stress [Angelidis et al., 1991; Li et al., 1991]. In humans, there is a dramatic increase in the incidence of heat stroke with age [Clowes and O'Donnell, 1974; Oechsli and Buechley, 1970; Kilbourne et al., 1982; and Jones et al., 1982]. Jones et al. [1982] reported that the rate of the incidence of heat stroke was 10-fold higher for persons 65 years or older than for younger individuals. Given this fact and the great deal of research showing that the expression of Hsp70 is reduced with age, it is ironic that there is little information on the effect of aging on thermotolerance at the cellular level. Luce and Cristofalo [1992] found that late passage human diploid fibroblasts, which had reduced levels of Hsp70 expression, were more sensitive to a temperature of 49°C than early passage cells. Thus, the decline in Hsp70 expression that occurs during *in vitro* aging in cultured cells is associated with a decrease in the ability of the cells to withstand hyperthermic stress. More recently Pahlavani et al. [1995] found that splenocytes isolated from old rats are likewise more sensitive to high temperatures than are splenocytes isolated from young rats. This decrease in thermotolerance is correlated with decreased expression of Hsp70.

Although it is clear that cells from old animals show a general decline in their ability to express Hsp70 in response to stress, it should be kept in mind that, in dealing with whole animals, other functional or metabolic changes that occur with aging could contribute to the loss of expression. For example, in examining the response to heat exposure *in vivo*, Blake et al [1991a] observed that the body temperature of aged animals was significantly lower than that of young animals. When young and old rats were exposed to an elevated ambient temperature, aged animals showed a slower rate of rise in body temperature than younger animals. The net result of this slower rise in temperature coupled with the lower basal temperatures of older animals was that, when exposed for a fixed period of time to the same ambient temperature, old rats did not show as great an increase in body temperature. It then predicted that the tissues of old animals would also show lesser induction of Hsp70 expression because they would receive less of the signal (heat) necessary to activate the heat shock response, and this is precisely what was observed. Interestingly, however, aged animals were more susceptible to the detrimental effects of hyperthermia and showed reduced survival compared with young animals.

Another example of a functional/metabolic change that could alter the ability of whole animals to respond to stress is shown in a recent study in which we examined the age-related decline in vascular Hsp70 expression in response to restraint. This response declines greater than 10-fold with aging, more than could likely be accounted for by the changes in HSF DNA-binding activity discussed above. Recent studies have provided evidence to indicate that restraint induces Hsp70 expression in the vasculature secondary to an elevation in blood pressure [Xu et al., 1995]. Interestingly, restraint results in a lower elevation in blood pressure in old animals than in young animals. The cause for this difference in blood pressure response in old and young animals is not clear, but, again, the net effect is that the aged animal receives less stimulus to initiate the response.

REFERENCES

Abravaya K, Myers MP, Murphy SP, Morimoto RI (1992): The human heat shock protein hsp70 interacts with HSF, the transcription factor that regulates heat shock gene expression. Genes Dev 6:1153–1164.

Amin J, Ananthan J, Voellmy R (1988): Key features of heat shock regulatory elements. Mol Cell Biol 8:3761–3769.

Ang D, Liberek K, Skowyra D, Zylicz M, Georgopoulos C (1991): Biological role and regulation of the universally conserved heat shock proteins. J Biol Chem 266:24233–24236.

Angelidis CE, Lazaridis I, Pagoulatos GN (1991): Constitutive expression of heat-shock protein 70 in mammalian cells confers thermoresistance. Eur J Biochem 199:35–39.

Baler R, Dahl G, Voellmy R (1993): Activation of human heat shock genes is accompanied by oligomerization, modification, and rapid translocation of heat shock transcription factor Hsf1. Mol Cell Biol 13:2486–2496.

Blake MJ, Fargnoli J, Gershon D, and Holbrook NJ (1991a): Concomitant decline in heat-induced hyperthermia and HSP70 mRNA expression in aged rats. Am J Physiol 260:663–667.

Blake MJ, Udelsman R, Feulner GJ, Norton DD, Holbrook NJ (1991b): Stress-induced heat shock protein 70 expression in adrenal cortex: an adrenocorticotropic hormone-sensitive, age-dependent response. Proc Natl Acad Sci USA 88:9873–9877.

Choi HS, Lin Z, Li B, Liu AYC (1990): Age-dependent decrease in the heat-inducible DNA sequence-specific binding activity in human diploid fibroblasts. J Biol Chem 265:18005–18011.

Clos J, Westwood JT, Becker PB, Wilson S, Lambert K, Wu C. (1990): Molecular cloning and expression of a hexameric drosophila heat shock factor subject to negative regulation. Cell 63:1085–1097.

Clowes CHA, O'Donnell TF (1974): Heat stroke. N Eng J Med 291:564–567.

Craig EA, Weissman JS, Horwich AL (1994): Heat shock proteins and chaperones. Cell 78:365–372.

Deguchi Y, Negoro S, Kishimoto S (1988): Age-related changes of heat shock protein gene transcription in human peripheral blood mononuclear cells. Biochem Biophys Res Commun. 157:580–584.

Effros RB, Zhu X, Walford RL (1994): Stress response of senescent T lymphocytes: Reduced hsp70 is independent of the proliferative block. J Gerontol 49:B65–B70.

Faassen AE, O'Leary JJ, Rodysill KJ, Bergh N, Hallgren HM (1989): Diminished heat-shock protein synthesis following mitogen stimulation of lymphocytes from aged donors. Exp Cell Res 183:326–334.

Fargnoli J, Kunisada T, Fornace AJ, Schneider EL, Holbrook NJ (1990): Decreased expression of heat shock protein 70 mRNA and protein after heat treatment in cells of aged rats. Proc Natl Acad Sci USA 87:846–850.

Fawcett TW, Sylvester SL, Sarge KD, Morimoto RI, Holbrook NJ (1995): Effects of neurohormonal stress and aging on the activation of mammalian heat shock factor 1. J Biol Chem 269:32272–32278.

Fernandes M, Xiao H, Lis JT (1994): Fine structure analysis of the *Drosophila* and *Saccharomyces* heat shock factor-heat shock element interactions. Nucleic Acids Res 22:167–173.

Fleming JE, Walton JK, Dubitsky R, Bensch KG (1988): Aging results in an unusual expression of *Drosophila* heat shock proteins. Proc Natl Acad Sci USA 85:4099–4103.

Gething MJ, Sambrook J (1992): Protein folding in the cell. Nature 355:33–44.

Heikkila JJ (1993): Heat shock gene expression and development. I. An overview of fungal, plant, and poikilothermic animal developmental systems. Dev Genet 14:1–5.

Hensold JO, Hunt CR, Calderwood SK, Housman DE, Kingston RE (1990): DNA binding of heat shock factor to the heat shock element is insufficient for transcriptional activation in murine erythroleukemia cells. Mol Cell Biol 10:1600–1608.

Heydari AR, Conrad CC, Richardson A (1995): Expression of heat shock genes in hepatocytes is affected by age and food restriction in rats. J Nutr 125:410–418.

Heydari AR, Richardson A (1992): Does gene expression play any role in the mechanism of the antiaging effect of dietary restriction. Ann NY Acad Sci 663:384–395.

Heydari AR, Takahashi Y, Gutsmann S, You S, Richardson A (1994): HSP70 and aging. Experientia 50:1092–1098.

Heydari AR, Wu B, Takahashi R, Strong R, Richardson A (1993): Expression of heat shock protein 70 is altered by age and diet at the level of transcription. Mol Cell Biol 13:2909–2918.

Johnston RN, Kucey BL (1988): Competitive inhibition of hsp70 gene expression causes thermosensitivity. Science 242:1551–1554.

Jones TS, Liang AP, Kilbourne EM, Griffin MR, Patriarca PA, Wassilak SGF, Mullan RJ, Herrick RF, Donnell HD, Choi K, Thacker SB (1982): Morbidity and mortality associated with the July 1980 heat wave in St Louis and Kansas City, Mo. JAMA 247:3327–3355.

Jurivich DA, Sistonen L, Kroes RA, Morimoto RI (1992) Effect of sodium salicylate on the human heat shock response. Science 255:1243–1245.

Kilbourne EM, Choi K, Jones TS, Thacker SB (1982): Risk factors for heatstroke. JAMA 247:3332–3336.

Li GC, Li L, Liu YK, Mak JY, Chen L, Lee WM (1991): Thermal response of rat fibroblasts stably transfected with the human 70-kDa heat shock protein-encoding gene. Proc Natl Acad Sci USA 88:1681–1685.

Lindquist S (1986): The heat-shock response. Annu Rev Biochem 55:1151–1191.

Lindquist S, Craig EA (1988): The heat-shock proteins. Annu Rev Genet 22:631–677.

Lis J, Wu C (1993): Protein traffic on the heat shock promoter: Parking, stalling, and trucking along. Cell 74:1–4.

Liu AY, Choi H, Lee Y, Chen KY (1991): Molecular events involved in transcriptional activation of heat shock genes become progressively refractory to heat stimulation during aging of human diploid fibroblasts. J Cell Physiol 149:560–566.

Liu AY, Lin Z, Choi H, Sorhage F, Li B (1989): Attenuated induction of heat shock gene expression in aging diploid fibroblasts. J Biol Chem 264:12037–12045.

Luce MC, Cristofalo VJ (1992): Reduction in heat shock gene expression correlates with increased thermosensitivity in senescent human fibroblasts. Exp Cell Res 202:9–16.

Martin GR, Danner DB, Holbrook NJ (1993): Aging-causes and defenses. Annu Rev Med 44:419–429.

Morimoto RI (1993): Cells in stress: Transcriptional activation of heat shock genes. Science 259:1409–1410.

Morimoto RI, Sarge KD, Abravaya K (1992): Transcriptional regulation of heat shock genes. A paradigm for inducible genomic responses. J Biol Chem 267:21987–21990.

Morimoto RI, Tissieres A, Georgopulous C (1990): Stress Proteins in Biology and Medicine. Cold Spring Harbor, NY: Cold Spring Harbor Laboratory Press.

Masoro EJ (1985): Nutrition and aging: A current assessment. J Nutr 115:842–848.

Masoro EJ (1988): Food restriction in rodents: An evaluation of its role in the study of aging. J Gerontol 43:B59–B64.

Mosser DD, Duchaine J, Massie B (1993) The DNA binding activity of the human heat shock transcription factor is regulated in vivo by hsp70. Mol Cell Biol 13:5427–5438.

Nakai A, Morimoto RI (1993): Characterization of a novel chicken heat shock transcription factor, heat shock factor 3, suggests a new regulatory pathway. Mol Cell Biol 13:1983–1997.

Niedzwiecki A, Kongpachith AM, Fleming JE (1991): Aging affects expression of 70-kDa heat shock proteins in *Drosophila*. J Biol Chem 266:9332–9338.

Nielson H, Hasendam JM, Pilegaard HK, Aalkjaer C, Mortensen FV (1992): Age-dependent changes in alpha-adrenoreceptor-mediated contractility of isolated human resistance arteries. Am J Physiol 263:H1190–H1196.

Oechsli FW, Buechley RW (1970): Excess mortality associated with three Los Angeles September hot spells. Environ Res 3:277–284.

O'Malley KA, Mauron JD, Barchas JD, Kedes L (1985): Constitutively expressed rat mRNA encoding a 70 kilodalton heat-shock-like protein. Mol Cell Biol 5:3476–3483.

Pahlavani MA, Denny M, Moore SA, Weindruch R,

Richardson A (1995): The expression of heat shock protein 70 decreases with age in lymphocytes from rats and rhesus monkeys. Exp Cell Res 218:310 318, 1995.

Pardue S, Groshan K, Raese JD, Morrison-Bogorad M (1992): Hsp70 mRNA induction is reduced in neurons of aged rat hippocampus after thermal stress. Neurobiol Aging 13:661–672.

Perisic O, Xiao H, Lis JT. (1989): Stable binding of *Drosophila* heat shock factor to head-to-head and tail-to-tail repeats of a conserved 5 bp recognition unit. Cell 59:797–806.

Rabindran SK, Giorgi G, Clos J, Wu C (1991): Molecular cloning and expression of a human heat shock factor, Hsf1. Proc Natl Acad Sci USA 88:6906–6910.

Rabindran SK, Haroun RI, Clos J, Wisniewski J, Wu C (1993): Regulation of heat shock factor trimer formation: Role of a conserved leucine zipper. Science 259:230–234.

Rabindran SK, Wisniewski J, Ligeng L, Li GC, Wu C. (1994): Interaction between heat shock factor and hsp70 is insufficient to suppress induction of DNA-binding activity in vivo. Mol Cell Biol 14:6552–6560.

Riabowol KT, Mizzen LA, Welch WJ (1988): Heat shock is lethal to fibroblasts microinjected with antibodies against hsp70. Science 242:433 436.

Richardson A (1985): The effect of age and nutrition on protein synthesis by cells and tissues from mammals. In Watson RR (ed): Handbook of Nutrition and Aging. Boca Raton, FL: CRC Press, pp 31–48.

Roberts J, Tumer (1987): Age-related changes in autonomic function of catecholamines. Rev Biol Res Aging 3:339–370.

Rothstein M (1979): The formation of altered enzymes in ageing animals. Mech Ageing Dev 9:197–202.

Rothstein M (1981): Posttranslational alteration of proteins. In Florini JR (ed): Handbook of Biochemistry in Aging. Boca Raton, FL: CRC Press, pp 103–111.

Rothstein M (1983): Enzymes, enzyme alteration, and protein turnover. In Rothstein M (ed): Review of Biological Research in Aging. New York: Alan R. Liss, Inc., pp 305–314.

Sapolsky RM (1990): The adrenocortical axis. In Schneider EL, Rowe JW (eds): The Handbook of the Biology of Aging, 3rd ed. New York: Academic Press, pp 330–346.

Sarge KD, Murphy SP, Morimoto RI (1993): Activation of heat shock gene transcription by heat shock factor 1 involves oligomerization, acquisition of DNA-binding activity, and nuclear localization and can occur in the absence of stress. Mol Cell Biol 13:1392–1407.

Sarge KD, Zimarino V, Holm K, Wu C, Morimoto RI (1991): Cloning and characterization of two mouse heat shock factors with distinct inducible and constitutive DNA-binding ability. Genes Dev 5:1902–1911.

Schuetz TJ, Gallo GJ, Sheldon L, Tempst P, and Kingston RE (1991): Isolation of a cDNA for HSF2: Evidence for two heat shock factor genes in humans. Proc Natl Acad Sci USA 88:6911–6915.

Shock NW (1977): Systems integration. In Finch CE, Hayflick L (eds): Handbook of the Biology of Aging. New York: Van Nostrand Reinhold Company, pp 639 665.

Shock NW, Greulich RC, Andres RA, Arenberg D, Costa PT, Jr, Lakatta EG, Tobin JD (1984): Normal human aging. The Baltimore Longitudinal Study of Aging, vol 84. Washington, DC: U.S. Government Printing Office, p 2450.

Sistonen L, Sarge KD, Morimoto RI (1994): Human heat shock factors 1 and 2 are differentially activated and can synergistically induce hsp70 gene transcription. Mol Cell Biol 14:2087–2099.

Smith, CD, Carney JM, Starke-Reed PE, Oliver CN, Stadtman ER, Floyd RA, Markesbery WR (1991): Excess brain protein oxidation and enzyme dysfunction in normal aging and in Alzheimer disease. Proc Natl Acad Sci USA 88:10540–10543.

Sorger PK (1991): Heat shock factor and the heat shock response. Cell 65:363–366.

Spindler SR, Crew MD, Mote PL, Grizzle JM, Walford RL (1990): Dietary energy restriction in mice reduces hepatic expression of glucose-regulated protein 78 (BiP) and 94 mRNA. J Nutr 120:1412–1417.

Stadtman ER (1992): Protein oxidation and aging. Science 257:1220 1224.

Starke-Reed PE, Oliver CN (1989): Protein oxidation and proteolysis during aging and oxidative stress. Arch Biochem Biophys 275:559–567.

Takahashi R, Heydari AR, Gutsmann A, Sabia M, and Richardson A (1994): The heat shock transcription factor in liver exists in a form that has DNA binding activity but no transcriptional activity. Biochem Biophys Res Commun 201:552–558.

Udelsman R, Blake MJ, Stagg CA, Li D, Putney DJ, Holbrook NJ (1993): Vascular heat shock protein expression in response to stress. Endocrine and autonomic regulation of this age-dependent response. J Clin Invest 91:465–473.

Udelsman R, Blake MJ, Holbrook NJ (1991): Molecular response to surgical stress: Specific and simultaneous heat shock protein induction in the adrenal cortex, aorta, and vena cava. Surgery 110:1125–1131.

Udelsman R, Blake MJ, Stagg CA, Holbrook NJ (1994): Endocrine control of stress-induced heat shock protein 70 expression in vivo. Surgery 115:611–616.

Udelsman R, Li DG, Stagg CA, Holbrook J (1995): Aortic co-transplantation between young and old rats: Effect upon the heat shock protein 70 stress response. J Gerontol 50A:B187–192.

Welch WJ (1993): Heat shock proteins functioning as molecular chaperones: their roles in normal and stressed cells. Philos Trans R Soc Lond 339:327 333.

Wu B, Gu MJ, Heydari AR, Richardson A (1993): The effect of age on the synthesis of two heat shock proteins in the HSP70 family. J Gerontol 48:B50–B56.

Xu Q, Li D, Holbrook NJ, Udelsman R (1995): Acute hypertension induces heat shock protein 70 expression in rat aorta. Circulation (in press).

ABOUT THE AUTHORS

ARLAN RICHARDSON is Professor of Physiology and Associate Director for Basic Biological Science at the Aging Research and Education Center at the University of Texas Health Science Center in San Antonio. He is also a Career Research Scientist at the Geriatric Research Education and Clinical Center at the Audie L. Murphy Memorial Veterans Hospital in San Antonio. Dr. Richardson's work has been published in such journals as the *Journal of Nutrition*, the *Annals of the New York Academy of Sciences, Molecular and Cellular Biology, Experimental Cell Research, Biochemical and Biophysical Research Communications*, and the *Journal of Gerontology*. He was also a contributor to the *Handbook of Nutrition and Aging*.

NIKKI J. HOLBROOK is a Senior Investigator and Chief of the Gene Expression and Aging Section at the National Institute on Aging in Baltimore. She earned a Ph.D. in 1980 from the University of South Florida Medical School in Tampa, studying glucocorticoid actions in cultured cells. She pursued further studies related to glucocorticoid actions and glucocorticoid receptor biochemistry as a postdoctoral fellow in the laboratory of Allan Munck at Dartmouth Medical School. She then undertook a second postdoctoral fellowship in the laboratory of Gerald Crabtree at the National Cancer Institute in Bethesda, Maryland, where she cloned and studied the regulation of the human interleukin 2 gene. In 1986 she moved to the National Institute on Aging where she developed a program focused on molecular and cellular responses to stress and the importance of these defenses to the aging process. Current areas of interest include the regulation and function of heat shock protein expression and signal transduction pathways controlling the cellular response to genotoxic stress. She is the author of more than 100 research articles and is an Associate Editor for the 4th edition of the *Handbook of the Biology of Aging*.

Cellular Aging and Cell Death: 81–107
© 1996 Wiley-Liss, Inc.

The Molecular and Cellular Biology of Aging in the Cardiovascular System

Michael T. Crow, Marvin O. Boluyt, and Edward G. Lakatta

I. INTRODUCTION

Cardiovascular disease is the leading cause of death and a major source of disability among older Americans. In the vasculature, advancing age is a strong risk factor for the development of atherosclerosis [Stout, 1987], while in the heart, changes in myocardial cell number and function lead to the loss of adaptive reserve with age [Lakatta, 1993]. An understanding of how aging affects the cardiovascular system requires a thorough knowledge of the physiology and biochemistry of vascular and cardiac cells in the young and the pathophysiologies of the disorders that affect them. Likewise, an understanding of the adaptive processes that occur during aging has already led in some instances to new insights into the pathogenesis of disease in both young and old. These insights can be expected to lead to the development of novel therapeutic strategies for the treatment of cardiac and vascular disorders.

The aim of this chapter is to concentrate on a selected number of issues related to the aging vasculature and myocardium, identifying, whenever possible, the molecular and cellular bases for the changes observed. We focus our discussion on four areas of cardiovascular aging: *(1)* the remodeling of the vasculature during aging: *(2)* the role that advanced glycation endproducts may play as extrinsic modifiers of vascular cell function with age; *(3)* changes in myocyte cell number, function, and microenvironment with age; and *(4)* the consequence of and potential mechanisms for the loss of β-adrenergic responsiveness in the aging cardiovascular system.

II. THE AGING VASCULATURE

II.A. Vascular Remodeling During Aging

II.A.1. Age-associated changes in vascular cell function. In both humans and many experimental animals, most large and medium-sized arteries become thick-walled, dilated, and less compliant with age. In addition, there is a relocation of smooth muscle cells from their "normal" position in the vessel where they control vascular tone to regions where they cause obstructions and disease. To understand the significance of these structural and compositional changes in the vasculature and the context in which these changes occur, the structure and function of nondiseased vessels in young to middle-aged animals are first reviewed.

Figure 1A shows a cross section of a carotid artery from a 6 month (young) adult rat. The blood vessel consists of well-defined layers of tissues, each of which plays an important role in the development of the vessel and the maintenance of its function. The basic structure of the vessel is that of a single endothelial layer separated from concentric, multiple layers of mesenchymal cells by a continuous basal lamina. The endothelial cell layer (stained dark brown in Fig. 1A) lines the vessel lumen and creates a nonthrombogenic surface. The mesenchymal cell layer is divided into sublayers by continuous elastic laminae (EL, Fig. 1). The layer of tissue between the vessel lumen and the first such membrane (the so-called internal elastic lamina, or IEL in Fig. 1) is the tunica intima. In most vessels, including the rat carotid artery shown in Figure 1, the tunica intima consists of only endothelial cells and

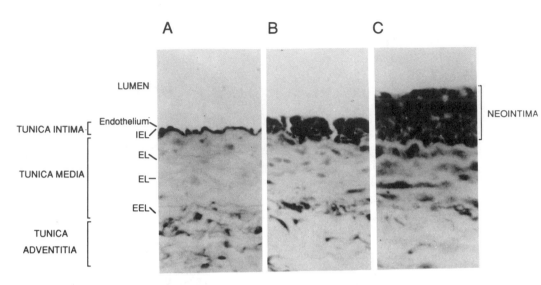

Fig. 1. *Cross sections of rat carotid arteries from an uninjured (control) animal* **(A)** *and from animals whose arteries had been balloon injured as previously described [Jenkins et al., 1994] and then processed for tissue sectioning 8 days* **(B)** *and 14 days* **(C)** *after the injury and immunostained with a rabbit antibody to an intimal cell marker. Detection was with a horseradish peroxidase–conjugated goat antibody to rabbit immunoglobulins and development with diaminobenzidine. SMC, smooth muscle cells; IEL, internal elastic lamina; EL; elastic laminae; EEL, external elastic lamina.*

the basal lamina, which is not visible at this magnification without specific staining. Mesenchymal cells that are likely to be undifferentiated smooth muscle cells, however, can be seen in the tunica intima of human vessels from birth and in many of the vessels of aged humans and rats. The outermost layer (with respect to the lumen) of concentrically arranged tissue beginning with the external elastic lamina (EEL, Fig. l) is the tunica adventitia. The adventitia is composed of fibroblasts and loose connective tissue and may contain small blood vessels and nerves. The area between internal and external elastic laminae is the tunica media and mesenchymal cells trapped between the layers of elastic laminae are medial vascular smooth muscle cells (VSMCs). The tunica media of the rat carotid artery contains many elastic laminae separating single layers of VSMGs (SMC in Fig. 1A).

The histology of the vessel wall is radically transformed by aging. Many vessels become dilated, thicker walled, and less compliant. Figure 2 shows data from a recent study of changes in

the aortic and carotid artery vessel walls of WAG/ Rij rats [Michel et al., 1994], although similar changes have been reported in other strains of rats [Wolinsky, 1970; Fornieri et al., 1992], other experimental animals [Stein et al., 1969], and humans [Spina et al., 1983]. The vessel itself dilates, with the lumenal diameter increasing by 50% in some studies. Both the tunica intima and the tunica media increase in width and undergo changes in their cellular and extracellular matrix compositions. The intima increases in width many fold. In the study presented in Figure 2A, there was an approximately twofold increase in the width of the intima, while in other studies, intimal thickness increased up to fivefold [Guyton et al., 1983; Li et al., 1995]. This increase is due, in part, to the recruitment of smooth muscle cells and monocytes to the intima [Guyton et al., 1983] and, in part, to the synthesis of interstitial extracellular matrix proteins by these cells. In Figure 2A, intimal collagen density, a measure of interstitial matrix production, increased sixfold.

The tunica media also increases in thickness with age (Fig. 2B), approximately doubling in

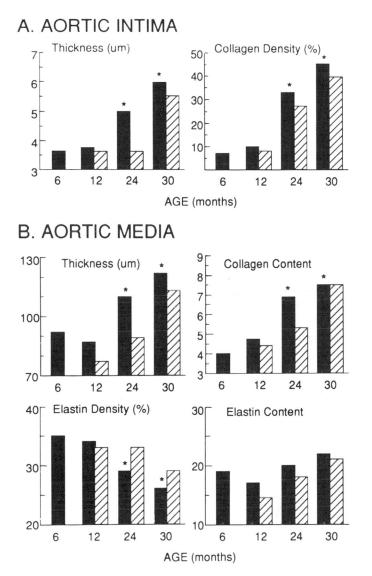

Fig. 2. *Changes in morphological measurements of the rat aorta with age. Graphs were derived from tabulated data [White et al., 1994]. Solid bars represent data from control (nontreated) animals, while cross-hatched bars represent data from animals chronically treated with the angiotensin–converting enzyme (ACE) inhibitor perindopril (1 mg/kg/day). Thickness is mea-sured in micrometers, collagen and elastin density as the percent area occupied by sirius red and orcein staining, respectively. Total elastin and collagen content were also calculated morphometrically as described by White et al., [1994]. Asterisk (*) indicates a statistically significant effect of age at p <0.001.*

size in some animals [Fornieri et al., 1992; Li et al., 1995]. This increase in thickness is due, in part, to an increase in VSMC size (hypertrophy) and, in part, to increased extracellular matrix deposition. The collagen content of the vessels in the study by Michel and colleagues

[1994] approximately doubled with age (Fig. 2B). In the same vessels, elastin content was unchanged, but the percentage density of elastin, a measure of its fibrillar organization, significantly decreased (Fig. 2B). In other studies, elastin fibers have been reported to be thin and disorganized in the aging vasculature [Cliff, 1970; Fornieri et al., 1992]. The increased thickness of the media is reminiscent of the vascular remodeling that is seen in hypertensive animals. While early studies [Stein et al., 1969] showed large decreases in the cellularity of the aging vessels, these measurements were reported as cell content per unit area. When adjusted for the fact that the area of the vessel increases with age, total cell number remains constant [Cliff, 1970; Fornieri et al., 1992] or decreases only slightly [Li et al., 1995]. Within the cells of the tunica media, the number of Golgi bodies and free ribosomes increases, changes that are consistent with the increased synthetic role that these cells play in altering the extracellular matrix. Interestingly, these cytological changes are similar to the morphological changes in VSMCs that have been characterized to occur in many vascular disorders [Chamley-Campbell and Campbell, 1981].

To understand the possible pathogenic significance of these changes, we need to understand the roles played by smooth muscle and endothelial cells in normal and diseased vessels. Vascular endothelial cells control the movement of macromolecules across the arterial wall [Belmin et al., 1993] and regulate the balance vasoconstriction and vasodilation in both an endocrine and paracrine fashion, maintaining vessel tone at a level appropriate for providing sufficient blood flow to downstream tissues [Furchgott and Zawadzki, 1980; Furchgott, 1983]. Vascular endothelial cells also produce both growth factors and growth antagonists. These growth modulators are likely to be involved in the repair of existing vessels and the development of new ones [Castellot et al., 1982; Collins et al., 1987]. Endothelial cells also tightly regulate the expression of surface adhesion molecules, so that inflammatory cells are targeted only to specific areas in response to vessel injury or underlying pathology [Collins, 1993]. Finally, endothelial cells normally provide an antithrombotic surface for blood flow, but can rapidly alter the expression of surface adhesion molecules to initiate a thrombogenic event by recruiting circulating platelets to sites of vessel injury [Radomski et al., 1987].

VSMCs directly control vessel tone through contraction and relaxation. The ability to contract is a differentiated function and depends on the activation of a differentiation-specific genetic program. Such genes include various components of the smooth muscle contractile apparatus, including smooth muscle α-actin, smooth muscle myosin heavy chain, smooth muscle–specific calponin and caldesmon, and myosin light chain kinase [Shanahan et al., 1993]. In the nondiseased vessels of young and middle-aged animals, the VSMCs present in the tunica media are, with few exceptions, in this differentiated state. In response to various signals, VSMCs can de-differentiate and display a proliferative, synthetic phenotype. There are a number of cytological changes that accompany this transition such that the cytoplasm no longer contains longitudinally organized contractile elements but extensive endoplasmic reticular networks and Golgi complexes [Mosse et al., 1985; Schwartz et al., 1990].

Because VSMCs in the synthetic state are more likely to proliferate [Chamley-Campbell et al., 1979, 1981], migrate [Pauly et al., 1994], and produce extracellular matrix components, the phenotypic modulation of VSMCs from the contractile to the synthetic state is an important component of most vascular disease. The development of lesions associated with both atherosclerosis and restenosis following angioplasty, for example, is the result of aberrant VSMC accumulation in the intima. These intimal VSMCs are de-differentiated [Gabbiani et al., 1984; Majesky et al., 1992], exhibit the synthetic phenotype, and are able to respond to growth factors secreted by other cells in the intima or those that are part of the inflammatory reactions that often accompany vascular diseases [Ross, 1993]. In the case of restenosis, VSMCs migrate across the extracellular matrix barriers into the intima and vessel lumen. Their subsequent proliferation in the intima creates what is referred to as the neointima. While the neointima may not on its own occlude blood flow through the vessel, it does create an environment for the development of numerous vascular complications, such as thrombosis. Figures 1B and 1C show the neointima that developed 8 and 14 days after balloon catheter injury to a rat carotid artery.

VSMCs are also found in the intima of vessels that exhibit no obvious pathology. These intimal VSMCs are often undifferentiated and exhibit cytological changes consistent with a transformation to the synthetic phenotype [Kocher and Gabbiani, 1986]. Their presence may predispose the area to atherosclerosis, since atherosclerotic lesions tend to occur in vessels in the arterial tree that "normally" contain such VSMCs in their tunica intima [Schwartz et al., 1990]. Careful kinetic studies indicate that atheromatous lesions in fat-fed pigs also originate in areas of pre-existing intimal thickening [Schwartz et al., 1985; Thomas et al., 1985].

With the above background, it is easy to see how the structural and compositional changes that take place in the aging vessel wall may play an important role in the vasculopathology of advancing age, particularly the increased incidence of atherosclerosis [Stout, 1987]. Atherosclerosis may be thought of as a disorder of the tunica intima [French, 1966]. It affects large and medium-sized muscular arteries, such as the coronary and carotid arteries, as well as elastic arteries, such as the aorta. Atherosclerosis is a predisposing factor to a number of acute vascular complications such as thrombosis, hemorrhage, and embolism. The initial step in the formation of the atherosclerotic lesion is likely to be the adherence of circulating monocytes to "dysfunctional" endothelium, expressing cell surface adhesion molecules. This is then followed by their subsequent migration into the intima. Oxidized low-density lipoproteins (OxLDL) are present in early atherosclerotic lesions [Steinberg et al., 1989] and *in vitro* studies have shown that OxLDL induces changes similar in the gene expression profile of endothelial cells to those seen in early lesions [Berliner et al., 1990], suggesting that it may be a factor in the conversion of the endothelium to its "dysfunctional" state. Once monocytes accumulate in the subendothelial space, they mature into macrophages, accumulate lipids, and become foam cells [Collins,

1993; Ross, 1993]. In this environment, VSMCs recruited from the tunica media proliferate, produce extracellular matrix components, and also accumulate lipids. The presence of both monocytes and VSMCs in the intima of many older vessels indicates that the initial steps in the development of the atherosclerotic lesion may already be complete.

The tunica media of the vessel wall also undergoes thickening with aging. This may be related to the hypertension that is often observed in aging animals and humans. Such thickening is a predictable adaptive response to increased blood pressure, and involves cell growth, death, migration, and the degradation and reorganization of the extracellular matrix. This remodeling of the vasculature is mediated by locally derived growth factors and vasoactive substances and is absolutely dependent on an intact endothelium, the source of many of these factors. The vasoactive substances and growth factors produced by the endothelium serve two purposes: as acute modifiers of vessel tone and as modulators of vascular smooth muscle cell growth [Gibbons and Dzau, 1994]. Are the changes in medial wall thickness with aging, therefore, just a chronic adaptation to the elevated systolic blood pressure that occurs during aging? There are a number of considerations that suggest that, while elevated blood pressure can affect the remodeling of the vascular wall during aging, it is unlikely to be the primary cause for the age-associated changes in the vasculature. First, changes in the arterial wall occur even in animals that exhibit little age-associated hypertension. The data shown in Figure 2, for example, are from a rat strain (WAG/Rij) that shows no detectable age-associated hypertension [Michel and colleagues, 1994]. The morphological and biochemical changes in the vessels occur despite any hypertension, although the magnitude of these changes are less than those seen in rat strains in which there is age-associated hypertension [Fornieri et al., 1992; Li et al., 1995]. Second, chronic administration of blood pressure–lowering drugs, such as angiotensin-converting enzyme (ACE) inhibitors delay but do not prevent the appearance of the age-associated changes in the vessel wall (Fig. 2). Finally, there is evidence for intrinsic changes in endothelial secretion of vasoactive substances with aging that persist independently of hemodynamic signals. In the intact animal, aging of the vasculature is associated with decreased production of vasodilatory substances, including prostacyclin and endothelium-derived relaxing factor nitric oxide (EDRF; NO) [Tokunaga et al., 1991; Egashira et al., 1993] and the increased production of vasoconstricting substances, such as endothelin-l (ET-l) [Luscher et al., 1992]. The decrease in EDRF production with age is likely to account, in part, for the reduced responsiveness of aging vessel to endothelium-dependent vasodilation [Yasue et al., 1990; Egashira et al., 1993]. These changes in the production of vasoactive substances and the reduced responsiveness of the vessel to vasodilators would be expected not only to increase active vessel stiffness due to increased VSMC contractility but also to promote VSMC growth. These effects would also be accentuated by the age-associated reduction in the response of the vessels to β-adrenergic agonists [Chin and Hoffman, 1991; Mader, 1992]. The age-associated changes in the pattern of vasoactive substance release by endothelial cells persists in cell culture [Tokunuga et al., 1991; Kumazaki et al., 1994], indicating that the changes are intrinsic to the endothelial cell and not merely a secondary response to increased blood pressure. Taken together, these observations indicate that, while hypertension can modify the vessel wall in the same direction as that observed during aging, there is an underlying age-associated pattern of change that may be independent of the changes in blood pressure and intrinsic to, at least, the endothelial cells.

The concept that age-associated changes are intrinsic to the cells in the vessels was addressed by experiments testing the responsiveness of the aged blood vessel to mechanical injury [Hariri et al., 1986]. Here vessels were treated with a nondistending coiled wire catheter that removed the endothelia but produced no observable damage to the underlying media. In response to this manipulation, intimal thickness was unchanged in young rats, while it increased many fold in older animals. The fact that mechanical injury was confined to the endothelium and there was no underlying medial damage is an important consideration in interpreting these results. As mentioned above and described in detail later (see Section II.A.2), aging blood vessels are less compliant so that a given injury may be expected to have a greater effect on older rather than younger vessels, a consideration not appreciated in earlier studies [Stemerman et al., 1982]. To eliminate the possibility that the differences in the response of old and young animals were due to changes in circulating factors, experiments were performed in which vessels from young and old animals were transplanted into syngeneic hosts of different ages and then injured (Fig. 3A). These results showed that the neointimal response was intrinsic to the vessel (donor)—transplantation of old vessels into young hosts produced the same response to injury that old vessels elaborated in old hosts, while young vessels failed to mount a response whether transplanted into either young or old hosts.

More recent studies have confirmed that there is little if any intimal growth in young animals in which the endothelium is removed without underlying medial cell damage, unless growth factors are infused after injury [Lindner et al., 1991]. It is not clear whether older VSMCs that migrate into the intima and proliferate produce their own growth/migratory factors or are resistant to growth antagonists present in the vessel wall. On the other hand, there are many structural changes that occur in the vessel wall with aging and it is equally likely that the factor(s) that antagonize growth are missing or modified in the aging vessel. To address this issue and extend the *in vivo* observations, VSMCs were isolated from young and old rat vessels and their growth properties examined in culture [McCaffrey et al., 1988]. These studies showed that VSMCs

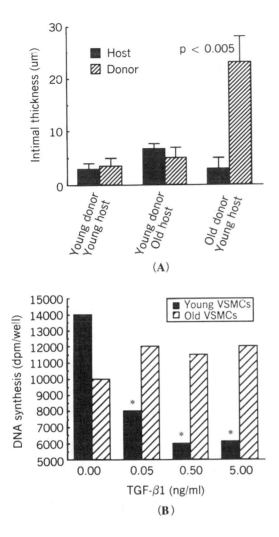

Fig. 3. *(A): Intimal response to gentle de-endothelialization of vessels following transplantation into young and old hosts [graph derived from tabulated data of Hariri et al., 1986]. The data show intimal thickness of injured donor vessels (crosshatched bars) and uninjured host controls (filled bars). Only old vessels transplanted into young hosts responded to gentle de-endothelialization with a significant increase in intimal thickness. Young vessels transplanted into either young or old host were nonresponsive. (B): Effect of TGF-β_1 on DNA synthesis in response to serum. VSMCs were isolated from medial explants of the rat thoracic aorta of young (<3 month) or old (>19 month) Fisher 344 rats and used in early (p <3) passage. Asterisk (*) indicates a statistically significant inhibition of DNA synthesis by TGF-β_1 at p <0.001. (Data from McCaffrey and Falcone [1993].)*

from older vessels had a growth advantage over VSMCs from younger animals and demonstrated autocrine production of growth factors that stimulated proliferation and that were inhibited by heparin. Further studies have suggested that an important component of this proliferative advantage is the resistance of older VSMCs to the growth-inhibitory effects of transforming growth factor, TGF-β_1 [McCaffrey and Falcone, 1993], a potent growth inhibitor of VSMCs from younger animals (Fig. 3B) and a possible key inhibitor of atherosclerosis [Grainger et al., 1993; 1994]. Indeed, recent evidence indicates that the serum concentration of activated TGF-β_1 is severely depressed in individuals with advanced atherosclerosis [Grainger et al., 1995]. Whether TGF-β_1 levels are reduced in older humans is not known, but VSMCs from both young and old rats produce equivalent amounts of total and activated TGF-β_1. Older VSMCs, however, express fewer receptors for TGF-β_1 and their growth is not inhibited even at TGF-β_1 concentrations that in the younger animals showed maximum inhibition of growth (Fig. 3B). In addition to reduced receptor binding, older VSMCs metabolized TGF-β_1 more rapidly than did cells from younger animals [McCaffrey and Falcone, 1993]. Given the changes that occur in extracellular matrix deposition in the medial wall of the blood vessel with age, it seems likely that some responses to TGF-β_1, such as increased collagen synthesis, must remain intact in the old VSMCs. Thus, while there are differences in TGF-β receptor number between young and old VSMCs, this difference alone seems unable to account for the resistance of older VSMCs to growth inhibition by TGF-β_1.

In addition to their resistance to growth antagonists, VSMCs from the media of older vessels also appear to exhibit an enhanced ability to migrate compared with cells from younger vessels. Experiments by Rebecca Pauly and Robert Monticone in our laboratory have tested the ability of VSMCs isolated from young (6 month) and old (20–24 month) Wistar rats to migrate toward the chemoattractant PDGF BB in a modified Boyden chamber [Pauly et al.,

1994]. Although VSMCs from young and old rats proliferate at comparable rates, medial VSMCs from older animals more readily migrate toward the chemoattractant than do VSMCs from younger animals. Since migration of VSMCs from the media to the intima is an obligate step in the pathogenesis of many different vascular diseases [Casscells, 1992], this increased ability to migrate may contribute to the increased susceptibility of older vessels to vascular disease.

The signaling pathways regulating cell migration are only beginning to be elucidated [Kundra et al., 1994]. For VSMCs, recent studies indicate that the regulation of calcium/calmodulin-dependent protein kinase II is a critical point in the control of migration [Pauly et al., 1995; Bilato et al., 1995]. Whether differences in the intracellular signaling pathways between young and old medial VSMCs account for the age-associated increases in migration and whether this is related to activation of calcium/calmodulin-dependent protein kinase II remain to be determined. The age-associated differences in migration may also be due to differences in the expression of differentiation-specific markers by young and old medial VSMCs. Surprisingly little is known about the expression of differentiation-specific markers by VSMCs in older vessels. From the description of these cells in the literature [Cliff, 1970; Michel et al., 1994], their increased expression or decreased degradation of interstitial type matrix components (e.g., fibronectin and collagen types I and III), and their behavior in cell culture, it would come as little surprise if their expression patterns were quite different from those of medial VSMC from younger animals and more like those of de-differentiated, neointimal VSMCs.

II.A.2. Changes in extracellular matrix composition of the vessel wall during aging. With aging, there is an increase in the thickness of the tunica media with little or no change in its cellularity but increased extracellular matrix (ECM) deposition, particularly that of collagen and fibronectin [Mamuya et al., 1992]. This increase in ECM deposition oc-

curs even though collagen I mRNA levels may be slightly depressed and tropoelastin mRNA expression virtually extinguished with aging [Quaglino et al., 1989]. In addition, other studies have reported that there is an increase in collagen insolubility and a decrease in elastin crosslinking with age [Reiser et al., 1987], as well as increased elastin degradation [Robert et al., 1984]. The expected result of these changes would be the creation of a stiffer, less compliant vessel wall. These changes in ECM deposition in the vessel wall are similar to the age-associated changes observed in other mesenchymal tissues in the body and reflect the overall increase in the stiffness of connective tissue throughout the body with aging [Sephel and Davidson, 1986; Martin et al., 1990; Quaglino et al., 1993].

These morphological changes fit well with a host of physiological and mechanical measurements that indicate that arterial wall stiffening occurs during both human [Bader, 1967; Avolio et al., 1985] and animal [Band et al., 1967; Cox, 1977; Michel et al., 1994] vascular aging. Stiffening of large arteries leads to increased aortic pulse-wave velocity and could account for increased arterial systolic and pulse pressure due to the summing of reflected and incident pulse waves [Lakatta, 1993]. Stiffening of the large arteries may also increase cardiac afterload by increasing the residual blood volume present in large arteries at the end of diastole and by increasing aortic impedance. These hemodynamic changes may also contribute to the age-associated hypertrophy of cardiomyocytes (see Section III.A).

While a decrease in total elastin content with aging may not be seen in all animal strains (see the data in Fig. 2B, for example), a redistribution of elastin in the vasculature does occur [Cliff, 1970; Robert et al., 1984]. That redistribution involves the loss of elastin from the laminae, a build up in the interlaminar tracts, and, in some cases, elastin condensation within the medial cells themselves [Cliff, 1970]. Understanding the molecular mechanisms underlying this phenomenon, which was initially described more than 25 years ago, has been a difficult problem, but recent insights into the nature of elastin metabolism and its assembly and turnover are providing new clues. One possible contributing factor is the decreased ECM crosslinking activity that has been observed in a number of studies [Fornieri et al., 1992; Quaglino et al., 1993], while another is a potential loss in elastin repair capacity with aging due to an alteration in elastin biosynthesis. This latter factor is an important consideration given the "wear and tear" that continuous pulsatile flow exerts on the vessel. Recent studies have shown that elastin biosynthesis involves a 67 kD "companion protein," the elastin-binding protein (EBP), that forms a complex with tropoelastin and acts as an intracellular chaperone, preventing the newly synthesized tropoelastin from undergoing self-aggregation and/or cleavage by intracellular proteolysis [Hinek and Rabinovitch, 1994]. EBP also acts as a cellular receptor for tropoelastin and is involved in elastin fiber assembly. In this context, EBP binds not only tropoelastin but also two transmembrane proteins on the surface of VSMCs. The "presentation" of elastin by VSMCs aligns the elastin monomer with the microfibrillar scaffold of the growing elastin fibril [Hinek et al., 1988]. EBP is also an integral component of the elastin fiber [Hinek et al., 1993]. Because EBP binds to the site on the elastin monomer that is cleaved by various elastin serine proteases, including that produced by vascular smooth muscle cells [Zhu et al., 1994], it may play a role in the mechanisms that are responsible for the extraordinary stability of the elastin fibril.

The role that EBP plays in vascular pathology is likely to be significant. VSMC migration into the tunica intima [Wight, 1989], the response to wounding [Benitz and Bernfield, 1990], systemic hypertension [Wolinsky, 1970], and experimental pulmonary vascular disease [Todorovich-Hunter et al., 1992] are all often accompanied by the disruption of existing elastin fibrils. Even more dramatic are the changes in elastogenesis that are seen in the ductus arteriosus (DA), a fetal vessel that

undergoes extensive remodeling during late gestation to produce intimal cushions in the vessel wall. These cushions are required to ensure that the vessel will close completely when it constricts after birth. The formation of these intimal cushions involves VSMC migration from the tunica media, a phenomenon that is accompanied and possibly even directed by defective elastin assembly in the vessel. An alteration in the DA smooth muscle cells that results in reduced expression of EBP is likely to be responsible for the defect in elastin assembly [Hinek et al., 1991]. Instead of a normal 72 kD tropoelastin monomer, DA cells secrete a 52 kD, serine protease–cleaved, truncated tropoelastin, which is unable to participate in fiber formation [Hinek and Rabinovitch, 1993].

Based on these and other studies, a number of hypotheses could be advanced to explain the age-associated alterations in elastin distribution and fiber structure. One is that there is a decrease in elastin production with age. Published data indicate that cultured VSMCs isolated from older animals do produce less tropoelastin than cells from younger animals [McMahon et al., 1985]. Other reports, however, suggest that elastin distribution, and not content, changes with age [Michel et al., 1994]. Another hypothesis would be that changes in EBP expression occur with age so that the elastin repair capacity of vessels decreases with age and is unable to keep up with the continuous "wear and tear" on the vessel caused by pulsatile blood flow. Yet another possible mechanism by which elastin repair capacity could be altered with aging relies on the ability of β-galacto sugars to decrease EBPs affinity for both elastin and its cell surface binding sites [Hinek et al., 1988, 1992]. EBP guides tropoelastin assembly through a highly coordinated mechanism in which binding of galactosylated proteins to the lectin-binding site of EBP releases tropoelastin onto the microfibrillar scaffold of the growing elastin fiber [Hinek et al., 1988, 1991]. The same effect of β-galactosylated sugars could also impair elastin fiber assembly and sensitize existing fibers to serine protease digestion if sugars were in excess and not associated with the microfibrillar scaffold used for elastin assembly [Hinek et al., 1992]. Large increases in such molecules occur in many pathogenetic circumstances. N-acetylgalactosamine–containing glycosaminoglycans (GAGs), especially chondroitin sulfate (CS) and dermatan sulfate (DS), accumulate during vessel injury, hypertension, and sclerotic disease. The appearance of CS/DS is also highly correlated with cell migration in cell culture [Kinsella and Wight, 1986]. As β-galacto sugars, CS/DS can cause the release of the EBP–tropoelastin complex from its cell surface binding site, thereby impairing new fiber assembly. They can also cause the release of EBP from the elastin fiber, thereby exposing a normally masked serine protease cleavage site. Although changes in EBP synthesis and extracellular localization with aging have not been characterized, arterial chondroitin sulfate content does increase with age in humans [Yla-Herttula et al., 1986].

Another potentially important mechanism controlling the redistribution of elastin with age may be the partial degradation of elastin by elastases. Elastases have long been thought to play an important role in degenerative arterial diseases [Hall, 1961]. Elastase activity has been shown to increase in atherosclerotic lesions [Dubick et al., 1988], aneurysms [Cohen et al., 1992], post-cardiac transplant arteriopathy [Oho and Rabinovitch, 1994], and in the rat pulmonary artery prior to the development of experimentally induced progressive pulmonary hypertension [Todorovich-Hunter et al., 1992]. Indeed, treatment with serine elastase inhibitors actually prevents pulmonary hypertension in these animals [Ye and Rabinovitch, 1991]. Elastase may be recruited from neutrophils or produced by VSMCs themselves [Zhu et al., 1994]. Expression of an endogenous vascular elastase (EVE) by VSMCs is stimulated by serum, although the serum factor(s) responsible are not known [Kobayashi et al., 1994]. It is known that various growth factors and cytokinases that are expressed in the vessel in association with various pathogenic states (e.g.,

PDGF, bFGF, TGF-β, IL1, TNF-α, and so forth) do not affect the expression of EVE. EVE shares its recognition site on elastin with EBP [Zhu et al., 1994], so that elastin degradation would be expected to be controlled by the relative levels of EVE, EBP, and various GAGs, such as CS and DS. EBP, however, will have no effect on metallo- or cysteinyl-enzymes with elastolytic activity. Both matrix metalloproteinase (MMP)-2 and MMP-9 (both type IV gelatinases) exhibit elastase activity [Senior et al., 1991], as does a metalloelastase cloned from macrophages [Shapiro et al., 1992]. While it has not been reported whether the metalloelastase is expressed in blood vessels, the activities of both MMP-2 and MMP-9 increase during the response to mechanical injury in the blood vessel [Jenkins et al., 1994].

Finally, it is worth considering whether elastin and EBP play an additional role in the maintenance of the differentiated phenotype in VSMCs. The 55 and 61 kD integral membrane proteins that bind EBP allow a transmembrane link between extracellular ligand and the cytoskeleton [Hinek et al., 1988]. As is the case with integrin receptors and their extracellular ligands, such a link may allow the transmission of information in both directions and may, therefore, be involved in regulating VSMC gene expression. That such regulation may include the maintenance of the differentiated phenotype is an attractive notion, based on the fact that VSMCs that are located *in vivo* between elastic laminae are almost always differentiated and that the disruption of these laminae is often associated with modulation of the VSMC to the synthetic phenotype.

II.B. Advanced Glycation Endproducts, Oxidant Stress, and Vascular Cell Gene Expression

When proteins or lipids are exposed to reducing sugars, they undergo nonenzymatic glycation and oxidation. Initially, this leads to the formation of reversible, early glycation products, such as the Schiff bases and Amadori products. Following further complex molecular rearrangements, however, the irreversible advanced glycation endproducts (AGEs) are formed. AGEs are a heterogeneous class of structures [Monnier, 1990]. They accumulate in the plasma and in the vessel wall during normal aging [Reiser, 1991; MacDonald et al., 1992] and at an accelerated rate in diabetes, where their presence has been linked to development of a variety of vascular complications [Brownlee et al., 1988]. One of the major ways in which AGEs exert their biological effects is through interaction with a specific cell surface receptor (receptor for AGEs [RAGE] [Neeper et al., 1992]. Molecular cloning of this receptor has identified it as a new member of the immunoglobulin superfamily of cell surface molecules. The extracellular region is comprised of one V (variable)-type domain and two C (constant)-type domains. There is a single transmembrane spanning domain followed by a short, highly charged, cytosolic tail [Neeper et al., 1992]. A number of studies, including many of those cited below, indicate that RAGE mediates the effects of AGEs on target cells.

AGE–RAGE interactions perturb normal cellular functions. They induce changes in the cell that are similar in many ways to those induced by minimally oxidized LDL. Infusion of AGEs, for example, can exacerbate atheromatous lesion formation in lipid–fed animals [Vlassara et al., 1994]. AGEs can also induce significant oxidant stress in endothelial cells, seen through the increased production of thiobarbituric acid–reactive substances, the induction of hemoxygenase mRNA, the activation of NF-kB, and the appearance of malondialdehyde epitopes [Yan et al., 1994]. By virtue of their ability to activate NF-kB, AGEs are likely to play a pivotal role the in pathogenesis of a variety of vascular diseases, since NF-kB activation appears to be a common pathogenetic mechanism for many of the important risk factors associated with vascular disease, including hyperlipidemia, smoking, diabetes, and hypertension [Collins, 1993].

AGEs may also play a significant role in the changes in intimal thickness and cellular composition that occur during aging and that involve increases in VSMC and monocyte number [Guyton et al., 1983; Haudenschild et al., 1981]. AGEs induce endothelial cells to express vascular cell adhesion molecule-1 (VCAM-1) [Schmidt et al., 1994], an endothelial–leukocyte adhesion molecule that captures circulating monocytes at the endothelial-vascular interface. Second, AGEs can stimulate expression of mRNA for monocyte chemoattractant protein-1 (MCP-1) by both cultured endothelial and vascular smooth muscle cells and in intact blood vessels as well [Fredman et al., 1994; Crow et al., 1995]. In the case of the cultured cells, this increase in mRNA is dependent on receptor binding, is due to transcriptional activation of the gene, and results in functionally active protein [Crow et al., 1995]. MCP-1 expression and secretion by intimal cells could attract migrating monocytes that have been activated by their binding to VCAM-1 on the surface of the endothelial cells and facilitate their transmigration across the endothelial barrier. Once in the intima, these monocytes could mature and form macrophages. AGEs also stimulate production of platelet-derived growth factor (PDGF) B chain by endothelial cells and VSMCs [Kirstein et al., 1990; Fredman et al., 1994]. PDGF BB is a potent chemoattractant for VSMCs and if expressed and secreted by endothelial or intimal smooth muscle cells would be expected to attract additional VSMCs to the intima.

There is little doubt then that vascular cells are targets for advanced glycation endproducts and that this interaction could increase smooth muscle cell migration and monocyte infiltration of the blood vessel intima, both of which are important prelesional events for atherogenesis. Because the receptor is also expressed in young animals in which there are no detectable levels of advanced glycation endproducts [Brett et al., 1993], additional ligands for the receptor are likely to exist. The documented association between receptor activation and oxidant stress suggests that the pathogenic effects of the RAGE/AGE/endogenous ligand system are likely to contribute to the progression of vascular disease in both young and old animals.

III. THE AGING HEART

III.A. Myocyte Hypertrophy and Myocardial Cell Loss With Aging

There are two major cell types in the heart: myocytes, which comprise some 75%–80% of the heart by volume, and fibroblasts, which outnumber myocytes by approximately 4 to 1 [Grove et al., 1969; Morkin and Ashford, 1968]. The contractile properties of the heart are largely dictated by individual myocyte function, although pump performance is also influenced by the viscoelastic properties of the extracellular matrix. As is the case for the vasculature and other mesenchymal tissues in the body [Martin et al., 1990; Quaglino et al., 1993], an increased collagen content has been reported in the aging heart. This change along with possible alterations in elastin assembly and distribution in the aging heart are likely to affect its passive mechanical properties.

The aging heart resembles the hypertrophic heart in younger mammals in a number of respects. The hearts of both humans and rats exhibit moderate (25%) left ventricular hypertrophy with age. The majority of the hypertrophy is due to myocardial cell enlargement, and it occurs in the absence of any overt sign of cardio-vascular disease [Yin et al., 1980; Anversa et al., 1986, 1990; Fraticelli et al., 1989; Olivetti et al., 1994]. There are also age-associated alterations in the expression of myocyte-specific genes that are similar to those seen in pressure-overloaded myocardium in young animals. Table I summarizes the changes in specific gene expression that occur in pressure-overloaded young and normotensive old animals as well as in cultured neonatal cardiomyocytes treated with various growth factors [Parker et al., 1990; Schneider and Parker, 1990].

Among the more dramatic of these changes is that of atrial natriuretic factor (ANF) gene expression. In the adult, ANF mRNA levels are highest in the atria, with the right and left ventricles of the heart expressing 50–100 fold less mRNA [Seidman et al., 1985; Gardner et al., 1986; Lewicki et al., 1986]. However, by virtue of their relatively large mass, the ventricles do significantly contribute to circulating plasma ANF and can have a considerable impact on systemic fluid balance [Thibault et al., 1989; Kinnunen et al., 1992]. Expression of ANF in adult ventricular tissues is increased when the heart is overloaded [Day et al., 1987; Mercadier et al., 1989]. ANF mRNA expression and peptide secretion is also significantly up-regulated during heart failure and during

TABLE I. Gene Expression in Various Models of Cardiac Hypertrophy

Gene	Aging	Hyper-tension	Heart failure	TGF-β_1	aFGF	bFGF
Atrial natriuretic factor (ANF)	↑↑	↑↑	↑↑	↑	↑↑	↑
Sarcoplasmic reticulum Ca^{2+} ATPase (SERCA)	↓↓	↓↓	↓↔*	↓↓	↓↓	↓↓
α-Myosin heavy chain (α-MHC)	↓↓	↓↓	↓↓	↓↓	↓↓	↓↓
β-Myosin heavy chain (β-MHC)	↑↑	↑↑	↔	↑	↑↑	↑
Skeletal α-actin (SK)	↓	↑↑**	↔	↑	↓↓	↑
Proenkephalin (PNK)	↑↑	↓↓	↔			
Fibronectin (FN)	↑↑	↑↑	↑↑			
Collagen type I	↑	↑↑	↑↑			
Collagen type III	↑	↑↑	↑↑			
Transforming growth factor-β_1 (TGF-β_1)	↔	↑	↑			

Directional information is based largely on studies of mRNA levels by Northern blot analysis and in a few cases protein concentration by Western blot analysis. ↑↑, Marked increase; ↑, increase; ↔, no detectable change; ↓↓, marked decrease; ↓, decrease; *, model dependent; **, transient change.

aging [Reckelhoff et al., 1992; Takahashi et al., 1992; Younes et al., 1995]. Recent work in cultured neonatal myocytes suggests that ANF may inhibit hypertrophic growth [Calderone et al., 1994]. The implications of this finding for both hypertrophying and aging hearts remain to be elucidated.

During aging and in many experimental models of cardiac hypertrophy, a shift occurs in myosin heavy chain (MHC) isoforms favoring expression of the β-MHC isoform over that of the α-isoform [O'Neill et al., 1991]. This switch in MHC gene expression may underlie, in part, the decline in the velocity of shortening exhibited by cardiac muscle during aging [Capasso et al., 1986], since the actin-activated ATPase activity of the β-isoform is less than that of the α-isoform. The change in MHC isoform expression occurs at both the mRNA and protein levels [Effron et al., 1987; O'Neill et al., 1991]. Most, but not all, studies have also shown a decrease in the mRNA levels for the cardiac sarcoplasmic reticulum calcium ATPase (SERCA) with advancing age [Buttrick et al., 1991; Lompre et al., 1991; Maciel et al., 1990]. This age-associated down-regulation of SERCA2 expression may contribute to diminished Ca^{2+} reuptake and be partially responsible for the age-related slowing of cardiac contraction. The expression of SERCA2 is downregulated in a number of models of hypertrophy, as well many models of heart failure (Table I). The recent identification of DNA sequences upstream of the SERCA2 gene that apparently regulate its transcription provide an approach to determining cis-acting elements responsible for the changes in expression of the SERCA2 gene during hypertrophy and aging [Fisher et al., 1993].

Comparison of the factors that regulate both the α- and β-MHC isoforms and SERCA2 may also provide clues to the mechanisms that regulate their expression during aging. Both the MHC and SERCA2 genes, for example, have in common the fact that the level of their respective mRNAs can be regulated by thyroid hormone. In the case of the α-MHC gene, thyroid hormone up-regulates its expression, while it inhibits transcription of the β-MHC [Gustafson et al., 1987; Edwards et al., 1994] and SERCA2 genes [Rohrer and Dillman, 1988; Rohrer et al., 1991]. The changes in these genes that occur during aging, therefore, mimic those observed in the hypothyroid state [Izumo et al., 1986; Arai et al., 1991]. Whether a relative hypothyroid state accompanies aging is uncertain. A small age-associated decline in plasma levels of thyroid hormones, 3,5,3'-triiodothyronine (T3) and thyroxine (T4), occurs in the rat [Effron et al., 1987; Schuyler and Yarbrough, 1990]. Replacement of sufficient T3 or T3 and T4 to restore plasma levels in older rats to levels equivalent to or greater than those in younger rats is incapable of completely reversing changes in the MHC isoform pattern and only partially abolishes the age-associated decline in myosin ATPase activity [Carter et al., 1987; Effron et al., 1987; Schuyler and Yarbrough, 1990]. These findings suggest that a decrease in the ability of myocardial cells to respond to thyroxine may occur with aging and underlie, in part, the pattern of changes depicted in Table I. Thyroid hormone activates transcription by first binding to thyroid hormone receptors, the products of the c-erbA protooncogene. Three such receptors are expressed in the heart and have been identified, including c-erbAα-l, c-erbAβ-l, and c-erbAα-2 [Weinberger et al., 1986; Koenig et al., 1988]. The first two gene products form functional receptors, while the c-erbAα-2 product does not bind thyroid hormone and may act to inhibit thyroid hormone action [Koenig et al., 1989]. Thyroid hormone binds to thyroid hormone receptors, and this complex then binds to thyroid response elements (TREs) along with other associated proteins in the transcriptional control regions of thyroid hormone–sensitive genes. The TRE is a DNA sequence exhibiting twofold symmetry of either direct or indirect repeats. The complex of thyroid hormone and its receptor actually binds to the

TRE as a heterodimer with another member of the steroid superfamily, the retinoid X receptors (RXRs) [Zhang and Pfahl, 1993]. Three RXR genes have been cloned (α, β, and γ), and each of these can and do form heterodimers with TRs [Leng et al., 1994; Mangelsdorf et al., 1992]. A decrease in the level of expression of either TR or RXR partner would, therefore, result in a functional "hypothyroid" state even in the face of elevated T3 or T4 plasma levels. Measurement of the mRNA for these different receptors indicates that there are no significant changes in any of the TR mRNA levels in the heart with age but that both RXRα and RXRγ mRNA expression in the heart are significantly reduced with age [X. Long, M. Boluyt, and M. Crow, unpublished data]. These preliminary observations suggest that a functional "hypothyroid" state capable of explaining the changes in MHC and SERCA2 gene expression may exist in the aged heart. These observations warrant further investigation, since they may also provide a partial explanation for changes in adrenergic responsiveness that are exhibited by the heart with aging (see Section III.B below).

While the age-associated changes in MHC and SERCA2 gene expression resemble those of the hypertrophied heart, the expression of a number of other genes in the myocardium is regulated differently in aging and in experimental hypertrophy. The skeletal isoform of the α-actin gene is expressed by the heart during late embryonic development but is partially replaced by the cardiac isoform soon after birth and long before the increase in circulating thyroid hormone that occurs postnatally in the rat [Carrier et al., 1992]. It is not regulated by manipulating the thyroid status of the animal and the expression of the skeletal α-actin gene is diminished with age, while its expression is transiently activated during pressure overload and other acute models of hypertrophy [Izumo et al., 1988; Boluyt et al., 1995]. Likewise, expression of the preproenkephalin (PNK) gene is down-regulated af-

ter either aortic coarctation (unpublished observations) or isoproterenol infusion in the rat [Boluyt et al., 1995], but up-regulated in cultured neonatal rat cardiomyocytes and fibroblasts in response to hypertrophy-inducing adrenergic agonists (unpublished observations). PNK mRNA expression is also up-regulated with age, increasing some five-fold over the life of the animal and resulting in increased accumulation of met- and leu-enkephalin [Boluyt et al., 1993; Caffrey et al., 1994]. The implications of this increase with age are unknown, but the negative effects of opioid peptides on cardiac contraction [Ventura et al., 1992] suggest that it may contribute to age-associated changes in cardiac performance, possibly even inhibition of the β-adrenergic augmentation of contraction [Xiao et al., 1991].

Although the patterns of change in gene expression observed in hypertensive young rats and growth factor–treated neonatal myocytes are not identical with the normotensive aging heart (compare, however, the effects of acidic FGF treatment of neonatal myocytes), they are so strikingly similar that it is tempting to speculate that such a common adaptive response reflects a common stimulus. This particular constellation of changes in gene expression appears to be adaptive in that it allows for an energy-efficient and prolonged contraction. In the hypertensive rodent heart, these changes are likely to represent a response to increased vascular afterload, dictated by a stiffer arterial system and an increase in arterial pressure and impedance. It could be argued that the mechanical and hormonal stimuli that initiate and maintain the cardiac hypertrophic response in the hypertensive heart are also present within the aging heart. While there is evidence for such changes in the arterial vasculature (see Section II.A), an additional stimulus could be the loss of myocardial cells that occurs with aging [Anversa et al., 1986]. Although indirect evidence for limited hyperplasia of ventricular myocytes has been reported [Anversa et al., 1990, 1991; Olivetti et al., 1994], the overwhelming preponderance of data indicates that the aging heart consists of fewer, but larger, myocytes [Anversa et al., 1986,1990; Fraticelli et al., 1989].

The mechanism responsible for myocyte loss in the heart is unknown, although recent evidence suggests that apoptosis of cardiomyocytes can occur under various pathophysiological circumstances and may, therefore, be occurring during normal aging. Apoptosis, or programmed cell death, is a process by which apparently healthy cells commit to a suicidal chain of events. Apoptosis can be initiated in various cell types through an assortment of signaling pathways, including incompatible growth signals [White, 1993]. While our current knowledge regarding apoptosis has derived almost exclusively from studies of nonmuscle cells, evidence for apoptotic myocyte cell death has recently emerged from work with myoctes in culture [Tanaka et al., 1994] and from a study of experimentally induced ischemia in the rabbit heart [Gottlieb et al., 1994]. DNA fragmentation and expression of Fas antigen mRNA, which is often associated with apoptosis, was observed in neonatal cardiomyocytes maintained under hypoxic conditions [Tanaka et al., 1994]. In contrast, evidence for apoptosis *in vivo* was observed in cardiomyocytes only during reperfusion from ischemia and not after even extended periods of ischemia/hypoxia [Gottlieb et al., 1994] Since it has been suggested that hypoxia due to reduced coronary blood flow reserves occurs with advancing age [Weisfeldt et al., 1971; Abu-Erreish et al., 1977], it is an intriguing possibility that apoptosis may account for myocyte loss observed in the heart during aging and in experimentally induced hypertrophy [Anversa et al., 1986, 1990; Olivetti et al., 1991]. Cell "dropout" could lead to augmented stretch of the remaining myocytes and stretch to their hypertrophy [Sadoshima et al., 1992]. One problem with this proposal is that the biophysical changes in the heart that occur during aging (i.e., the increased action po-

tential and calcium transient and the prolonged contracture) as well as the specific changes in gene expression shown in Table I occur in both the right and left ventricles, even though myocyte number in the right ventricle may not be reduced with aging [Anversa et al., 1986].

While it remains unclear whether cell loss is responsible for any of the biophysical or biochemical changes observed in the aging heart, cell loss is likely to be an important factor in the development of fibrosis in the heart. Fibrosis is a well-documented correlate of aging in the heart that is associated with increased cardiac muscle stiffness [Lakatta, 1993]. An early step in the elaboration of additional extracellular matrix and a signal that reparative or reactive fibrosis is underway is the production and secretion of fibronectin [Raghow, 1994; Weber and Brilla, 1991]. Fibronectin is a 440 kD glycoprotein, existing as a dimer of two nearly identical polypeptide chains harboring domains that interact with collagen, heparin, and fibrin. Increased expression of fibronectin is a common feature of cardiac hypertrophy models, such as aortic constriction [Villareal and Dillman, 1992], spontaneous hypertension [Mamuya and Brecher, 1992], isoproterenol infusion [Boluyt et al., 1995], and heart failure [Boluyt et al., 1994]. In each case, an elevation in the "embryonic" fibronectin pattern of alternatively spliced EIIIA and EIIIB segments accounts for at least a portion of the increased expression. In the case of heart failure, elevated fibronectin levels are accompanied by elevated levels of TGF-β_1 mRNA [Boluyt et al., 1994]. Elevated EIIIA- and EIIIB-containing fibronectins were first observed as a hallmark of the wound healing process [ffrench-Constant et al., 1989]. While initially high after birth, fibronectin mRNA and protein levels fall between 2 and 10 months of age in the rat heart [Mamuya et al., 1992]. Between 6 and 24 months of age, however, there is a five-fold increase in fibronectin expression in the left ventricle and a slightly smaller increase in the atrium (Fig. 4). Thus, just as in

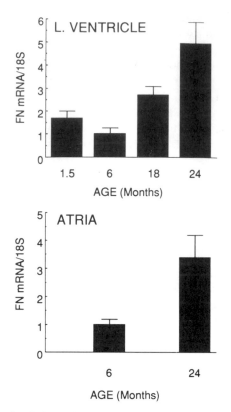

Fig. 4. *Relative levels of fibronectin (FN) mRNA in the left ventricles and atria of Wistar rats during aging. Total RNA was probed with a cDNA specific for the coding region of the fibronectin gene as previously described [Boluyt et al., 1994] and normalized for loading differences to 18S rRNA. Levels of FN mRNA were elevated in atria of 24 month compared with 6 month rats (p <0.05, t test) and in the left ventricles of 24 month compared with 1.5 and 6 month rats (p <0.05; Tukey procedure).*

experimental hypertension and cardiac hypertrophy, the elevated level of fibronectin mRNA that occurs with aging is consistent with remodeling of the heart.

Collagens I and III comprise the vast majority of extracellular matrix protein in the heart. As is the case for fibronectin, expression of the genes encoding types I and III collagen decreases from youth to middle age and then increases with advancing age [Mamuya et al., 1992; Besse et al., 1994]. Whereas the relatively high level of collagen gene expression in the young adult heart does not result in ex-

cess accumulation of collagen, the relatively small increases in collagen types I and III mRNAs that occur between late middle age and senescence coincide with a nearly two-fold increase in accumulation of connective tissue as assessed by hydroxyproline content [Besse et al., 1994]. While translational regulation could account for these age-associated differences in protein accumulation, changes in collagen turnover with age could also affect collagen accumulation. Degradation of collagen is initiated by an interstitial collagenase belonging to the matrix metalloproteinase (MMP) family. Interstitial collagenase or MMP-1 cleaves native collagens I and III, rendering them susceptible to further digestion by gelatinases such as MMP-2, which is consititutively expressed in many tissues [Matrisian, 1992]. MMP-1 is generally regulated at the level of gene transcription by cytokines and growth factors through an AP-1 site located in its transcriptional control region [Angel et al., 1987]. Given the fact that numerous physiological manipulations are less effective in aging animals at activating immediate early gene responses, such as AP-1 activation [Riabowol et al., 1992; Takahashi et al., 1992; Burke et al., 1994], MMP-1 activation might be expected to decrease with age.

A host of growth and humoral factors have been implicated in changes in myocardial extracellular matrix gene expression during experimental hypertrophy, heart failure, and aging, but little attention has been given to endothelin-1 (ET-1) in the heart, even though the effects of ET in extracellular matrix changes in the aging vascular system have been described [Kumazaki et al., 1991, 1993]. ET-1 is expressed by both myocytes and fibroblasts in the heart [Ito et al., 1993; Fujisaki et al., 1993] and is a potent hypertrophic agent for neonatal myocytes [Ito et al., 1991]. There is also evidence that the local production of ET-1 in myocytes may mediate the hypertropic effects of aortic constriction *in vivo* [Ito et al, 1994] and of angiotensin II *in vitro* [Ito et al,

1992]. Both ET-l mRNA and protein secretion are increased in the aging blood vessel, where it is likely to be at least partly responsible for the relatively large increase in fibronectin deposition in the aging blood vessel [Kumazaki et al., 1994]. Similar effects might be expected in the heart.

III.B. β-Adrenergic Responsiveness Is Lost by Cardiomyocytes During Aging

One of the most consistent cross-species effects of age on the cardiovascular system is the decline in the response of cardiovascular tissue to β-adrenergic agonists. In humans, this is manifested as an age-associated decrease in heart rate augmentation, left ventricular dilatation, and diminished left ventricular ejection fraction during exercise [Rodeheffer et al., 1984]. This decline in the efficacy of the β-adrenergic receptor (β-AR) system would seem to be postsynaptic, in that there is no deficit in the ability of older individuals to increase plasma levels of both norepinephrine and epinephrine in response to exercise; in fact, these levels are in general greater than those achieved by younger individuals [Fleg et al., 1985]. Likewise, infusion of β-AR agonists elicits a lesser augmentation of heart rate in old than in young individuals [Yin et al., 1979] and stimulation of isolated myocytes from older animals with β-AR agonists elicits less of a contractile response [Guarnieri et al., 1980] and change in the calcium transient [Xiao et al., 1994] than in cells from younger animals.

Attempts to characterize the biochemical alterations associated with the effects of aging on the β-AR signal transduction pathway have indicated that multiple steps in the signaling pathway seem to be affected. Both β1-ARs and β2-ARs are present in the heart. Upon agonist stimulation, both β-ARs activate adenylate cyclase to increase cAMP levels and activate protein kinase A (PKA) (see Fig. 5). In the heart, the likely targets of PKA phosphorylation are troponin I, C-protein, the L-type calcium channel, and phospholamban. Phosphorylation of

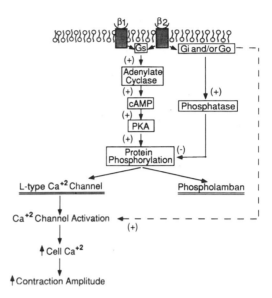

Fig. 5. *Hypothetical β1- and β2-adrenergic receptor signaling cascade in the heart. PKA, protein kinase A; β1, β1-adrenergic receptor, β2, β2-adrenergic receptor.*

the latter two proteins is, in fact, associated with accelerated calcium handling. Limited data are available on the relative ability of β-AR agonists to induce the phosphorylation of these proteins in aging hearts, showing that the phosphorylation of both troponin I and C-protein following β1-AR stimulation are significantly impaired with aging [Sakai et al., 1989]. In humans, but not rats, there is a substantial decrease in β1-AR density with aging [White et al., 1994; Lakatta, 1993], while in both species there is a decrease in high-affinity binding states of the β1-AR [White et al., 1994; Scarpace and Abrass, 1986]. Guanine nucleotide-binding (G)-proteins act as coupling factors to transduce the signal from the receptor to the catalytic subunit of adenylate cyclase. Studies in both humans and rats indicate that, in addition to receptor loss or desensitization, there are also abnormalities in the G$_s$-protein–mediated activation of adenylate cyclase. A recent study in humans indicates that while direct G$_s$-protein–mediated adenylate cyclase activity was decreased with age, there is no change in adenylate cyclase activity induced by manganese, a stimulus of adenylate cyclase that does not involve G-proteins [White et al., 1994]. Similar studies using reagents to distinguish between G-protein–mediated and direct adenylate cyclase activity have not been performed in the rat.

A similar lack of β-AR responsiveness is observed in heart failure and in some experimental models of hypertension, such as the rat aortic constriction model [Chevelier et al., 1989]. In heart failure, plasma norepinephrine levels are elevated, and activation of the sympathetic nervous system is increased. These changes represent an attempt by the organism to maintain blood pressure and flow in the face of a progressively failing "pump" and reduced cardiac output. While increased sympathetic activation may initially be helpful to the chronically failing heart, it is likely to trigger desensitization of the β-ARs and actually accelerate failure. In both animal [Calderone et al., 1991] and human [Bristow et al., 1989] heart failure, total β-AR density, particularly that of the β_1-AR, is decreased. In some animal models of heart failure, β_2-AR density is either unaltered or actually increased [Bristow et al., 1986, 1989; Kiuchi et al., 1993]. It has been hypothesized that selective desensitization/down-regulation of β_1-AR vs β_2-AR in heart failure is caused directly by the elevated plasma levels of norepinephrine, a relatively selective β_1-adrenergic agonist. A similar mechanism may be at work during aging, because plasma catecholamine levels may also be chronically elevated with aging based on the excessive neural discharge during stress [Fleg et al., 1985] and the reduced clearance capacity of the kidneys with age.

Recent data indicate that β_2-AR density in humans is also not altered by aging [Bohm et al., 1993; White et al., 1994], although coupling to adenylate cyclase is impaired, as would be expected from the fact that the ability of G_s to activate adenylate cyclase is diminished in the aging heart [White et al., 1994]. The selectivity of β_1 vs. β_2-AR down-regulation in both heart failure and aging is of interest for a num-

ber of reasons. Recent studies on the signal transduction pathway initiated by β_2-AR stimulation indicate that while stimulation with β_2-specific agonists can increase cAMP levels, this is unlikely to be the mechanism by which calcium movement into the cell and the intracellular calcium transient are increased. The dose–response curves for cAMP production and calcium movement in response to β_2-specific stimulation are dissociated and, in some species (e.g., the dog), β_2-specific stimulation does not increase cAMP at all, even though it does affect calcium mobilization [Xiao et al., 1994; Xiao and Lakatta, 1993]. In fact, in contrast to β_1-AR stimulation, β2-AR stimulation does not increase the phosphorylation of phospholamban [Xiao et al., 1994]. This failure to phosphorylate phospholamban may be related to the fact that β_2-ARs are likely to associate with at least two and possibly three different G-proteins: G_s, to stimulate adenylate cyclase, a pertussis toxin–sensitive G-protein that may be involved in antagonizing G_s-mediated changes in the cell [Xiao et al., 1995], and a G-protein that may directly link activation of the β_2-AR to calcium channel activation. In both heart failure and aging, therefore, it is possible that a pathway to stimulate calcium movements following β_2-AR stimulation may remain open, despite the fact that receptor activation of cAMP is substantially reduced.

On the surface, the selectivity of β_1-AR down-regulation versus that of β_2-AR with aging seems at odds with the fact that desensitization/down-regulation following receptor occupancy has clearly been shown to affect β_2-ARs much more extensively and rapidly than β_1-ARs [Muntz et al., 1994]. This unexpected pattern may be related to the relative selectivity of norepinephrine for β_1-ARs, although epinephrine (which is a potent β_2-AR agonist) is also increased in older individuals, particularly during stress [Fleg et al., 1985]. Another possible mechanism for the selective age-related decline in β_1-AR density may be related to the "functionally hypothyroid" state associated with age (see Section II.A above). It has been shown in neonatal cardiomycoyte cultures that the administration of thyroid hormone regulates β_1-AR levels but has no effect on β_2-ARs [Bahouth, 1991]. If such a mechanism were effective in the mature heart, the development of a "functionally hypothyroid" state would be expected to cause the selective decline in β_1-AR density with age.

Is the loss of β-adrenergic responsiveness likely to be of significance in other tissues? Adrenergic modulation of isolated blood vessels, particularly β-adrenergic–mediated vasodilation, has been found to decrease with age [O'Donnell and Wanstall, 1984; Tsujimoto et al., 1986]. Attempts to determine if this loss is due to changes in receptor density, which are predominantly $\beta2$-AR, however, have produced inconsistent results. It is clear, however, that there is an age-associated deficit in cAMP production in response to β-AR stimulation in most vessels [Deisher et al., 1989]. The nature of β_2-AR receptor coupling in the vasculature is unknown and may differ from that in the heart, relying predominantly on G_s-mediated, cAMP-dependent, postreceptor signal transduction. It is also interesting to note that the decline in β-AR-mediated relaxation in the vasculature can be partially restored by treatment with thyroid hormone [O'Donnell and Wanstall, 1986].

ACKNOWLEDGMENTS

We wish to acknowledge Marlene Rabinovitch (Hospital for Sick Children, Toronto) and Jeff Froehlich (NIA-NIH) for helpful discussions and critical reading of the manuscript, Linda Cheng and Mark Jenkins (NIA-NIH) for the tissue sections shown in Figure 1, and Lydia O'Neill and Andrea Meredith for data shown in Figure 4.

REFERENCES

Abu-Erreish GM, Neely JR, Whitmer JT, Whitman V, Sanadi DR (1977): Fatty acid oxidation by isolated perfused working hearts of aged rats. Am J Physiol 232:E258–E262.

Angel P, Baumann I, Stein B, Delius H, Rahrnsdorf HJ, Herrlich P (1987): 12-O-Tetradecanoyl-phorbol-acetate induction of the human collagenase gene is mediated by an inducible enhancer element located in the 5´ flanking region. Mol Cell Biol 7:2256–2266.

Anversa P, Fitzpatrick D, Argani S, Capasso JM (1991): Myocyte mitotic division in the aging mammalian rat heart. Circ Res 69:1159–1164.

Anversa P, Hiler B, Ricci R, Guideri G, Olivetti G (1986): Myocyte cell loss and myocyte hypertrophy in the aging rat heart. J Am Coll Cardiol 8:1441–1448.

Anversa P, Palackal T, Sonnenblick EH, Olivetti G, Meggs LG, Capasso JM (1990): Myocyte cell loss and myocyte hyperplasia in the hypertrophied aging rat heart. Circ Res 67:871–885.

Arai M, Otsu K, MacLennan DH, Alpert NR, Periasamy M (1991): Effect of thyroid hormone on the expression of mRNA encoding sarcoplasmic reticulum proteins. Circ Res 69:266–276.

Avolio AP, Fa-Quan D, Wei-Qiang L, Yao-Fei L, Zhen-Jong H, Lian-Fen X, O'Rourke MF (1985): Effects of aging on arterial distensibility in populations with high and low prevalence of hypertension: Comparison between urban and rural communities in China. Circulation 71:202–210.

Bader H (1967): Dependence of wall stress in the human thoracic aorta on age and pressure. Circ Res 20:354–361.

Bahouth S (1991): Thyroid hormones transcriptionally regulate the β_1-adrenergic receptor gene in cultured ventricular myocytes. J Biol Chem 266:15863–15869.

Band W, Goedhard WJA, Knoop AA (1967): Effects of aging on dynamic viscoelastic properties of the rat's thoracic aorta. Pfluegers Arch 331:357–364.

Belmin J, Corman B, Merval R, Tedgui A (1993): Age-related changes in endothelial permeability and distribution volume of albumin in rat aorta. Am J Physiol 264:H679–H685.

Benitz WE, Bernfield M (1990): Endothelial cell proteoglycans: Possible mediators of vascular responses to injury. Am J Respir Cell Mol Biol 2:407–408.

Berliner JA, Territo MC, Sevenian A, Ramin S, Kim JA, Bahshad B, Esterson M, Fogelman MA (1990): Minimally modified low density lipoprotein stimulates monocyte endothelial interreactions. J Clin Invest 85:1260–1266.

Besse S, Robert V, Assayag P, Delcayre C, Swynghedauw B (1994): Non synchronous changes in myocardial collagen mRNA and protein during aging. Effect of DOCA-salt hypertension. Am J Physiol 267: H2237–H2244.

Bilato C, Pauly RR, Monticone R, Gorelick-Feldman D, Gluzband YA, Sollot SJ, Ziman B, Crow MT (1995): Intracellular signalling pathways required for rat vascular smooth muscle cell migration: Interaction between basic fibroblast growth factor and platelet-derived growth factor. J Clin Invest 96 (in press).

Bohm M, Dorner H, Htun P, Lenche H, Platt D, Erdmann E (1993): Effects of exercise on myocardial adenylate cyclase and G_{α} expression in senescence. Am J Physiol 264:H805–H814.

Boluyt MO, Long X, Eschenhagen T, Mende U, Schmitz U, Crow MT, Lakatta EG (1995): Isoprenaline-induced alterations in expression of hypertrophy-associated genes in rat heart. Am J Physiol 269 (in press).

Boluyt MO, O'Neill L, Meredith AL, Bing OHL, Brooks WW, Conrad CH, Crow MT, Lakatta EG (1994): Alterations in cardiac gene expression during the transition from stable hypertrophy to heart failure: Marked upregulation of genes encoding extracellular matrix proteins. Circ Res 75:23–32.

Boluyt MO, Younes A, Caffrey JL, O'Neill L, Barron BA, Crow MT, Lakatta EG (1993): Age-associated increase in rat cardiac opioid production. Am J Physiol 265:H212–H218.

Brett J, Schmidt AM, Yan SD, Zou YS, Weidman E, Pinsky D, Nowygrod R, Neeper M, Przysiecki C, Shaw A, Migheli A, Stern D (1993): Survey of the distribution of a newly characterized receptor for advanced glycation end products in tissues. Am J Pathol 143:1699–1712.

Bristow MR, Ginsburg R, Minobe W, Cubiciotti RS, Sageman WS, Lurie K, Billingham ME, Harrison DC, Stinson EB (1992): Decreased catecholamine sensitivity of beta-adrenergic receptor density on failing human hearts. N Engl J Med 307:205–211.

Bristow MR, Ginsburg R, Uman V, Fowler M, Minobe W, Rasmussen R, Zera P, Menlove R, Shah P, Jamieson S, Stinson EB (1986) β_1- and β_2-adrenergic receptor subpopulations in nonfailing and failing human ventricular myocardium: Coupling of both receptor subtypes to muscle contraction and selective β_1-receptor downregulation in heart failure. Circ Res 59:297–309.

Bristow MR, Hershberger RE, Port JD, Minobe W, Rasmussen R (1989): β_1- and β_2-adrenergic receptor mediated adenylate cyclase stimulation in nonfailing and failing human ventricular myocardium. Mol Pharmacol 35:295–303.

Brownlee M, Cerami A, Vlassara H (1988) Advanced dycadon endproducts in tissues and the biochemical basis of diabetic complications. N Engl J Med 318:1315–1320.

Burke EM, Horton WE, Pearson JD, Crow MT, Martin GR (1994): Altered transcriptional regulation of human interstitial collagenase in cultured skin fibroblasts from old donors. Exp Gerontol 29:37–53.

Buttrick P, Malhotra A, Factor S, Greenen D, Leinwand L, Scheuer J (1991) Effect of aging and hypertension on myosin biochemistry and gene expression in the rat heart. Circ Res 68:645–652.

Caffrey JL, Boluyt MO, Younes A, Barron BA, O'Neill L, Crow MT, Lakatta EG (1994): Aging, cardiac proenkephalin mRNA and enkephalin peptides in the Fischer 344 rat. J Moll Cell Cardiol 26:701–711.

Calderone A, Bouvier M, Li K, Juneau C, De Champlain J, Rouleau JL (1991): Dysfunction of the beta- and alpha-adrenergic systems in a model of congestive heart failure. The pacing-overdrive dog. Circ Res 69:332–343.

Calderone A, Takahashi N, Thaik CM, Colucci WS (1994): Atrial natriuretic factor and cyclic guanosine monophosphate modulate cardiac myocyte growth and phenotype. Circulation 90:I317 (abstract).

Capasso JM, Malhotra A, Scheuer J, Sonnenblick EH (1986): Myocardial biochemical, contractile and electrical performance after imposition of hypertension in young and old rats. Circ Res 58:445–460.

Carrier L, Boheler KR, Chassagne C, de la Bastie D, Wisnewsky C, Lakatta EG, Schwartz K (1992): Expression of the sarcomeric actin isogenes in the rat heart with development and senescence. Circ Res 70:999–1005.

Carter WJ, Kelly WF, Fass FH, Lynch ME, Perry CA (1987) Effect of graded doses of tri-iodothyronine on ventricular ATPase activity and isomyosin profile in young and old rats. Biochem J 247:329–334.

Casscells W (1992): Endothelial and smooth muscle cell migration: Critical factors in restenosis. Circulation 86:723–729.

Castellot JJ Jr, Favreau LV, Kamovsky MJ, Rosenberg RD (1982): Inhibition of vascular smooth muscle cell growth by endothelial cell-derived heparin. J Biol Chem 257:11256-11260.

Chamley-Campbell JH, Campbell GR (1981): What controls smooth muscle phenotype? Atherosclerosis 40:347–357.

Chamley-Campbell JH, Campbell GR, Ross R (1979): The smooth muscle cell in culture. Physiol Rev 59:1–61.

Chamley-Campbell JH, Campbell GR, Ross R (1981): Phenotype-dependent response of cultured aortic smooth muscle to serum mitogens. J Cell Biol 89:379–383.

Chevalier B, Mansier P, Callens el Amrani F, Swynghedauw B (1989): β-Adrenergic system is modified in compensatory pressure cardiac overload in rats: physiological and biochemical evidence. J Cardiol Pharmacol 13:412–420.

Chin JH, Hoffman BB (1991): Beta-adrenergic responsiveness in cultured vascular smooth muscle cells and fibroblasts: effect of age. Mech Ageing Dev 57:259–273.

Cliff WJ (1970): The aortic tunica media in aging rats. Exp Mol Pathol 13:172–189.

Cohen JR, Sarfati I, Danna D, Wise L (1992): Smooth muscle cell elastase, atherosclerosis, and abdominal aortic aneurysms. Ann Surg 216:327–332.

Collins T (1993): Biology of disease. Endothelial nuclear factor-κβ and the initiation of the atherosclerotic lesion. Lab Invest 68:499–508.

Collins T, Pober JS, Gimbrone JA Jr, Hammacher A, Betscholtz C, Westermark B, Heldin CH (1987): Cultured human endothelial cells express platelet-derived growth factor A chain. Am J Pathol 126:7–12.

Cox RH (1977): Effect of age on the mechanical properties of rat carotid artery. Am J Physiol 233:H256–H263.

Crow MT, Juhasz O, Pauly RR, Yamamoto H, Schmidt AM, Kinsella J, Monticone R, Stern D (1995) Advanced glycation endproducts stimulate monocyte chemoattractant production by vascular cells through a receptor-mediated mechanism involving an NFκB-dependent oxidant stress pathway. J Clin Invest (in review).

Day ML, Schwartz D, Wiegand RC, Stockman PT, Brunnert SR, Tolunay HE, Currie MG, Standaert DG, Needleman P (1987): Ventricular atriopeptin. Unmasking of messenger RNA and peptide synthesis by hypertrophy or dexamethasone. Hypertension 9:485–491.

Deisher TA, Mankani S, Hoffman BB (1989): Role of cyclic AMP-dependent protein kinase in the diminished beta adrenergic responsiveness of vascular smooth muscle with increasing age. J Pharmacol Exp Ther 249:812–819.

Dubick MA, Hunter GC, Perez-Lizano E, Mar G, Goekas MC (1988): Assessment of the role of pancreatic proteases in human abdominal aortic aneurysms and occlusive disease. Clin Chim Acta 177:1–10.

Edwards JG, Bahl JJ, Flink IL, Cheng SY, Morkin E (1994): Thyroid hormone influences β-myosin heavy chain (β-MHC) expression. Biochem Biophys Res Commun 199:1482–1488.

Effron MB, Bhatnagar GM, Spurgeon HA, Ruano-Arroyo G, Lakatta EG (1987): Changes in myosin isoenzymes, ATPase activity, and contraction duration in rat cardiac muscle with aging can be modulated by thyroxine. Circ Res 60:238–245.

Egashira K, Inou T, Hirooka Y, Kai H, Sugimachi M, Suzuki S, Kuga T, Urabe Y, Takeshita A (1993): Effects of age on endothelium-dependent vasodilation of resistance coronary artery by acetylcholine in humans. Circulation 88:77–81.

ffrench-Constant C, Van De Water L, Dvorak HF, Hynes RO (1989): Reappearance of an embryonic pattern of fibronectin splicing during wound healing in the adult rat. J Cell Biol 109:903–914.

Fisher SA, Buttrick PM, Sukovich D, Periasamy M (1993): Characterization of promoter elements of the rabbit cardiac sarcoplasmic reticulum Ca²⁺-ATPase gene required for expression in cardiac muscle cells. Circ Res 73:622–628.

Fleg JL, Tzankoff SP, Lakatta EG (1985): Age-related augmentation of plasma catecholamines during dynamic exercise in healthy males. J Appl Physiol 59:1033–1039.

Fornieri C, Quaglino D, Mori G (1992): Role of the extracellular matrix in age-related modifications of the rat aorta. Arterioscler Thromb 12:1008–1016.

Fraticelli A, Josephson R, Danziger R, Lakatta EG, Spurgeon HA (1989): Morphological and contractile characteristics of rat cardiac myocytes from maturation to senescence. Am J Physiol 257:H259–H265.

Fredman J, Pauly RR, Stern D, Schmidt AM, Yan SD, Brett J, Monticone R, Crow MT (1994): Advanced glycation endproducts interacting with their receptors activate vascular smooth muscle cells: Induction of chemoattractants for smooth muscle cells and monocytes. Circulation 90:I-291 (abstract).

French JE (1966): Atherosclerosis in relation to the structure and function of the arterial intima, with special reference to the endothelium. Int Rev Exp Pathol 5:253–354.

Fujisaki H, Ito H, Kimoto H, Adachi S, Tanaka M, Marumo F, Hiroe M (1993): Expression of endothelin-1 mRNA and its inhibitory regulation by atrial and brain natriuretic peptides in cardiac fibroblasts. Circulation 88:I–141 (abstract).

Furchgott RF (1983): Role of endothelium in responses of vascular smooth muscle. Circ Res 53:557–573.

Furchgott RF, Zawadzki JV (1980): The obligatory role of endothelial cells in the relaxation of arterial smooth muscle by acetylcholine. Nature 288:373–376.

Gabbiani G, Kocher O, Bloom WS, Vandekerckhove J, Weber K (1984): Actin expression in smooth muscle cells of rat aortic intimal thickening, human atheromatous placque, and cultured rat aortic media. J Clin Invest 73:148–152.

Gardner DG, Deschepper CG, Ganong WF, Hane S, Fiddes J, Baxter DJ, Lewicki J (1986): Extra-atrial expression of the gene for atrial natriuretic factor. Proc Natl Acad Sci USA 83:6697–6701.

Gibbons, GH, Dzau VJ (1994): The emerging concept of vascular remodelling. New Engl J Med 330:1431–1438.

Gottlieb RA, Burleson KO, Kloner RA, Babior BM, Ender RL (1994): Reperfusion injury induces apoptosis in rabbit cardiomyocytes. J Clin Invest 94:1621–1628.

Grainger DJ, Kemp PR, Liu AC, Lawn RM, Metcalfe JL (1994): Activation of transforming growth factor β is inhibited in apolipoprotein(a) transgenic mice. Nature 370:460–462.

Grainger DJ, Kemp PR, Metcalfe JC, Liu AC, Lawn RM, Williams NR, Grace AA, Schofield PM, Chauhan A (1995): The serum concentration of active transforming growth factor-β is severely depressed in advanced atherosclerosis. Nature Med 1:74–79.

Grainger DJ, Kirshenlohr HL, Metcalfe JL, Weissberg PL, Wade DP, Lawn RM (1993): Proliferation of human smooth muscle cells promoted by lipoprotein(a). Science 260:1655–1658.

Grove D, Zak R, Nair KG, Aschenbrenner V (1969): Biochemical correlates of cardiac hypertrophy. IV. Observations on the cellular organization of growth during myocardial hypertrophy in the rat. Circ Res 25:473–485.

Guarnieri T, Filburn CR, Zitnik G, Roth GS, Lakatta EG (1980): Contractile and biochemical correlates of β-adrenergic stimulation of the aged heart. Am J Physiol 238:H501–H508.

Gustafson TA, Markham BE, Bahl JJ, Morkin E (1987): Thyroid hormone regulates expression of a transfected α-myosin heavy chain fusion gene in fetal heart cells. Proc Natl Acad Sci USA 84:3122–3126.

Guyton JR, Lindsay KL, Dao DT (1983): Comparison of aortic intima and inner media in young adult versus aging rats. Am J Pathol 111:234–246.

Hall DA (1961): The possible implications of enzymes of the elastase complex in atherosclerosis. J Atheroscler Res 1:173–183.

Hariri RJ, Alonso DR, Hajjar DP, Coletti D, Weksler ME (1986): Aging and arteriosclerosis. I. Development of myointimal hyperplasia after endothelial injury. J Exp Med 164:1171–1178.

Haudenschild CC, Prescott MF, Chobanian AV (1981): Aortic endothelial and subendothelial cells in experimental hypertension and aging. Hypertension 3(Suppl I): I-148–I-153.

Hinek A, Boyle J, Rabinovitch M (1992): Vascular smooth muscle cell detachment from elastin and migration through elastic laminae is promoted by chondroitin sulfate-induced "shedding" of the 67-kDA cell surface elastin binding protein. Exp Cell Res 203:344–353.

Hinek A, Mecham RP, Keeley FW, Rabinovitch M (1991): Impaired elastin fiber assembly related to reduced 67-kD elastin-binding protein in fetal lamb ductus arteriosus and in cultured aortic smooth muscle cells treated with chondroitin sulfate. J Clin Invest 88:2083–2094.

Hinek A, Rabinovitch M (1994): 67-kD elastin-binding protein is a protective "companion" of extracellular insoluble elastin and intracellular tropoelastin. J Cell Biol 126:563–574.

Hinek A, Rabinovitch M, Keeley FW, Callahan J (1993): The 67-kD elastin/laminin binding protein is related to an alternatively spliced β-galactosidase. J Clin Invest 91:1198–1205.

Hinek A, Rabinovitch M (1993) The ductus arteriosus migratory smooth muscle cell phenotype processes tropoelastin to a 52-kDa product associated with impaired assembly of elastic laminae. J Biol Chem 268:1405–1413.

Hinek A, Wrenn DS, Mecham RP, Barondes SH (1988): The elastin receptor: A galactoside-binding protein. Science 239:1539–1541.

Ito H, Hirata Y, Adachi S, Tanaka M, Tsujino M, Koike A, Nogami A, Marumo F, Hiroe M (1993): Endothelin-l is an autocrine/paracrine factor in the mechanism of andotensin II–induced hypertrophy in cultured rat cardiomyocytes. J Clin Invest 92:398–403.

Ito H, Hirata Y, Hiroe M, Tsujino M, Adachi S, Takamoto T, Nitta M, Taniguchi K, Marumo F (1991): Endothelin-l induces hypertrophy with enhanced expression of muscle specific genes in cultured neonatal rat cardiomyocytes. Circ Res 69:209–215.

Ito H, Hiroe M, Hirata Y, Fujisaki H, Adachi S, Akimoto H, Ohta Y, Marumo F (1994): Endothelin ET_A receptor antagonist blocks cardiac hypertrophy provoked by hemodynamic overload. Circulation 89:2198–2203.

Izumo S, Nadal-Ginard B, Mahdavi V (1986): All members of the MHC family respond to thyroid hormone in a highly tissue-specific manner. Science 231:597–600.

Izumo S, Nadal-Ginard B, Mahdavi V (1988): Proto-oncogene induction and reprogramming of cardiac gene expression produced by pressure overload. Proc Natl Acad Sci USA 85:339–343.

Jenkins GM, Crow MT, Bilato C, Ryu W-S, Li Z, Stetler-Stevenson WG, Nater C, Froehlich J, Lakatta EG, Cheng L (1994): The 72kD type IV collagenase (matrix metalloproteinase-2) is preferentially expressed in the neointima of the rat carotid artery following balloon injury. FASEB J 8(4):Pt. I, A51.

Kinnunen P, Vulteenaho O, Usima P, Ruskoaho H (1992): Passive mechanical stretch releases atrial natriuretic peptide from rat ventricular myocardium. Circ Res 70:1244–1253.

Kinsella MG, Wight TN (1986): Modulation of sulfated proteoglycan synthesis by bovine aortic endothelial cells during migration. J Cell Biol 102:679–687.

Kirstein M, Brett J, Radoff S, Ogawa S, Stern D, Vlassara H (1990): Advanced protein glycosylation induces transendothelial human monocyte chemotaxis and secretion of platelet-derived growth factor: role in vascular disease of diabetes and aging. Proc Natl Acad Sci USA 87:9010–9014.

Kiuchi K, Shannon RP, Komamura K, Cohen DJ, Bianchi C, Homcy CJ, Vatner SF, Vatner DE (1993): Myocardial β-adrenergic receptor function during the development of pacing induced heart failure. J Clin Invest 91:907–914.

Kobayashi J, Wigle D, Childs T, Zhu L, Keeley FW, Rabinovitch M (1994): Serum-induced vascular smooth muscle cell elastolytic activity through tyrosine kinase intracellular signalling. J Cell Physiol 160:121–131.

Kocher O, Gabbiani G (1986): Expression of actin mRNAs in rat aortic smooth muscle cells during development, experimental intimal thickening, and culture. Differentiation 32:245–251.

Koenig RJ, Lazar MA, Hodin RA, Brent GA, Larsen PR, Chin WW, Moore DD (1989): Inhibition of thyroid hormone action by a non-hormone binding c-erbA protein generated by alternative mRNA splicing. Nature 337:659–661.

Koenig RJ, Warne RL, Brent GA, Harney JW, Larsen PR, Moore DD (1988): Isolation of a cDNA clone encoding a biologically active thyroid hormone receptor. Proc Natl Acad Sci USA 85:5031–5035.

Kumazaki T, Fujii T, Kobayashi M, Mitsui Y (1994): Aging-and growth-dependent modulation of endothelin-1´-gene expression in human vascular endothelial cells. Exp Cell Res 211:6–11.

Kumazaki T, Kobayashi M, Mitsui Y (1993): Enhanced expression of fibronectin during in vivo cellular aging of human vascular endothelial cells and skin fibroblasts. Exp Cell Res 205:396–402.

Kumazaki T, Robetorye RS, Robetorye SC, Smith JR (1991): Fibronectin expression increases during in vitro cellular senescence: Correlation with increased cell area. Exp Cell Res 195:13–19.

Kundra V, Escobedo JA, Kasluskas A, Kim HK, Rhee SG, Williams LT, Zetter BR (1994): Regulation of chemotaxis by the platelet-derived growth factor-β. Nature 367:474–476.

Lakatta EG (1993): Cardiovascular regulatory mechanisms in advanced age. Physiol Rev 73:413–467.

Leng X, Blanco J, Tsai SY, Ozato K, O'Malley BW, Tsai M-J (1994): Mechanisms for synergistic activation of thyroid hormone receptor and retinoid X receptor on different response elements. J Biol Chem 269:31436–31442.

Lewicki J Greenberg B, Yamanaka M, Vlasuk G, Brewer M, Gardner D, Baxter J, Johnson LK, Fiddes JC (1986): Cloning sequence analysis and processing of the rat and human atrial natriuretic peptide precursors. Fed Proc 45:2086–2090.

Li Z, Miyashita Y, Cheng L, Lakatta E, Froehlich J (1995): Remodeling of the rat aortic wall during aging. FASEB J 9:A606.

Lindner V, Lappi DA, Baird A, Majack RA, Reidy MA (1991): Role of basic fibroblast growth factor in vascular lesion formation. Circ Res 68:106–113.

Lompre A-M, Lambert F, Lakatta EG, Schwartz K (1991): Expression of sarcoplasmic reticulum Ca^{2+}-ATPase and calsequestrin genes in rat heart during ontogenic development and aging. Circ Res 69:1380–1388.

Luscher TF, Tanner FC, Dohi Y (1992): Age, hypertension and hypercholesterolemia alter endothelium-dependent vascular regulation. Phamacol Toxicol 70:S32–S39.

MacDonald E, Lee WK, Hepburn S, Bell J, Scott PJW, Dominiczak MH (1992): Advanced glycosylation end products in the mesenteric artery. Clin Chem 38:530–533.

Maciel LMZ, Polikar R, Rohrer D, Popovich BK, Dillman WH (1990): Age-induced decreases in the messenger RNA coding for the sarcoplasmic reticulum Ca^{2+}-ATPase of the rat heart. Circ Res 67:230–243.

Mader SL (1992): Influence of animal age on the beta-adrenergic system in cultured rat aortic and mesenteric artery smooth muscle cells. J Gerontol 47:B32–B36.

Majesky MW, Gianchelli C, Reidy MA, Schwartz SM (1992): Rat carotid neointimal smooth muscle cells reexpress a developmentally regulated mRNA phenotype during repair of arterial injury. Circ Res 71:759–768.

Mamuya WS, Brecher P (1992): Fibronectin expression in the normal and hypertrophic rat heart. J Clin Invest 89:392–401.

Mamuya WS, Chobanian A, Brecher P (1992): Age-related changes in fibronectin expression in spontaneously hypertensive, Wistar-Kyoto, and Wistar rat hearts. Circ Res 71:1341–1350.

Mangelsdorf DJ, Borgmeyer U, Heyman RA, Zhou IY, Ong ES, Oro AE, Kakizuka A, Evans RM (1992): Characterization of three RXR genes that mediate the action of 9-cis retinoic acid. Genes Dev 6:329–344.

Martin M, Einabout R, Lafuma C, Crechet F, Remy J (1990): Fibronectin and collagen gene expression during in vitro aging of pig skin fibroblasts. Exp Cell Res 191:8–13.

Matrsian LM (1992): The matrix-degrading metalloproteinases. BioEssays 14:455–465.

McCaffrey TA, Falcon DJ (1993): Evidence for an age-related dysfunction in the antiproliferative response to transforming growth factor-β in vascular smooth muscle cells. Mol Biol Cell 4:315–322.

McCaffrey TA, Nichlson AC, Szabo PE, Weksler ME, Weksler BB (1988): Aging and atherosclerosis: the increased proliferation of arterial smooth muscle cells isolated from old rats is associated with increased platelet derived growth factor-like activity. J Exp Med 167:163–174.

McMahon MP, Faris B, Wolfe BL, Brown KE, Pratt CA, Toselli P, Franzblau C (1985): Aging effects on the elastin composition in the extracellular matrix of cultured rat aortic smooth muscle cells. In Vitro Cell Dev Biol 21:674–680.

Mercadier JJ, Samuel JL, Michel JB, Zongazo MA, de la Bastie D, Lompre AM, Wisnewski C, Rappaport L, Levy B, Schwartz K (1989): Atrial natriuretic factor gene expression in rat ventricle during experimental hypertension. Am J Physiol 257:H979–H987.

Michel JB, Heudes D, Michel O, Poitevin P, Phillipp M, Scalbert E, Corman B, Levy BI (1994): Effect of chronic ANG I-converting enzyme inhibition on aging processes. II. Large arteries. Am J Physiol 267:R124–R135.

Monnier VM (1990): Nonenzymatic glycosylation, the Maillard reaction, and the aging process. J Gerontol 45:B105–B111.

Morkin E, Ashford TP (1968): Myocardial DNA synthesis in experimental cardiac hypertrophy. Am J Physiol 215:1409–1413.

Mosse PRL, Campbell GR, Wang ZL, Campbell JH (1985): Smooth muscle phenotypic expression in human carotid arteries, I: Comparison of cells from diffuse intimal thickenings adjacent to atheromatous placques with those of the media. Lab Invest 53:556–562.

Muntz KH, Zhao M, Miller JC (1994): Downregulation of myocardial β-adrenergic receptors. Receptor subtype selectivity. Circ Res 74:369–375.

Neeper M, Schmidt A-M, Brett J, Yan S-D, Wang F, Pan Y-C, Elliston K, Stern D, Shaw A (1992): Cloning and expression of RAGE: A cell surface receptor for advanced glycosylation endproducts of proteins. J Biol Chem 267:14998–15004.

O'Donnell SR, Wanstall JC (1984): Beta-1 and beta-2 adrenoceptor mediated responses in preparations of pulmonary artery and aorta from young and aged rats. J Pharmacol Exp Ther 228:733–738.

O'Donnell SR, Wanstall JC (1986): Thyroxine treatment of aged or young rats demonstrates that vascular responses mediated by β-adrenoceptor subtypes can be differentially regulated. Br J Pharmacol 88:41–49.

Oho S, Rabinovitch M (1994): Post-cardiac transplant arteropathy in piglets is associated with fragmentation of elastin and increased activity of a serine elastase. Am J Pathol 145:202–210.

Olivetti G, Melissari M, Balbi T, Quaini F, Sonnenblick EH, Anversa P (1994): Myocyte nuclear and possible cellular hyperplasia contribute to ventricular remodeling in the hypertrophic senescent heart in humans. J Am Coll Cardiol 24:140–149.

Olivetti G, Melissari M, Capasso JM, Anversa P (1991): Cardiomyopathy of the aging human heart. Myocyte loss and reactive cellular hypertrophy. Circ Res 68:1560–1568.

O'Neill L, Holbrook NJ, Fargnoli J, Lakatta EG (1991): Progressive changes from young adult age to senescence in mRNA for rat cardiac myosin heavy chain genes. Cardioscience 2:1–5.

Parker TG, Chow KL, Schwartz RJ, Schneider MD (1990): Differential regulation of skeletal α-actin transcription in cardiac muscle by two fibroblast growth factors. Proc Natl Acad Sci USA 87:7066–7070.

Pauly RR, Bilato C, Sollot SJ, Monticone R, Kelly PT, Lakatta EG, Crow MT (1995): The role of calcium/calmodulin-dependent protein kinase II in the regulation of vascular smooth muscle cell migration. Circulation (in press).

Pauly RR, Passaniti A, Bilato C, Monticone R, Cheng L, Papadopoulos N, Gluzband YA, Smith L, Weinstein C, Lakatta EG, Crow MT (1994): Migration of cultured vascular smooth muscle cells through a basement membrane barrier requires type IV collagenase activity and is inhibited by cellular differentiation. Circ Res 75:41–54.

Quaglino D, Fornieri C, Nannay LB, Davidson JM (1993): Extracellular matrix modifications in rat tissues of different ages. Matrix 13:481–490.

Quaglino D, Kennedy R, Fornieri C, Nanney LB, Pasquali Ronchetti I, Davidson JM (1989): Matrix gene expression during the aging process revealed by in situ hybridization. J Histochem Cytochem 37:933–941.

Radomski MW, Palmer RMJ, Moncada S (1987): The anti-aggregating properties of vascular endothelium: Interactions between prostacyclin and nitric oxide. Br J Pharmacol 92:639–646.

Raghow R (1994): The role of extracellular matrix in postinflammatory wound healing and fibrosis. FASEB J 8:823–831.

Reckelhoff JF, Morris M, Bayliss C (1992): Basal and stimulated plasma atrial natriuretic peptide (ANP) concentrations and cardiac ANP contents in old and young rats. Mech Ageing Dev 63:177–181.

Reiser KM (1991): Nonenzymatic glycation of collagen in aging and diabetes. Proc Soc Exp Biol. Med. 196:17–29.

Reiser KM, Hennessy SM, Last JA (1987): Analysis of age-associated changes in collagen crosslinking in the skin and lung in monkeys and rats. Biochim Biophys Acta 926:339–348.

Riabowol K, Schiff J, Gilman MZ (1992): Transcription factor AP-1 activity is required for initiation of DNA synthesis and is lost during cellular aging. Proc Natl Acad Sci USA 89:157–161.

Robert L, Jacobs MP, Frabces C, Godeau G, Hornebeck W (1984): Interaction between elastin and elastases and its role in the aging of the arterial wall, skin and other connective tissues. Mech Ageing Dev 28:155–166.

Rodeheffer RJ, Gerstenblith G, Becker LC, Fleg JL, Weisfeldt ML, Lakatta EG (1984): Exercise cardiac output is maintained with advancing age in healthy human subjects: Cardiac dilatation and increased stroke volume compensate for a diminished heart rate. Circulation 69:203–213.

Rohrer DK, Dillman WH (1988): Thyroid hormone markedly increases the mRNA coding for sarcoplasmic reticulum Ca^{2+}-ATPase in the rat heart. J Biol Chem 263:6941–6944.

Rohrer DK, Hartong R, Dillman WH (1991): Influence of thyroid hormone and retinoic acid on slow sarcoplasmic reticulum Ca^{2+}-ATPase and myosin heavy chain α gene expression in cardiac myocytes. J Biol Chem 266:8638–8646.

Ross R (1993): The pathogenesis of atherosclerosis: A perspective for the 1990s. Nature 362:801–809.

Sadoshima J, Jahn L, Takahashi T, Kuhlik TJ, Izumo S (1992): Molecular characterization of the stretch-induced adaptation of cultured cardiac cells. J Biol Chem 267:10551–10560.

Sakai M, Danziger RS, Staddon JM, Lakatta EG, Hansford RG (1989): Decrease with senescence in the norepinephrine-induced phosphorylation of myofilament proteins in isolated rat cardiac myocytes. J Mol Cell Cardiol 21:1327–1336.

Scarpace PJ, Abrass IB (1986): Beta-adrenergic agonist mediated desensitization in senescent rats. Mech Ageing Dev 35:255–264.

Schmidt AM, Zhang JH, Crandall J, Cao R, Yan SD, Brett J, Stern D (1994): Interaction of advanced glycation endproducts with their endothelial cell receptor leads to enhanced expression of VCAM-1: a mechanism of augmented monocyte-vessel wall interactions in diabetes. FASEB J 8:3841 (abstract A662).

Schneider MD, Parker TG (1990): Cardiac myocytes as targets for the action of peptide growth factors. Circulation 81:1443–1456.

Schuyler GT, Yarbrough LR (1990): comparison of myosin and creatine kinase isoforms in left ventricles of young and senescent Fischer 344 rats after treatment with triiodothyronine. Mech Ageing Dev 56:39–48.

Schwartz SM, Heimark RL, Majesky MW (1990): Developmental mechanisms underlying pathology of arteries. Physiol Rev 70:1177–1209.

Schwartz SM, Reidy MA, Clowes AW (1985): Kinetics of atherosclerosis: A stem cell model. Ann N Y Acad Sci 454:292–304.

Seidman CE, Bloch KD, Zifsin JB, Smith JA, Haber E, Homcy C, Duby AD, Choi E, Graham RM, Seidman JG (1885): Molecular studies of the atrial natriuretic factor gene. Hypertension 7:31–34.

Senior RM, Griffin GL, Fliszar CJ, Shapiro SD, Goldberg GI, Welgus HG (1991): Human 92- and 72-kilodalton type IV collagenases are elastases. J Biol Chem 266:7870–7875.

Sephel GC, Davidson JM (1986): Elastin production in human skin fibroblast cultures and its decline with age. J Invest Dermatol 86:279–285.

Shanahan CM, Weissberg PL, Metcalfe JC (1993): Isolation of gene markers of differentiated and proliferating vascular smooth muscle cells. Circ Res 73:193–204.

Shapiro SD, Griffin GL, Gilbert DJ, Jenkins NA, Copeland NG, Welgus HG, Senior RM, Ley TJ (1992): Molecular cloning, chromosomal localization, and bacterial expression of a murine macrophage metalloelastase. J Biol Chem 267:4664–4671.

Spina M, Garbisa S, Hinnie J, Hunter JC, Serafini-Fracassini A (1983): Age-related changes in composition and mechanical properties of the tunica media of the upper thoracic human aorta. Arteriosclerosis 3:67–76.

Stein O, Eisenberg S, Stein Y (1969): Aging of aortic smooth muscle cells in rats and rabbits. Lab Invest 21:386–397.

Steinberg D, PArthasarathy S, Carew TA, Khoo JC, Witzum JL (1989): Beyond cholesterol: Modifica-

tions of low-density lipoprotein that increase its artherogenicity. N Engl J Med 320:915–924.

Stemerman MB, Weinstein R, Rowe JW, Maciag T, Fuhro R, Gardner R (1982): Vascular smooth muscle cell growth kinetics in vivo in aged rats. Proc Natl Acad Sci USA 79:3863–3866.

Stout RW (1987): Ageing and atherosclerosis. Age Ageing 16:65–72.

Takahashi T, Schunkert H, Isoyama S, Wei JY, Nadal-Ginard B, Grossman W, Izumo S (1992): Age-related differences in the expression of proto-oncogene and contractile protein genes in response to pressure overload in the rat myocardium. J Clin Invest 89:939–946.

Tanaka M, Ito H, Adachi S, Akimoto H, Nishiwaka T, Kasajima T, Marumo F, Hiroe M (1994): Hypoxia induces apoptosis with enhanced expression of Fas antigen messenger RNA in cultured neonatal rat cardiac myocytes. Circ Res 75:426–433.

Thibault G, Nemer M, Crouin J, Levigne JP, Ding J, Charbonneau C, Garcia R, Genest J, Jasmin G, Sole M, Cantin M (1989): Ventricles as a major site of atrial natriuretic factor synthesis and release in cardiomyopathic hamsters with heart failure. Circ Res 65:71–82.

Thomas WA, Lee KT, Kim DN (1985): Cell population kinetics in atherogenesis. Cell births and losses in intimal cell mass-derived lesions in the abdominal aorta of swine. Ann NY Acad Sci 454:305–315.

Todorovich-Hunter L, Johnson DJ, Ranger P, Keeley FW, Rabinovitch M (1992): Increased pulmonary artery elastolytic activity and monocrotaline-induced progressive hypertensive pulmonary vascular disease in adult rats compared to infant rats with non-progressive disease. Am Rev Respir Dis 146:213–223.

Tokunaga O, Yamada T, Fan J, Watanabe T (1991): Age-related decline in prostacyclin synthesis by human aortic endothelial cells. Am J Pathol 138:941–949.

Tsujimoto G, Lee C-H, Hoffman BB (1986): Age-related decrease in beta adrenergic receptor-mediated vascular smooth muscle relaxation. J Pharmacol Exp Ther 239:411–415.

Ventura C, Spurgeon H, Lakatta EG, Guarnieri C, Capogrossi M (1992): Kappa and delta opioid receptor stimulation affects cardiac myocyte function and Ca^{2+} release from an intracellular pool in myocytes and neurons. Circ Res 70:66–81.

Villareal FJ, Dillman WH (1992): Cardiac hypertrophy-induced changes in mRNA levels for $TGF\beta_1$, fibronectin, and collagen. Am J Physiol 31:H1861–H1866.

Vlassara H, Fuh H, Cybulsky M (1994): Adhesion molecules and atheroma in rabbits injected with advanced glycosylation products. Circulation 90:I–84 (abstract).

Weber KT, Brilla CG (1991): Pathological hypertrophy and

cardiac interstitium. Fibrosis and renin–angio-tensin–aldosterone system. Circulation 83:1849–1865.

Weinberger CC, Thompson CC, Ong ES, Lebo R, Gruol DJ (1986): The c-erbA gene encodes a thyroid hormone receptor. Nature 324:641–646.

Weisfeldt ML, Wright JR, Shreiner DP, Lakatta E, Shock NW (1971): Coronary flow and oxygen extraction in the perfused heart of senescent male rats. J Appl Physiol 30:44–49.

White E (1993): Death-defying acts: A meeting review on apoptosis. Genes Dev 7:2277–2284.

White M, Roden R, Minobe W, Farid Khan M, Larrabee P, Wollmering M, Port D, Anderson F, Campbell D, Feldman AM, Bristow MR (1994): Age-related changes in β-adrenergic neuroeffector systems in the human heart. Circulation 90:1225–1238.

Wight TN (1989): Cell biology of arterial proteoglycans. Arteriosclerosis 9:1–20.

Wolinsky H (1970): Response of the rat aortic media to hypertension: Morphological and chemical studies. Circ Res 26:507–522.

Xiao R-P, Capogrossi MC, Spurgeon HA, Lakatta EG (1991): Stimulation of delta opioid receptors in single heart cells blocks β-adrenergic receptor mediated increase in calcium and contraction. J Mol Cell Cardiol 23 (Suppl III):S83 (abstract).

Xiao R-P, Hohl C, Altschuld R, Jones L, Livingston B, Ziman B, Tantini B, Lakatta EG (1994): β₂-adrenergic receptor-stimulated increase in cAMP in rat heart cells is not coupled to changes in Ca^{2+} dynamics, contractility, or phospholamban phosphorylation. J Biol Chem 269:19151–19156.

Xiao R-P, Ji X, Lakatta EG (1995): Functional coupling of the β2-adrenoceptor to a pertussis toxin sensitive G protein in cardiac myocytes. Journal (in press).

Xiao R-P, Lakatta EG (1993): β₁-adrenoceptor stimulation and β₂-adrenoceptor stimulation differ in their effects on contraction, cytosolic Ca^{2+}, and Ca^{2+} current in single rat ventricular cells. Circ Res 73:286–300.

Xiao R-P, Spurgeon HA, O'Connor F, Lakatta EG (1994): Age-associated changes in β-adrenergic modulation of rat cardiac excitation-contraction coupling. J Clin Invest 94:2051–2059.

Yan SD, Schmidt AM, Anderson GM, Zhang J, Brett J, Zou YS, Pinsky D, Stern D (1994): Enhanced cellular oxidant stress by the interaction of advanced glycation end products with their receptors/binding proteins. J Biol Chem 269:9889–9897.

Yasue H, Matsuyama K, Okumura K, Morikami Y, Ogawa H (1990): Responses of angiographically normal coronary arteries to intracoronary injection of acetylcholine by age and segment. Circulation 81:482–490.

Ye C, Rabinovitch M (1991): Inhibition of elastolysis by SC37698 (Searle) reduced development and progression of monocrotaline pulmonary hypertension. Am J Physiol 261: H1255–H1267.

Yin FCP, Spurgeon HA, Greene NL, Lakatta EG, Weisfeldt ML (1979): Age-associated decrease in heart rate response to isoproterenol in dogs. Mech Ageing Dev 10:17–25.

Yin FCP, Spurgeon HA, Weisfeldt ML, Lakatta EG (1980): Mechanical properties of myocardium from hypertrophied rat hearts. A comparison between hypertrophy induced by senescence and by aortic banding. Circ Res 46:292–300.

Yla-Herttula S, Sumuvuori H, Karkola K, Mottonen M, Nikkari T (1986): Glycosaminoglycans in normal and atherosclerotic human coronary arteries. Lab. Invest. 54:402–407.

Younes A, Boluyt MO, O'Neill L, Crow MT, Lakatta EG (1995): Age-associated increase in ANP mRNA and peptide levels in rat left ventricle. Am J Physiol 269 (in press).

Zhang X, Pfahl M (1993): Regulation of retinoid and thyroid hormone action through homodimeric and heterodimeric receptors. Trends Endocrinol Metab 4:156–162.

Zhu L, Wigle D, Hinek A, Kobayashi J. Ye C, Zuker M, Dodod H, Keeley FW, Rabinovitch M (1994): The endogenous vascular elastase that governs development and progression of monocrotaline-induced pulmonary hypertension in rats is a novel enzyme related to the serine proteinase adipsin. J Clin Invest 94:1163–1171.

ABOUT THE AUTHORS

MICHAEL T. CROW received his doctoral degree in Physiology/Biophysics from Harvard University and did postdoctoral training in molecular and developmental biology at Stanford University. He is currently Head of the Vascular Biology Unit in the Laboratory of Cardiovascular Science at the National Institutes of Health's Gerontology Research Center in Baltimore. Dr. Crow's area of expertise is in signal transduction mechanisms related to vascular cell migration and differentiation.

MARVIN O. BOLUYT earned his doctoral degree in kinesiology from the University of Michigan in 1990. He then joined the Laboratory of Cardiovascular Science at the National Institutes of Health's Gerontology Research Center in Baltimore as a postdoctoral fellow. He is currently a Senior Staff Fellow at NIH.

Dr. Boluyt's areas of interest/expertise include signal transduction mechanisms involved in cardiac myocyte hypertrophy, age-associated alterations in cardiac gene expression, and mechanisms regulating the transition from stable hypertrophy to heart failure.

EDWARD G. LAKATTA received his M.D. degree from Georgetown University School of Medicine, his post-doctoral training in medicine at Strong Memorial Hospital, University of Rochester, his cardiology fellowship at Georgetown University Hospital and Johns Hopkins School of Medicine, and physiology training at University College, London, and the National Institutes of Health, Bethesda. Dr. Lakatta is presently the Director of the Laboratory of Cardiovascular Science at the National Institute on Aging.

Cellular Aging and Cell Death: 109–121
© 1996 Wiley-Liss, Inc.

Molecular Genetics of *In Vitro* Cellular Senescence

Cynthia A. Afshari and J. Carl Barrett

I. INTRODUCTION

Senescence is a term that describes the physiological process that an organism or a cell undergoes when it reaches the end of its proliferative life span. The aging, or senescence, of whole organisms is a complex process involving biological changes in many organ systems, the eventual decline of which ultimately leads to death of the organism. Since the study of aging in the whole organism is prohibitively complex, models of *in vitro* aging have been established as tools to investigate the genetic and physiological changes that occur at the cellular level as the aging process develops. *In vitro* models of cellular senescence have been developed for a variety of cell types, including fibroblasts [Hayflick and Moorhead, 1961], endothelial cells [Mueller et al., 1980], keratinocytes [Rheinwald and Green, 1975], and lymphocytes [Tice et al., 1979]. In culture, normal cells undergo a finite number of population doublings after which both DNA and cellular division cease, but metabolic activity and cellular viability may be maintained for an extended period of time. The loss of proliferative potential in these *in vitro* systems has been termed *cellular senescence*.

Extensive studies of both genetic and physiological changes involved in cellular aging have been conducted in fibroblasts derived from a variety of species, including human, chicken, mouse, and hamster. Early studies by Hayflick and Moorhead [1961] showed that the finite life span in cultures of human fibroblasts was not related to artifacts of tissue culture conditions. Furthermore, when normal cells were transplanted serially *in vivo*, the cells exhibited a similar life span to those that were maintained in cell culture, further indicating that *in vitro* senescence is not due to the artificial cell culture environment [Daniel et al., 1968]. Thus it seems clear that it is the number of cell doublings a cell population has undergone that correlates with the population's proliferative potential, not the chronological age of the culture [Hayflick, 1976]. This is further supported by the observation that human embryo fibroblasts can be maintained for 50–60 population doublings before they achieve senescence, regardless of whether the culture remains static for a period of time, such as by cryopreservation or contact inhibition [Dell'Orco et al., 1974; Roberts and Smith, 1980].

There is evidence that *in vitro* cellular senescence may directly correlate with aging *in vivo*. An inverse relationship has been observed between the age of a donor and the proliferative potential of their fibroblasts *in vitro* [Schneider and Mitsui, 1976]. Fibroblast cultures established from young donors grow more rapidly and proliferate for many more population doublings than fibroblasts from older organisms [Hayflick, 1976; Dell'Orco et al., 1974].

In addition to the inverse correlation that exists between donor age and proliferative potential of that donor's cells in culture, other evidence suggests a relationship between aging *in vitro* and aging of the organism. There is a correlation between the proliferative potential of cells from a certain species and the average life span of that species. For example, while human fibroblasts can undergo 50–60 population doublings in culture with a maximum organismal life span of 120 years, rodent fibroblasts can only undergo 20–40 population doublings, correlating with a shorter maximum organismal life span of 2–5 years [Goldstein, 1974].

Studies of premature aging disorders such as Hutchinson-Gilford (progeria) and Werner's syndromes also show a relationship between aging of the organism and *in vitro* cellular senescence. Individuals with progeria begin to age prematurely very early in life and by the end of their first decade show phenotypic signs of aging that are characteristic of the seventh decade of life. Werner's syndrome patients begin to show signs of accelerated aging typically in the third or fourth decade of life. Fibroblasts derived from individuals afflicted with these illnesses have a shortened life span in culture compared with age-matched normal controls [Goldstein, 1969; Goldstein et al., 1989; Brown et al., 1985].

While the combination of these observations correlating *in vitro* life span with *in vivo* life span, do not prove that *in vitro* models of senescence represent aging at the organismal level, they do provide strong support for the use of these simple cellular models to investigate the basic mechanisms of aging.

II. DEFINITION OF CELLULAR SENESCENCE

Senescent fibroblasts, in culture, display an enlarged, flattened morphology that is accompanied by the failure of the cells to replicate their DNA in response to normal growth stimuli [Cristofalo and Pignolo, 1993]. As a culture of normal human fibroblasts reaches the end of its life span, significant cell death occurs along with the emergence of a stable, nonproliferative population that is viable, metabolically active, and, by definition, senescent [Matsumura et al., 1979]. Senescent cells may be maintained in culture for a prolonged period of time [Hayflick, 1976].

Normal, untransformed cells are unique in their ability to achieve a permanently arrested, postmitotic state. Many cell lines that have been established from tumors are immortal, i.e., capable of proliferation in culture indefinitely. The progression of normal human cells to acquisition of immortality occurs very rarely. In the study of human tumor models it has been shown that multiple mutagenic events are required for this progression [Vojta and Barrett, 1995]. However, it has been shown that treatment of human fibroblast cultures with the DNA tumor virus SV40 T antigen can lead to a period of crisis that is followed by growth of a small number of cells that escape from cellular senescence [Tsuji et al., 1983; Shay and Wright, 1989; Shay et al., 1993]. While some cells become immortal, most respond by entering into a round of DNA synthesis that is followed by a permanent growth arrest [Ide et al., 1983]. A few other agents, such as cytomegalovirus [Ide et al., 1984], phosphatase inhibitors [Afshari and Barrett, 1994], and transfection of c-*fos* [Phillips et al., 1992], can induce DNA synthesis in senescent cells, but all fail to reverse permanent division arrest, showing that some, but not all, arrest signals may be overcome in these cells.

Analysis of the DNA content of senescent cells shows that the majority of these cells have a near diploid DNA content resembling cells that are arrested in G_1 [Yanishevsky et al., 1974]. Young cells that are arrested in G_0 by nutrient deprivation can be induced to enter the cell cycle by the addition of mitogens. While senescent cells require mitogens for continued viability, they do not progress through the cell cycle as a response. Recent analyses of senescent cells have focused on the signaling pathways that remain intact, versus those that appear to be blocked, in response to mitogens. The information learned from these studies provides insights into the events that are ultimately required for irreversible growth inhibition that may lead to potential cancer therapy strategies.

One approach investigators have used to understand the difference in signaling pathways between young and senescent cells has been to compare gene expression in the two cell states. The expression of some genes, such as histone, fibronectin [Seshadri and Campisi, 1989], and insulin-like growth factor–binding protein 3 [Goldstein et al., 1991] are higher in senescent cells than in their younger cell counterparts. In addition, the ornithine decarboxylase [Seshadri and Campisi, 1989], c-*myc*, H-*ras*, and thymidine kinase genes [Rittling et al., 1986] are still mitogenic responsive in stimulated senescent cells, while the c-*fos* gene is down-regulated [Seshadri and Campisi, 1990]. The loss of c-*fos* gene expression in senescent cells in turn leads to loss of *fos–jun* heterodimers, a complex of the AP-1 transcription factor family. AP-1 activity is still present in senescent cells through the formation of *jun* homodimers [Riabowol et al., 1992]. However, the consequences of these changes on the expression of genes regulated by AP-1 is unclear. In addition to changes in AP-1, CREBP and CTF transcription factor complexes arc also reduced in senescent cells. In contrast, TFIID, CREBP, NFkappaB, and Sp-1 complexes are unchanged [Dimri and Campisi, 1994].

In addition to the study of gene expression and transcription regulatory events in senescent cells, the discovery of mechanisms of cell cycle progression has produced another avenue of investigation for putative mediators of senescence (Table I, Fig. 1). These studies began with focus on the retinoblastoma (Rb) protein, a tumor suppressor gene product . The retinoblastoma gene is lost in a variety of human tumors [Friend et al., 1986; Fung et al., 1987], and reintroduction of the gene into Rb-deficient tumor cell lines causes suppression of tumorigenicity [Huang et al., 1988; Sumegi et al., 1990]. Observations on the interactions of the Rb protein with other cellular proteins has led to a strong working hypothesis for the function of this gene product. It is believed that Rb functions primarily in the G_1 phase of the cell cycle to negatively regulate growth. The Rb protein is unphosphorylated in G_1, and, as cells progress through the G_1/S transition, the Rb protein becomes phosphorylated until the cells exit mitosis. It is hypothesized that phosphorylation inactivates the negative functions of Rb [Chen et al., 1989; Cooper and Whyte, 1989; Mihara et al., 1989; DeCaprio et al., 1989]. Consistent with this hypothesis, the unphosphorylated form of Rb binds the products of the transforming DNA tumor viruses, SV40 T antigen, adenovirus E1A, and papilloma virus E7 [DeCaprio et al., 1988; Ludlow et al., 1989; Dyson et al., 1989; Ewen et al., 1989], which presumably inactivate the Rb protein.

Studies of the Rb protein in senescent cells show that Rb is present in the unphosphorylated

TABLE I. Regulation of Cell Cycle Components in Stimulated Senescent and Quiescent Young Cells

Cell cycle regulatory components	Quiescent	Senescent	References
c-*fos* mRNA	Increases early G1	Not expressed	Seshadri and Campisi [1990]
c-*myc* mRNA	Increases early G1	Expressed	Rittling et al. [1986]
cdk2 mRNA	Expressed	Not expressed	Stein et al. [1991]
Cyclin A mRNA	Cell cycle regulated	Not expressed	Afshari et al. [1993]
Cyclin C, D, E mRNA	Cell cycle regulated	Expressed	Afshari et al. [1993]
E2F1 mRNA	Increases through G1	Present at low level	Afshari et al. [unpublished data]
p21 mRNA	Expressed	Elevated 10-fold	Noda et al. [1994]
Rb protein	Phosphorylated at G1/S	Unphosphorylated	Futreal and Barrett [1991], Stein et al. [1990]
Cyclin A, B proteins	Cell cycle regulated	Down-regulated	Stein et al. [1991]
MAP kinase	Expressed	Expressed	Afshari et al. [1993]
cdk2 protein	Expressed	Expressed	Dulic et al. [1993]
cdc2 protein	Expressed	Down-regulated	Richter et al. [1991]
Cyclins D, E proteins	Expressed	Expressed	Dulic et al. [1993]
AP-1	Expressed	Expressed	Riabowol et al. [1992]
p53	Expressed	Expressed	Afshari et al. [1993]

112   Afshari and Barrett

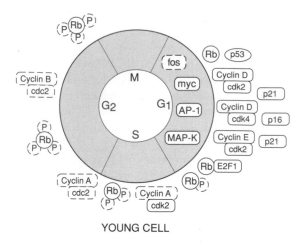 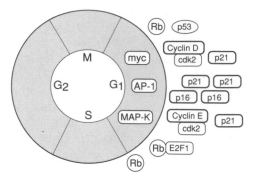

YOUNG CELL   SENESCENT CELL

Fig. 1. *Regulation of cell cycle molecules in young and senescent fibroblasts. Cell cycle components are depicted in young and senescent cells. Components that decrease during senescence are indicated by dotted lines; components that increase in senescent cells are indicated by bold lines.*

form and is incapable of becoming phosphorylated upon stimulation by the addition of mitogens [Stein et al., 1990; Futreal and Barrett, 1991]. Therefore, it is possible that Rb may contribute to the arrest of senescent cells by remaining in its active, negative growth regulatory state. One mechanism by which Rb function may be mediated is through its interaction with the transcription factor E2F. E2F has been shown to positively regulate many of the genes required for DNA synthesis [Mudryj et al., 1990], and E2F has also been shown to form a complex with the unphosphorylated form of Rb [Chellappan et al., 1991], leading to downregulation of E2F activity [Hiebert et al., 1992; Weintraub et al., 1992]. An Rb–E2F interaction has been detected in senescent cells [Afshari et al., unpublished observation]. Therefore, it is likely that part of the growth arrest mechanism in senescent cells involves the interaction between Rb and E2F, which consequently leads to lack of transcription of genes required for DNA synthesis. In addition, it has been shown that the RB–E2F interaction is disrupted by binding by SV40 T antigen, E1A, and E7. Therefore, the mechanism by which these transforming proteins function may be to disrupt the Rb–E2F complex in arrested cells, releasing free E2F, which in turn

allows transcription of the genes required for DNA synthesis, and unregulated entry into the cell cycle, and transformation.

Following the discovery that Rb was permanently unphosphorylated in senescent cells, investigation of the events regulating RB phosphorylation began. It was postulated that either the phosphatase that dephosphorylated Rb is constitutively activated or the kinase that phosphorylated Rb is permanently down-regulated in senescent cells. It was shown that the cdc2 kinase complex was capable of phosphorylating Rb *in vitro* [Lin et al., 1991; Lees et al., 1991; Hu et al., 1992]. Studies of the cdc2 protein in young, serum-starved, quiescent versus senescent cells revealed that the cdc2 protein is down-regulated in senescent cells [Richter et al., 1991]. In addition, it was shown that the cyclin binding partners of cdc2 are also down-regulated in senescent cells [Stein et al., 1991]. However, while the down-regulation of cdc2 kinase activity may be an important regulatory event in senescence, it is not sufficient for the senescence arrest, because reintroduction of cdc2 and cyclin A into cells is insufficient to allow escape from senescence [Afshari et al., 1993].

Other cell cycle regulatory components have been investigated in senescent cells. Cdk2 mRNA is down-regulated in senescent cells [Stein et al., 1991; Afshari et al., 1993; Lucibello et al., 1993], but the protein is still expressed [Dulic et al., 1993; Afshari, unpublished observation]. In addition, two cdk2-binding partners, cyclin E and cyclin D1, are elevated in senescent cells [Dulic et al., 1993; Lucibello et al., 1993]. The level of expression of these two cyclins increases throughout G_1, reaching their peak prior to the G_1/S border. It is possible that their elevation during senescence may be because senescent cells are arrested in G_1 at the precise stage where these cyclins are normally elevated. However, in spite of the presence of cdk2 and these cyclin proteins, the cdk2 kinase complexes have very low enzymatic activity [Dulic et al., 1993]. It has been shown that cyclin A–cdk kinase activity is required for progression of cells into DNA synthesis [Walker et al., 1991; Pagano et al., 1992]. Therefore, it is possible that another mechanism of growth arrest in senescent cells may be responsible for down-regulation of the cdk kinase activity that in turn leads to cell cycle arrest.

Recently, a new class of cell cycle regulatory molecules was discovered. These proteins, known as *cdk-inhibitory proteins* (cips), were first discovered independently by several laboratories using different cloning approaches. Harper and coworkers [1993], screening for proteins that bind to cdk–cyclin complexes, found a small, 21 kD protein that was shown to inhibit the kinase activity of these complexes. The same 21 kD protein was also found in a screen for genes that are regulated by p53 [El-Diery et al., 1993]. In addition, the gene for this 21 kD protein was cloned because of its ability to induce a senescence-like growth arrest. Moreover, the mRNA of this gene is elevated in senescent fibroblasts [Noda et al., 1994]. The discovery of this gene product, known as p21, provides an explanation for the lack of an enzymatically inactive cdk–cyclin complexes in senescent cells.

p21 has been shown to be regulated by p53 [El-Diery et al., 1993]; however, it is likely that other mechanisms are involved in the regulation of p21 expression because p21 levels increase in the absence of induced, wild-type p53 activity [Johnson et al., 1994]. Interestingly, p53 may also play an important role in senescence-induced arrest. It has been shown that transfection of both antisense Rb and p53 into cells can induce an extension of life span of normal cells [Hara et al., 1991]. However, blocking the expression of these two genes is insufficient to induce immortalization; therefore, at least one other event must be required for escape from senescence. Conversely, introduction of both Rb and p53 into immortal and transformed lines leads to growth arrest with a senescent phenotype [Huang et al., 1988; Finlay et al., 1989; Shay et al., 1991]. Further evidence that supports a role for Rb and p53 in the senescence arrest mechanism comes from experiments where it has been shown that the transforming protein SV40 T antigen, which binds both Rb and p53, can lead to escape from senescence at a low frequency [Tsuji et al., 1983; Shay and Wright, 1989; Shay et al., 1993]. In addition, the DNA transforming viral gene products E1B and E6 (which bind p53) cooperate with E1A and E7 (which bind Rb) to induce immortalization of rodent fibroblasts [Ruley et al., 1983; Munger et al., 1989].

While the role of p53 in normal cell cycle progression is unknown, it has been shown that p53 levels are markedly increased in cells in response to DNA damage. Cells that lack wild-type p53 function fail to growth arrest in response to DNA damage and thus have decreased DNA repair before entering into the replicative phase of the cell cycle [Kastan et al, 1991; Lu and Lane, 1993]. In senescent fibroblasts, p53 protein is present, although it does not appear to be elevated [Afshari et al., 1993]. Therefore, it is likely that p53 plays a role in cellular senescence. Further indirect evidence for involvement of p53 in cellular life span control is the observation that a majority of human tumors and immortal cell lines contain p53 mutations, indicating that loss of wild-type p53 function may be required to overcome the signals of the senescence program [Nigro et al., 1989]. It is possible that p53 may sense, by an unknown mechanism, an increase in DNA damage or shortened telomeric sequences in senescent cells, which then leads to induction of a certain genetic program, including p21, which in turn leads to cell cycle arrest. It has also been shown that p21, through an interaction with E2F, can lead to decreased expression of E2F-regulated genes and thus may contribute to the failure of senescent cells to enter into DNA synthesis [Afshari et al., unpublished data].

III. CAUSES OF SENESCENCE

It is known that a finite number of population doublings exists for any normal cell type in culture. Therefore, the only known cause of senescence is cell growth and division for a predetermined number of divisions. However, the mechanisms by which a cell ultimately defines the end of its life span and ceases to proliferate are unknown. Several theories on the cause of cellular senescence exist. These include a genetic model, a programmed clock model, and a damage model. It is likely that the true mechanism of senescence arrest is a combination of some or all of these scenarios. These models are discussed below.

III.A. Genetic Model

Recently, a major focus of senescence research has been on the genetic program required for this process. There are several lines of evidence that suggest that senescence is a genetically controlled process. Initial studies in this area were conducted using heterokaryon fusion experiments where the fusion of senescent and young cell fibroblasts showed that DNA synthesis within the young cell nucleus is inhibited by the nucleus of the senescent cell [Norwood et al., 1974]. This inhibition only occurs in young cell nuclei that are in the G_1 or G_2 phases of the cell cycle. Nuclei that were already committed to DNA synthesis completed the current round of replication before arresting [Yanishevsky and Stein, 1980; Rabinovitch and Norwood, 1980]. These results indicate that the signals for growth arrest, carried within the nucleus, are dominant

over cellular signals to proliferate. Similar results were obtained when fusions were made between young quiescent and young proliferating cells, suggesting that some of the mechanisms involved in quiescence–arrest are similar to senescence–arrest [Stein, 1983; Norwood et al., 1990]. However, the quiescence–arrest of young cells is reversible whereas senescence–arrest is not, so differences between the two processes also exist.

In addition to the heterokaryon fusion studies, somatic cell fusions provided strong evidence for the existence of a genetic program for senescence. Pereira-Smith and Smith [1983, 1988] performed somatic cell fusions between different immortal cell lines and found that the fusions of certain transformed lines result in senescence of the hybrids. Using this technique, 31 human cell lines were assigned to four senescence complementation groups. When a fusion between two cell lines results in a hybrid with an indefinite proliferative capacity (greater than 100 population doublings), the hybrid is termed *immortal*, and the two cell lines used to form the hybrid are assigned to the same complementation group. These results indicate that different genetic changes may occur as cells become immortal and that the defect in one cell may be complemented by a cell with a different defect (giving rise to a senescent phenotype), while a similar defect may have occurred in cells that are assigned to the same complementation group. These data provides evidence for the existence of at least four senescence genes or pathways.

A new somatic cell genetics technique provided direct evidence of the existence of senescence genes by mapping the chromosomal location of these genes. The technique of microcell-mediated chromosomal transfer allows the transfer of a single human chromosome into a cell line that may be deficient in a gene product from that particular chromosome. Sugawara and coworkers [1990] first used the microcell fusion technique to show the existence of a senescence gene on human chromosome 1. A single human chromosome 1 was introduced into an immortal hamster cell line, and clones were selected for successful fusion by growth in G418, since the introduced chromosome 1 carried a gene for neomycin resistance. Drug-resistant colonies ceased proliferation and appeared to have a senescent phenotype after a few population doublings. Control fusions using human chromosome 11 produced colonies that did not senesce. These data indicated the presence of a senescence gene on chromosome 1. Chromosome 1 also induces senescence of a human endometrial carcinoma cell line [Yamada et al., 1990], and Hensler et al. [1994] have assigned chromosome 1 to the senescence complementation group C.

To date senescence genes have been assigned to chromosomes 1 [Sugawara et al., 1990] 2, 3 [M. Oshimura, personal communication], 4 [Ning et al., 1991], 6 [Hubbard-Smith et al., 1992], 7 [Ogata et al., 1993], 9 [M. Diaz and R. Newbold, personal communication], 11 [Koi et al., 1993], 18 [Sasaki et al., 1994], and X [Klein et al., 1991]. The genes on these chromosomes that are responsible for inducing the senescent phenotype in cells have not yet been cloned, but hybrids that lose only part of the transferred chromosome and escape from senescence are being used to narrow the active regions further . In addition, neo-tagged chromosome fragments are now being used for transfer to define further the regions of the genes. Cloning of these genes will not only provide an important breakthrough into understanding how senescence occurs, but it is possible that introduction of these genes will revert the transformed phenotype and may be utilized in gene therapy strategies for the treatment of human tumors.

III.B. Programmed Clock

Because the proliferative potential of normal cells in culture is correlated with number of population doublings and not chronological time, it is likely that some type of intracellular counting or clock mechanism exists within cells. One intriguing mechanism for performing this function is the shortening of chromosomal telomeric length. As human fibroblasts age, the telomere length shortens [Harley et al., 1990].

Further support of this mechanism is provided by the analysis of human immortal lines. Immortal tumor lines contain a telomerase enzymatic activity that adds repeat sequences to the ends of chromosomes and maintains telomere length. In cell lines with active telomerase, telomere length does not decrease with repeated population doublings [Morin, 1991; Counter et al., 1994]. Interestingly, the expression of telomerase activity can only be detected during development and is turned off in all adult tissues except germline tissues, unless it is reactivated in malignant cancers [Counter et al., 1994].

A decreased telomere length may regulate gene expression by causing the activation or repression of certain genes. Telomere shortening could cause certain genes to be moved into, or out of, transcriptionally inactive heterochromatin regions [Wright and Shay, 1992]. Recent evidence in further support for a role of telomerase down-regulation in senescence is provided by microcell fusion experiments where human chromosome 3 was transferred into a renal carcinoma cell line. The transfer of chromosome 3 not only restored a finite life span to the cell line, but it was also shown that these clones lost telomerase activity and had shortened telomeres. Therefore, chromosome 3 may harbor a gene that is important for the regulation of telomerase activity [M. Oshimura, personal communication] and cellular senescence.

III.C. Damage Model of Cellular Senescence

There are several models that hypothesize that cellular senescence is due to cellular damage: mutations, changes in DNA methylation, defects in cellular proteins, or oxidative damage. There is evidence that supports each one of these theories, but there are also exceptions that can contradict each model. Thus, the role that DNA damage plays in the initiation of senescence is highly controversial [Cristofalo and Pignolo, 1993].

DNA methylation is one form of genetic modification that changes as cells age. It has been reported that methylation of cytosine decreases in fibroblasts that are near the end of their life span [Wilson and Jones, 1983; Fairweather et al., 1985; Gray et al., 1991], and investigations into the effect of methylation on gene expression suggest that methylation is associated with a repression of gene expression [Cedar, 1988]. Conversely, immortal cell lines have been shown to contain higher levels of methylated DNA than cells with finite life spans [Fairweather et al., 1985; Gray et al., 1991; Matsumura et al., 1989]. However, treatment of normal fibroblasts with 5-azacytidine, a chemical that inhibits DNA methyltransferase, can shorten the life span of the cells but does not induce growth arrest, even though the level of DNA methylation is lower than that found in senescent cells [Gray et al., 1991; Holliday, 1986; Fairweather et al., 1987; Honda and Matsuo, 1987]. These data indicate that, while decreased DNA methylation may play a role in induction of the senescence program, it may not be sufficient. On the other hand, the pattern of methylation required may be specific and cannot be mimicked by nonspecific chemical treatments.

The somatic mutation model of senescence initiation proposes that the accumulation of mutational events over the lifecourse of a cell eventually leads to a decline of proliferative capacity and cell death [Szilard, 1959; Morley, 1982]. Inherent within this model is the implication that repair mechanisms may also become impaired as cells age. The evidence supporting this model is highly controversial. It has been reported that senescent cells have an increase in chromosomal abnormalities, including chromosomal and chromatid breaks and gaps [Sherwood et al., 1988]. In addition, investigation of mutation frequency of the hypoxanthine-guanine phosphoribosyl-transferase (HGPRT) locus in lymphocytes from young versus old donors shows that the mutation frequency in this locus was approximately twice as high in the cells from the older donors (aged 65–69) versus cells from the younger donors (aged 35–39 and 50–54). In addition, an increase in chromosomal aberrations was observed in the older population

[King et al., 1994]. Another study analyzing fibroblasts from young and elderly donors, as well as cells from patients with premature aging disorders, showed a correlation between the reduction of maximal life span and elevated levels of chromosomal damage and micronuclei formation, as well as decreased repair of a transfected, damaged, DNA template [Weirich-Schwaiger et al., 1994]. Conversely, Gaubatz and Tan [1994] investigated the level of repair induced after damage incurred by methylnitrosourea (MNU) treatment of young and old mice. Analysis of repair in liver, kidney, and brain showed that initial repair rates were similar between the two age groups. However, the older animals had higher levels of persistent adducts, suggesting that the accumulation of damage in older tissues is not due to lack of repair enzymes but to other alterations that may occur in older chromatin that may prevent efficient repair.

A third model of damage-induced senescence involves free radical damage [Harman, 1956, 1987]. This model suggests that damage by oxygen free radicals increases as cells age due to a decrease in scavenging enzymes such as superoxide dismutase (SOD). However, it has been shown that there is no significant decrease in SOD levels in fibroblasts from young versus old donors [Allen and Bailin, 1988]. In addition, treatment of fibroblasts with antioxidants did not lead to an extension of life span [Balin et al., 1976,1977]. Similar results were reported after analysis of mouse dermal tissue from young and old animals. There was no decrease in the levels of catalase, SOD, or glutathione reductase in the tissues of the older animals. There was, however, a decrease in glutathione peroxidase, but in general there was no difference in the scavenging activity between the mice of different ages [Lopez-Torres et al., 1994].

In spite of the evidence against the theory of free radical–induced aging, some compelling data exist that appear to support this theory. It has been shown that dietary restriction in rodents, which protects against oxidative damage, also retards the aging process [Masoro,

1993]. In addition, age-1 mutants of *Caenorhabditis elegans* have an extended life span and a resistance to oxidative stress. Analyses of SOD and catalase enzymatic activity show an age-dependent increase in age-1 mutants versus parental controls [Larsen, 1993]. In addition, treatment of human fibroblasts with hydrogen peroxide, which induces oxidative damage, results in a growth arrest state that is similar to senescence [Chen and Ames, 1994]. While these data appear to be controversial, it is possible that a decrease in oxidative damage repair may not occur with aging, but perhaps the exposure to free radicals over a lifetime may still result in the accumulation of some damage, as no repair activity is completely efficient, without fail. This argument would also be supported by evidence that suggests that a decreased exposure to damaging agents, such as dietary restriction, would decrease the overall amount of unrepaired damage.

IV. CONCLUSION

We have discussed the recent theories of cellular aging and senescence based on *in vitro* studies. It is likely that the program of aging is controlled by both physical as well as genetic factors, the combination of which result in the ultimate phenotype of cessation of replicative capacity. It is likely that physical damage to DNA, by decreased methylation, shortened telomeres, or general unrepaired damage could contribute to activation of some senescent pathways. Genetic data indicate that there are at least 13 genes responsible for senescent activity. When these genes are cloned and their mechanism of activation defined, it will be possible to understand the mechanism of cellular aging more completely.

REFERENCES

Afshari CA, Barrett JC (1994): Disruption of G_0-G_1 arrest in quiescent and senescent cells treated with phosphatase inhibitors. Cancer Res. 54:2317–2321.

Afshari CA, Vojta PJ, Annab LA, Futreal PA, Willard TB, Barrett, JC (1993): Investigation of the role of G_1/S cell cycle mediators in cellular senescence. Exp Cell Res 209:231–237.

Allen RG, Balin AK (1988): Developmental changes in the superoxide dismutase activity of human skin fibroblasts are maintained *in vitro* and are not caused by oxygen. J Clin Invest 82:731–734.

Balin AK, Goodman, DBP, Rasmussen H, Cristofalo VJ (1976): The effect of oxygen tension on the growth and metabolism of WI-38 cells. J Cell Physiol 89:235–250.

Balin AK, Goodman DBP, Rasmussen H, Cristofalo VJ (1977): The effect of oxygen and vitamin E on the life span of human diploid cells *in vitro*. J Cell Biol 74:58–67.

Brown WT, Kieras FJ, Houck GE Jr., Dutkowski R, Jenkins EC (1985): Comparison of adult and childhood progerias: Werner syndrome and Hutchinson-Guilford progeria syndrome. In Salk D, Fujiwara Y, Martin GM (eds): Werner's Syndrome and Human Aging. New York: Plenum, pp 229–244.

Cedar H (1988): DNA methylation and gene activity. Cell 53:3–4.

Chellappan SP, Hiebert S, Mudryj M, Horowitz JM, Nevins JR (1991): The E2F transcription factor is a cellular target for the RB protein. Cell 65:1053–1061.

Chen P-L, Scully P, Shew J-Y, Wang J, Lee W-H (1989): Phosphorylation of the retinoblastoma gene product is modulated during the cell cycle and cellular differentiation. Cell 58:1193–1198.

Chen Q, Ames BN (1994): Senescence-like growth arrest induced by hydrogen peroxide in human diploid fibroblast F65 cells. Proc Natl Acad Sci USA 91:4130–4134.

Cooper JA, Whyte P (1989): RB and the cell cycle: entrance or exit? Cell 58:1009–1011.

Counter CM, Hirte HW, Bacchetti S, Harley CB (1994): Telomerase activity in human ovarian carcinoma. Proc Natl Acad Sci USA 91:2900–2904.

Cristofalo VJ, Pignolo RJ (1993): Replicative senescence of human fibroblast-like cells in culture. Physiol Rev 73:617–638.

Daniel CW, Deohme KB, Young JT, Blair PB, Faulkin LJ (1968): The in vivo life span of normal and preneoplastic mouse mammary glands: A serial transplantation study. Proc Natl Acad Sci USA 61:52–60.

DeCaprio JA, Ludlow JW, Figge J, Shew J-Y, Huang C-M, Lee W-H, Marsillio E, Paucha E, Livingston DM (1988): SV40 large tumor antigen forms a specific complex with the product of the retinoblastoma susceptibility gene. Cell 54:275–283.

DeCaprio JA, Ludlow JW, Lynch D, Furukawa J, Griffin Y, Piwnica-Worms H, Huang C-M, Livingston D-M (1989): The product of the retinoblastoma susceptibility gene has properties of a cell cycle regulatory element. Cell 58:1085–1095.

Dell'Orco RT, Mertens GB, Kruse PF (1974): Doubling

potential and calender time of human diploid cells in culture. Exp Cell Res 84:363–366.

Dimri GP, Campisi J (1994): Altered profile of transcription factor–binding activities in senescent human fibroblasts. Exp Cell Res 212:132–140.

Dulic V, Drullinger LF, Lees E, Reed SI, Stein GH (1993): Altered regulation of Gl cyclins in senescent human diploid fibroblasts: Accumulation of inactive cyclinE–cdk2 and cyclin Dl–cdk2 complexes. Proc Natl Acad Sci USA 90:11034–11038.

Dyson N, Howley PM, Munger K, Harlow, E (1989): The human papilloma viurs–16 E7 oncoprotein is able to bind to the retinoblastoma gene product. Science 243:934–937.

El-Deiry WS, Tokino T, Velculescu VE, Levy DB, Parsons R, Trent JM, Lin D, Mercer WE, Kinzler KW, Vogelstein B (1993): WAF1, a potential mediator of p53 tumor suppression. Cell 75:817–825.

Ewen ME, Ludlow JW, Marsillio E, DeCaprio JA, Millikan RC, Cheng SH, Paucha E, Livingston, DM (1989): An N-terminal transformation-governing sequence of SV40 large T antigen contributes to the binding of both p110Rb and a second cellular protein, p120. Cell 58:257–267.

Fairweather DS, Fox M, Margison G (1985): DNA hypomethylation: A new theory on aging. Clin Sci 69:53.

Fairweather S, Fox M, Margison BP (1987): The in vitro lifespan of MRC-5 cells is shortened by 5-azacytidine-induced demethylation. Exp Cell Res 168:153–159.

Finlay CA, Hinds PW, Levine AJ (1989): The p53 proto-oncogene can act as a suppressor of transformation. Cell 57:1083–1093.

Friend SH, Bernards R, Rogelj S, Weinberg RA, Rapaport JM, Albert DM, Dryja TP (1986): A human DNA segment with properties of the gene that predisposes to retinoblastoma and osteosarcoma. Nature 323:643–646.

Fung Y-K T, Murphree AL, T'Ang A, Qian J, Hinrichs SH, Benedict WF (1987): Structural evidence for the authenticity of the human retinoblastoma gene. Science 236:1657–1661.

Futreal PA, Barrett JC (1991): Failure of senescent cells to phosphorylate the RB protein. Oncogene 6:1109–1113.

Gaubatz JW, Tan BH (1994): Aging affects the levels of DNA damage in postmitotic cells. Ann NY Acad Sci 719:97–107.

Goldstein S (1969): Life span of cultured cells in progeria. Lancet 1:424.

Goldstein S (1974): Aging in vitro. Growth of cultured cells from the Galapagos turtle. Exp Cell Res 83:297–302.

Goldstein S, Moerman EJ, Jones RA, Baxter RC (1991): Insulin-like growth factor binding protein 3 accumulates to high levels in culture medium of senescent and quiescent human fibroblasts. Proc Natl Acad Sci USA 88:9680–9684.

Goldstein S, Murano S, Benes H, Moerman EJ, Jones RA, Thewatt R, Shmookler Reis RJ, Howard BH (1989): Studies on the molecular genetic basis of replicative senescence in Werner syndrome and normal fibroblasts. Exp Gerontol 24:461–468.

Gray MD, Jesch SA, Stein GH (1991): 5-Azacytidine-induced demethylation of DNA to senescent level does not block proliferation of human fibroblasts. J Cell Physiol 149:477–484.

Hara E, Tsurui H, Shinozaki A, Nakada S, Oda K (1991): Cooperative effect of antisense-Rb and antisense-p53 oligomers on the extension of life span in human diploid fibroblasts, TIG-1. Biochem Biophys Res Commun 179:528–534.

Harley CB, Futcher AB, Greider CW (1990): Telomeres shorten during aging of human fibroblasts. Nature 345:458–460.

Harman D (1956): Aging: A theory based on free radical and radiation chemistry. J Gerontol 11:298–300.

Harman D (1987): The free radical theory of aging. In: Warner HR, Butler RN, Sprott RL, Schneider EL, (eds): Modern Biological Theories of Aging. New York: Raven pp 81–87.

Harper JW, Adami GR, Wei N, Keyomarsi K, Elledge SJ (1993): The p21 cdk-interacting protein cip1 is a potent inhibitor of G_1 cyclin-dependent kinases. Cell 75:805–816.

Hayflick L (1976): The cell biology of human aging. N Engl J Med 295:1302–1308.

Hayflick L, Moorhead PS (1961): The serial cultivation of human diploid strains. Exp Cell Res 25:585–621.

Hensler PJ, Annab LA, Barrett JC, Pereira-Smith OM (1994): A gene involved in the control of human cellular senescence localized to human chromosome 1q. Mol Cell Biol 14:2291–2297.

Hiebert SW, Chellappan SP, Horowitz JM, Nevins JR (1992): The interaction of RB with E2F coincides with an inhibition of the transcriptional activity of E2F. Genes Dev 6:177–185.

Holliday R (1986): Strong effects of 5-azacytidine on the in vitro life span of human diploid fibroblasts. Exp Cell Res 166:543–552.

Honda S, Matsuo M (1988): Relationship between the cellular glutathione level and in vitro life span of human diploid fibroblasts. Exp Gerontol 23:81–86.

Hu Q, Lees JA, Buchkovich KJ, Harlow E (1992): The retinoblastoma protein physically associates with the human cdc2 kinase. Mol Cell Biol 12:971–980.

Huang H-JS, Lee J-K, Shew J-Y, Chen P-L, Bookstein R, Friedmann T, Lee EY-HP, Lee W-H (1988): Suppression of the neoplastic phenotype by replacement of the RB gene in human cancer cells. Science 242:1563–1566.

Hubbard-Smith K, Patsalis P, Pardinas JR, Jha KK, Henderson AS, Ozer HL (1992): Altered chromosome 6 in immortal human fibroblasts. Mol Cell Biol 12:2273–2281.

Ide T, Tsuji Y, Ishibashi S, Mitsui Y (1983): Reinitiation of host DNA synthesis in senescent human diploid cells by infection with Simian virus 40. Exp Cell Res 143:343–349.

Ide T, Tsuji Y, Ishibashi S, Mitsui Y, Toba M (1984): Induction of host DNA synthesis in senescent human diploid fibroblasts by infection with human cytomegalovirus. Mech Ageing Dev 25:227–235.

Johnson N, Dimitrov D, Vojta PJ, Barrett JC, Noda A, Pereira-Smith OM, Smith JR (1994): Evidence for a p53-independent pathway for upregulation of SDI1/CIP1/WAF1/p21 RNA in human cells. Mol Carc 11:59–64.

Kastan MB, Onyekwere O, Sidransky D, Vogelstein B, Craig RW (1991): Participation of p53 protein in the cellular response to DNA damage. Cancer Res 51:6304–6311.

King CM, Gillespie ES, Mckenna PG, Barnett YA (1994): An investigation of mutation as a function of age in humans. Mutat Res 316:79–90.

Klein CB, Conway K, Wang XW, Bhamra RK, Lin X, Cohen MD, Annab L, Barrett JC, Costa M (1991): Senescence of nickel-transformed cells by a mammalian X chromosome: Possible epigenetic control. Science 251:796–799.

Koi M, Johnson LA, Kalikin LM, Little PFR, Nakamura Y, Feinberg AP (1993): Tumor cell growth arrest caused by subchromosomal transferable DNA fragments from chromosome 11. Science 260:361–364.

Larsen PL (1993): Aging and resistance to oxidative damage in *Caenorhabditis elegans*. Proc Natl Acad Sci USA 90:8905–8909.

Lees JA, Buchkovich KJ, Marshak DR, Anderson CW, Harlow E (1991): The retinoblastoma protein is phosphorylated on multiple sites by human cdc2. EMBO J 10:4279–4290.

Lin BT-Y, Gruenwal S, Morla AO, Lee W-H, Wang JYJ (1991): Retinoblastoma cancer suppressor gene product is a substrate of the cell cycle regulator cdc2 kinase. EMBO J 10:857–864.

Lopez-Torres M, Shindo Y, Packer L (1994): Effect of age on antioxidants and molecular markers of oxidative damage in murine epidermis and dermis. J Invest Dermatol 102:476–480.

Lu X, Lane DP (1993): Differential induction of transcriptionally active p53 following UV or ionizing radiation: Defects in chromosome instability syndromes? Cell 75:765–778.

Lucibello FC, Sewing A, Brusselbach S, Burger C, Muller R (1993): Deregulation of cyclins D1 and E and suppression of CDK2 and cdk4 in senescent human fibroblasts. J Cell Sci 105:123–133.

Ludlow JW, DeCaprio JA, Huang C-M, Lee W-H, Livingston DM (1989): SV40 large T antigen binds preferentially to an underphosphorylated member of the retinoblastoma susceptibility gene product family. Cell 56:57–65.

Masoro EJ (1993): Dietary restriction and aging. J Am Geriatr Soc 41:994–999.

Matsumura T, Malik F, Holliday R (1989): Levels of DNA methylation in diploid and SV40 transformed human fibroblasts. Exp Gerontol 24:477–481.

Matsumura T, Zerrudo Z, Hayflick L (1979): Senescent human diploid cells in culture: Survival, DNA synthesis and morphology. J Gerontol 34:328–334.

Mihara K, Cao X-R, Yen A, Chandler S, Driscoll B, Murphree AL, T'Ang A, Fung Y-KT (1989): Cell cycle-dependent regulation of phosphorylation of the human retinoblastoma gene product. Science 246:1300–1303.

Morin GB (1989): The human telomere terminal transferase is a ribonucleoprotein that synthesizes TTAGGG repeats. Cell 59:521–529.

Morley AA (1982): Is ageing the result of dominant or co-dominant mutations? J Theor Biol 98:469–474.

Mudryj M, Hiebert SW, Nevins JR (1990): A role for the adenovirus inducible E2F transcription factor in a proliferation dependent signal transduction pathway. EMBO J 9:2179–2184.

Mueller SN, Rosen EM, Levine EM (1980): Cellular senescence in a cloned strain of bovine fetal aortic endothelial cells. Science 207:889–891.

Munger K, Werness BA, Dyson N, Phelps WC, Howley PM (1989): The E6 and E7 genes of the human papillomavirus type 16 together are necessary and sufficient for transformation of primary human keratinocytes. J Virol 63:4417–4421.

Nigro JM, Baker SJ, Preisinger AC, Jessup JM, Hostetter R, Cleary K, Bigner SH, Davidson N, Baylin S, Devilee P, Glover T, Collins FS, Weston A, Modali R, Harris CC, Vogelstein B (1989): Mutations in the p53 gene occur in diverse human tumor types. Nature 342:705–708.

Ning Y, Weber JL, Killary AM, Ledbetter DH, Smith JR, Pereira-Smith OM (1991): Genetic analysis of indefinite division in human cells: Evidence for a senescence-related gene(s) on human chromosome 4. Proc Natl Acad Sci USA 88:5635–5639.

Noda A, Ning Y, Venable SF, Pereira-Smith OM, Smith JR (1994): Cloning of senescent cell-derived inhibitors of DNA synthesis using an expression screen. Exp Cell Res 211:90–98.

Norwood TH, Pendergrass WR, Sprague CA, Martin GM (1974): Dominance of the senescent phenotypes in heterokaryons between replicative and post-replicative human fibroblast-like cells. Proc Natl Acad Sci USA 73:2223–2236.

Norwood TH, Smith JR, Stein GH (1990): Aging at the cellular level: the human fibroblastlike cell model. In Schneider EL, Rowe JR, (eds): Handbook of the Biology of Aging, 3rd ed. San Diego: Academic Press, pp 131–154.

Ogata T, Ayusawa D, Namba M, Takahashi E, Oshimura M, Oishi M (1993): Chromosome 7 suppresses indefinite division of nontumorigenic immortalized human fibroblast cell lines KMST-6 and SUSM-1. Mol Cell Biol 13:6036–6043.

Pagano M, Pepperkok R, Verde F, Ansorge W, Draetta G (1992): Cyclin A is required at two points in the human cell cycle. EMBO J 11:961–971.

Pereira-Smith OM, Smith JR (1983): Evidence for the recessive nature of cellular immortality. Science 221:964–966.

Pereira-Smith OM, Smith JR (1988): Genetic analysis of indefinite division in human cells: Identification of four complementation groups. Proc Natl Acad Sci USA 85: 6042–6046.

Phillips PD, Pignolo RJ, Nishikura K, Cristofalo VJ (1992): Renewed DNA synthesis in senescent WI-38 cells by expression of an inducible chimeric c-*fos* construct. J Cell Physiol 151:206–212.

Rabinovitch PS, Norwood TH (1980): Comparative heterokaryon study of cellular senescence and the serum deprived state. Exp Cell Res 130:101–109.

Rheinwald JG, Green H (1975): Serial cultivation of strains of human epidermal keratinocytes: The formation of keratinizing colonies from single cells. Cell 6:331–334.

Riabowol K, Schiff J, Gilman MZ (1992): Transcription factor AP-1 activity is required for initiation of DNA synthesis and is lost during cellular aging. Proc Natl Acad Sci USA 89:157–161.

Richter KH, Afshari CA, Annab LA, Burkhart BA, Owen RD, Boyd J, Barrett JC (1991): Down-regulation of cdc2 in senescent human and hamster cells. Cancer Res 51:6010–6013.

Rittling SR, Brooks KM, Cristofalo VJ, Baserga R (1986): Expression of cell cycle–dependent genes in young and senescent WI-38 fibroblasts. Proc Natl Acad Sci USA 83:3316–3320.

Roberts TW, Smith JR (1980): The proliferative potential of chick embryo fibroblasts: Population doubling vs. time in culture. Cell Biol Int Rep 4:1057–1063.

Ruley HE (1983): Adenovirus early region 1A enables viral and cellular transforming genes to transform primary cells in culture. Nature 304:602–606.

Sasaki M, Honda T, Yamada H, Wake N, Barrett JC, Oshimura M (1994): Evidence for multiple pathways to cellular senescence. Cancer Res 54:6090–6093.

Schneider EL, Mitsui Y (1976): The relationship between *in vitro* cellular aging and *in vivo* human age. Proc Natl Acad Sci USA 73:3548–3588.

Seshadri T, Campisi J (1989): Growth-factor–inducible gene expression in senescent human fibroblasts. Exp Gerontol 24:515–522.

Seshadri T, Campisi J (1990): Repression of c-*fos* transcription and an altered genetic program in senescent human fibroblasts. Science 247:205–209.

Shay JW, Pereira-Smith OM, Wright WE (1991): A role for both RB and p53 in the regulation of human cellular senescence. Exp Cell Res 196:33–39.

Shay JW, van der Haegen BA, Ying Y, Wright W (1993): The frequency of immortalization of human fibroblasts and mammary epithelial cells transfected with SV40 large T-antigen. Exp Cell Res 209:45–52.

Shay JW, Wright WE (1989): Quantitation of the frequency of immortalization of normal human diploid fibroblasts by SV40 large T-antigen. Exp Cell Res 184:109–118.

Sherwood SW, Rush D., Ellsworth JL, Schimke RT (1988): Defining cellular senescence in IMR-90 cells: A flow cytometric analysis. Proc Natl Acad Sci USA 85:9086–9090.

Stein GH (1983): Human diploid fibroblasts (HDF) can induce DNA synthesis in cycling HDF but not quiescent HDF or senescent HDF. Exp Cell Res 144:468–471.

Stein GH, Beeson M, Gordon L (1990): Failure to phosphorylate the retinoblastoma gene product in senescent human fibroblasts. Science 249:666–669.

Stein GH, Drullinger LF, Robetorye RS, Pereira-Smith OM, Smith JR (1991): Senescent cells fail to express cdc2, cycA, and cycB in response to mitogen stimulation. Proc Natl Acad Sci USA 88:11012–11016.

Sugawara OM, Oshimura M, Koi M, Annab L, Barrett JC (1990): Induction of cellular senescence in immortalized cells by human chromosome 1. Science 247:707–710.

Sumegi J, Uzvolgyi E, Klein G (1990): Expression of the RB gene under the control of MULV-LTR suppresses tumorigenicity of WERI-Rb-27 retinoblastoma cells in immunodefective mice. Cell Growth Differ 1:247–250.

Szilard L (1959): On the nature of the aging process. Proc Natl Acad Sci USA 45:30–45.

Tice RR, Schneider EL, Kram D, Thorne P (1979): Cytokinetic analysis of impaired proliferative response of peripheral lymphocytes from aged humans to phytohemagglutinin. Exp Med 149:1029–1041.

Tsuji Y, Ide T, Ishibashi S (1983): Correlation between the presence of T-antigen and the reinitiation of host DNA synthesis in senescent human diploid fibroblasts after SV40 infection. Exp Cell Res 144:165–169.

Vojta PJ, Barrett JC (1995): Genetic analysis of cellular senescence. Biochim Biophys Acta (in press).

Walker DH, Maller JL (1991): Role for cyclin A in the dependence of mitosis on completion of DNA replication. Nature 354:314–317.

Weintraub SJ, Prater CA, Dean DC (1992): Retinoblastoma protein switches the E2F site from positive to negative element. Nature 358:259–261.

Weirich-Schwaiger H, Weirich HG, Gruber B, Schweiger M, Hirsch-Kauffmann M (1994): Correlation between senescence and DNA repair in cells from young and old individuals and in premature aging syndromes. Mutat Res 316:37–48.

Wilson VL, Jones PA (1983): DNA methylation decreases in aging but not in immortal cells. Science 220:1055–1057.

Wright WE, Shay JW (1992): Telomere positional effects and the regulation of cellular senescence. Trends Genet 8:193–197.

Yamada H, Wake N, Fujimoto S, Barrett JC, Oshimura M (1990): Multiple chromosomes carrying tumor suppressor activity for a uterine endometrial carcinoma cell line identified by microcell mediated chromosome transfer. Oncogene 5:1141–1147.

Yanishevsky R, Mendelsohn ML, Mayall BH, Cristofalo VJ (1974): Proliferative capacity and DNA content of aging human diploid cells in culture: A cytophotometric and autoradiographic analysis. J Cell Physiol 84:165–170.

Yanishevsky RM, Stein GH (1980): On-going DNA synthesis continues in young human diploid cells (HDC) fused to senescent HDC, but entry into S phase is inhibited. Exp Cell Res 126:469–472.

ABOUT THE AUTHORS

CYNTHIA A. AFSHARI is currently a fellow at the National Institute of Environmental Health Sciences in Research Triangle Park, North Carolina. She received her B.S. in biochemistry/biophysics from the University of Pittsburgh in Pennsylvania. Following this she pursued doctoral research at the University of North Carolina in the Curriculum of Toxicology under the guidance of Dr. J. Carl Barrett. Her thesis study focused on the loss of mechanisms of growth control in cells undergoing neoplastic progression. After receiving her degree in 1992, she began postdoctoral training at the Center for the Study of Aging and Human Development at Duke University, North Carolina. While conducting research in the laboratory of Dr. Maria Mudryj, Dr. Afshari investigated mechanisms of transcriptional regulation during cellular senescence. Now at NIEHS, she is continuing these studies and focusing on regulatory changes that occur early in the neoplastic process, including the function of tumor suppressor genes in cellular senescence.

J. CARL BARRETT is currently Director, Environmental Carcinogenesis Program, and Chief, Laboratory of Molecular Carcinogenesis of the Division of Intramural Research at the National Institute of Environmental Health Sciences (NIEHS) in Research Triangle Park, North Carolina. He received his B.S. in chemistry from the College of William and Mary in Virginia. He followed this with completion of a Ph.D. in biophysical chemistry in 1974 from Johns Hopkins University. Dr. P.O.P Ts'o served as his Ph.D. mentor. His thesis research focused on the utilization of oligonucleotide ethyl phosphotriesters as probes for the study of nucleic acid function. He continued his research career with postdoctoral work in the Laboratory of Pulmonary Function and Toxicology at the NIEHS, where he made significant contributions to the understanding of diethylstilbestrol and asbestos-induced carcinogenesis. In 1977 he became head of his own group and since has focused on the genetic mechanisms of multistage neoplastic progression and cellular senescence. Dr. Barrett also serves as an adjunct faculty member to four departments at the University of North Carolina in Chapel Hill. In addition, he is editor-in-chief of *Molecular Carcinogenesis* and has served on the editorial boards of 11 other journals. He has received numerous awards and has lectured all over the world. Currently, Dr. Barrett has authored over 260 manuscripts.

Cellular Aging and Cell Death: 123–138
© 1996 Wiley-Liss, Inc.

Telomeres and Telomerase in Cell Senescence and Immortalization

Carol W. Greider and Calvin B. Harley

I. TELOMERES AND TELOMERASE

Chromosome stability is essential for cell viability. Eukaryotes have linear chromosomes, and the telomeres that cap the ends protect chromosomes from degradation and recombination. In the 1930s, Muller and McClintock recognized that broken chromosomes that lacked telomeres were unstable [reviewed by Greider, l991a]. Subsequently, experiments in protozoa, yeast, and mammalian cells demonstrated the essential nature of telomeres for chromosome structure and function. The recent finding that telomeres shorten with age in normal human somatic cells but are stabilized in tumor cells has suggested that the regulation of telomere length may play a role in cellular senescence and immortalization. In this chapter, we first review the discovery and the function of telomeres and telomerase in lower eukaryotes. We then discuss the data that suggest telomeres play a critical role in the replicative life span of human cells and that telomerase may be important for the indefinite growth of immortalized cells.

I.A. Identification of *Tetrahymena* Telomerase

Telomeric DNA sequences are highly conserved in diverse eukaryotes from ciliates to humans [reviewed by Blackburn, 1991]. The sequences consist of simple tandem repeats of specific GT-rich motifs. The exact sequences are characteristic of a particular organism, for example, d(TTGGGG) in *Tetrahymena*, d(TTTTGGGG) in *Oxytricha*, and d(TTAGGG) in humans and mice. The number of repeats

on any given chromosome end may vary, giving telomeres a characteristic heterogeneous or "fuzzy" appearance on Southern blots. In addition to sequence conservation, telomere function is also conserved in eukaryotes. *Tetrahymena* and human telomeres function in the yeast *Saccharomyces cerevisiae* [Brown, 1989; Cross et al., 1990; Riethman et al., 1989; Szostak and Blackburn, 1982]. Thus the mechanisms for maintaining a stable end must share essential features in diverse organisms.

The properties of DNA polymerases predict that sequences will be lost from chromosome ends during replication [Olovnikov, 1973; Watson, 1972]. Although leading strand synthesis can proceed to the molecular end, lagging strand synthesis may result in a loss of the region where an RNA primer was used to prime DNA synthesis (see Fig. 1). Cells must have a mechanism to overcome this end replication problem and to maintain telomere length. The heterogeneous size of terminal restriction fragments (TRFs) and the growth of telomeres in Trypanosomes and *Tetrahymena* suggested that loss of sequences during replication may be compensated by *de novo* sequence addition [Bernards et al., 1983; Larson et al., 1987; Shampay et al., 1984]. Thus experiments were initiated to look for an enzyme activity that would add telomeric sequences *de novo*. *Tetrahymena* extracts were initially used because a single *Tetrahymena* cell contains over 10,000 telomeres; thus we expected the enzymes needed to synthesized telomeres might also be abundant. Using these extracts, a highly specific telomere repeat addition ac-

Left Telomere   Right Telomere

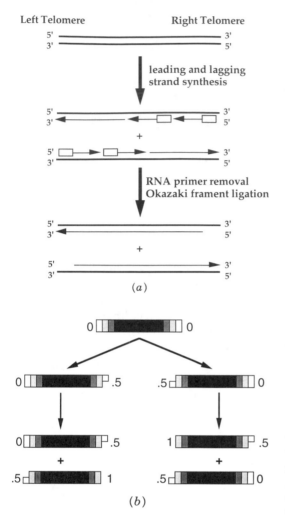

(a)

(b)

Fig. 1. *The end-replication problem. (a): Replication of a linear parent duplex (heavy lines) is shown. Lagging strand synthesis is discontinuous in Okazaki fragments, which initiate with a labile RNA primer (box). After RNA primer removal and Okazaki fragment extension and ligation, the most 5′ Okazaki fragment will remain incomplete since the RNA primer cannot be replaced. If the 5′ terminal Okazaki fragment does not initiate directly opposite the 3′ end of the template DNA, there will be additional bases incompletely replicated. (b): After two rounds of division, the four daughters of a single chromosome have a distribution of predicted deletion events as shown. Numbers represent "units" of incomplete replication based on the distance from the terminus of the chromosome to the 5′ most DNA of the terminal Okazaki fragment. Half-values represent single-strand deletions. [Adapted from Levy et al., 1992, with permission of the publisher.]*

tivity was identified. This activity, termed *telomere terminal transferase* (later shortened to *telomerase*), synthesized *Tetrahymena* telomeric d(TTGGGG) repeats *de novo* onto synthetic single-stranded d(TTGGGG) primers. The nucleotides were added one base at a time without any apparent template DNA [Greider and Blackburn, 1985].

To determine how a polymerase synthesizes a specific sequence *de novo*, telomerase extracts were treated with nucleases to determine if there was a nucleic acid component. RNase treatment abolished the telomere primer elongation activity, suggesting that the enzyme contained an essential RNA. Telomerase was then partially purified, and a 159 nt RNA component containing the sequence CAACCCCAA, complementary to the telomere d(TTGGGG) repeats, was identified [Greider and Blackburn, 1989]. When mutations were made in the template region and the gene was reintroduced into *Tetrahymena* cells, mutant telomeres were generated [Yu et al., 1990]. This demonstrated that telomerase is responsible for telomere synthesis *in vivo*.

Based on the RNA sequence, the synthesis of long stretches of d(TTGGGG)n and the ability of telomerase always to add the correct next nucleotide onto primers with different 3′ ends, a model was proposed for telomerase elongation [Greider and Blackburn, 1989] (Fig. 2). Telomerase first binds the telomeric primer and aligns the primer with the RNA sequence. Elongation then fills out the sequence to the end of the template. Translocation then repo-

Fig. 2. *Model for elongation of telomeres by telomerase. (1) Telomerase (represented by two gray ovals) recognizes the telomere substrate and the terminal TTGGGG repeat is allowed to base pair with the CAACCCCAA region of the RNA component. (2) The RNA is copied to the end of the template region. (3) Translocation repositions the terminal TTGGGGTTG sequence and exposes additional template sequences for a second round of repeat synthesis. (4) Another round of template copying produces additional TTGGGG repeats. [Model adapted from Greider and Blackburn, 1989, with permission of the publisher.]*

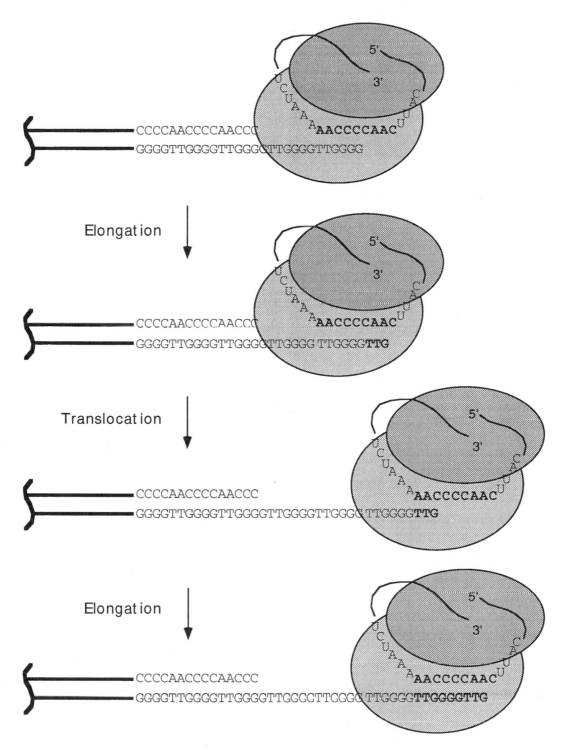

Figure 2.

sitions the primer on the template to allow a second round of d(TTGGGG) synthesis. The general concepts of this model have been verified, and recent data have allowed us to refine the model of telomerase action further [Collins and Greider, 1993; Lee and Blackburn, 1993; for review, see Greider, 1995].

I.B. Telomerase Activities

A typical telomerase reaction requires telomeric primer, buffer (Tris-acetate, Mg^{2+}, and glycerol), dTTP, and dGTP. The dGTP is usually labeled with ^{32}P to visualize the products. Labeled dTTP can also be used, however; because the K_m for dGTP is much lower than for dTTP, a higher specific activity is obtained with ^{32}P-dGTP [Greider and Blackburn, 1987]. End-labeled primers are a substrate, however; for enzymes from several species, most of the input primer is not elongated and thus the signal at the input primer size is very strong and the products are more difficult to detect. When the products of the elongation reaction are analyzed on a sequencing gel, a banding pattern with a six base periodicity is detected. This six base periodicity corresponds to the addition of each d(TTGGGG) repeat [Greider, 1991b; Greider and Blackburn, 1987]. A more sensitive polymerase chain reaction (PCR) assay for telomerase elongation was recently developed [Kim et al., 1994] (see below).

In addition to *Tetrahymena*, telomerase activity has been identified in the ciliates *Oxytricha* and *Euplotes*, as well as in immortalized human cell lines and in various tissues from mouse (see below), rat, and *Xenopus* [Mantell and Greider, 1994; Morin, 1989; Prowse et al., 1993; Shippen-Lentz and Blackburn, 1989; Zahler and Prescott, 1988; Mantell and Greider, unpublished data]. A telomerase RNA component has additionally been cloned from two different yeast species [Singer and Gottschling, 1994; McEachern and Blackburn, personal communication] (see below). All of the telomerase enzymes identified have properties similar to *Tetrahymena* telomerase. They are RNase sensitive and elongate only telomeric primers with the addition

of one nucleotide at a time. The sequence synthesized, however, is specific to the given organism and corresponds to the telomere repeats in that organism.

The products generated by telomerase from different organisms are similar; tandem repeats of the telomeric sequence are generated. Telomerase activity isolated from mouse and *Xenopus* cells, however, synthesizes much shorter products than those synthesized by the ciliate and human telomerases [Mantell and Greider, 1994; Prowse et al., 1993]. Under conditions where the human enzyme synthesizes many repeats, the mouse enzyme will generate one predominant band. The position of this strong stop varies with the sequence of the input oligonucleotide, similar to both the human and the *Tetrahymena* enzymes [Greider and Blackburn, 1987; Morin, 1989; Prowse et al., 1993].

Recent biochemical evidence suggested a two-site model for primer binding and elongation. Evidence for this model came from experiments on telomerase processivity. First, with human telomerase, primers of different lengths generated different length products [Morin, 1989]. The length of products primed by d(TTAGGGTTA) were shorter than those generated with (TTAGGG)2. Since elongation of d(TTAGGGTTA) by the addition of only three G residues will convert it to d(TTAGGG)2, it was suggested that the initial recognition of the shorter primer must affect processivity. Early experiments showing that *Tetrahymena* telomerase is highly processive used d(TTGGGG)3 and d(TTGGGG)4 as primers [Greider, 1991b]. Later experiments using shorter oligonucleotides showed that the length of the input primers greatly affects the processivity of the enzyme [Collins and Greider, 1993; Lee and Blackburn, 1993]. The primers (TTGGGG) and d(TTGGGGTT) are extended by addition of only one repeat corresponding to the sequence of the RNA template region. Further elongation does not occur, suggesting that the product dissociates after elongation to the end of the template. These experiments suggested that primers longer than

10 nucleotides are processively elongated because they bind to both the template and a second site or "anchor site." Binding to the anchor site prevents product dissociation [Collins and Greider, 1993; Harrington and Greider, 1991; Lee and Blackburn, 1993; Morin, 1989]. The two-site model is based on the proposed mechanism for processive elongation by RNA polymerase. RNA polymerase binds the elongating RNA chain both at the site of polymerization (site l) and at a second site at least eight nucleotides away from the 3′ end (site 2). This two-site mechanism allows 3′ end release during translocation while maintaining contact with the growing product for the next round of sequence addition [Chamberlin, 1993].

Initial experiments with mouse and *Xenopus* telomerase suggested that the products synthesized were predominantly one or two repeats long. Recent experiments showed that *Xenopus*, mouse, and rat telomerase enzymes can be processive under certain conditions. Increasing the concentration of dGTP 100-fold over that used with the human enzyme showed that *Xenopus* telomerase will synthesize long products analogous to those seen with human telomerase [Mantell and Greider, unpublished data]. Subsequent experiments with mouse and rat enzymes demonstrated that they also will generate long products at high dGTP concentrations. Thus, the processivity difference may be due in part to a difference in affinity for dGTP between telomerases from different organisms.

I.C. Ciliate and Yeast Telomerase RNAs

The similar biochemical properties of telomerases suggest a similar mechanism of *de novo* repeat synthesis. To test this for the *Euplotes* telomerase, the RNA component was cloned by looking for an RNA containing the sequence complementary to the telomere repeats [Shippen-Lentz and Blackburn, 1990]. RNase H cleavage of nuclear RNA after addition of a d(TTTTGGGG) oligonucleotide identified a 190 nt RNA as a candidate for telomerase RNA. This RNA contained the sequence CAAAACCCCAAAA that was shown

to provide template function using antisense oligonucleotides inhibition and elongation experiments similar to those used for *Tetrahymena* telomerase. The most striking thing about the sequences of the initial *Tetrahymena* and *Euplotes* RNAs is that they are not well conserved. Although they both contain template regions, the two will not cross-hybridize on blots, and very little similarity is found in sequence comparisons [Shippen-Lentz and Blackburn, 1990]. Thus cloning of telomerase RNA from additional organisms required a number of special tricks.

Lingner et al. [1994] cloned the telomerase RNAs from a variety of hypotrichous ciliate species. Taking advantage of the conserved upstream promoter elements in ciliate RNAs and the large template region identified in the *Euplotes crassus* sequence, they used PCR to clone telomerase-specific products from several new ciliate species. The remainder of the gene was then cloned from a genomic library [Lingner et al., 1994]. The *Oxytricha nova* RNA was also identified by direct sequencing of a CCCCAAAA containing RNA and then cloning of the gene from a library [Melek et al., 1994]. Although the hypotrich telomerase RNAs are longer than the *Tetrahymena*, (190 vs. 160 nt), the secondary structure predicted from nucleotide co-variation was very similar to that proposed for the *Tetrahymena* enzyme. The one exception was stem II, which was not found in the hypotrich RNA structure.

Two very different approaches allowed the identification of telomerase RNAs from two yeast species. The first approach was a systematic search for potential yeast telomerase RNAs. Several species of budding yeast have unusually long (26 bp) telomere repeats [McEachern and Hicks, 1993]. The dynamic behavior of these repeated sequences suggested that they were synthesized by telomerase. Thus it was reasoned that a gene for the telomerase RNA would contain at least one repeat of this telomeric sequence. A chromosomal internal fragment was cloned from *Kluveromyces lactis* that contained the telomeric repeat plus a repeated region of five nucleotides at each end similar to the template sequence redundancy in the ciliate telomerase RNAs. To determine if this cloned chromosomal region was involved in telomerase, a 2 bp change to introduce a Bgl II restriction site was made in the potential template region, and the wild-type locus was replaced with the mutant copy. After growing mutant cells for several generations, the mutant sequence was detected at the telomeres by oligonucleotide hybridization. Digestion with Bgl II removed the mutant repeats and shortened the telomere tracts. Thus, changing the chromosome internal copy of the telomerase RNA gene altered the sequence of the telomere repeats, formal proof that this gene represented the telomerase RNA component. Northern analysis showed that an RNA encoded by this locus is 1.3 kb in length, much larger than the RNAs in ciliate telomerases [McEachern and Blackburn, 1995].

A different approach resulted in the cloning of the *S. cerevisiae* telomerase RNA gene. Genes placed near telomeres in yeast are subject to metastable transcriptional repression [Gottschling et al., 1990; for review, see Shore, 1995]. This transcriptional silencing, referred to as *telomere position effect* (TPE), may be similar to the position effects seen in heterochromatin in *Drosophila* [Henikoff, 1990]. To understand both telomere silencing and telomere function in general, a genetic screen was set up to identify genes that alleviated TPE. Ten genes were identified and were initially screened for general effects on telomere metabolism [Singer and Gottschling, 1994]. One of the genes, *TLC1*, had shortened telomeres. The gene hybridized to a 1.3 kb RNA on a Northern blot; however, no open reading frame was found within the cloned fragment. Careful analysis of the DNA sequence revealed a 17 nucleotide region with homology to *S. cerevisiae* telomeres. Although telomere repeats in this yeast are irregular, the sequence in this gene was exactly that predicted for a telomerase RNA from analysis of many chromosome healing sites in *S. cerevisiae* [Kramer and Haber, 1993]. To verify that this gene was the yeast telomerase RNA component, the se-

quence of the potential template region was altered to generate a restriction site. Yeast expressing this mutant RNA generated mutant sequence telomere repeats onto newly introduced telomeres *in vivo*, confirming the predicted templating function of the RNA [Singer and Gottschling, 1994]. Like the *K. lactis* RNA, the *S. cerevisiae* telomerase RNA is >1.3 kb.

The large size of the telomerase RNAs in both *K. lactis* and *S. cerevisiae* was not predicted from the cloned ciliate telomerase RNAs. It is not yet clear what structure these RNAs will form or what function the "extra" sequences might have for these telomerase RNAs. The yeast U2 RNA is also much larger than U2 RNAs from many other species. Yet this larger RNA can form the same secondary structure as the shorter mammalian counterparts after the deletion of a large internal region of the RNA [Ares, 1986; Igel and Ares, 1988]. Deletional analysis of the yeast telomerase RNAs may be required before either a structural model can be derived or the function of the various regions of the RNA can be determined.

I.D. Mammalian Telomerase RNAs

Because of the lack of sequence similarities in the telomerase RNAs of single cell eukaryotes, strategies that did not rely on cross-hybridization were designed to clone the mammalian RNA component. A cDNA library was prepared and subjected to several rounds of selection for RNAs containing the potential template sequence CTAACCCTAA [Feng et al., in preparation]. A number of candidate RNAs containing this sequence were characterized to determine if the RNA co-purified with telomerase activity and whether oligonucleotides complementary to the template region would inhibit activity similar to what was found with ciliate telomerases [Greider and Blackburn, 1989]. A candidate for the human telomerase RNA (hTR) was identified and shown to be a single copy gene in the human genome. To demonstrate this functionally mutants in the telomerase RNA gene were used. The putative template region was altered to generate a telomerase that would synthesize TTTGGG or TTGGGG repeats. Using both transient transfection and stable transformants, the mutant telomerase activity was detected in extracts from these cells. Thus, like the *Tetrahymena* and yeast telomerase RNAs, altering the template region of the telomerase RNA altered the sequence of the repeats synthesized. Like the yeast telomerase RNAs, the human RNA component is much larger than the ciliate RNA. The RNA is approximately 560 nt in length.

The mouse telomerase RNA gene was cloned from a genomic library using the human telomerase gene as a probe [Blasco et al., in preparation]. Surprisingly, the mouse RNA has only 58% similarity to the human RNA gene. In contrast, other small RNAs are highly conserved between human and mouse. The U-RNA species range from 100% identity for U6 to 85% identity for U7. The mouse and human RNase P RNA and MRP RNAs are 86% and 78% identical, respectively. The low degree of homology in telomerase RNAs extends through the template region of the human and mouse RNAs. Interestingly the mouse template region is shorter than that in the human RNA. This shortened template may play a role in the decreased processivity of the mouse telomerase. The probability of dissociation may be greater for the mouse than the human telomerase, because the degree of complementarity between the template and the telomere primer is less. Mutational analysis of the human and mouse RNAs will allow a greater understanding of the functional consequences of the sequence divergence between these species.

Because the mouse RNA sequence is so different from the human RNA, functional evidence was obtained to demonstrate that this RNA is required for telomerase activity. Oligonucleotides complementary to the RNA that covered the template region specifically inhibited elongation, while an oligonucleotide with its 3′ end immediately adjacent to the template was elongated with the sequence predicted from the template. These results are exactly what was found with analogous oligonucleotides against the *Tetrahymena* telomerase [Greider and Blackburn, 1989].

I.E. Chromosome Healing

In addition to telomere replication, telomerase is also involved in chromosome healing. The phenomenon of healing was first observed by McClintock [1942] in maize. She found that chromosomes undergoing repeated cycles of breakage and fusion would sometimes spontaneously heal as if they had regained a telomere. The ability to heal depended on the type of tissue in which the broken chromosome was located When a single broken chromatid was generated at meiosis, rounds of breakage and fusion would occur in the following mitotic divisions. After fertilization the breakage cycle continued in the endosperm tissues but was healed in the zygotic tissues [reviewed by Greider, 1991a]. Spontaneous chromosome healing has also been described in *Plasmodium*, yeast, and human cells. In ciliates and *Ascaris*, chromosome healing is a developmentally regulated event. The chromatin diminution that occurs in the somatic nuclei of these organisms generates fragmented chromosomes onto which telomeres are added. The DNA sequence organization of healed chromosomes suggests *de novo* telomere repeat addition by telomerase. Healed sites usually lack recombination junctions and homology to either telomere sequences or to telomere-associated sequences [reviewed by Greider, 1991a].

Direct experiments in *Tetrahymena* demonstrated that telomerase is involved in developmentally programmed chromosome healing. A mutation was introduced into the telomerase RNA template region, and the gene was transformed into *Tetrahymena* prior to chromosome fragmentation. The mutant telomere sequences appeared as the first repeat added onto the newly generated end, demonstrating that telomerase was responsible for telomere sequence repeat addition [Yu and Blackburn, 1991]. The structure of healed chromosomes in organisms such as *Plasmodium*, yeast, and *Paramecium*, where telomerase has not yet been identified, implies that telomerase is active in these cells. Healing has recently been demonstrated in human cells. Two independent cases of α-thalassemia both contain terminal truncations of chromosome 16 where telomeric d(TTAGGG) repeats were added *de novo* within the wild-type α-globin locus [Wilkie et al., 1990]. Presumably a broken chromosome generated in the germline was healed by telomerase. This healed chromosome was subsequently stably segregated in meiosis and mitosis. Chromosome healing in mammalian tissue culture cells has also been used as a tool to generate large chromosomal deletions to study chromosome structure and function [Farr et al., 1991].

II. TELOMERE LENGTH REGULATION IN IMMORTAL EUKARYOTIC CELLS

The model for telomerase in telomere replication suggests an equilibrium is established between telomere shortening and telomere lengthening. DNA replication leads to telomere shortening because DNA polymerase cannot replicate the very end of a DNA molecule. Telomerase elongates the chromosome through *de novo* sequence addition. DNA polymerase and primers then fill in the complementary C-rich strand. In immortal single cell eukaryotes this model appears to be valid; evidence suggests that a number of genes are involved in regulating the equilibrium between telomere shortening and lengthening [reviewed by Greider, 1993].

Although telomerase can lengthen telomeres by adding telomeric sequences, the overall length of the repeat tract is regulated by additional factors. In yeast, telomere tracts are heterogeneous in size, and different strains maintain this heterogeneity over a wide range of average sizes. Within one strain the number of repeats on all chromosomes is approximately the same; between strains, the tract length can vary from around 100 bp to over 500 bp. When strains containing long telomeres are crossed to strains with short telomeres, the heterozygote telomeres are intermediate in length. Genetic analysis suggests that multiple genes determine the length at which telomere tracts are maintained [Walmsley and Petes, 1985].

Maize telomere repeat tract length is also maintained by a number of different genes. Us-

ing parental strains with very different telomere tract lengths, Burr et al. [1992] studied the tract length in recombinant inbred plants. Telomere length in the recombinants was maintained at different lengths intermediate between the two parental extremes. Using quantitative trait analysis, they showed that three genes are responsible for 50% of the variation in tract length. These data indicate that multiple factors independently contribute to telomere length maintenance.

Mutational analysis in yeast provides information about some of the functions that may influence telomere length. The mutations *tel-1*, *tel-2*, and *est-1* were identified using screens for telomere length mutants [Lundblad and Szostak, 1989; Lustig and Petes, 1986]. Each mutation resulted in cells with shortened telomere sequence tracts. Two other genes, *CDC17* and *RAP1*, which were initially identified for their cell cycle and silencing phenotypes, respectively, also have dramatic effects on telomere length [Carson and Hartwell, 1985; Shore and Naysmyth, 1987]. *CDC17* encodes DNA polymerase α, indicating that a mutant in the DNA replication machinery can affect telomere length. RAP1 (repressor activator protein) is a transcriptional silencer and activator that binds tightly to telomeric sequences *in vitro* and *in vivo* [Buchman et al., 1988; Conrad et al., 1990; Longtine et al., 1989]. Temperature sensitive alleles of *rap-1* can generate progressively shorter telomeres, while overexpression of the C-terminal region and some point mutations results in very long, heterogeneous telomere tracts [Conrad et al., 1990; Kyrion et al., 1993; Lustig et al., 1990]. Alleles of *rap-1* have been isolated that appear not to affect transcription but do affect silencing and telomere lengths, indicating that the telomere length effects may be due to direct action of RAP1. [Sussel and Shore, 1991]. RIF1 (RAP1 interacting factor), a protein identified by its ability to bind RAP1, can also disrupt normal telomere length regulation [Hardy et al., 1992]. The *EST1* (ever shorter telomeres) gene has been suggested as a candidate for encoding a telomerase component in yeast. In *est-1* null alleles telomeres shorten progressively with each round of cell division [Lundblad and Szostak, 1989]. In *rad-52* recombination-deficient backgrounds the cells senesce and die. However, if recombination is allowed to occur, pseudorevertants or "bypass" mutants arise that recombine telomeric repeats onto the ends of shortening chromosomes through addition of subtelomeric Y′ elements [Lundblad and Blackburn, 1993]. To determine the role of EST1 in telomerase, it will be necessary to assay telomerase directly in yeast.

With the cloning of the yeast telomerase RNA components (see above), it was possible to detect the fate of cells deleted for a specific essential telomerase component. Interestingly, the phenotypes of the telomerase RNA gene deletions from both *K. lactis* and *S. cerevisiae* are very similar to the phenotype seen previously for the *est-1* deleted strains. In both cases telomeres shorten progressively with increasing rounds of cell division, the cells get very sick, and most cells die after the telomeres are shortened. However, at a very low rate "survivors" are generated that restore telomere length by a recombinational mechanism [Singer and Gottschling, 1994; McEachern and Blackburn, 1995]. These experiments are excellent models for what might happen to immortalized human cells if telomerase is deleted or inhibited (see below).

III. TELOMERE LENGTH REGULATION IN MAMMALIAN CELLS
III.A. Normal and Immortalized Human Somatic Cells

Unlike the equilibrium that is established to maintain telomere length in single cell eukaryotes, telomere length is not maintained in primary human somatic cells that have a limited life span. In primary human cells and tissues telomeres shorten with replicative age [Harley et al., 1990; Hastie et al., 1990; Lindsey et al., 1991]. As in the simpler eukaryotes, telomere dynamics in mammalian cells are typically examined by Southern analysis of the TRFs. The TRFs contain both subtelomeric DNA and terminal d(TTAGGG) repeats; the functional

telomere is likely constrained to the terminal d(TTAGGG) sequences. Estimates of the length of the subtelomeric non-d(TTAGGG) DNA in human TRFs range from 2 to 5 kbp [Allsopp et al., 1992; Counter et al., 1992]. Some d(TTAGGG) repeats may not be contiguous with the terminal d(TTAGGG) array; thus the functional telomere may represent only a fraction of the total TRF length.

At birth, the TRFs of normal human somatic cell chromosomes are ≈10–12 kbp [Allshire et al., 1989; Allsopp et al., 1992; de Lange et al., 1990; Vaziri et al., 1993]. The TRFs gradually shorten with cell division in culture and as a function of donor age *in vivo* in human somatic cells but not germline or cancer cells [Allsopp et al., 1992; Counter et al., 1992; Harley, 1991; Harley et al., 1990; Hastie et al., 1990]. Shortening of the TRFs represents telomere loss, since the signal strength of the hybridization pattern detected with a telomeric probe also decreased with age. The rate of telomere loss in proliferating cells or tissues is typically 50–150 bp per cell doubling *in vitro* and 15–50 bp per year *in vivo*. When DNA was examined from cells that were kept quiescent in culture, TRFs did not decrease as a function of time. Similarly, TRFs from DNA isolated from brain tissue did not decrease as a function of donor age [Allsopp et al., submitted]. Sperm cell TRFs remain constant with age of the donor and are longer than that seen for somatic cells at birth (≈12–15 kbp vs. ≈10–12 kbp) [Allsopp et al., 1992]. Tumor cells and cells transformed and immortalized *in vitro* had stable but generally short TRFs, as did tumor tissues [Harley et al., 1994; Kim et al., 1994]

These observations are consistent with the hypothesis that telomerase is not active in somatic tissues to reduce the probability of cancer in long-lived organisms such as humans [reviewed by Harley et al., 1994]. Thus, telomerase would be active in some cells in the reproductive lineage to maintain chromosome length between generations and might be aberrantly reactivated in cancer (generally late in life). This hypothesis received substantial support through use of a highly sensitive and reproducible PCR-based assay for telomerase activity. Kim and coworkers confirmed and extended earlier reports that telomerase activity is not detected in primary human fibroblasts, embryonic kidney cells (in culture), lymphocytes, and epithelial cells (Table I) [Counter et al., 1992, 1994a,b; Kim et al., 1994; Shay et al., 1993; Vaziri et al., 1993]. In this study, telomerase was active in 98 of 100 immortal cell lines, 90 of 101 tumor samples representing 12 tumor types, and in normal testes and ovaries. In contrast, telomerase activity was not detected in 22 of 22 normal somatic cell cultures and in 50 of 50 normal or benign biopsies representing 18 different tissues.

Questions concerning cell immortality, telomere regulation, and telomerase expression in true somatic stem cells in humans have not been fully answered. Although telomere loss as a function of replicative aging in culture or as a function of donor age *in vivo* has been documented in a variety of human cells and tissues that undergo renewal, including candidate hematopoietic stem cells [Vaziri et al., 1994], weak telomerase activity has been detected with the PCR-based assay in some preparations of candidate stem cells stimulated for growth and differentiation *in vitro* and in peripheral blood lymphocytes [C.-P. Chiu, N.W. Kim, C.B. Harley, P.M. Lansdorp, S. Bacchetti, C. Counter, J. Shay, unpublished data]. Given that telomeric DNA is lost with age *in vitro* and *in vivo* in these cells [Vaziri et al., 1993, 1994], the biological significance of this activity is uncertain.

The telomere hypothesis of cell aging and immortalization suggests that a causal link may exist between telomere loss and aging [Harley, 1991; Harley et al., 1990] (Fig. 3). In addition, cell immortality and cancer may be linked to telomerase activation. Critical telomere loss on one or more chromosomes may signal cell cycle arrest by a checkpoint mechanism. This mechanism may be similar to or share components with the p53-mediated pathway that monitors DNA integrity and can arrest cells if damage is present [Goldstein, 1990; Harley et al., 1990] see also Wright and Shay, this vol-

ume; Lowe and Ruley, this volume]. In the absence of telomerase or other mechanisms to repair telomere loss, human somatic cells might irreversibly arrest in a metabolically active state at senescence. There are, in fact, similarities in morphology and gene expression between fibroblasts that are senescent and fibroblasts that harbor as few as one double strand DNA break induced by radiation [DiLeonardo et al., 1994].

The second part of the hypothesis is that telomerase reactivation may be required for the growth of immortalized cells. This need not necessarily be linked to the role of short telomeres in signaling senescence. Telomerase reactivation may be necessary to stabilize chromosomes with shortened telomeres and permit continued cell proliferation. Although telomerase activity is associated with immortal cells, telomerase activation does not necessarily associate with malignancy. Telomerase is expressed in normal germline cells. In addition, some malignant tumors have little or no detectable telomerase [Kim et al., 1994; Shay et al., unpublished data]. The malignant and metastatic potential of tumors may be primarily a function of mutations in genes involved directly or indirectly in cell–cell communication in the context of growth regulation. Telomerase activation may be increasingly likely as tumors evolve and progress toward metastasis, since strong selection for telomerase activation may occur after many cellular divisions when telomeres become critically short. Without telomerase activation, many advanced tumors may regress due to telomere loss and possibly massive chromosome instability.

To determine ultimately whether telomere loss and telomerase activity are simply biomarkers of aging and cancer, or one of their fundamental causes, we await results from experiments designed to establish if manipulation of telomere length or telomerase activity in human cells affects cell life span and diseases of aging, including cancer. As discussed above, a precedent for such a causal relationship exists in ciliates and yeast, since critical telomere loss and a "senescent" phenotype can

TABLE I. Studies of Telomerase Activity*

Source	Phenotype	Tissue origin	Telomere dynamics	Telomerase activity	References
Cultured cells	Mortal (dividing)	Various (skin, lung, vascular, hemopoietic, ovary)	Shorten or n.t.[§]	0 of 25+[¶]	Allsopp et al. [1992], Counter et al.[1992], Harley et al. [1990], Vaziri et al. [1993]
	Mortal (nondividing)	Connective (skin, lung)	Stable	0 of 3+	————**
	Mortal (extended life span)	Emb. kidney Connective Blood	Shorten Shorten Shorten	0 of 5+	Counter et al. [1992, 1994a]
	Immortal (transformed lines)[†]	Various (lung, kidney, prostate, retina)	Stable or n.t.	8 of 10+	Counter et al. [1992, 1994a], Kim et al. [1994], Shay et al. [1993a]
	Immortal (tumor lines)	Various (14 different tissue origins)	Stable or n.t.	0 of 70+	Counter et al. [1992, 1994a] Kim et al. [1994]
Normal tissues	Immortal[‡]	Testes	Stable	2 of 2+	Allsopp et al. [1992], Kim et al. [1994]
		Connective	Shorten		Allsopp et al. [1992], Harley et al. [1990], Kim et al. [1994]
		Epidermis	Shorten		
		Blood	Shorten		Lindsey et al. [1991][††]
	Mortal[‖]	Vascular intima	Shorten	0 of >60+	Counter et al. [1994a,b], Hastie et al. [1990], Vaziri et al. [1993]
		Brain	Stable		
		Others (breast, prostate, uterus, intestine, kidney, liver, lung, muscle, spleen)	n.t.		Kim et al. [1994]**
Tumor tissues		Breast	n.t.		Kim et al. [1994]
		Prim. node neg.		1 of 4+	
		Duct node pos.		14 of 15+	
		Ovarian carcinoma	Stable	7 of 7+	Counter et al. [1994b]
		Prostate	n.t.		Kim et al. [1994]
		BPH		1 of 10+	
		PIN3		3 of 5+	
		Adenocarcinoma		2 of 2+	Kim et al. [1994]
		Neuroblastoma	Shorten	5 of 5+	Kim et al. [1994]
		Head and neck	n.t.	14 of 16+	
		Colon			Kim et al. [1994], Vaziri et al. [1993]
		Polyp	n.t.	0 of 1+	
		Tubular adenoma	Shorten or n.t.	0 of 1+	
		Carcinoma	Shorten or n.t.	8 of 8+	Kim et al. [1994]
		Uterine			
		Fibroids		0 of 11+	
		Sarcoma		3 of 3+	

*Adapted from Harley et al. [1994], with permission.

[†]Sublines of T-antigen transformed and immortalized cells may become telomerase negative, perhaps through genetic instability [Kim et al., 1994]. It is not known whether these subclones have an indefinite replicative capacity.

[‡]The germline lineage is immortal, even if specific cell types may not be.

[‖]Whether normal tissues contain rare immortal stem cells that are not detected by current telomerase assays has not been determined.

[§]n.t., Not tested.

[¶]Some cultured hemopoietic cells may display weak telomerase activity detected with the PCR-based assay [Kim et al., 1994; C.-P. Chiu, N.W. Kim, P.M. Landsdorp and C.B. Harley, unpublished data].

**R.C. Allsopp and C.B. Harley (unpublished data).

[††]K.R. Prowse and C.B. Harley (unpublished data).

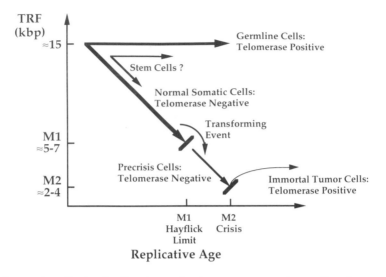

Fig. 3. *The telomere hypothesis of cell aging and immortalization. Telomerase is active in germline cells, maintaining long stable telomeres, but is repressed in most normal somatic cells, resulting in telomere loss in dividing cells. Telomerase activity and telomere length have not been well characterized during embryonic and fetal development or for true somatic stem cells. At mortality phase 1 (M1, or the Hayflick limit), there is presumed critical telomere loss on one or perhaps a few chromosomes signaling irreversible cell cycle arrest. Transformation events may allow somatic cells to bypass M1 without activating telomerase. When telomeres become critically short on a large number of chromosomes, cells enter crisis (M2). Rare clones that activate telomerase escape M2, stabilize chromosomes, and acquire an indefinite growth capacity. [Adapted from Harley, et al., 1994, with permission of the publisher.]*

be induced by mutations in the telomerase RNA component or other genes involved in telomere maintenance [Lundblad and Szostak, 1989; Singer and Gottschling, 1994; McEachern and Blackburn, 1995; Yu et al., 1990].

III.B. Are Mice Different?

Compared with humans, many laboratory strains of mice and other rodents have relatively long arrays of d(TTAGGG) repeats. Thus, it has been difficult to assess the significance of reports that telomere loss does not occur as a function of age in mice [Kipling and Cooke, 1990], since loss of true telomeric DNA may be obscured by the very large size of the TRFs. Loss of 5–10 kb might not be detected on the 100–200 kb terminal restriction fragments. *Mus spretus* has TRFs comparable in size with those of humans, and in this species we have demonstrated telomere loss as a function of replicative aging of fibroblasts in culture. Moreover, it appears that telomeres may

be lost with age *in vivo* in some somatic tissues [Prowse and Greider, 1995].

However, it is not clear whether murine cells have a stringent control of replicative capacity analogous to that seen in human somatic cells. Rodent fibroblasts spontaneously immortalize in culture at a relatively high frequency, concomitant with telomerase activation [Prowse and Greider, 1995]. In addition, telomerase activity can readily be detected in a variety of normal rodent tissues in vivo [Prowse and Greider, 1995; C. Chadeneau, W. Muller, C. Harley, and S. Bacchetti, 1995]. Incomplete repression of telomerase in rodent cells may account for their high frequency of spontaneous immortalization and the extremely high frequency of cancer in mice *in vivo* on a per cell, per year basis. The effects of experimental manipulation of telomere length on age-related disease and cancer in mice, for example, by knocking out the telomerase RNA gene, will help address these issues.

IV. IMPLICATIONS IN MEDICINE

Replicative senescence, possibly at focal sites of high cell turnover, likely occurs with increasing probability with age in a variety of tissues. Thus, cell senescence, reflected not only in the inability to divide but also in the aberrant pattern of gene expression, may contribute to many age-related pathologies. Examples include reduced wound healing in skin (dermal fibroblast aging), immune dysfunction (T-cell aging), and cardiovascular diseases (endothelial cell aging). It also seems clear that escape from replicative senescence permits many tumors to progress to a malignant or metastatic state. Since telomere loss and telomerase activation may be fundamentally involved in these processes, telomeres and telomerase are exciting targets for discovering new diagnostics and therapeutics. For example, delaying critical telomere loss through transient reactivation of telomerase or slowing the rate of telomere loss might extend cell life span and delay the onset of replicative senescence. This could be of clinical use in *ex vivo* applications as well as diseases *in vivo*. In contrast, telomerase inhibition could provide a safe and effective therapy for many, if not all, cancers. Before these ideas can be tested in humans we need to discover and develop specific methods to modulate telomerase or telomere length and analyze the effects of such modulation on normal growth, development, and disease progression in animals.

V. CONCLUSION

Telomeres are the dynamic structures at chromosome ends and are essential for chromosome stability and replication. Thus, understanding the dynamics of telomere maintenance is central to our understanding of cell biology and may lead to fundamentally new insights into the complex processes of cancer and aging. The cloning of the RNA component of telomerase from ciliates, yeast, humans, and rodents should permit critical testing of hypotheses presented here. We hope to

know soon if telomere loss is causally linked to replicative senescence and age-related diseases and if telomerase reactivation is causally linked to cell immortality and cancer. If these associations are formally established, there may be important implications for novel drug discovery and benefits to human health.

ACKNOWLEDGMENTS

We thank many of our colleagues in our own labs and in the fields of telomeres, aging, and cancer for critical discussions and their contributions to ideas and concepts described in this chapter. We thank Chantal Autexier for critical reading of the manuscript. Some of the work described here was supported by grants from the NIH (AG 09383 to C.W.G. and C.B.H.), by the Allied Signal Outstanding Research Award to C.B.H. and C.W.G., and by Geron Corporation.

REFERENCES

Allshire RC, Dempster M, Hastie ND (1989): Human telomeres contain at least three types of G-rich repeats distributed non-randomly. Nucleic Acids Res 17:4611–4627.

Allsopp RC, Vaziri H, Patterson C, Goldstein S, Younglai EV, Futcher AB, Greider CW, Harley CB (1992): Telomere length predicts the replicative capacity of human fibroblasts. Proc Natl Acad Sci USA 89:10114–10118.

Ares M (1986): U2 RNA from yeast is unexpectedly large and contains homology to vertebrate U4, U5 and U6 small nuclear RNAs. Cell 47:49–59.

Bernards A, Michels PAM, Lincke CR, Borst P (1983): Growth of chromosomal ends in multiplying trypanosomes. Nature 303:592–597.

Blackburn EH (1991): Structure and function of telomeres. Nature 350:569–573.

Brown WRA (1989): Molecular cloning of human telomeres in yeast. Nature 338:774–776.

Buchman AR, Kimmerly WJ, Rine J, Kornberg RD (1988): Two DNA binding factors recognize specific sequences at silencers, upstream activating sequences, autonomously replicating sequences, and telomeres in *S. cerevisiae*. Mol Cell Biol 8:210–225.

Burr B, Burr F, Matz EC, Romero-Severson J (1992): Pinning down their loose ends: Mapping telomeres and factors effecting their length. Plant Cell 4:953–960.

Carson M, Hartwell L (1985): CDC 17: An essential gene that prevents telomere elongation in yeast. Cell 42:249–257.

Chamberlin MJ (1995): New models for the mechanism of transcription elongation and its regulation. Harvey Lectures. Series 88 (1992–1993). New York: John Wiley & Sons, pp 1–21.

Collins K, Greider CW (1993): *Tetrahymena* telomerase catalyzes nucleolytic cleavage and non-processive elongation. Genes Dev 7:1364–1376.

Conrad MN, Wright JH, Wolf AJ, Zakian VA (1990): RAPl protein interacts with yeast telomeres *in vivo*: Overproduction alters telomere structure and decreases chromosome stability. Cell 63:739–750.

Counter CM, Avilion AA, LeFeuvre CE, Stewart NG, Greider CW, Harley CB, Bacchetti S (1992): Telomere shortening associated with chromosome instability is arrested in immortal cells which express telomerase activity. EMBO J 11:1921–1929.

Counter CM, Botelho FM, Wang P, Harley CB, Bacchetti S (1994a): Stabilization of short telomeres and telomerase activity accompany immortalization of Epstein-Barr virus transformed human B lymphocytes. J Virol 68:3410–3414.

Counter GM, Hirte HW, Bacchetti S, Harley CB (1994b): Telomerase activity in human ovarian carcinoma. Proc Natl Acad Sci USA 91:2900–2904.

Cross S, Lindsey J, Fantes J, McKay S, Cooke H (1990): The structure of a subterminal repeated sequence present on many human chromosomes. Nucleic Acids Res 18:6649–6657.

de Lange T, Shiue L, Myers R, Cox DR, Naylor SL, Killery AM, Varmus HE (1990): Structure and variability of human chromosome ends. Mol Cell Biol 10:518–527.

DiLeonardo A, Linke S, Clarkin K, Whal G (1994): DNA Damage triggers a prolonged p53 dependent G_1 arrest and long term induction of Cipl in normal fibroblasts. Genes Dev 8:2540–2551.

Farr C, Fantes J, Goodfellow P, Cooke H (1991): Functional reintroduction of human telomeres into mammalian cells. Proc Natl Acad Sci USA 88:7006–7010.

Goldstein S (1990): Replicative senescence: The human fibroblast comes of age. Science 249: 1129–1133.

Gottschling DE, Aparicio OM, Billington BL, Zakian VA (1990): Position effect at *S. cerevisae* telomeres: reversible repression of PolII transcription. Cell 63:751–762.

Greider C (1995): Telomerase biochemistry and regulation. In Blackburn EH, Greider CW (eds): Telomeres. Cold Spring Harbor, NY: Cold Spring Harbor Laboratory, pp 35–68.

Greider CW (1991a): Chromosome first aid. Cell 67:645–647.

Greider CW (1991b): Telomerase is processive. Mol Cell Biol 11, 4572–4580.

Greider CW (1993): Telomerase and telomere length regulation: Lessons from small eukaryotes to mammals. Cold Spring Harbor, Symp Quant Biol 58:719–723.

Greider CW, Blackburn EH (1985): Identification of a specific telomere terminal transferase activity in *Tetrahymena* extracts. Cell 43: 405–413.

Greider CW, Blackburn EH (1987): The telomere terminal transferase of *Tetrahymena* is a ribonucleoprotein enzyme with two kinds of primer specificity. Cell 51:887–898.

Greider CW, Blackburn EH (1989): A telomeric sequence in the RNA of *Tetrahymena* telomerase required for telomere repeat synthesis. Nature 337:331–337.

Hardy CFJ, Sussel L, Shore D (1992): A RAPl-interaction protein involved in transcriptional silencing and telomere length regulation. Genes Dev 6:801–814.

Harley CB (1991): Telomere loss: Mitotic clock or genetic time bomb? Mutat Res 256:271–282.

Harley CB, Futcher AB, Greider CW (1990b): Telomeres shorten during ageing of human fibroblasts. Nature 345:458–460.

Harley CB, Kim NW, Prowse KR, Weinrich SL, Hirsch KS, West MD, Bacchetti S, Hirte HW, Counter CM, Greider CW, Wright WE, Shay JW (1994): Telomerase, cell immortality and cancer. Cold Spring Harbor Symp Quant Biol 59:307–315.

Harrington LA, Greider CW (1991): Telomerase primer specificity and chromosome healing. Nature 353:451–454.

Hastie ND, Dempster M, Dunlop MG, Thompson AM, Green DK, Allshire RC (1990): Telomere reduction in human colorectal carcinoma and with ageing. Nature 346:866–868.

Henikoff S (1990): Position-effect variegation after 60 years. Trends Genet 6:422–426.

Igel AH, Ares M (1988): Internal sequences that distinguish yeast from metazoan U2 snRNA are unnecessary for pre-RNA splicing. Nature 334:450–453.

Kim NW, Piatyszek MA, Prowse KR, Harley CB, West MD, Ho PL, Coviello GM, Wright WE, Weinrich SL, Shay JW (1994): Specific association of human telomerase activity with immortal cells and cancer. Science 266:2011–2014.

Kipling D, Cooke HJ (1990): Hypervariable ultra-long telomeres in mice. Nature 347:400–402.

Kramer KM, Haber JE (1993): New telomeres in yeast are initiated with a highly selected subset of TG1-3 repeats. Genes Dev 7:2345–2356.

Kyrion G, Boakye K, Lustig A (1993): C-terminal truncation of RAPl results in deregulation of telomere size, stability and function in *Saccharomyces cerevisiae*. Mol Cell Biol 12:5159–5173.

Larson DD, Spangler EA, Blackburn EH (1987): Dynamics of telomere length variation in *Tetrahymena thermophila*. Cell 50:477–483.

Lee MS, Blackburn EH (1993): Sequence-specific DNA primer affects on telomerase polymerization activity. Mol Cell Biol 13:6586–6599.

Levy MZ, Allsopp RC, Futcher AB, Greider CW, Harley CB (1992): Telomere end-replication problem and cell aging. J Mol Biol 225:951–960.

Lindsey J, McGill NI, Lindsey LA, Green DK, Cooke HJ (1991): *In vivo* loss of telomeric repeats with age in humans. Mutat Res 256:45–48.

Lingner J, Hendrick LL, Cech TR (1994): Telomerase RNAs of different ciliates have a common secondary structure and a permuted template. Genes Dev 8:1984–1998.

Longtine M, Wilson N, Petracek M, Berman J (1989): A yeast telomere binding activity binds to two related telomere sequence motifs and is indistinguishable from RAPl. Curr Genet 16:225–239.

Lundblad V, Blackburn EH (1993): An alternative pathway for yeast telomere maintenance rescues estl senescence. Cell 73:347–360.

Lundblad V, Szostak JW (1989): A mutant with a defect in telomere elongation leads to senescence in yeast. Cell 57:633–643.

Lustig AJ, Kurtz S, Shore D (1990): Involvement of the silencer and UAS binding protein RAPl in regulation of telomere length. Science 250:549–552.

Lustig AL, Petes TD (1986): Identification of yeast mutants with altered telomere structure. Proc Natl Acad Sci USA 83:398–1402.

Mantell LL, Greider CW (1994): Telomerase activity in germline and embryonic cells of *Xenopus*. EMBO J 13:3211–3217.

McClintock B (1942): The fusion of broken ends of chromosomes following nuclear fission. Proc Natl Acad Sci USA 28:458–463.

McEachern MJ, Hicks JB (1993): Unusually large telomeric repeats in the yeast *Candida albacans*. Mol Cell Biol 13:551–560.

Melek M, Davis B, Shippen DE (1994): Oligonucleotides complementary to the *Oxytricha nova* telomerase RNA delineate the template domain and uncover a novel mode of primer utilization. Mol Cell Biol 14:7827–7838.

Morin GB (1989): The human telomere terminal transferase enzyme is a ribonucleoprotein that synthesizes TTAGGG repeats. Cell 59:521–529.

Olovnikov AM (1973): A theory of marginotomy. J Theor Biol 41:181–190.

Prowse KR, Avilion AA, Greider CW (1993): Identification of a nonprocessive telomerase activity from mouse cells. Proc Natl Acad Sci USA 90:1493–1497.

Prowse KR, Greider CW (1995): Developmental and tissue specific regulation of mouse telomerase and telomere length. Proc Natl Acad Sci USA 92:4818–4822.

Riethman HC, Moyzis RK, Meyne J, Burke DT, Olson MV (1989): Cloning human telomeric DNA fragments into *Saccharomyces cerevisiae* using a yeast-artificial-chromosome vector. Proc Natl Acad Sci USA 86:6240–6244.

Shampay J, Szostak JW, Blackburn EH (1984): DNA sequences of telomeres maintained in yeast. Nature 310:154–157.

Shay JW, Wright WE, Werbin H (1993): Toward a molecular understanding of human breast cancer: A hypothesis. Breast Cancer Res Treat 25:83–94.

Shippen-Lentz D, Blackburn EH (1989): Telomere terminal transferase activity from *Euplotes crassus* adds large numbers of TTTTGGGG repeats onto telomeric primers. Mol Cell Biol 9:2761–2764.

Shippen-Lentz D, Blackburn EH (1990): Functional evidence for an RNA template in telomerase. Science 247:546–552.

Shore D, Naysmyth K (1987): Purification and cloning of a DNA binding protein from yeast that binds to both silencer and activator elements. Cell 51:721–732.

Shore DA (1995): Telomere positions effects. In Blackburn EH, Greider CW (eds): Telomeres. CSH Press, pp 139–199.

Singer MS, Gottschling DE (1994): TLCl: Template RNA component of *Saccharomyces cerevisiae* telomerase. Science 266:404–409.

Sussel L, Shore D (1991): Separation of transcriptional activation and silencing functions of the RAPl-encoded repressor/activator protein: Isolation of viable mutants affecting both silencing and telomere length. Proc Natl Acad Sci USA 88:7749–7753.

Szostak JW, Blackburn EH (1982): Cloning yeast telomeres on linear plasmid vectors. Cell 29:245–255.

Vaziri H, Dragowska W, Allsopp RC, Thomas TE, Harley CB, Lansdorp PM (1994): Evidence for a mitotic clock in human hematopoietic stem cells: Loss of telomeric DNA with age. Proc Natl Acad Sci USA 91:9857–9860.

Vaziri H, Schaechter F, Uchida I, Wei L, Xiaoming Z, Effros R, Choen D, Harley CB (1993): Loss of telomeric DNA during aging of normal and trisomy 21 human lymphocytes. Am J Hum Genet 52:661–667.

Walmsley RM, Petes TD (1985): Genetic control of chromosome length in yeast. Proc Natl Acad Sci USA 82:506–510.

Watson JD (1972): Origin of concatameric T4 DNA. Nature New Biol 239:197–201.

Wilkie AO, Lamb J Harris PC, Finney RD, Higgs DR (1990): A truncated human chromosome 16 associated with alpha thalassaemia is stabilized by addition of telomeric repeat (TTAGGG)n. Nature 346:868–871.

Yu GL, Blackburn EH (1991): Developmentally programmed healing of chromosomes by telomerase in *Tetrahymena*. Cell 67:823–832.

Yu GL, Bradley JD, Attardi LD, Blackburn EH (1990): *In vivo* alteration of telomere sequences and senescence caused by mutated *Tetrahymena* telomerase RNAs. Nature 344:126–132.

Zahler AM, Prescott DM (1988): Telomere terminal transferase activity in the hypotrichous ciliate *Oxytricha nova* and a model for replication of the ends of linear DNA molecules. Nucleic Acids Res 16:6953–6972.

ABOUT THE AUTHORS

CAROL W. GREIDER is a Senior Staff Scientist At Cold Spring Harbor Laboratory in New York. She also holds a position as an Adjunct Professor in the Genetics program and the Department of Molecular and Cellular Biology at the State University of New York at Stony Brook. Dr. Greider obtained her B.A. degree in Biology at the University of California at Santa Barbara in 1983. She then began her graduate work at the University of California at Berkeley working in Dr. Elizabeth Blackburn's laboratory. Dr. Blackburn's group had then recently proposed that telomere replication might involve the *de novo* addition of sequences onto chromosome ends. To test this idea, Dr. Greider initiated biochemical experiments designed to identify a DNA polymerase that was specific to telomeres. In 1985 Drs. Greider and Blackburn reported the first identification of telomerase in cell extracts from the ciliate *Tetrahymena*. They subsequently showed that telomerase required RNA for enzyme function. Dr. Greider received her Ph.D. in 1987, and in 1988 she went to Cold Spring Harbor Laboratory as an independent Cold Spring Harbor Fellow, where she identified and cloned the *Tetrahymena* telomerase RNA component. In 1990 Dr. Greider was appointed as a Staff Investigator at Cold Spring Harbor Laboratory and was awarded a Pew Scholarship in the Biomedical Sciences. In this year in a collaborative paper, Dr. Greider and Dr. Harley (then at McMaster University) reported that human telomeres shorten with age *in vitro* and *in vivo*. In 1992 Dr. Greider was appointed as Senior Staff Investigator and in 1994 as a Senior Staff Scientist at Cold Spring Harbor Laboratory. Dr. Greider currently directs a group of scientists studying both the biochemical mechanism of *Tetrahymena* and mammalian telomerase and the role of telomerase and telomere length regulation in aging and cancer.

CALVIN B. HARLEY is Vice President of Research at Geron Corporation, Menlo Park, California. He also holds a part-time appointment with the Department of Biochemistry at McMaster University, Ontario. Dr. Harley completed his B.Sc. in Honors Science at the University of Waterloo, Ontario, in 1975. He held a National Research Council, Canada, Centennial Scholarship while conducting his graduate work in Sam Goldstein's laboratory at McMaster University (1975–1979). His Ph.D. thesis was entitled "Aging, protein synthesis, and mistranslation in human cells." He did postdoctoral training in evolution with John Maynard Smith, University of Sussex, England, and in molecular biology with Herbert W. Boyer, University of California, San Francisco, before returning to McMaster University in 1982 as a faculty member and Medical Research Council Scholar. He was Chair-elect (1987–1989) and Chairman (1989–1991) of the Canadian Association on Gerontology Division of Biological Sciences, and Associate Editor of the *Canadian Journal on Aging* (1991–1993). Dr. Harley joined Geron Corporation, a biotechnology company focusing on the discovery and development of solutions for health care problems associated with aging, in 1993. The primary focus of research at Geron is the role of replicative senescence in age-related diseases, including cancer, and the development of methods to discover novel drugs to treat these diseases.

Cellular Aging and Cell Death: 139–151
© 1996 Wiley-Liss, Inc.

EF-1$_\alpha$–S1 Gene Family and Regulation of Protein Synthesis During Aging

Stephen Lee, Atanu Duttaroy, and Eugenia Wang

I. INTRODUCTION

The relationship between protein synthesis and aging has been extensively studied in the past three decades. An astonishing number of reports have led today's investigators in the aging field to believe generally that cellular and organismal aging is accompanied by a slowing down of protein synthesis. The exact consequence of such decrease is still a mystery, although it has been suggested that decreased protein synthesis may contribute to the deterioration of normal cellular machinery in senescent cells and organisms. Since decreased protein synthesis is implicated in the attainment of the senescent state *in vivo* and in culture, therefore it is necessary to elucidate the exact mechanisms leading to the down-regulation of translation.

Protein translation takes place in three major steps, initiation, elongation, and termination, with each step being promoted by soluble protein factors, known as initiation factors, elongation factors, and termination factors, and their subsequent interaction with ribosomes, mRNA, and aminoacyl-tRNAs. All of these factors may be implicated or may be the direct cause of impaired protein synthesis during aging and therefore have been studied individually for their possible role in age-related changes in protein synthesis [see Rattan, 1991, for a detailed review of age-related changes in the components of translation]. Such observations suggest that the elongation step of protein synthesis is most affected during aging. For example, in the fruit fly *Drosophila melanogaster*, peptide chain elongation is decreased significantly with increasing age [Webster, 1985]. Similar decreases in peptide elongation rate have been reported in rat and mouse, in which case the binding of aminoacyl-tRNA to the ribosome was reduced, causing a drop in elongation rate. It is known that peptide elongation essentially requires elongation factor 1$_\alpha$ (EF-1$_\alpha$), which helps to bind charge tRNA molecules to the ribosome. This key role of EF-1$_\alpha$ in protein chain elongation makes EF-1$_\alpha$ a prime candidate for defiants in protein synthesis with age.

EF-1$_\alpha$ is a ubiquitous, highly conserved protein that has been the focus of intensive investigation for the past 40 years. Several facts emerged from these studies: *(1)* EF-1$_\alpha$ is an abundant protein expressed in every tissue and cell, *(2)* it is encoded by more than one gene and they are expressed differentially during development, and *(3)* EF-1$_\alpha$ controls the translational rate. In addition to these functions, a casual correlation has been repeatedly observed between EF-1$_\alpha$ expression and protein synthesis with cellular and organismal aging, although the exact role of EF-1$_\alpha$ in senescence is still highly debatable. In recent years, a more direct relationship between EF-1$_\alpha$ levels, translational fidelity, and cellular longevity is beginning to emerge. Differentially expressed and cell type–specific EF-1$_\alpha$ isotypes have been uncovered, including the more recent reports of a terminal differentiation–specific EF-1$_\alpha$ isotype.

This chapter is an effort to relate EF-1$_\alpha$ to cellular and organismal aging and to summarize the vast amount of literature pertaining to EF-1$_\alpha$. In doing so we first survey the literature on the biochemical and molecular characterization of EF-1$_\alpha$ and related genes. Second, the expression pattern of EF-1$_\alpha$ gene(s) during development and aging is considered. We conclude with an up-to-date analysis of the two mammalian EF-1$_\alpha$ genes and speculate on their role in the maintenance of cellular survival during mammalian aging.

II. EF-1$_\alpha$ BIOCHEMISTRY

The first step in peptide elongation consists of cyclic addition of aminoacyl-tRNAs to the acceptor site of the 80S ribosomal complex. The EF-1$_\alpha$ subunit of the EF-1 complex performs this function. EF-1$_\alpha$ is a 50 kD molecule that binds GTP and aminoacyl-tRNA to form the EF-1$_\alpha$–GTP–aminoacyl-tRNA ternary complex [Jurnak, 1985; LaCour et al., 1985; Lauer et al., 1984; van Hemert et al., 1984; Kohno et al., 1986]. Its principal function is to deliver the charged tRNA to the A site of the ribosome in a codon-specific manner. The binding of EF-1$_\alpha$ to the ribosome promotes GTP hydrolysis and release of the EF-1$_\alpha$–GDP binary complex that accumulates in the assay mixture, indicating that another factor must be present in the EF-1 complex to permit the guanosine nucleotide exchange reaction to occur [Weissbach et al., 1973; Lin et al., 1969; Slobin and Moller, 1976; Ejiri et al., 1977; Slobin and Moller, 1978; Grasmuk et al., 1978; Motoyoshi et al., 1977]. It is the task of two other EF-1 polypeptides, EF-1β and 1-γ, to facilitate the guanyl nucleotide exchange on EF-1$_\alpha$ [Riis et al., 1990; Ryazanov and Davydoka, 1989; Janssen and Moller, 1988; Iwasaki et al., 1976] permitting the formation of a new ternary complex EF-1$_\alpha$–GTP-aminoacyl-tRNA. The γ-subunit has been reported to be sufficient for this activity [Iwasaki et al., 1976; Slobin and Moller, 1976]. After the correct match is made, an unknown signal triggers GTP hydrolysis, and EF-1$_\alpha$ is released from the ribosomal complex linked to GDP as a binary complex. Formation of the peptide bond is catalyzed by the peptidyltransferase center of the large subunit of the ribosome [Nygard and Nilssen, 1990].

After the formation of the peptidyl bond, the ribosome shifts one codon downstream, leaving the acceptor site free for another EF-1$_\alpha$–GTP–aminoacyl-tRNA complex. The shifting of ribosome is presumably performed by EF-2, a single polypeptide of 9 kD with a GTP-dependent activity [Hershey, 1991; Merrick, 1992]. The exact mechanism by which EF-2 translocates the mRNA on the surface of the ribosome is still obscure, although the EF-2–GTP complex alone appears to be able to translocate the mRNA in the elongating ribosome. The addition of aminoacyl-tRNA is ensued by the formation of peptide bond and translocation of mRNA within the active site of the ribosome. This cycle will terminate only when a stop codon (usually TGA) is positioned in the A site of ribosome.

II.A. Analysis of the EF-1$_\alpha$ Functional Domains

Comparison of the amino acid sequence between different EF-1$_\alpha$s revealed several interesting features. One is the presence of GTP-binding domains, known as G-domains in the EF-1$_\alpha$–Tu primary amino acid sequence [Jurnak, 1985; LaCour et al., 1985; Kohno et al., 1986], which are totally conserved from yeast to man, indicating a strong selection pressure relating to the proper GTPase activity in all EF-1$_\alpha$s [Valencia et al., 1991a,b]. Striking sequence and structural similarities were reported between the G-domains of ras p21 and members of the RAS family and EF–Tu/EF-1$_\alpha$ [van Damme et al., 1995; Jurnak, 1985; Valencia et al., 1991a,b]. So far, three independent G-domains and one putative domain have been characterized in the EF–Tu/EF-1$_\alpha$ peptide [Dever et al., 1987].

Other domains of EF-1$_\alpha$, including the putative tRNA- and EF-1$\beta\gamma$ binding domains are also very highly conserved. Work done with *Escherichia coli* has shown that certain key amino acids are responsible for binding aminoacyl-tRNA [Moller et al., 1987; Kinzy et al., 1992; van Damme et al., 1992], and such sequences are conserved in all EF–Tu/EF-1$_\alpha$ protein sequences [van Noort et al., 1984, 1985].

The prokaryotic EF–Tu can function in peptide chain elongation using the eukaryotic tRNAs, further indicating that the conserved elements play a role in tRNA attachment [Metz-Boutigue et al., 1989]. There are also several other highly conserved domains in eukaryotic EF-1$_\alpha$s whose exact function is not clear, but could involve binding to mRNA, to ribosomes, or to the nucleotide exchange factor EF–Ts/EF-1$\beta\gamma$ [van Damme et al., 1992]. The structure of the EF-1$_\alpha$ protein is therefore in good agreement with the known biochemical characteristics of EF-1$_\alpha$, including the formation of the ternary structure.

II.B. Post-Translational Modification

The EF-1$_\alpha$ protein from three different sources has been chemically sequenced [Merrick, 1992]. This has brought about the identification of novel post-translational modifications, including the possible modification of glutamic acid to glutamate, and in methylation of certain lysines [Dever et al., 1989; Toledo and Jerez, 1989; Merrick, 1992]. While the significance of the glutamic acid modification is unclear, methylation of the lysine at position 56 has been shown to play a role in controlling EF–Tu activity in bacteria [Toledo and Jerez, 1989]. A second lysine residue at position 79 is also methylated, and, contrary to the lysine at position 56, this modification has been shown to be present in all species [Merrick, 1992]. In fact, four out of five lysines that are methylated are the same (positions 36, 55, 79, 318) between *Artemia salina* and rabbit. It was found in *Mucor racemous* that increased methylation is correlated with observed elevation in overall elongation rate in the yeast-to-hyphac and spore germination stages [Hiatt et al., 1982]. In contrast, hypo- and hypermethylated forms of EF-1$_\alpha$ are equally active in *in vitro* assays [Sherman and Sypherd, 1989]. A more complex modification consisting of addition of glycerylphosphorylethanolamine at position 301 has been identified in both mouse and rabbit [Dever et al., 1989; Whiteheart et al., 1989] is probably present at position 374 in *Artemia,* but is absent from the yeast protein [Merrick, 1992]. The exact function for this post-translational modification remains unknown.

It has also been reported that EF–Tu is phosphorylated at position 382 [Lipmann et al., 1993], and Venema et al. [1991a,b] have shown that mammalian EF-1$_\alpha$ is phosphorylated *in vivo* and *in vitro* at not yet identified sites by PMA and protein kinase C, respectively. They concluded that this modification activates poly(U)-dependent poly(Phe) synthesis. However, sequence alignment from all known EF–Tu and EF-1$_\alpha$ has revealed that the identified position corresponding to 382-The in *E. coli* is completely conserved in all species.

III. MOLECULAR CLONING OF EF-1$_\alpha$
III.A. Isolation of EF-1$_\alpha$ cDNAs

Our comprehension of EF-1$_\alpha$s role in protein synthesis was enhanced with the cloning of EF-1$_\alpha$ cDNA. The first eukaryotic species cDNA corresponding to EF-1$_\alpha$ was cloned in the brine shrimp *A. salina* [van Hemert et al., 1983, 1984]. It was quickly followed by the isolation of the yeast *Saccharomyces cerevisiae*'s EF-1$_\alpha$ cDNA [Nagata et al., 1984; Cottrelle et al., 1985a]. The amino acid sequence homology between these two very divergent organisms and a prokaryotic species EF–Tu was found to be strikingly high and was in agreement with previous analysis that indicated functional similarity between all EF-1$_\alpha$s. Since then cDNAs encoding for EF-1$_\alpha$ have been isolated from a multitude of eukaryotic organisms.

Among mammalian species, human EF-1$_\alpha$ cDNA was the first to be isolated, followed by the cloning of its corresponding gene [Brands et al., 1986; Uetsuki et al., 1989]. cDNA sequences from mouse [Lu and Werner, 1989], rat [Ann et al., 1992; Shirasawa et al., 1992], and rabbit [Cavallius and Merrick, 1992] EF-1$_\alpha$ have been reported in the last 5 years. The predicted amino acid sequence of mammalian cDNA has indicated that the EF-1$_\alpha$ protein is extremely conserved in all mammalian species (97%–100% conservation) [Riis et al., 1990].

Furthermore, in comparison with human EF-1_α, yeast and tomato EF-1_αs are about 80% conserved. Another interesting feature of the mammalian EF-1_α cDNAs is its highly conserved 3′ untranslated region (UTR). The nucleotide sequence of this 3′ UTR is even more conserved than the coding region, while the 5′ UTR in EF-1_α mRNA is divergent between different species.

III.B. EF-1_α Gene Copy Number

Multiple active EF-1_α genes have been found in prokaryotes [An and Friesen, 1980], different yeast species [Schirmaier and Philippsen, 1984; Sundstrom et al., 1990], the fungus *M. racemous* [Linz et al., 1986b], *Artemia* [Lenstra et al., 1986], *Drosophila* [Hovemann et al., 1988], and *Xenopus* [Dje et al., 1990]. Some nine EF-1_α genes have been identified in a plant species, while in animals the number of EF-1_α genes varies between two and four.

III.C. Isolation of a Second EF-1_α Isotype in Mammals—The S1 Gene

Our research is focused on the identification and characterization of factors that control cell growth, quiescence, and senescence. In specific studies mice were injected with the Triton-insoluble extract of *in vitro* aged human fibroblasts. Two monoclonal antibodies (mAB) obtained were referred to as *S-30* and *S-44,* which recognized an antigen named *statin*, that is localized in the nuclei of nonproliferating, but not in proliferating, fibroblasts [Wang, 1985a,b].

In screening a λgt11 expression library with the mAB30 and mAB44, we identified a cDNA clone referred to as *S1* [Wang et al., 1989]. The nucleotide and predicted amino acid sequence of S1 cDNA revealed that it is similar to mouse and human EF-1_α [Ann et al., 1991]. The predicted coding region of S1 is 89% similar to that of both mouse and human EF-1_α. Interestingly, the 5′ and 3′ UTRs of S1 mRNA are totally divergent from those of the EF-1_α mRNA, suggesting that S1 is not the rat homolog of mouse and human EF-1_α. Furthermore, a comparison of the amino acid analysis

of purified statin and the deduced sequence of S1 shows that statin and S1 are distinctly different. In depth, immunoblotting characterization shows that the monoclonal antibodies (S-30 and S-44) do not recognize S1 protein product, suggesting that the two proteins do not share common antigenic epitopes and the original selection of S1 clone from the λgt11 library is the product of fortuitous crossreaction.

The exact function of the EF-1_α isotypes is not known. In yeast, both EF-1_α genes can complement each other, although only one EF-1_α gene is enough for normal metabolic activity [Cotrelle et al., 1985b]. This is also true in bacteria, where the expression of a single EF–Tu gene is necessary for survival [Hughes, 1990]. The amphibian *Xenopus laevis* contains three active EF-1_αs [Dje et al., 1990] of which EF-1S and 42Sp48 are 63% similar, but both are capable of peptide elongation in an *in vitro* system [Mattaj et al., 1987]. The third EF-1_α gene in *Xenopus*, EF-1_αO, is 90% homologous to EF-1S. There is no evidence to support EF-1_αO's involvement in peptide elongation, even though it has been shown to bind to GTP and tRNA *in vitro* [Viel et al., 1987; Bourne et al., 1991].

IV. EF-1_α GENE EXPRESSION DURING DEVELOPMENT AND AGING

The occurrence of different EF-1_α isotypes in various organisms are discussed above. Interestingly some of these EF-1_α isotypes have shown spatial and temporal restrictions with respect to their expression pattern. Such differential expression of the EF-1_α isotypes implies either that different tissues require distinct regulation of the unique isotypes of the EF-1_α family or that the activity of protein synthesis among the various tissues needs the involvement of each specific EF-1_α isotype. These suggestions have arisen from the results gathered from various systems such as unicellular organisms, *Drosophila*, rat, mouse, and human species. The main conclusions derived from these studies on developmental regulation of protein synthesis, translational accuracy, cellular longevity, and aging, are treated separately.

IV.A. Unicellular Organisms

In yeast, EF-1$_\alpha$ is encoded by two unlinked genes, known as TEF-1 and TEF-2. Both *tef* genes encode an identical protein, but their mRNAs differ in the 5´ and 3´ UTRs [Nagata et al., 1984; Cotrelle et al., 1985a; Schirmaier and Philippsen, 1984]. Insertional inactivation of either TEF-1 or TEF-2 suggests that either one of them is sufficient for normal growth [Cotrelle et al., 1985b]. In *Salmonella typhimurium* only one *tef* gene was found to be sufficient for survival [Hughes, 1990]. While these results in yeast and in *Salmonella* suggest a functional redundancy for the EF-1$_\alpha$ gene, an attractive alternate proposal suggests a requirement for multiple EF-1$_\alpha$ genes to permit accumulation of adequately synthesized proteins [Song et al., 1987]. In other words, an enhanced translational accuracy is achieved owing to an increase in the cellular pool of EF-1$_\alpha$. This proposal agrees well with the computer simulation of protein translation data [Pingoud et al., 1990].

As in yeast, three highly homologous EF-1$_\alpha$ genes have been isolated from the fungus *M. racemous* [Linz et al., 1986a,b]. These genes, TEF-1, -2, and -3, differ in their flanking UTRs, encoding for a protein that is nearly identical except for a lysine instead of a glutamate at amino acid position 41 in TEF-2 and TEF-3. The mRNAs of TEF-1, -2 and -3 do not accumulate to the same levels [Linz and Sypherd, 1987]. It is now hypothesized that in *M. racemous* the modulation of EF-1$_\alpha$ activity is a result of post-translational modification, since a concomitant increase in lysine methylation of EF-1$_\alpha$ occurs during this period [Linz and Sypherd, 1987; Hiatt et al., 1982; Sherman and Sypherd, 1989; Fonzi et al., 1985]. Although this does not explain why *M. racemous* and yeast contain three and two copies of exactly similar EF-1$_\alpha$ genes, respectively. It is unlikely that multiple genes could have a quantitative effect, since yeast can grow normally with either of its two EF-1$_\alpha$ genes deleted. So far, no function has been attributed to the extra EF-1$_\alpha$ copies, except for the possible effect of EF-1$_\alpha$ gene dosage on translation fidelity.

Unicellular organisms have yielded important information on EF-1$_\alpha$'s role in peptide chain elongation and how it is possibly related to growth and development. However, the most striking report linking translational accuracy and cellular longevity was recently described in the unicellular ascomycete fungus *Podospora anserina* by Silar and Picard [1994]. These investigators have shown by complementation analysis that a mutant AS4 gene that codes for EF-1$_\alpha$ is functionally linked to translational fidelity. Most importantly, increased accuracy in AS4 expression results in enhanced longevity in this organism. No such mutational evidence is available for the EF-1$_\alpha$ isotype in higher organisms. But the translational accuracy model in lower organisms provides an attractive working hypothesis that may explain why higher organisms also contain more than one EF-1$_\alpha$ gene.

IV.B. *Drosophila melanogaster*

Drosophila contains two actively transcribed EF-1$_\alpha$ genes, F1 and F2 [Hovemann et al., 1988; Walldorf et al., 1985]. Contrary to bacteria and yeast, these two genes differ in their expression profile and coding potential, since the F1 and F2 proteins, while similar, are not exactly identical (92% similarity). With regard to their expression profile, F1 is the housekeeping gene that encodes for an EF-1$_\alpha$ that is abundantly expressed in all cells. F2 gene expression, on the other hand, is temporally restricted to the pupal stage of development. Hovemann et al. [1988] have suggested that F2 exerts a specific function by preferring its own aminoacyl-tRNA pool and thereby controls certain aspects of the elongation step during development. Another possibility is that expression of F2 will increase the pool of active EF-1$_\alpha$ in tissues already expressing F1. This might cause an enhanced translational accuracy in F1–F2-positive cells [Song et al., 1987].

Most of the attempts made to elucidate the mechanisms involved in the age related reduction of protein synthesis were carried out in *Drosophila*. The step most impaired during *Drosophila* aging is the binding of aminoacyl-tRNA to the ribosome, which is the primary function of EF-1$_\alpha$ [Webster and Webster, 1982]. Subsequently, starting in early life there is a sharp decline in the rate of EF-1$_\alpha$ polypeptide synthesis and in EF-1$_\alpha$ mRNA levels [Webster and Webster, 1983, 1984; Webster, 1985]. This phenomenon precedes by a few days a decline in protein synthesis in this organism, implying a direct relationship between the decline in protein synthesis, EF-1$_\alpha$ activity, and aging [Webster and Webster, 1983]. If EF-1$_\alpha$ is directly responsible for a decline in protein synthesis in aged *Drosophila*, then restoring the EF-1$_\alpha$ activity in flies should have an impact on the aging process. With this rationale, transgenic *Drosophila* flies were produced with an extra copy of the F1 gene, the *Drosophila* homolog for EF-1$_\alpha$ [Shepherd et al., 1989], under the control of the Hsp70 promoter, which should drive F1 expression in all cells in *Drosophila*. Two other F1-expressing transgenic strains have longer life spans than the nontransgenic control flies. Interestingly, transgenic strains incubated at 25.0°C did not live as long as strains incubated at 29.5°C, probably due to increased transcription of the transgene due to temperature affects on the Hsp70 promoter. These experiments suggest that increased amounts of EF-1$_\alpha$ can extend the life expectancy in *Drosophila*. Shikama et al. [1994] performed a similar analysis of the transgenic EF-1$_\alpha$ *Drosophila* and concluded that transgenic EF-1$_\alpha$ flies do live longer but not due to the EF-1$_\alpha$ transgene. These authors have also shown that EF-1$_\alpha$ mRNA and protein levels do not dramatically fluctuate during *Drosophila* aging, suggesting that the slow elongation rate may not result from a diminution of the pool of active EF-1$_\alpha$. Hence the role of EF-1$_\alpha$ in *Drosophila* aging remains unclear.

IV.C. Mammalian Species

EF-1$_\alpha$ cDNAs have been isolated from several mammalian species, including rabbit, man, mouse, and rat, and in man, mouse, and rat both EF-1$_\alpha$ and the highly homologous S1 genes are found [Ann et al., 1991, 1992; Shirasawa et al., 1992; Knudson et al., 1993; Lee et al., 1994].

EF-1$_\alpha$ mRNA accumulates at a higher level in rat muscle than in liver, and transformed cells in culture contain more EF-1$_\alpha$ mRNA than cells in normal tissues [Sanders et al., 1992]. Tumor cells also accumulate higher levels of EF-1$_\alpha$ mRNA *in vivo* [Grant et al., 1992]. Such an accumulation of EF-1$_\alpha$ mRNA in transformed cells has been related to the overexpression of EF-1$_\alpha$ in mouse 3T3 cells, which has been suggested to render these cells more susceptible to oncogenic transformation [Tatsuka et al., 1992]. Studies on the post-transcriptional and translational control of EF-1$_\alpha$ mRNA and proteins [Slobin and Jordan, 1984; Rao and Slobin, 1987] have shown that EF-1$_\alpha$ mRNA is more stable in nonproliferating Friend erythroleukemia cells than in their proliferating counterparts. In a later paper, the same authors showed that EF-1$_\alpha$ mRNA, isolated from growth-arrested murine erytholeukemia cells, is not translatable in an *in vitro* assay [Slobin and Rao, 1993] in contrast to most RNPs, which are readily translatable in the same assays. This inhibition may be caused by trans-acting factor(s) that bind to EF-1$_\alpha$ mRNA *in vivo*. These results, along with the post-translational modification discussed above, suggest that in mammalian cells the level and activity of EF-1$_\alpha$ are controlled at post-transcriptional, translational, and post-translational levels.

Both mammalian EF-1$_\alpha$ genes show differential regulation in their expression. It has been shown by this laboratory and by others that S1 gene expression is limited to brain, heart, and muscle in rat, mouse, and man [Lee et al., 1992]. On the other hand, EF-1$_\alpha$ mRNA can be detected in all tissues. When S1 gene expression is examined in the brain by *in situ* hybridization in sections, only certain neurons were found to be S1 positive. This led us to hypothesize that S1 is a terminal differentiation-specific EF-1$_\alpha$. This hypothesis is further strengthened by the fact that S1 expression is a late event in brain, heart, and muscle development, with accumulation of S1 mRNA at a time when cells are permanently leaving the cell cycle [Olson, 1993; Sassoon, 1993; Lee et al., 1993a]. Also in agreement with the terminal differentiation hypothesis is the fact that S1 expression is dependent on the formation of multinucleated myotubes during rat L8 myogenic cell myogenesis in culture [Lee et al., 1993a]. Finally, S1 expression is detected only in primary cultures of terminally differentiated cortical neurons, whereas astrocytes and microglia were shown to be S1 negative. Taken together, these results strongly suggest that mammalian cells contain a second EF-1$_\alpha$ gene whose expression is strictly limited to terminally differentiated tissues such as neurons, cardiomyocytes, and myocytes.

Along the same line of investigation, we noticed that EF-1$_\alpha$ mRNA is maintained at a steady level during the development of S1-negative tissues, such as liver, spleen, and lung, while the EF-1$_\alpha$ level drops down late during brain, heart, and muscle development. This later observation was a surprise considering the general presumption that the EF-1$_\alpha$ gene is constitutively expressed during cell growth and de-

velopment and suggests that EF-1$_\alpha$ gene expression is partially suppressed in postmitotic neurons and myocytes [Lee et al., 1993a]. This phenomenon is less appreciable in total brain, but may reflect the high cellular heterogeneity of this organ. Further studies analyzing different regions of rat brain have clearly shown that EF-1$_\alpha$ mRNA is down-regulated, most dramatically in the cerebral cortex [Lee et al., in press]. Therefore, these results suggest that there is a change in the quantitative and qualitative expression of EF-1$_\alpha$ and S1 in mammalian cells during terminal differentiation (Fig. 1).

This terminal differentiation-dependent expression of the S1 gene led us to evaluate S1 and EF-1$_\alpha$ gene expression as a function of aging. Northern analysis and RNase protection assays have shown no significant change in the S1 and EF-1$_\alpha$ expression levels during the adult life span in rat. But modulation in expression of both genes occurs during development, and we presume that this phenomenon is related to

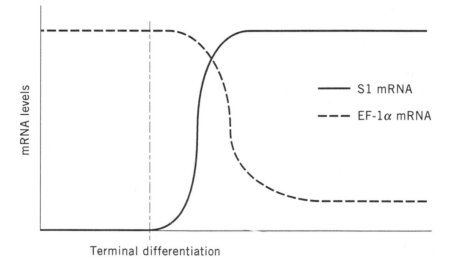

Terminal differentiation

Life span of organisms

Fig. 1. *A model depicting possible post-translational differentiation dependent switch governing the expression of EF-1$_\alpha$ to S1. We suggest that S1 replaces, either partially or totally, EF-1$_\alpha$ in mature, terminally differentiated neurons, cardiomyocytes, and myocytes. The switch only occurs after these cell types have per-* *manently withdrawn from the cell cycle (terminal differentiation). In other cell types, EF-1$_\alpha$ mRNA levels stay stable, independent of cell growth and differentiation. Neurons, cardiomyocytes, and myocytes are the only cell types whose life span is similar to that of the organism.*

the terminal differentiation process in S1-positive tissues. If members of the EF-1$_\alpha$ gene family are implicated in mammalian aging, such changes are not regulated at the mRNA level since EF-1$_\alpha$ message level stays constant from 9 to 26 months of adult life.

IV.D. EF-1$_\alpha$ and Cellular Senescence

EF-1$_\alpha$ expression and activity level have been studied in normal human fibroblast cells during *in vitro* cellular senescence. Cavallius et al. [1986] have reported a 45% decrease in active EF-1$_\alpha$ protein in late passage senescent human fibroblasts (MRC-S). These results concur with work by Cavallius et al. [1989], who have shown a decline in EF-1$_\alpha$ mRNA in senescent MRC-5 fibroblasts. On the other hand, senescent WI-38 fibroblasts were reported to have a 50-fold accumulation of EF-1$_\alpha$ transcript [Giordano and Foster, 1989; Giordano et al., 1989]. More recently, a direct correlation between an increased level of active EF-1$_\alpha$ and stimulation of protein synthesis in phorbol ester–treated aged human fibroblast was reported. However, no major difference in EF-1$_\alpha$ protein level was observed in young, old, or transformed human cells [Rattan et al., 1986]. As for this laboratory, S1 mRNA has not been detected in every fibroblast cell line that we have analyzed. More precise studies are necessary to investigate whether in some of the fibroblast cell lines EF-1$_\alpha$ causes the observed decline in protein synthesis during cellular aging.

IV.E. S1, EF-1$_\alpha$, and Cellular Longevity

We have inferred a role for S1 in the elongation step of translation mainly because of its high sequence similarity to EF-1$_\alpha$. The S1 putative primary amino acid sequence shows total conservation of the functional domains, such as the GTP- and tRNA-binding domains when compared with those of EF-1$_\alpha$ [Dever et al., 1987; Valencia et al., 1991; van Damme et al., 1992; Kinzy et al., 1992]. Yet these arguments do not necessarily imply functional redundancy, as seems to be the case in lower species [Linz et al., 1986; Cotrelle et al., 1985; Hughes et al., 1990]. The EF-1$_\alpha$ protein sequence is almost totally conserved in mammals. This fact alone suggests that any mutations occurring in the primary amino acid sequence of EF-1$_\alpha$ would lead to functional aberrations and eventually to the death of the organism. S1 does have about 40 amino acid substitutions compared with EF-1$_\alpha$. These are not random events, as the substitutions have been conserved from rat to man. If the amino acid modifications had not resulted in the modification of function, they would not have been so conserved between the two species. We presume that the above-mentioned changes in the primary amino acid sequence has led to distinct functional changes between EF-1$_\alpha$ and S1. Hence, our data imply that terminally differentiated cells contain a specific translation elongation program in which S1 is probably implicated.

We have observed a change in the abundance of expression from EF-1$_\alpha$ to S1 after terminal differentiation in three well-defined cells: neurons, cardiomyocytes, and myocytes [Lee et al., 1993a,b]. These three cell types share one common characteristic, that is, they are irreversibly growth arrested and long lived, having no proliferative capacity. This later characteristic differentiates neurons, cardiomyocytes, and myocytes from other terminally differentiated cells such as keratinocytes, which have a relatively short life span in the organism. It would then be logical to assume that some still unknown molecular mechanism exists for maintaining the integrity of highly differentiated cells whose life span is similar to that of the individual organisms. Does S1 play a role in the putative function involving the maintenance of the long-lived phenotype of neurons, cardiomyocytes, and myocytes? One plausible hypothesis is that it is part of a mechanism to increase translational fidelity, which may be necessary to prevent the accumulation of incorrectly synthesized proteins, detrimental to cell function and survival. Proponents of the aging error theory still consider that translation error may be an important mechanism by which cellular longevity or transformation may occur [Dice, 1993]. As discussed above, there is ample evidence that links EF-1$_\alpha$ gene mutation, modification of primary amino acid structure, and alteration in the expression of different EF-1$_\alpha$ isotypes to cellular transformation and longevity. Until recently it has been firmly established that certain modifications in the primary amino acid sequence of EF-1$_\alpha$ result in a significant increase in translation fidelity and cellular longevity in *P. anserina* [Silar and Picard, 1994]. One possibility is that the increase in fidelity of EF-1$_\alpha$ mutants is a result of a slowing down of the elongation rate, leading to a longer life span of the organism [a concept reviewed by Thompson, 1988]. Another possibility is that the production of erroneously synthesized proteins would result in cellular malfunction and eventual death. Nevertheless, it is feasible that a highly accurate EF-1$_\alpha$ isotype like S1 has been acquired for specialized cells by neurons, myocytes, and cardiomyocytes for vital organs such as brain, muscle, and heart.

V. CONCLUSION

There is no doubt that EF-1$_\alpha$ is the center piece involved in the regulation of translation rate and accuracy. The presence of multiple EF-1$_\alpha$ genes in all but a few species and the fact that the expression of these genes is so highly regulated suggest its importance in cell type–specific fine tuning of protein synthesis at the elongation step. The evidence pointing toward a direct relationship between EF-1$_\alpha$ and aging is less clear. One of the reasons may be that investigators are looking at gross differences in either activity or level of translation factor as evidence for age-related problems with the protein synthesis apparatus. Even the assumption that reduced elongation rate is necessarily detrimental to the well-being of aged cells is uncertain. Translation elongation rate has been shown to be linked with cellular growth and translational accuracy. It is clear that the three cell types in mammalian organisms that are long lived acquire, in addition to the ubiquitous EF-1$_\alpha$, the specialized S1 after they have

permanently withdrawn from the cell cycle traverse. To us, this suggests that the elongation step of protein synthesis may play a crucial role in the maintenance of cellular longevity, at least in mammalian species. It will be the next challenge to elucidate how exactly EF-1$_\alpha$ and related genes such as its sister gene S1 help to accomplish the formidable task of keeping cells long lived.

ACKNOWLEDGMENTS

The authors express their gratitude to Andrea LeBlanc for many helpful discussions. This work was supported by an operative grant (MT-12112) from the Medical Research Council of Canada to E.W. and a student fellowship from the Medical Research Council of Canada to S.L.

REFERENCES

An G, Friesen JD (1980): The nucleotide sequence of tufB and four nearby tRNA structural genes of *E. coli*. Gene 12:33–39.

Ann DK, Lin HH, Lee S, Tu ZJ, Wang E (1992): Characterization of the statin-like S1 and rat elongation factor 1$_\alpha$ as two distinctly expressed messages in rat. J Biol Chem 267:699–702.

Ann DK, Moutsatsos IK, Nakamura T, Lin HH, Mao PL, Lee MJ, Chin S, Liem RKH, Wang E (1991): Isolation and characterization of the rat chromosomal gene for a polypeptide (pSl) antigenically related to statin. J Biol Chem 266:10429–10437.

Bourne HR, Sanders DA, McCormick F (1991): The GTPase superfamily: Conserved structure and molecular mechanism. Nature 349:117–127.

Brands JH, Maassen JA, van Hemert FJ, Amons R, Moller W (1986): The primary structure of the alpha subunit of human elongation factor 1. Structural aspect of guanine-nucleotide–binding sites. Eur J Biochem 155:167–171.

Cavallius J, Merrick WC (1992): Nucleotide sequence of rabbit elongation factor-1 alpha cDNA. Nucleic Acids Res 20:1422.

Cavallius J, Rattan SIS, Clark BF (1986): Changes in activity and amount of active elongation factor 1 alpha in aging and immortal human fibroblast cultures. Exp Gerontol 21:149–157.

Cavallius J, Rattan SIS, Riis B, Clark BF (1989): A decrease in levels of mRNA for elongation factor-1 alpha accompanies the decline in its activity and the amounts of active enzyme in rat livers during aging. Topics Aging Res Eur 13:125–132.

Cottrelle P, Thiele D, Price VL, Memet S, Micouin JY, Marck C, Buhler JM, Senenac A, Fromageot P (1985a): Cloning, nucleotide sequence and expression of one of two genes coding for yeast elongation factor 1_α. J Biol Chem 260:3090–3096.

Cottrelle P, Cool M, Thuriaux P, Price VL, Thiele D, Buhler JM, Fromageot P (1985b): Either one of the two yeast EF-1_α alpha genes is required for cell viability. Curr Genet 9:693–697.

Dever TE, Costello CE, Owens CL, Rosenberry TL, Merrick WC (1989): Location of seven post-translational modifications in rabbit EF-1_α including dimethyllysine, trimethyllysine, and glycerylphosphorylethanolamine. J Biol Chem 264:20518–20525.

Dever TE, Ravel JM, Merrick WC (1987): The GTP-binding domain: Three consensus sequence elements with distinct spacing. Proc Natl Acad Sci USA 84:1814–1818.

Dice JF (1993): Cellular and molecular mechanism of aging. Physiol Rev 73:149–159.

Dje MK, Mazabraud A, Viel A, Maire ML, Denis H, Crawford E, Brown DD (1990): Three genes under different developmental control encode elongation factor-1 alpha in *Xenopus laevis*. Nucleic Acids Res 18:3489–3493.

Ejiri S, Murakami K, Katsumoyo T (1977): Elongation factor 1 from the silk gland of silkworm. FEBS Lett 82:111–114.

Fonzi WA, Katayama C, Leathers T, Sypherd PS (1985). Regulation of protein synthesis elongation factor-1 alpha in *Mucor racemous*. Mol Cell Biol 5:1100–1103.

Giordano T, Foster DN (1989): Identification of a highly abundant cDNA isolated from senescent WI-38 cells. Exp Cell Res 185:399–406.

Giordano T, Kleinsek D, Foster DN (1989): Increase in the abundance of a transcript hybridizing to elongation factor 1 alpha during cellular senescence and quiescence. Exp Gerontol 24:501–513.

Grant AG, Flomen RM, Tizzard ML, Grant DA (1992): Differential screening of a human pancreatic adenocarcinoma lambda gtll expression library has identified increased transcription of elongation factor EF-1_α in tumor cells. Int J Cancer 50:740–745.

Grasmuk H, Nolan RD, Drews J (1978): The isolation and characterization of elongation factor eEF-Ts from Krebs II mouse ascites tumour cells and its role in the elongation process. Eur J Biochem 92:479–490.

Hershey JWB (1991): Translational control in mammalian cells. Annu Rev Biochem 6:717–755.

Hiatt WR, Garcia R, Merrick WC, Sypherd PS (1982). Methylation of elongation factor 1 alpha from the fungus *Mucor*. Proc Natl Acad Sci USA 79:3433–3437.

Hovemann B, Richter S, Waldorf U, Cziepluch C (1988): Two genes encode related cytoplasmic elongation factor 1 alpha in *Drosophila melanogaster* with continuous and stage specific expression. Nucleic Acids Res 16:3175–3194.

Hughes D (1990): Both genes for EF-Tu in *Salmonella typhimurium* are individually dispensible for growth. J Mol Biol 215:41–51.

Iwasaki K, Motoyoshi K, Nagata S, Karizo Y (1976): Purification and properties of a new polypeptide chain elongation factor, EF-1_α, from pig liver. J Biol Chem 251:1843–1845.

Janssen GMC, Moller W (1988): Kinetic studies on the role of elongation factors 1_β and 1_γ in protein synthesis. J Biol Chem 263:1773–1778

Jurnak F (1985): Structure of the GDP domain of EF-Tu and location of the amino acids homologous to *ras* oncogene proteins. Science 230:32–36.

Kinzy TG, Freeman JP, Johnson AE, Merrick WC (1992): A model for the aminoacyl-tRNA binding site of eucaryotic elongation factor 1 alpha. J Biol Chem 267:1623–1632.

Knudson SM, Poulsen K, Dahl O, Clark BFC, Hjorth IP (1993): Tissue dependent variation in the expression of elongation factor-1 alpha isoforms: Isolation and characterization of a cDNa encoding a novel variant of human elongation factor-1 alpha. Eur J Biochem 215:549–554.

Kohno K, Uchida T, Ohkubo H, Nakanishi S, Nakanishi T, Fukui T, Ohtsuka E, Ikehara M, Okada Y (1986): Amino acid sequence of mammalian EF-2 deduced from the cDNA sequence: Homology with GTP-binding proteins. Proc Natl Acad Sci USA 83:4978–4982.

LaCour TFM, Nyborg J, Thirup S, Clark RFC (1985): Structural details of the binding of guanosine diphosphate to elongation factor Tu from *E. coli* as studied by X-ray crystallography. EMBO J 4:2385–2388.

Lauer SJ, Burks E, Irvin JD, Ravel JM (1984): Purification and characterization of three elongation factors EF-1_α, EF$_{\beta\gamma}$, and EF-2 from wheat germ. J Biol Chem 259:1644–1648.

Lee S, Wolfraim LA, Wang E (1993a): Differential expression of S1 and elongation factor-1_α during rat development. J Biol Chem 268:24453–24459.

Lee S, Stollar E, Wang E (1993b): Localization of S1 and EF-1_α mRNA in rat brain and liver by nonradioactive *in situ* hybridization. J Histochem Cytochem 41:1093–1098.

Lee S, Francoeur A-M, Liu S, Wang E (1992): Tissue specific expression in mammalian brain, heart, and muscle of Sl, a member of the elongation factor-1_α gene family. J Biol Chem 267:24064–2468.

Lee S, Ann D, Wang E (1994): Cloning of human and Mouse cDNAs for Sl, the second member of the mammalian elongation factor-1 alpha gene family: Analysis of a possible evolutionary pathway. Biochem Biophys Res Commun 203:1371–1377.

Lee S, LeBlanc A, Duttaroy A, Wang E (1995): Terminal differentiation dependent alteration in the expression of translation elongation factor-1_α and its sister gene, S1, in neurons. Exp Cell Res (in press).

Lenstra JA, Van Vliet A, Arnberg AC, Van Hemert FJ,

Moller W (1986): Genes coding for the elongation factor EF-1$_\alpha$ in *Artemia*. Eur J Biochem 155:475–483.

Lin SY, McKeehan WL, Culp W, Hardesty B (1969): Partial characterization of the enzymatic properties of the aminoacyl transfer ribonucleic acid binding enzyme. J Biol Chem 244:4340–4350.

Linz JE, Sypherd PS (1987): Expression of three genes for elongation factor 1 alpha during morphogenesis of *Mucor racemous*. Mol Cell Biol 7:1925–1932.

Linz JE, Katayama C, Sypherd PS (1986a): Three genes for the elongation factor-1 alpha in *Mucor racemous*. Mol Cell Biol 6:593–600.

Linz JE, Lira LM, Sypherd PS (1986b): The primary structure and the functional domains of an elongation factor-1 alpha from *Mucor racemous*. J Biol Chem 261:15022–15029.

Lipmann C, Lindschau C, Vijgenboom E, Schroder W, Bosch L, Erdmann VA (1993): Prokaryotic elongation factor Tu is phosphorylated *in vivo*. J Biol Chem 268:601–607.

Lu XA, Werner D (1989): The complete sequence of mouse elongation factor-1 alpha (EF-1 alpha) mRNA. Nucleic Acids Res 17:442.

Mattaj IW, Coppard NJ, Brown RS, Clark BFC, DeRobertis EM (1987): 42Sp48—the most abundant protein in previtellogenic *Xenopus* oocytes—resembles elongation factor 1$_\alpha$ structurally and functionally. EMBO J 6:2409–2413.

Merrick WC (1992): Mechanisms and regulation of eukaryotic protein synthesis. Microbiol Rev 56:291–315.

Metz-Boutigue M-H, Reinbolt J, Ebel J-P, Ehresmann C, Ehresmann B (1989): Crosslinking of elongation factor Tu to tRNA-Phe by trans-diaminedichloroplatinum (11). FEBS Lett 245:194–200.

Moller W, Schipper A, Amons R (1987): A conserved amino acid sequence around ARG-68 of *Artemia* elongation factor 1 alpha is involved in the binding of guanine nucleotides and aminoacyl transfer RNAs. Biochimie 69:983–989.

Motoyoshi K, Iwasaki K, Kaziro Y (1977): Purification and properties of polypeptide chain elongation factor-1$_{\beta\gamma}$ from pig liver. J Biochem 82:145–155.

Nagata S, Nagashima K, Tsunetsugu-Yokota Y, Fujirama K, Miyazaki M, Kaziro Y (1984): Polypeptide chain elongation factor 1 alpha from yeast, nucleotide sequence of one of the two genes for EF-1$_\alpha$ from *Sacharomyces cerevisiae*. EMBO J 3:1835–1830.

Nygard O, Nilsson L (1990): Translational dynamics: Interactions between the translational factors, tRNA and ribosomes during eukaryotic protein synthesis. Eur J Biochem 191:1–17.

Olson EN (1992): Interplay between proliferation and differentiation within the myogenic lineage. Dev Biol 154:261–272.

Pingoud A, Gast FU, Peters F (1990): The influence of the concentration of elongation factors on the dynamics and accuracy of protein biosynthesis. Biochem Biophys Acta 1050:252–258.

Rao T, Slobin LI (1987): Regulation of the utilization of mRNA for eucaryotic elongation factor Tu in Friend erythroleukemia cells. Mol Cell Biol 7:687–697.

Rattan SIS (1991): Protein synthesis and the components of protein synthetic machinery during cellular aging. Mutat Res 256:115–125.

Rattan SIS, Cavallius J, Hartvigsen G, Clark BF (1986): Activity and amounts of active elongation factor 1 alpha in mouse livers during ageing. Trends Aging Res 147:135–140.

Riis B, Rattan SIS, Clark BFC, Merrick WC (1990): Eukaryotic protein elongation factors. Trends Biochem Sci 15:420–424.

Ryazanov AG, Davydoka EK (1989): Mechanism of elongation factor 2 (EF-2) inactivation upon phosphorylation. FEBS Lett 251:187–190.

Sanders J, Maassen JA, Moller W (1992): Elongation factor-1 messenger RNA levels in cultured cells are high compared to tissue and are not drastically affected further by oncogenic transformation. Nucleic Acids Res 20:5907–5910.

Sassoon DA (1993): Myogenic regulatory factors: Dissecting their role and regulation during vertebrate embryogenesis. Dev Biol 156:11–23.

Schirmaier F, Philippsen P (1984): Identification of two genes coding for the translational elongation factor EF-1$_\alpha$ of *S. cerevisiae*. EMBO J 3:3311–3315.

Shepherd IC, Walldorf U, Hug P, Gehring WJ (1989): Fruit flies with additional expression of the elongation factor-1 alpha live longer. Proc Natl Acad Sci USA 86:7527–7521.

Sherman M, Sypherd PS (1989): Role of lysine methylation in the activities of elongation factor 1 alpha Arch Biochem Biophys 275:371–378.

Shikama N, Ackerman R, Back C (1994): Protein synthesis elongation factor EF-1 alpha expression and longevity in *Drosophila melanogaster*. Proc Natl Acad Sci USA 91:4199–4203.

Shirasawa T, Sakamoto K, Akashi T, Takahashi H, A Kawashima (1992): Nucleotide sequence of rat elongation factor-1 alpha. Nucleic Acids Res 20:909.

Silar P, Picard M (1994): Increased longevity of EF-1 alpha high-fidelity mutants in *Podospora anserina*. J Mol Biol 235:231–236.

Slobin LI, Rao MN (1993): Translational repression of EF-1$_\alpha$ mRNA *in vitro*. Eur J Biochem 213:919–926.

Slobin LI, Jordan P (1984): Translational repression of mRNA for eucaryotic elongation factors in Friend erythroleukemia cells. Eur J Biochem 145:143–150.

Slobin LI, Moller W (1976): Characterization of developmentally regulated forms of elongation factors in *Artemia salina*. I. Purification and structural properties of the enzymes. Eur J Biochem 69:351–375.

Slobin LI, Moller W (1978): Purification and properties of an elongation factor functionally analogous to bacterial elongation factor Ts from embryos of *Artemia salina*. Eur J Biochem 84:69–77.

Song JM, Picologlou S, Grant CM, Firoozan M, Tuite GM, Liebman S (1989): Elongation factor-l alpha gene dosage alters translational fidelity in *Saccharomyces cerevisiae*. Mol Cell Biol 9:4571–4575.

Sundstrom P, Irwin M, Smith D, Sypherd PS (1991): Both genes for EF-1 alpha in *Candida albicans* are translated. Mol Microbiol 5:1703–1706.

Sundstrom P, Deborah S, Sypherd PS (1990): Sequence analysis and expression of two gene for elongation factor-l alpha from the dimorphic yeast *Candida albicans*. J Bacteriol 172:2036–2045.

Tatsuka M, Mitsui H, Wada M, Naga A, Nojima H, Okayama H (1992): Elongation factor-l alpha gene determines susceptibility to transformation. Nature 359:333–336.

Thompson RC (1988: EFTu provides an internal kinetic standard for translational accuracy. Trends Biochem Sci 3:91–93.

Toledo CA, Jerez CA (1989): Methylation of elongation factor Tu affects rate of trypsin degradation and tRNA-dependent GTP hydrolysis. FEBS Lett 252:37–41.

Uetsuki T, Naito A, Nagata S, Kazir Y (1989): Isolation and characterization of the human chromosomal gene for polypeptide chain elongation factor-l alpha J Biol Chem 264:5791–5798.

Valencia A, Kjeldgaa M, Pai EF, Sander C (1991a): GTPe domain of *ras* p21 oncogene and elongation factor Tu: Analysis of three dimensional structures, sequence families, and functional sites. Proc Natl Acad Sci USA 88:5443–5447.

Valencia A, Chardin P, Whittenghofer A, Sander C (1991b): The *ras* protein family: Evolutionary tree and role of conserved amino acids. Biochemistry 30:4637–4648.

van Damme HT, Amons R, Moller W (1992): Identification of the sites in the eukaryotic elongation factor 1 alpha involved in the binding of elongation factor 1 beta and aminoacyl-tRNA. Eur J Biochem 207:1025–1034.

van Hemert FJ, van Ormond H, Moller W (1983): A bacterial clone carrying sequences coding for elongation factor-l alpha from mia. FEBS Lett 157:289–293.

van Hemert FJ, Amons R, Pluijms WJM, van Ormondt H, Moller W (1984): The primary structure of elongation factor EF-1$_\alpha$ from the brine shrimp *Artemia*. EMBO J 3:1109–1113.

van Noort JM, Kraal B, Bosch L (1985): A second tRNA binding site on elongation factor Tu is induced by a factor that is bound to the ribosome. Proc Natl Acad Sci USA 82:3212–3216.

van Noort, JM, Kraal B, Bosch L, LaCour TFM, Nyborg J, Clark BFC (1984): Cross-linking of tRNA at two different sites of the elongation factor Tu. Proc Natl Acad Sci USA 81:3969–3972.

Venema RC, Peters HI, Traugh JA (1991a): Phosphory- lation of elongation factor 1 (EF-1) and valyl-tRNA synthetase by protein kinase C and stimulation of EF-1 activity. J Biol Chem 266:12574–12580.

Venema RC, Peters HI, Traugh JA (1991b): Phosphorylation of valyl-tRNA synthetase and elongation factor 1 in response to phorbol esters is associated with stimulation of both activities. J Biol Chem 266:11993–11998.

Viel A, Dje MK, Mazabraud A, Denis H, Maire Mle (1987): Thesaurin a, the major protein of *Xenopus laevis* previtellogenic oocytes, present in the 42S particles, is homologous to elongation factor EF-1 alpha. FEBS Lett 223:232–236.

Walldorf U, Hovemann D, Bautz EK (1985): F1 and F2: two similar genes regulated differentially during development of *Drosophila melanogaster*. Proc Natl Acad Sci USA 82:5795–5799.

Wang E (1985a): A 57,000-mol-wt protein uniquely present in nonproliferating cells and senescent human fibroblasts. J Cell Biol 100:545–551.

Wang E (1985b): Rapid disappearance of statin, a nonproliferating and senescent cell specific protein, upon reentering the process of cell cycling. J Cell Biol 101:1695–1702.

Wang E, Moutsatos IK, Nakamura T (1989): Cloning and molecular characterization of a cDNA clone to statin, a protein specifically expressed in nonproliferating quiescent and senescent fibroblasts. Exp Gerontol 24:485–499.

Webster GC (1985): Protein synthesis in aging organisms. In Sohal RS, Birnbaum LS, Cutler RG (eds): Molecular biology of aging: gene stability and gene expression. Raven, New York: pp 263–289.

Webster GC, Webster SL (1982): Effects of age on post-initiation stages of protein synthesis. Mech Ageing Dev 18:369–378.

Webster GC, Webster SL (1983): Decline in synthesis of elongation factor one (EF-1) precedes the decreased synthesis of total protein in aging *Drosophila melanogaster*. Mech Ageing Dev 22:121–128.

Webster GC, Webster SL (1984): Specific disappearance of translatable messenger RNA for elongation factor one in aging *Drosophila melanogaster*. Mech Ageing Dev 24:335–342.

Webster GC (1985): Protein synthesis in aging organisms. In Sohal RS, Birnbaum LS, Cutler RG (eds): Molecular biology of aging: Gene stability and gene expression. Raven, New York: pp 263–289.

Weissbach H, Redfield B, Moon HM (1973): Further studies on the interactions of elongation factor 1 from animal tissues. Arch Biochem Biophys 156:267–275.

Whiteheart SW, Shenbagamurthi P, Chen L, Cotter RJ, Hart GW (1989): Murine EF-1$_\alpha$ is post-translationally modified by novel amide-linked ethanolamine-phosphoglycerol moieties. J Biol Chem 264:14334–14341.

ABOUT THE AUTHORS

STEPHEN LEE is a Visiting Fellow working with Dr. Richard D. Klausner, Chief of the Cell Biology and Metabolism Branch of the National Institute of Child Health and Human Development, National Institutes of Health in Bethesda, Maryland. After receiving a B.Sc. from the Université du Québec à Montréal, in Montreal, Canada, he obtained an M.Sc., working with Dr. Carl Séguin, from the Université Laval in Quebec City, Canada. He then earned a Ph.D. with Dean Honor mention in 1994, from McGill University, working on the molecular characterization of the S1 gene, the second member of the mammalian elongation factor-1 alpha gene family, in Dr. Eugenia Wang's laboratory, Bloomfield Centre for Research in Aging, The Lady Davis Institute for Medical Research, Sir Mortimer B. Davis—Jewish General Hospital, Montreal, Canada. Dr. Lee's research is now focused on the characterization of the von Hippel-Lindau tumor suppressor gene that is responsible for the familial form of human kidney cancer.

ATANU DUTTAROY is a postdoctoral researcher in Prof. Eugenia Wang's laboratory at the Bloomfield Centre For Research On Aging, McGill University. He pursued his doctoral research in the field of *Drosophila* genetics at the University of Calcutta, India, under the supervision of Prof. Asish K. Duttagupta. During his postdoctoral career he has been involved studying the biology of a transposable element in *Drosophila* and also cloning and characterization of the manganese superoxidase dismutase gene from *Drosophila*. He is currently engaged in studying the mechanism of programmed cell death.

EUGENIA WANG is a Professor of Medicine at McGill University in Montreal, Quebec, Canada, where she directs the Bloomfield Centre for Research in Aging of the Lady Davis Institute for Medical Research of the Sir Mortimer B. Davis—Jewish General Hospital; she also teaches undergraduate and graduate courses dealing with the cell and molecular biology of aging. She received her B.Sc. degree from the National Taiwan University, her M.A. from Northern Michigan University, followed by her Ph.D. in 1974 for work on the cell biology of intermediate filaments structures in the laboratory of Dr. Robert D. Goldman at Case Western Reserve University in Ohio. This was followed by postdoctoral research at the Rockefeller University in New York City, where she worked with Drs. Allan R. Goldberg, Purnell Choppin, and Igor Tamm on the characterization of cytoskeletal structures in virally infected and interferon-treated cells. Since 1981, Dr. Wang's research has involved the investigation of the molecular mechanisms controlling the cessation of proliferation in senescent human fibroblasts; her work along this line has allowed the identification of a protein, statin, whose presence is used as a marker for the nonproliferating state. She went on to identify another protein, terminin, whose unique presence in the 30 kD, form has been used to mark the commitment to programmed cell death (apoptosis). Development of monoclonal antibodies to statin and terminin has contributed to the mechanistic understanding of cellular aging and apoptosis.

Cellular Aging and Cell Death: 153–166
© 1996 Wiley-Liss, Inc.

Mechanisms of Escaping Senescence in Human Diploid Cells

Woodring E. Wright and Jerry W. Shay

I. INTRODUCTION

The elegant work of Greider and Harley (this volume) has shown that telomeres (the ends of the chromosomes) shorten progressively with each division of normal diploid cells. This has led to the hypothesis that telomere shortening is the clock that times cellular senescence. However, the detailed molecular mechanisms by which telomere shortening regulates cellular senescence and the pathways by which cells escape these mechanisms to become immortal still need to be established. This chapter reviews several theories concerning the molecules and pathways involved in the different stages of this process.

II. THE M1/M2 MODEL OF CELLULAR SENESCENCE

The two-stage model of cellular senescence, in which M1 (mortality stage 1) and M2 (mortality stage 2) represent independent mechanisms limiting the proliferative capacity of normal cells, was derived to explain the behavior of cells transfected with an inducible SV40 large T-antigen [Wright et al., 1989]. Proteins from a variety of DNA tumor viruses (T-antigen from SV40, E6/E7 from the high-risk strains of human papilloma viruses, E1A/E1B from adenovirus type 5) can extend the cultured life span of normal human fibroblasts significantly, but do not directly immortalize the cells. Rather than entering a period of prolonged quiescence as do normal cells at the limit of their proliferative capacity, cells expressing such proteins enter crisis. Crisis is characterized by an ongoing process of cell division and cell death so that the population size initially ceases to increase and eventually decreases so that the culture is lost. Occasionally, as a very rare event, an immortal cell emerges from crisis. This is often observed as a focus of small cells surrounded by the larger cells still in crisis (Fig. 1). Figure 2 shows the behavior of IDH4 cells, IMR90 human embryonic lung fibroblasts immortalized by a dexamethasone-inducible large T-antigen. The immortal cells stopped dividing when T-antigen was de-induced by the removal of dexamethasone, suggesting that the continued expression of T-antigen was necessary for the maintenance of the immortal state. However, the cells also stopped dividing when T-antigen was de-induced during the period of extended life span prior to crisis, when the cells were not yet immortal.

The model we proposed to explain these observations postulated that senescence is initially caused by the first of two independent mechanisms, mortality stage 1 (or the M1 mechanism). We now know that the M1 mechanism is executed by the cellular proteins p53 and either pRb or an Rb-like activity [Shay et al., 1991; Wright et al., 1989]. Tumor virus proteins like SV40 T-antigen bind to both p53 and pRb and block their activity, thus bypassing M1 and extending the life span of the entire population of cells until an independent second mechanism, M2, is activated. M2 causes crisis and is not directly affected by the presence of T-antigen or similar viral proteins. If a critical M2 gene becomes inactivated, the cells

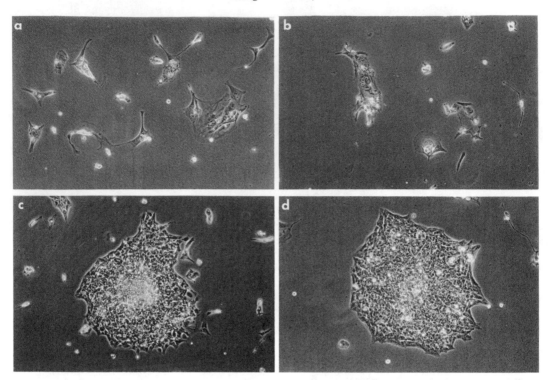

Fig. 1. *Phase contrast photomicrographs of SV40 T-antigen–expressing IMR-90 fibroblasts at crisis. While transforming proteins of small DNA tumor viruses can significantly extend the cultured life span of normal human fibroblasts, they do not directly immortalize the cells. Instead, cells stop proliferating at mortality stage 2 (M2) (a,b). The population of cells in crisis initially ceases to increase (likely due to a* *balance between successful division of cells not yet in crisis and cells in crisis undergoing cell death). In most instances the population eventually decreases so that the culture is lost. In rare instances, a cluster of immortal cells emerges from crisis that can be seen as a focus of small cells surrounded by larger cells still in crisis (c,d).*

can escape crisis and become immortal. However, these immortal cells have a perfectly functional M1 mechanism, which was never mutated but simply blocked by the presence of T-antigen. Consequently, if T-antigen is removed any time after the M1 mechanism has been induced, the cells stop dividing due to the actions of M1-activated p53 and pRb.

This model can be recast as follows to integrate its explanatory power with the observations that telomeres shorten during cellular proliferation. Telomere shortening would be the mitotic clock that counts cell divisions and induces the M1 mechanism when there are still several kilobases of telomeric repeats remaining. Later in the chapter, two models for this induction are discussed. If the M1 mechanism

is blocked by agents such as T-antigen, then the cells would continue dividing. Telomerase, a ribonucleoprotein that synthesizes telomere repeats and thus maintains telomere length in tumor cells, is not induced by T-antigen, and telomeres have been found to continue to shorten during the period of extended life span prior to crisis [Counter et al., 1992; Shay et al., 1993b]. The M2 mechanism would represent terminal telomere shortening, when there were so few repeats remaining that the direct consequences of the lack of repeats (end-to-end chromosome fusions, potential detachment of the telomeres from the nuclear matrix, and so forth) would cause cells to stop dividing and eventually die. Escape from M2 might represent the inactivation of the pathway by which

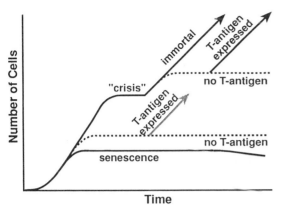

Fig. 2. *The behavior of T-antigen dependent immortal human fibroblasts. IMR90 human diploid fibroblasts transfected with an MMTV promoter-driven SV40 T-antigen and maintained in the presence of dexamethasone express T-antigen, which causes cells to bypass or neutralize M1 and proliferate until the M2 mechanism causes crisis. After inactivation of M2 by spontaneous mutation/nondisjunctional events, an immortal line appears. Since the M1 mechanism is still present, such immortal cells require the continuous presence of T-antigen in order to proliferate. De-induction of T-antigen either in postcrisis cells or during the extended life span phase causes the cells to re-express the M1 mechanism, stop dividing, and exhibit the senescent phenotype (M1). [Reproduced from Shay et al., 1994b, with permission of the publisher.]*

TABLE I. Both Rb- and p53-Binding Domains Are Needed to Overcome M1

Protein	Growth without steroids	Binds Rb	Binds p53
SV40 large T-antigen	+	+	+
Adenovirus E1A	−	+	−
Adenovirus E1B	−	−	+
E1A + E1B	+	+	+
Papillomavirus E6	−	−	+
Papillomavirus E7	−	+	−
E6 + E7	+	+	+

Small DNA tumor viruses such as SV40, adenovirus 5, and the high-risk strains of human papilloma virus appear to exert some of their proliferative and oncogenic effects on the host cell by encoding factors that interact with cellular proteins such as pRb and p53. This table demonstrates a relationship between the proliferation of IDH4 cells following de-induction of T-antigen (in steroid-depleted medium) and the presence of viral proteins that bind to pRb and p53. These results establish that two distinct classes of activity are required to overcome the M1 mechanism of human cellular senescence. Neither the class represented by adenovirus E1A or human papilloma virus E7 nor that represented by adenovirus E1B or human papilloma virus E6 is sufficient to replace intact T-antigen and permit the growth of IDH4 cells in the absence of steroids, while combinations of factors from each class can do so.

telomerase is repressed. The re-expression of telomerase would then permit the cell to add telomeric repeats to the denuded chromosome ends, maintain a stable telomere length, and become immortal.

III. P53, PRB AND THE M1 MECHANISM

The initial evidence that p53 and an Rb like activity were key intermediates in the M1 mechanism came from studies that determined which viral proteins were able to replace SV40 T-antigen and maintain the reversibly immortalized IDH4 in a proliferative state after T-antigen had been de-induced. Although T-antigen has the ability to bind both p53 and pRb, this binding is present on two different molecules in the corresponding proteins from the papilloma viruses and adenoviruses. Table I

shows that molecules capable of binding either p53 (papilloma virus E6 protein, adenovirus E1B protein) or pRb (papilloma virus E7 protein, adenovirus E1A protein) failed as single factors to permit growth once T-antigen had been de-induced by the removal of steroids. However, combinations of proteins in which both p53- and pRb-binding activities were present did permit cell proliferation. Similarly, T-antigen mutants defective in p53- or pRb-binding were able to replace the wild-type T-antigen only when both mutants were provided at the same time [Shay et al., 1991].

Viral proteins are multifunctional, and the above studies could at best suggest that p53 and pRb were the functional targets responsible for the M1 mechanism. Direct evidence for their involvement first came from studies by Hara et al. [1991], who showed that antisense pRb oligonucleotides extended the life span of human TIG-1 fibroblasts by about 10 doublings, while antisense pRb plus antisense p53 oligonucleotides extended their life span by about 20 doublings. One curious aspect of their study was that antisense p53 alone had no effect on the cell's proliferative capacity: Its effect was only observed in conjunction with antisense pRb. One interpretation of their observations is that different parts of the M1 mechanism become activated at different times. The pRb pathway would be activated first, so that antisense p53 treatment would be ineffective unless the pRb pathway was first blocked by antisense pRb. Antisense p53 would have a phenotype only if the cells were able to continue to divide deeper into M1 to the point where the p53 arm of the M1 mechanism was activated.

Although this interpretation may apply to the TIG-1 fibroblasts used by Hara et al. [1991], it is inconsistent with the behavior of IMR90 human lung fibroblasts. Retroviral vectors expressing either human papilloma virus E6 (which binds p53) or E7 (which binds pRb) are individually able to extend the life span of IMR90 by approximately 15–20 doublings. However, this represents only a partial M1 block, since the life span is extended by about 30–40 doublings when both E6 and E7 are present [Shay et al., 1993]. A similar partial extension of the life span of human fibroblasts by a dominant negative mutant p53 has also been reported [Bond et al., 1994]. These results suggest that the p53 arm of the M1 mechanism is not delayed with respect to the Rb-like M1 pathway.

To avoid confusion, we would like to point out some subtleties about the data concerning the number of doublings during the extended life span period. There is a great deal of clonal heterogeneity in a population of cells [Smith and Hayflick, 1974], and the life span of the population tends to reflect a clonal succession [Martin et al., 1974] as individual clones senesce and the culture is taken over by the longest lived clones. Thus, individual clones generally have a proliferative capacity signifi-

cantly less than that of the population. The original statements that T-antigen extended the life span of fibroblasts by about 15–20 doublings compared the proliferation of clones of cells transfected with a T-antigen expression vector to the population of nontransfected cells and thus is an underestimate of the true extension of life span. Most clones behaved like the clone that gave rise to IDH4 cells (Fig. 2) and entered crisis at about population doubling level 70 (PDL 70), although occasional clones entered crisis at PDL 80–85. The parental populations of IMR90 senesced at PDL 50–55, leading to the conclusion that the average extension of life span was 15–20 doublings. The more recent use of retroviral vectors containing HPV16 E6, E7, or E6/E7 has led to the comparison of the life span of large populations of cells infected with experimental versus control constructs. As described above, these experiments suggest that the actual prolongation of life span between the onset of M1 and the full expression of M2 in IMR90 fibroblasts is closer to 30–40 doublings. The absolute number of divisions between M1 and M2 may vary between different lineages.

Additional evidence for the specific involvement of p53 in the M1 mechanism comes from studies of the immortalization of human mammary epithelial cells (HME cells). In contrast to fibroblasts, Rb-like functions do not appear to play a role in HME cellular senescence. The initial evidence for this difference came from comparing the effects of the papilloma virus proteins E6 and E7. Although both E6 (which binds p53) and E7 (which binds Rb-like proteins) were required for the complete extension of life span and eventual immortalization of fibroblasts, E7 had no effect on HME, and the prolongation of life span and immortalization could be obtained with E6 alone (Table II). This observation was confirmed by showing that transfection with vectors expressing dominant-negative p53 mutants was sufficient to extend the life span of HME and lead to their eventual immortalization (unpublished observations).

Human cells spontaneously immortalize with a frequency too low to be observed repro-

TABLE II. E6 Alone Is Sufficient to Overcome M1 Stage of Senescence in Human Mammary Epithelial Cells but Not in Fibroblasts

Papillomavirus 16 protein expressed in pLXSN plasmid	IMR90 lung fibroblasts		Human mammary epithelial cells	
	PDL*	Immortal	PDL	Immortal
Nil	55–60	–	37–42	–
E6	66–69	–	42–62	+
E7	69–71	–	38–42	–
E6 + E7	75–95	+	42–65	+

The control IMR90 fibroblast cells (nil) senesce at approximately population doubling (PDL) 55–60, but those in which both HPV16 E6 and E7 were expressed have extended life (overcome the M1 stage of immortalization), enter crisis (M2) at approximately PDL 75–95, and some escape crisis and immortalize. The IMR90 fibroblasts containing either E6 or E7 alone display limited extended life (E6, PDL 66–69; E7, PDL 69–71) but do not immortalize. However, the same defective retroviral constructs containing either HPV16 E6 or E7 produces a very different behavior in human mammary epithelial cells. While HPV16 E7 infections does not suppress the M1 mechanism (the human mammary epithelial cells senescing at approximately PDL 38–42), most cells infected with HPV16 E6/E7 or E6 alone not only go through extended life (up to PDL 6265), but many escape from crisis and immortalize. [Reproduced from Shay et al., 1993a, with permission of the publisher.]

ducibly. However, patients with the Li-Fraumeni syndrome often have a dominant-negative mutant p53 allele. Evidence for the specific involvement of p53 in the M1 mechanism in the absence of any experimental manipulation comes from the observation that HME cells from Li-Fraumeni patients spontaneously immortalize with a detectable frequency [Shay et al., 1995]. Additional evidence for the different roles of pRb in HME and fibroblastoid cells is that mammary stromal fibroblasts from Li-Fraumeni patients do not reproducibly spontaneously immortalize unless provided with a pRb-binding factor like papilloma virus E7 [Shay et al., 1995]. Although there have been reports of the spontaneous immortalization of Li-Fraumeni fibroblasts [Bischoff et al., 1990; Rogan et al., 1995], these were extraordinarily rare events that likely required second mutations in the pRb arm of M1 in addition to the rare events inactivating the M2 mechanism.

There is now considerable evidence specifically implicating p53 in the M1 mechanism. However, aside from the single report of antisense pRbB being able to extend the life span of human fibroblasts, all of the evidence implicating pRb is indirect and relies on the use of multifunctional viral proteins. Since a variety of known factors (e.g., p107 and p300 [Whyte et al., 1989]) are bound by the same viral domains that bind

pRb, we believe it prudent at present to attribute the pRb arm of the M1 mechanism to an Rb-like activity rather than to pRb itself.

Although we describe the antiproliferative aspects of M1 as being due to the induction or "constitutive activation" of pRb and p53, the specific details of this process are unknown. Protein levels for these factors do not change significantly as a function of cellular senescence in fibroblasts [Afshari et al., 1993; W.E. Wright and J.W. Shay, unpublished observations], although, as expected for nonproliferating cells pRb, is mainly in a hypophosphorylated state [Stein et al., 1990]. There could be direct M1-regulated post-translational modifications of these proteins or changes in the factors with which pRb and p53 interact (such as E2F or the D cyclins with pRb [Hiebert et al., 1992; Ewen et al., 1993; Dowdy et al., 1993]) or proteins like $p34^{cdc2}$ or mdm2 with p53 [Sturzbecher et al., 1990; Momand et al., 1992; Moll et al., 1992] that alter either the activities or targets of pRb and p53 action.

IV. HYPOTHESES FOR THE INDUCTION OF M1

There are at least two working models for the mechanism by which M1 could be induced while there are still several kilobases of telomeric repeats remaining. One postulates the production of DNA damage signals, while the other invokes changes in expression of regulatory genes located in subtelomeric heterochromatin. These theories are not mutually exclusive, and both could easily contribute to the phenotype of senescent cells.

The DNA damage hypothesis is derived from calculations made for the distributions of telomeric repeats between daughter DNA stands during progressive cell divisions in the absence of telomerase [Levy et al., 1992]. This analysis indicated that a relatively stable distribution of sizes should be rapidly achieved. For cells like IMR90 human diploid fibroblasts, which lose approximately 50 bp of repeats per division, this led to the prediction of a standard deviation of about 500 bp in the size of telomeres. However, since a diploid cell with 23 chromosome pairs has 92 telomeres, it is probable that at least a few of these telomeres would have few or no repeats once the mean size of the repeats has decreased to several kilobases. The consequences of having a chromosome lacking repeats at one end could be the production of an ongoing DNA damage signal, which could induce the M1 mechanism [Levy et al., 1992].

One very attractive feature of the telomeric DNA damage hypothesis is that a molecular pathway consistent with some of its predictions has begun to emerge. It is known that DNA damage such as UV and X-irradiation can induce p53 [Maltzman and Czyzyk, 1984; Kastan et al., 1991; Fritsche et al., 1993]. Although p53 protein levels are not elevated in senescent cells [Afshari et al., 1993], the increased p53 protein in irradiation studies was transient, and it is not yet known what the p53 response to a chronic low level DNA damage signal would be. p53 has been found to induce a molecule variously called CIP1/Waf1/p21/SDI [Harper et al., 1993; El-Deiry et al., 1993; Xiong et al., 1993; Noda et al., 1994]. SDI ("senescent DNA synthesis Inhibitor") levels have been found to be elevated in senescent human fibroblasts [Noda et al., 1994]. This factor binds to cdk4–cyclin complexes and blocks the

ability of cdk4 to phosphorylate and activate the cyclins. Although both cyclins D and E accumulate to normal levels in senescent cells [Afshari et al., 1993], they appear to be in inactive forms since they fail to phosphorylate substrates *in vitro* [Lucibello et al., 1993; Stein et al., 1990]. The outline of an M1 pathway that derives from these observations is as follows: Telomere shortening leads to a DNA damage signal that induces p53, which in turn induces CIP1/Waf1/p21/SDI. This molecule then blocks the ability of cdk4 and other kinases to phosphorylate the cyclins and activate their ability to phosphorylate downstream products (including pRb) necessary for progression through the cell cycle.

Although this model has many attractive features, there are a variety of outstanding issues that it does not yet adequately explain. For example, telomerase is present in the reversibly immortalized IDH4 cells [Shay et al., 1994a]. One would predict that telomerase would add repeats to the denuded chromosome ends and thus remove the DNA damage signal in these immortal cells. However, the M1 mechanism is still active in IDH4 cells since the cells stop dividing once T-antigen is deinduced (Fig. 2). It is of course possible that the telomeres in IDH4 cells are stable but sufficiently short that they still produce an ongoing DNA damage signal. However, the concept that telomerase is unable to produce a good functional telomeric end that eliminates DNA damage signals is an unsatisfying one. The demonstration that CIP1/Waf1/p21/SDI mRNA falls back to levels comparable to those in young cells following the escape from crisis [Rubelj and Pereira-Smith, 1994] suggests that the DNA damage signal from denuded telomeres has in fact been eliminated following immortalization.

According to the simple outline of the pathway described above, pRb would be downstream of p53 in the M1 mechanism. However, blocking p53 with viral proteins such as papilloma virus E6, while sufficient to extend the life span of fibroblasts partially, is not sufficient to block the Rb arm of the M1 mechanism. To our knowledge, there is little evidence for pRb-like functions being induced during the response to DNA damage.

A second hypothesis for the induction of the M1 mechanism invokes the presence of regulatory genes located near telomeres whose activity is controlled by telomere length. Studies in yeast have demonstrated that the artificial positioning of reporter genes adjacent to telomeres results in an unstable phenotype [Gottschling et al., 1990]. Gene expression periodically switched between on and off states, similar to position–effect variation in *Drosophila*. Telomeres are thought to be heterochromatic, and the results were interpreted as indicating that the reporter gene was expressed only when it could

temporarily overcome the silencing effects of telomeric heterochromatin and establish a local euchromatic domain. We then hypothesized that similar mechanisms could regulate the M1 mechanism [Wright and Shay, 1992]. If the amount of heterochromatin induced by telomeres were proportional to the length of the TTAGGG repeats, then young cells with long telomeres would have a large amount of subtelomeric DNA silenced by being heterochromatic. Any M1 regulatory factors located in this region would not be expressed (Fig. 3). As the cells divided and lost telomeric repeats, the stimulus for heterochromatin formation would become progressively reduced. Once the telomeres had become sufficiently short, pre-

Telomere & Heterochromatin Shortening

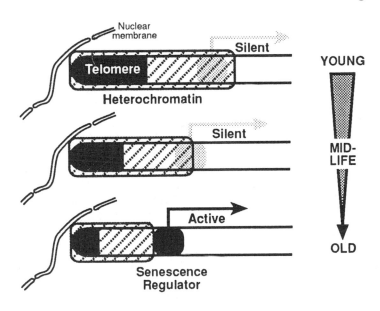

Fig. 3. *Model implicating telomere positional effects in the regulation of human cellular senescence. In young cells telomeres are long and heterochromatic. Gene(s) regulating M1 are postulated to be located in the subtelomeric region and are silent in young cells since they are inside the heterochromatic telomere region. As cells divide, the telomeres become progressively shorter because somatic cells lack telomerase activity and thus fail to replicate the very ends of their chromosomes. As the telomeres shorten, the strength of the signal to form heterochromatin* *declines, and the amount of heterochromatin in the subtelomeric DNA decreases. Once the telomeres become very short and the cells are near senescence, the silent M1 regulatory locus becomes active since it is now in a euchromatic region. When expressed it is proposed that this M1 regulatory region activates the M1 stage of cellular senescence, which likely involves a cascade of other downstream regulatory (e.g. Rb and p53) and structural genes. [Reproduced from Shay et al., 1994b, with permission of the publisher.]*

viously heterochromatic subtelomeric M1 regulatory genes would be able to establish euchromatic domains and become expressed, inducing the M1 mechanism. Consistent with this model, it has now been found that the strength of telomeric position effects in yeast is proportional to telomere length [Kyrion et al., 1993]. There is at present no evidence that addresses the validity of such effects in mammalian cells.

Both of the above models predict that the onset of the M1 mechanism could be a progressive event that might blend in over time. The strength of a DNA damage signal would be expected to increase if cells were forced to divide (by viral oncoproteins) beyond the point at which the first telomere was denuded of repeats. Similarly, there may be multiple M1 regulatory factors located in the subtelomeric DNA of different chromosomes so that the cell would be able to monitor average telomere length rather than just the length of a single telomere. Under these circumstances, the fraction of chromosomes with sufficiently short telomeres to permit the subtelomeric locus to be expressed would increase with increasing cell divisions. In either case, the cell's decision on whether to divide would be based on the balance between positive signals (the usual growth factor–cell cycle stimuli) and negative ones (the progressive induction of M1). Agents such as the papilloma E7 protein might partially extend the life span by both contributing positive growth signals and blocking the negative contribution by the Rb arm of the M1 mechanism. The cells would then continue to divide until the strength of the negative signal from the p53-arm of the pathway became sufficiently strong to counterbalance the positive signals. Only when both arms of the M1 mechanism had been blocked would the cells be able to divide until M2 was reached and they became competent to immortalize.

SDI mRNA levels have been found to increase during the period of extended life span in T-antigen–expressing cells and then drop dramatically following immortalization [Rubelj and Pereira-Smith, 1994]. These results

have been interpreted as indicating that there is no difference between the M1 and M2 mechanisms, since SDI would be responsible for both effects. These authors hypothesized that the down-regulation of the pathway by which SDI was induced allowed immortalization. Our interpretation of these results is that SDI levels increase during the extended life span period because of a progressively increasing strength of the p53 arm of the M1 mechanism and that SDI levels fall after immortalization because the expression of telomerase has resulted in the repair of telomeric ends and consequently reduced one of the signals for SDI expression. We find it difficult to understand why the reversibly immortalized IDH4 cells should stop dividing following the removal of T-antigen if SDI were responsible for both M1 and M2 and its down-regulation represented the means by which cells escaped crisis and became immortal.

V. THE M2 MECHANISM AND THE ESCAPE FROM CRISIS

A working hypothesis for the M2 mechanism is that it represents the consequences of terminal telomere shortening. The inactivation of a component of the pathway by which telomerase was repressed would then permit the re-expression of telomerase, the repair of the telomeres, and the escape from crisis. There are two principle reasons for assuming that this escape involves a recessive event. Cell hybrids between normal and immortal cells have a limited life span [Periera-Smith and Smith, 1983]. This is true even when the immortal cell line expresses T-antigen, which would be expected to inactivate in trans any M1 functions contributed by the normal parent [Periera-Smith and Smith, 1987]. Telomerase activity has recently been measured and found to be repressed in such cell hybrids [Wright et al., 1996]. These results suggest that the normal cell is contributing factors that repress telomerase that were lacking in the immortal cell.

A second reason for suspecting that the escape from M2 is a recessive event is the inability of mutagenic treatments to immortalize human cells directly with a frequency consistent with a dominant effect. The premature reactivation of telomerase would be expected to block the telomere shortening that normally activates the M1 mechanism, and it should thus directly immortalize cells. Mutagenesis should inactivate single alleles in the pathway by which telomerase is repressed, and if this were sufficient to derepress telomerase then immortalization should occur. Although other interpretations are possible (for example, that the telomerase gene is heavily methylated and thus inaccessible for derepression prior to changes that occur after M1), the lack of direct immortalization of normal pre-senescent human cells following mutagenesis is consistent with the requirement for a recessive event.

The probable requirement for a recessive mutation in order to escape from crisis predicts that the inactivation of the M2 mechanism should involve two events. The first would be the mutational inactivation of one allele of a telomerase repressor gene, and the second would be one or more loss-of heterozygosity events (such as nondisjunction and the elimination of a chromosome or gene conversion) that would eliminate the remaining wild-type allele(s).

There are many observations that are consistent with this interpretation. The frequency of immortalization of T-antigen–transfected human fibroblasts is considerably less than would be expected for a simple mutational event. This frequency was determined by what amounts to a fluctuation analysis. Cells were divided into a large number of independent series, for example, where 1 million cells were seeded into one daughter dish at each passage for each series. The fraction of series that eventually immortalized was then determined. In a typical experiment, 30% of the series passaged at 10^6 cells/dish immortalized. The frequency of immortalization was then calculated as three events per 10^7 cells or 3×10^{-7} [Shay and Wright, 1989]. Mutagenic rates are normally expressed per cell division. Since the independent series were established approximately 10

doublings prior to crisis, this frequency would be approximately 3×10^{-8} if expressed per cell division, which is lower than the approximate frequency of 10^{-6}/cell division for spontaneous mutations at haploid loci.

SV40 T-antigen–transfected human fibroblasts are aneuploid, with most cells often having approximately 70–85 chromosomes. There should thus be three to four copies of most chromosomes. The hypothesis that one needs both a mutation in one allele and the elimination of most or all of the remaining wild-type alleles predicts that those cells with only three copies of the relevant chromosome might be able to eliminate the remaining two wild-type chromosomes. However, the probability of successfully eliminating the wild-type chromosomes would be compromised if additional copies were present. Only approximately 10%–20% of the clones of IMR90 lung fibroblasts isolated following transfection with SV40 T-antigen were able to immortalize [Shay and Wright, 1989; Shay et al., 1993a]. This is consistent with the hypothesis that in general only those clones with the smallest number of target chromosomes are able to eliminate the remaining wild-type copies following the mutational inactivation of one allele.

This hypothesis was considerably strengthened when the behavior of human fibroblasts was compared with HME cells. HME cells expressing SV40 large T-antigen or HPV 16 E6/E7 were found to immortalize with a frequency of approximately 3×10^{-5}, at least 30- to 100-fold more frequent than that for fibroblasts [Shay et al., 1993a]. Figure 4 presents a sample chromosome distribution for both cell types before and after crisis. The fibroblasts had pseudotetraploid chromosome numbers both before and after crisis. In contrast to this, the HME cells maintained a subpopulation of pseudodiploid cells before crisis. It is likely that the cells that immortalized came from this subpopulation since they were uniformly pseudodiploid. A simple explanation for the higher frequency of escape from M2 in the mammary epithelial cells is that the pseudodiploid subpopulation contained only two cop-

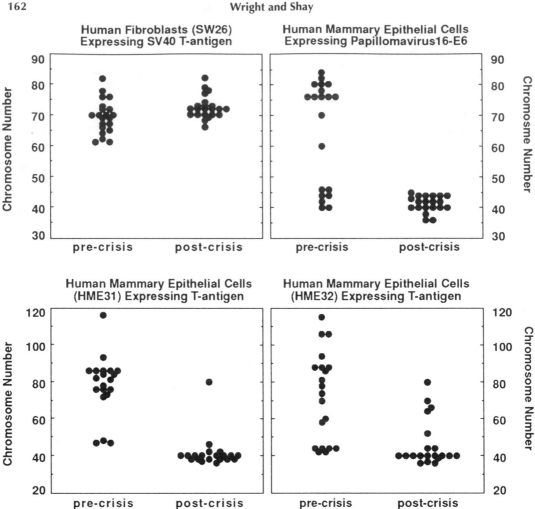

Fig. 4. *Chromosome distribution in immortaliza-tion-competent human fibroblasts and human mammary epithelial cells expressing viral oncoproteins. Human IMR90 fibroblast clones expressing SV40 large T-antigen that are capable of immortalizing generally have a pseudotetraploid distribution of chromosomes both pre- and postcrisis (top left) and immortalize with a frequency of approximately 3 × 10⁻⁷. Human mammary epithelial cells expressing* HPV16 E6 (top right) or SV40 large T-antigen (bottom) that are immortalization-competent generally have a population of both pseudodiploid and pseudotetraploid chromosomes precrisis. These immortalization-competent cells escape crisis at a much higher frequency (10⁻⁵) than fibroblasts and contain a distribution of chromosomes that is primarily pseudodiploid (aneuploid). [Reproduced from Shay et al., 1993a, with permission of the publisher.]

ies of the target chromosome so that following the mutational inactivation of one allele there was only a single wild-type allele that needed to be eliminated.

If pseudodiploid T-antigen expressing HME are able to escape M2 and reactivate telomerase with a high frequency, why do not normal diploid cells also reactivate telomerase with a high frequency? The answer is almost certainly that the chromosomal instability produced by T-antigen or similar agents contributes to the ability of the cells to eliminate the remaining wild-type alleles after one allele has been inactivated by mutation. Thus, although we have stated that M2 is independent of T-antigen (which is able to inactivate only the M1 mechanism), the indirect effects of T-antigen likely increase the ability of the cells to undergo the nondisjunctional/chromosome conversion events needed for the recessive escape from crisis. It is possible that this chromosomal instability may permit the elimination of wild-type alleles prior to crisis. However, we also believe that the end-to-end fusion of chromosomes that occurs during crisis due to the lack of protective telomere repeats contributes to the nondisjunctional events that are needed. Thus, although the escape from M2 could theoretically occur at any time, we suspect that in most cases it requires both chromosomal instability and the increased frequency of nondisjunctional events that occurs during crisis.

Evidence for a mutational component to the escape from M2 comes from retroviral insertional mutagenesis experiments. T-antigen blocks the M1 mechanism so that the only barrier to immortalization in fibroblasts transfected with T-antigen is the M2 mechanism. A clone of such fibroblasts was expanded, and 200 million cells were infected with a defective retrovirus or the supernatant from a control packaging line. The cells were then divided into a large number of independent series. Table III shows that the frequency of immortalization was increased by approximately two- to threefold following insertional mutagenesis [Shay et al., 1993a]. Although it is difficult to get reliable numbers from rare events, the consistency of the frequency of immortalization of the controls

(between 3.0 and 4.5 × 10^{-7} in four experiments conducted over a 3 year period) suggests that the elevated level found in the insertionally mutagenized cells represents a real increase.

VI. CONCLUSION

Cellular immortalization represents a multistep process, the molecular details of which are beginning to emerge. Our working hypothesis is that telomere shortening is the molecular clock that counts cell divisions and ultimately limits the proliferative capacity of normal cells. Once there are only several kilobases of telomeric repeats remaining, the M1 mechanism is induced. This could be mediated by DNA damage signals from rare telomeres that lack repeats, by regulatory genes previously trapped in subtelomeric heterochromatin that are now able to establish local euchromatic domains, or by both or unknown additional processes. The M1 mechanism is carried out through the actions of p53 in human mammary epithelial cells and both p53 and an pRb-like activity in fibroblasts. The downstream effects of p53 are likely to include such factors as CIP1/Waf1/SDI. If the actions of p53 and pRb are blocked, then the cells can continue to divide and telomeres continue to shorten until the M2 mechanism causes crisis. Although mutations in an M2 gene could occur at any time,

TABLE III. Mutagenesis Increases the Frequency of Immortalization of SW26 Cells

Experiment	Addition	Immortalization	Frequency
1	Control	1/10	3.0×10^{-7}
	Bleomycin sulfate	1/4	7.6×10^{-7}
2	Control	10/68	4.5×10^{-7}
	Bleomycin sulfate	7/27	7.9×10^{-7}
	LNL6 retrovirus	36/99	11.0×10^{-7}

SW26 fibroblasts were plated at 0.33×10^6 cells per 50 cm^2 dish at each passage and were split four to six times before crisis. Cells were treated with bleomycin sulfate (50 μg/ml) for 2 hours or were infected with the LNL6 retrovirus after the initial plating into different dishes. Immortalization is expressed as the number of immortal lines obtained per number of culture series, each series being derived from a single dish at the initiation of the experiment. Frequency is expressed as the probability of obtaining an immortal cell line based on the number of cells plated at each passage (nor per cell division). These results indicate that mutagenesis increases the frequency of escape from crisis approximately two- to threefold.

the nondisjunctional events that are produced by terminal telomere shortening probably contribute to the ability to eliminate wild-type alleles and retain extra copies of chromosomes containing critical mutant alleles. Once a sufficient loss of heterozygosity has occurred, the repression of telomerase is abolished, telomerase is re-expressed, the ends of the telomeres are repaired, and the cell becomes immortal.

ACKNOWLEDGMENTS

These studies were supported by research grant (AG07992) from the National Institutes of Health.

REFERENCES

Afshari CA, Vojta PJ, Annab LA, Futreal PA, Willard TB, Barrett JC (1993): Investigation of the role of G1/S cell cycle mediators in cellular senescence. Exp Cell Res 209:231–237.

Bischoff FZ, Yim SO, Pathak S. Grant G. Sciciliano MJ, Giovanella BC, Strong LC, Tainsky MA (1990): Spontaneous abnormalities in normal fibroblasts from patients with Li-Fraumeni cancer syndrome: Aneuploidy and immortalization. Cancer Res 50:7979–7984.

Bond JA, Wyllie FS, Wyndford-Thomas D (1994): Escape from senescence in human diploid fibroblasts induced directly by mutant p53. Oncogene 9: 1885–1889.

Counter CM, Avillon AA, LeFeuvre E, Stewart NG, Greider CW, Harley CB, Bacchetti S (1992): Telomere shortening associated with chromosome instability is arrested in immortal cells which express telomerase activity. EMBO J 11:1921–1929.

Dowdy SF, Hinds PW, Louie K, Reed SI, Arnold A, Weinberg RA (1993): Physical interaction of the retinoblastoma protein with human D cyclins. Cell 73:499–511.

Dulic V, Drullinger LF, Lees E, Reed SI, Stein GH (1993): Altered regulation of G1 cyclins in senescent human diploid fibroblasts: accumulation of inactive cyclin E–Cdk2 and cylin D1–Cdk2 complexes. Proc Natl Acad Sci USA 90:11304–11038.

El-Deiry WS, Tokino T, Velculescu VE, Levy DB, Parsons R. Trent JM, Lin D, Mercer WE, Kinzler KW, Vogelstein B (1993): WAF1, a potential mediator of p53 tumor suppression. Cell 75:817–825.

Ewen ME, Sluss HK, Sherr CJ, Matsushime H, Kato J, Livingston DM (1993): Functional interactions of retinoblastoma protein with mammalian D-type cyclins. Cell 73:487–497.

Fritsche M, Haessler C, Brandner G (1993): Induction of nuclear accumulation of the tumorsuppressor protein p53 by DNA-damaging agents. Oncogene 8:307–318.

Gottschling DE, Aparicio OM, Billington BL, Zakian VA (1990): Position effect at S. cerevisiae telomeres: Reversible repression of PolII transcription. Cell 63:751–762.

Hara E, Tsurai H, Shinozaki A, Nakada S, Oda K (1991): Cooperative effect of antisense-Rb and antisense-p53 oligomers on the extension of life span in human diploid fibroblasts, TIG-1. Biochem Biophys Res Commun 179:528–534.

Harper JW, Adami GR, Wei N, Keyomarsi K, Elledge SJ (1993): The p21 Cdk-interacting protein Cip1 is a potent inhibitor of G1 cyclin-dependent kinases. Cell 75:805–816.

Hiebert SW, Chellappan SP, Horowitz JM, Nevins JR (1992): The interaction of RB with E2F coincides with inhibition of the transcriptional activity of E2F. Genes Dev 6: 177–185.

Kastan MB, Onyekwere O. Sidransky D, Vogelstein B, Craig RW (1991): Participation of p53 protein in the cellular response to DNA damage. Cancer Res 51:6304–6311.

Kyrion G, Liu K, Liu C, Lustig AJ (1993): RAP1 and telomere structure regulate telomere position effects in Saccharomyces cerevisiae. Genes Devel 7:1146–1159.

Levy MA, Allsopp RC, Futcher AB, Greider CW, Harley CB (1992): Telomere end-replication problem and cell aging. J Mol Biol 225:951–960.

Lucibello FC, Sewing A, Brusselbach S. Burger C, Müller R (1993): Deregulation and suppression of cdk2 and cdk4 in senescent human fibroblasts. J Cell Sci 105:123–133.

Maltzman W, Czyzyk L (1984): UV irradiation stimulates levels of p53 cellular tumor antigen in non-transformed mouse cells. Mol Cell Biol 4:1689–1694.

Martin GM, Sprague CA, Norwood TH, Pendergrass WR (1974): Clonal selection, attenuation and differentiation in an in vitro model of hyperplasia. Am J Pathol 74:137–153.

Moll UM, Riou G, Levine AJ (1992): Two distinct mechanisms alter p53 in breast cancer: Mutation and nuclear exclusion. Proc Natl Acad Sci USA 89:7262–7266.

Momand J, Zambetti GP, Olson DC, George D, Levine AJ (1992): The mdm-2 oncogene product forms a complex with the p53 protein and inhibits p53-mediated transactivation. Cell 69: 1237–1245.

Noda A, Ning Y, Venable SF, Pereira-Smith OM, Smith JR (1994): Cloning of senescent cell–derived inhibitors of DNA synthesis using an expression screen. Exp Cell Res 211:90–98.

Periera-Smith OM, Smith JR (1983): Evidence for the recessive nature of cellular immortality. Science 221:964–966.

Pereira-Smith OM, Smith JR (1987): Functional SV40

T-antigen is expressed in hybrid cells having finite proliferative potential. Mol Cell Biol 7: 1541–1547.

Rogan EM, Bryan TM, Hakku B, Maclean K, Chang AC-M, Moy EL, Englezou A, Warneford SG, Dalla-Pozza L, Reddel RR (1995): Changes in p53 status and telomere length during spontaneous immortalization of Li-Fraumeni syndrome fibroblasts. Mol Cell Biol (in press).

Rubelj I, Pereira-Smith OM (1994): SV40-transformed human cells in crisis exhibit changes that occur in normal cellular senescence. Exp Cell Res 211:82–89.

Shay JW, Brasiskyte D, Ouellete M, Piatyszek MA, Werbin H, Ying Y, Wright WE (1994a): Methods for analysis of telomerase and telomeres. Methods Mol Genet 5:263–280.

Shay JW, Pereira-Smith OM, Wright WE (1991): A role for both Rb and p53 in the regulation of human cellular senescence. Exp Cell Res 196:33–39.

Shay JW, Tomlinson G. Piatyszek MA, Gollahon LS (1995): Spontaneous *in vitro* immortalization of breast epithelial cells from a Li-Fraumeni patient. Mol Cell Biol 15:425–432.

Shay JW, Van Der Haegen VA, Ying Y. Wright WE (1993a): The frequency of immortalization of human fibroblasts and mammary epithelial cells transfected with SV40 large T-antigen. Exp Cell Res 209:45–52.

Shay JW, Werbin H, Wright WE (1994b): Telomere shortening may contribute to aging and cancer: A perspective. Mol Cell Differ 2:1–21.

Shay JW, Wright WE (1989): Quantitation of the frequency of immortalization of normal human diploid fibroblasts by SV40 large T-antigen. Exp Cell Res 184:109–118.

Shay JW, Wright WE, Brasiskyte D, Van Der Haegen BA (1993b): E6 of human papillomavirus 16 can overcome the M1 stage of immortalization in human mammary epithelial cells but not in human fibroblasts. Oncogene 8:1407–1413.

Smith JR, Hayflick L (1974): Variation in the life-span of clones derived from human diploid cell strains. J Cell Biol 62:48–53.

Stein GH, Beeson M, Gordon L (1990): Failure to phosphorylate the retinoblastoma gene product in senescent human fibroblasts. Science 249:666–668.

Sturzbecher HW, Maimets T, Chumaker P, Brain R, Addison C, Simianis V, Rudge K, Philip R. Grumaldi M, Court W, Jenkins JR (1990): p53 interacts with p34^{cdc2} in mammalian cells: Implication for cell cycle control and oncogenesis. Oncogene 5:795–801.

Whyte P, Williamson NM, Harlow E (1989): Cellular targets for transformation by the adenovirus E1A proteins. Cell 56:67–75.

Wright WE, Brasiskyte D, Piatyszek MA, Shay JW (1996): Manipulating telomere length in immortal human cells determines the life span of immortal × normal cell hybrids (submitted).

Wright WE, Periera-Smith OM, Shay JW (1989): Reversible cellular senescence: Implications for immortalization of normal human diploid fibroblasts. Mol Cell Biol 9:3088–3092.

Wright WE, Shay JW (1992): Telomere positional effects and the regulation of cellular senescence. Trends Genet 8:193–197.

Xiong Y, Hannon G, Zhang H, Casso D, Kobayashi R, Beach D (1993): p21 is a universal inhibitor of the cyclin kinases and a potential effector of the p53 checkpoint pathway. Nature 366:701–704.

ABOUT THE AUTHORS

WOODRING E. WRIGHT is Professor of Cell Biology and Neuroscience at the University of Texas Southwestern Medical Center in Dallas, Texas. He received a B.A. *summa cum laude* from Harvard College in Cambridge, Massachusetts, where he worked on the visual pigment rhodopsin with George Wald. He then completed an M.D./Ph.D. program at Stanford University in California, where he earned his Ph.D. in 1974 for research in the laboratory of Leonard Hayflick showing that the control of cellular senescence resided in the nucleus rather than the cytoplasm. After receiving his M.D. in 1975, he pursued postdoctoral studies at the Pasteur Institute in Paris with François Gros on the molecular biology of muscle differentiation. Since joining the faculty of Southwestern Medical Center in 1978, his research has focused on the discovery and characterization of myogenin, one of the key bHLH proteins that regulates myogenesis, as well as the mechanisms controlling cellular senescence and immortalization. In 1985 he began what has become a very close and productive collaboration with Jerry Shay that led to the development of the two-stage model for cellular senescence. He has received a Research Award from the Texas Affiliate of the American Heart Association, a Career Development Award from the National Institute on Aging, and a MERIT award from the National Institute on Aging. He has served on the Molecular Cytology NIH study section and is on the Scientific Research Board of the American Foundation for Aging Research (AFAR) and the Scientific Advisory Board for the biotechnology company Geron.

JERRY W. SHAY is Professor of Cell Biology and Neuroscience at the University of Texas Southwestern Medical Center in Dallas, Texas. After receiving a B.A. and M.A. from the University of Texas in Austin, he pursued doctoral research at the University of Kansas at Lawrence, where he earned a Ph.D. in 1972 in Physiology and Cell Biology. This was followed by postdoctoral research at the University of Colorado in Boulder, where he worked with Keith Porter and David Prescott on nuclear–cytoplasmic interactions. Since 1975, Dr. Shay has been in Dallas and has been interested in the relationship between aging and cancer. In collaboration with Dr. Woodring E. Wright, he proposed a two stage model of cellular senescence and has obtained evidence that the tumor suppressor genes p53 and Rb are part of the first stage and terminal telomere shortening and reactivation of telomerase are part of the second stage. Dr. Shay has been a member of the Mammalian Genetics NIH study section and is on the Scientific Advisory Board of Geron Corporation (a biotechnology company that studies aging). In addition, he is currently on the editorial board of several journals, including *Cellular and Molecular Differentiation, In Vitro, Cellular and Developmental Biology, Methods in Cell Science*, and the *International Journal of Oncology*. Both Dr. Shay and Dr. Wright are the recipients of the 1995 Allied Signal Award for Research on Aging.

Cellular Aging and Cell Death: 167–180
© 1996 Wiley-Liss, Inc.

The Biology of Cell Death and Its Relationship to Aging

Richard A. Lockshin and Zahra Zakeri

I. DEFINITIONS OF CELL DEATH

I.A. Senescence In Culture

The topic of cell death in aging evokes different interpretations in different contexts. Senescence of cells in culture, referring to the limited number of population doublings that most diploid cells express when maintained in culture, is one version. Though the phenomenon is real and has led to excellent studies in cell biology, as discussed elsewhere in this book, its relevance to aging of the organism is frequently challenged. The most sturdy arguments for its relevance relate to the failure of the immune system later in life and the potential that the exit from mitotic cycle reflects an underlying physiological change in cell function.

I.B. Physiological Cell Death

A perhaps entirely independent phenomenon is that known as *physiological, programmed,* or *active cell death* and apoptosis. Physiological cell death, as distinguished from traumatic or necrotic cell death, is likewise a well-documented biological process of unclear relevance to aging. As is discussed below, it is clear that cells under specific circumstances will commit suicide and may be pushed into suicide by specific adverse conditions. However, individuals do not routinely die by running out of cells as a process of normal aging. The relevance to gerontology more likely lies in the following: Throughout life various pathologies or stresses undoubtedly lead to the physiological loss of substantial numbers of critical cells.

These losses lower the ability of the individual to resist various insults. Particularly in the reticuloendothelial system and in the central nervous system, these losses may well compromise the individual. Another consideration is that failure of the apparently universal cell death mechanism is one basis for the expansion of a cell lineage in malignancy. If we understand the signaling mechanism and the physiological events that lead to commitment to die, we may learn to recognize signals of danger or to intervene appropriately. The chapters that follow therefore explore the control and response to cell death signals.

The terms *programmed cell death* and *apoptosis* have slightly differing connotations, and confusion of these connotations has led to some sterile argumentation. Therefore, several laboratories have adopted more general and less confining terms such as *physiological* or *active* cell death. Ultimately the world will generate its own sense of the meanings of these terms irrespective of the wishes of linguistic purists, but it is useful for the sake of clarity to highlight the differing implications of the terms.

I.B.1. Programmed cell death.

Programmed cell death was first used in a developmental context and therefore carried with it the implication of genetic control of undefined directness [Lockshin and Williams, 1964, 1965a; Saunders, 1966; Saunders and Fallon, 1966]. At defined developmental periods, specific cells destroyed themselves in a noninflammatory process [Saunders, 1966], and the cells destined for this fate could be identified as progressing along this pathway well before the process became irreversible [Lockshin and Williams, 1964, 1965b]. As antibiotic inhibitors became generally available, three laboratories independently documented that death could be prevented by blocking mRNA or protein synthesis, thus establishing the theoretical potential for genes controlling cell death [Tata, 1966; Lockshin, 1969a; Munck, 1971]. As is discussed below, these studies led to the specific identification of genes causing and preventing cell death in invertebrates and to the search for genes regulating cell death in many different systems [Ellis et al., 1991b]. Nevertheless, there is no compelling reason to expect that there will be specific and universal *cell death genes*. Indeed, the proliferation of candidates for this role has led several laboratories to adumbrate the concept of *cell confusion,* or death from conflicting stimuli.

I.B.2. Apoptosis. *Apoptosis,* derived from Greek terms referring to the dropping of leaves in the autumn or to male-pattern balding, described the morphology of cells disappearing in a noninflammatory manner [Kerr et al., 1972]. Originally there was no suggestion of a requirement for protein synthesis, and the striking morphology, particularly the condensation and coagulation of chromatin, led rather quickly to the realization that DNA was degraded in a specific manner. The emphasis on the degradation of DNA led over the years to a complete loss of an earlier primary focus of this research, the recognition of the substantial involvement of lysosomes in the destruction of cytoplasm of vertebrate secretory cells and many invertebrate cells [Lockshin, 1969b; Strange et al., 1992]. Although most of the early researchers understood these distinctions, only a few made any effort to emphasize them [Schweichel and Merker, 1973; Beaulaton and Lockshin, 1977, 1982]. In a prescient article, Levi-Montalcini and Aloe [1981] emphasized the argument that the metabolic and developmental status of a cell might determine the morphology of its death. Thus there are variants of the patterns of cell death, and insistence on universality has in fact hindered research (Table I).

TABLE I. Patterns of Physiological Cell Death

Characteristic	Type I (apoptosis)	Type II
Nucleus	Early collapse	Late collapse
Chromatin	Intact → 500 kp → ~200 bp ladders (early step); marginated early	Intact → 500 kp → ~200 bp ladders (degradation to ladders is late); condensed, electron dense (late)
Cell membrane	Intact until late	Intact until late
Lysosomes	No major role	Primary destruction of cytoplasm
Respiration	Intact until late	Intact until late
Phagocytosis	By phagocytes or neighboring cells	Absent until late
Cell shape	Fragments	Ultimately fragments
Inflammation	No	No
Protein synthesis required	Sometimes	Often
Specific genes up-regulated	Sometimes	Yes
Provoked by	Modest pathologies; "confusion" among stimuli; viruses	Developmental stimuli (hormones or interactions among cells)
Earliest manifestation	DNA degradation?	Decreased protein synthesis? Increased lysosomes?
Examples	Thymocytes; lymphocytes; various cell lines	Metamorphosing insect or tadpole tissues; mammary gland

I.C. Description Of Type I Cell Death (Apoptosis)

Type I cell death, as defined by Schweichel and Merker [1973], is what is now recognized as classic apoptosis. Although early embryologists clearly described non-necrotic cell death and various structures such as Councilman bodies may have been apoptotic bodies, the description of apoptosis as a generalizable phenomenon dates to 1972 [Kerr et al., 1972]. The primary image of apoptosis is that of the dying thymocyte: fusion of chromatin into one mass, which binds to the nuclear membrane (marginates), while the cytoplasm remains apparently intact before beginning to condense (Fig.1a). The nuclear change is one of the earliest visible processes; the conversion to the condensed state occurs rapidly and is accompanied by endonucleolytic degradation of DNA between nucleosomes. Once the chromatin has condensed, electrophoresis of the DNA demonstrates a ladder of fragments differing in size by 180 bp, generated by an enzymatic activity resembling that of DNase I (Fig. 2). The condensation of the cytoplasm, by unknown means [Lockshin and Beaulaton, 1981] appears to be secondary and is frequently followed by budding and fragmentation into several pieces [Wyllie, 1993]. Intracellular organelles maintain a relatively normal morphology, including also little visible change in lysosomes, and the cell fragments are ultimately consumed by phagocytes or by neighboring cells that become phagocytic [Kerr et al., 1972]. The plasmalemma remains intact throughout this process, and in the absence of phagocytosis the apoptotic cell continues to exclude trypan blue through the earlier phases of death.

This type of cell death is seen in many varieties of cells, especially those that, like lymphocytes or thymocytes, have relatively little cytoplasm and are highly mitotic or derive from highly mitotic lines. In this situation, in which mitotic cells are likely to face challenges by mutagens (viruses, toxins), an appropriate biological imperative would be to destroy the DNA rapidly and effectively. Thus this type of cell death is particularly dramatic among hematopoietic cells and their derivatives.

I.D. Description Of Type II Cell Death

Unlike the situation in apoptosis, cells affected by type II death are characterized by an elaborate cytoplasm, are frequently secretory, and eliminate their cytoplasm by lysosomal means. Perhaps because these cells typically do not die as isolated individuals but rather die en masse following a developmental stimulus, phagocytosis appears to be a rather late and minor aspect of their death (Fig. 1b).

In the situations that have been studied, the first detectable changes are rather subtle: an increase in intact lysosomes [Lockshin and Zakeri, 1994; Zakeri et al., 1995; Halaby et al., 1995], a developing erosion of rough endoplasmic reticulum, and a substantial drop in protein synthesis [Zakeri et al., 1995]. The cytoplasmic changes are conspicuous, while the nucleus at low magnification appears relatively normal. (At high magnification there may be a change in the granularity of the nuclear matrix and chromatin, but this has not yet been well studied (in preparation). During this period, respiration and intracellular ATP are normal [Halaby et al., 1995], as might be expected since the cells do not undergo osmotic swelling; the resting potential and contractility of muscle is normal [Lockshin and Beaulaton, 1979], and there is little evidence that cell function is compromised. Cyclic nucleotides and other potential mediators change little or equivocally [Zakeri et al., 1995], and interference with their function appears to have little impact. Pronounced collapse of cellular function is quite late. Other measures of cell function and trafficking, such as alteration of cytoskeleton, are suggestive, but have not been studied extensively. Both type I and type II are physiological, and the dying cell asserts control for a considerable part of the process. It may be that the differences result from different priorities, such as the danger of a highly mitotic cell being damaged by a mutagen; different amounts of cytoplasm; or differences in the availability of phagocytes.

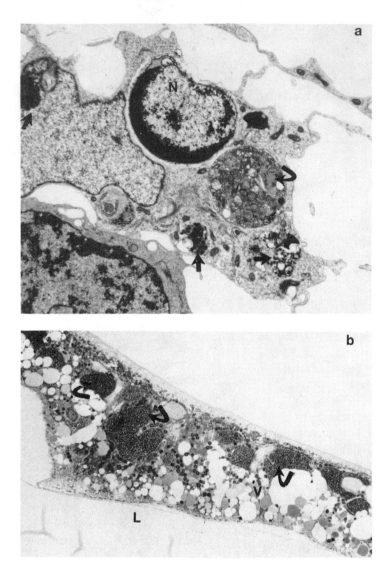

Fig. 1. *(a): Electron micrograph of apoptotic (type I) cell death in the hand palate of a day 14.5 postcoitus mouse embryo Several of the characteristics of apoptotic cell death are evident: phagocytosis of the fragments of a dead cell by a phagocytic cell (arrows); condensation of the cytoplasm of the dead cell (curved arrow) and coalesced and marginated chromatin in the phagocytosed nucleus (N). ×10,000. (b): Type II cell death, as seen in labial gland of* Manduca sexta, *the tobacco hornworm, on day 3 of metamorphosis. Semithin section, ×780. The lumenal side of the gland is indicated by L. The cytoplasm is mark-* *edly vacuolated, and the nuclei (curved arrows) are just beginning to undergo condensation. Though not shown here, lysosomal activity is high. Many of the vacuoles (V) are secretory, but some may be recognized by histochemical staining to be lysosomal, and, at later stages, lysosomes constitute the bulk of the cytoplasmic content. Type II apoptosis is also evident in insect muscle in which full but late nuclear condensation has been demonstrated. [Beaulaton and Lockshin, 1977, 1982; Lockshin and Beaulaton, 1979], as well as in vertebrate tissues [Zakeri et al., 1994].*

C   E

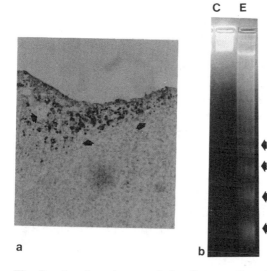

Fig. 2. *Another characteristic of apoptotic cell death. (a): End labeling of single- and double-stranded breaks in DNA in dying cells in the cell death region, indicated by arrows, of a mouse limb, as measured by the ApopTag kit. (b): Ladder of DNA fragments produced when DNA is extracted from glucocorticoid-treated thymocytes, electrophoresed in 2% agar, and stained with ethidium bromide. C, control; E, experimental (1 μM dexamethasone for 6 hours). The arrows at the right indicate from the bottom (anodal) end successive fragments of DNA differing in size by 180 bp (one nucleosome).*

Our search focuses on the common factors, which may include the initiation of the process.

II. THE NUCLEUS AND CELL DEATH

The ascendancy of the lymphocyte/thymocyte model, abetted by the relative ease of electrophoresing DNA extracts, created the impression that calcium-activated DNA degradation both caused and acted as a hallmark of apoptosis. This argument was inherently unreasonable. Complete inactivation of the genome, by enucleation, inhibition of transcription, or radiolytic breakage of DNA, does not lead to collapse of the cytoplasm with a morphology or speed approaching that of apoptosis, and there is no reason to believe that fragmented DNA is toxic. More recently, continual disagreement about the mechanism of activation

of the endonuclease, coupled with several lines of evidence pointing to earlier events, have forced researchers to re-examine the question. Several laboratories have now documented a temporal or other discordance between generation of ladders and morphological apoptosis [Oberhammer et al., 1992, 1993b, 1994; Brown et al., 1993; Zakeri et al., 1993; Enright et al., 1994] or cell death itself [Oberhammer et al., 1993a]. High-sensitivity techniques, such as end labeling of DNA or *in situ* end labeling, have demonstrated that, in type II cell death, a ladder of fragmented DNA may be generated, but only after much of the cytoplasm has already been destroyed [Zakeri et al., 1993; Woo et al., 1994], and one laboratory reports that ladders may be generated as artifacts [Enright et al., 1994]. Other laboratories, using techniques such as pulsed-field gel electrophoresis, which can resolve very high molecular weight fragments, have demonstrated that the first steps appear to be fragmentation of DNA into 300 kbp pieces, followed by digestion to sizes 50–100 kbp, before generation of the ladders [Oberhammer et al., 1993b; Walker et al., 1993, 1994; Woo et al., 1994]. These early steps, unlike the final steps, may not be dependent on Ca^{2+} [Walker et al., 1994], and may even involve proteolytic separation of DNA from its attachment to the nuclear membrane [Weaver et al., 1993]. These earlier steps may be more universal in that they have been detected in type II cell death. In the labial glands of the tobacco hornworm, which undergo type II apoptosis and very late DNA fragmentation, the early, large-sized fragmentation occurs well before the gland actually collapses, suggesting that the gland is compromised long before the initiating event.

III. LYSOSOMES AND CELL DEATH

The report of an early proteolytic step, coupled with a recent observation that a gene up-regulated in the death of both rat prostate and rat mammary epithelium is cathepsin B. brings to mind some older work that perhaps deserves re-examination. Using fluorescence labeling of lysosomes, Szego and coworkers argued that an early response of sensitive tissues to estrogenic stimulus was migration of lysosomes to the nuclear membrane and fusion of the lysosomes with the nuclear membrane, allowing lysosomal enzymes to enter the nucleus [Szego et al., 1971; Szego and Seeler, 1973; Szego, 1974; Nazareno et al., 1981]. The argument that proteolytic liberation of the chromosomes from their attachment sites is brought about by lysosomal enzymes is bizarre and does not appear probable given the dual nature of the nuclear membrane but, in light of recent findings, may require re-investigation.

IV. LESSONS FROM CAENORHABDITIS AND DROSOPHILA (GENES THAT CONTROL CELL DEATH)

That cell death in development is under genetic control is obvious, since any characteristic reproducible among individuals of a species is ultimately a product of selection. The assumption of a genetics of cell death was implicit in the earlier studies by Saunders and colleagues, who noted that the webbing between the digits of aquatic birds and the absence of webbing in other birds could be explained by an absence of cell death in the former situation [Saunders, 1966; Saunders and Fallon, 1966]. The control may, however, be indirect, for the interdigital cell death seen in the development of vertebrate limbs may be regulated by distribution of morphogens or morphogen-like substances such as, perhaps, retinoic acid [Zakeri and Ahuja, 1995]. Likewise the ability to prevent programmed cell death by inhibiting protein synthesis suggests but does not prove that there are specific cell death genes. The most direct proof for a genetics of cell death derives primarily from Caenorhabditis (see Lints and Driscoll, this volume) and secondarily from Drosophila, in which the existence of genes required for cell death and capable of blocking it has been documented [Horvitz et al., 1983; Ellis and Horvitz, 1986; Yuan and Horvitz, 1990; Ellis et al.,

1991a,b; Hengartner et al., 1992; Yuan and Horvitz, 1992; Hengartner and Horvitz, 1994a,b; Steller et al., 1994; White et al., 1994]. Interestingly, as is discussed below, these genes may have homologs in mammals. Another important observation from this work, however, is a generality that is highly relevant to the overall question of aging: In Caenorhabditis and Drosophila, it appears that all cells are genetically capable of self-destruction and that they are usually impeded from doing so by genes that inhibit cell death [Ellis et al., 1991b; Hengartner et al., 1992; Hengartner and Horvitz, 1994b]. The cell death inhibitor gene, ced-9 in Caenorhabditis, has a mammalian homolog that is also an inhibitor of cell death, bcl-2 [Vaux et al., 1992; Hengartner and Horvitz, 1994a]. The implication for gerontology is profound: Cells that fail to receive the proper support, or that for any reason are incapable of responding to signals for suppression of cell death, will die (see Tolkovsky et al., this volume). It therefore becomes extremely important to recognize the situations in which cells will escape from the suppression of cell death. It should be noted, however, that the existence of genes that clearly control cell death has been documented presently only for invertebrates. Whether this limitation relates to the ease of genetic analysis in invertebrates or to a more fundamental difference is not currently resolved.

V. CELL DEATH GENES

Other than the direct genetic evidence for cell death genes as detected in invertebrate development, most of the argument for cell death genes is indirect: temporary protection of cells by inhibitors of protein synthesis such as cycloheximide. Many interpretations of this result other than the existence of "thanatogenes" [Guenette et al., 1994] are possible. Most of these other interpretations can be categorized into two groups.

One group of interpretations suggests that a short-lived destructive element, such as a protease or nuclease, is continuously synthesized but is kept under control (inhibited) by factors relating to the maintenance of the cell. The disappearance of these factors allows the hypothetical enzyme to act, perhaps by alteration of configuration of substrate, and the enzyme is the effector of cell death. Blockage of protein synthesis leads to depletion of this enzyme, thereby protecting the cell. In this scenario, there is no requirement for specificity of the proteins synthesized by the dying cell.

A second group of interpretations suggests that the act of synthesis drains from the cell a limiting resource and that blockage of synthesis spares this resource. Several suggestions for such a resource have come forth: energy, such as ATP [Halaby et al., 1995] (but in many instances cells appear to be adequately supplied with energy during the commitment phase of cell death [Carson et al., 1986, 1988], as is predictable from the fact that they do not lyse); and glutathione, cysteine, or other traps for free radicals or reactive oxygen groups [Latour et al., 1992; Shen et al., 1992; Flomerfelt et al., 1993; Hockenbcry et al., 1993; Pan and Perez-Polo, 1993; Cheng et al., 1994; Fernandes and Cotter, 1994; Ratan et al., 1994; Whittemore et al., 1994; see also Wolozin et al., this volume]. All of these intermediaries of course may change as a result of changing metabolism in a dying cell. A more theoretical and explicit expansion of this general concept in the interpretation of the kinetics of death from NGF deprivation as reflecting a nonconvergent "death set" is given by Tolkovsky et al. (this volume).

These two suggestions underscore the point that "killer" genes or "thanatogenes" have not yet been identified for vertebrates, and the function of the invertebrate genes, as deduced from sequence motifs, has not yet suggested specifically destructive mechanisms [Yuan and Horvitz, 1992; Driscoll, 1992; Hengartner and Horvitz, 1994a,b; Kumar et al., 1994]. Indeed, many of the genes known to be differentially expressed are oncogenes or cell cycle genes [Buttyan et al., 1988; Bursch et al., 1990; O'Connor et al., 1991; Bissonnette et al., 1992; Colotta et al., 1992; Evan et al., 1992; González-Martín et al., 1992; Grassilli et al., 1992; Korsmeyer, 1992; Ryan et al., 1993; Fesus, 1993; Furuya and Isaacs, 1993; Heintz, 1993; Manfredini et al., 1993; Quinlan, 1993; Smeyne et al., 1993; Vaux and Weissman, 1993; Yonish-Rouach et al., 1993; Chiou et al., 1994; Freeman et al., 1994; Harrington et al., 1994; Kerr et al., 1994]. For most of these, the data are controversial. To take *c-myc* as an example, various means of up-regulating its transcription may protect the cell [Bennett et al., 1994; Toren and Rechavi, 1994] or induce apoptosis [Amati et al., 1993; Davidoff and Mendelow, 1993; Milner et al., 1993; Evan et al., 1994].

These contradictory findings have spawned a host of interpretations, mostly focusing on apoptosis as an abortive mitosis [Colombel et al., 1992]. It is possible to generalize these findings based on the frequent association of cell death with dysregulation of growth factors, growth factor receptors, oncogenes, and cell cycle genes [Evan et al., 1994; Kerr et al., 1994]. The generalization derives from a crude analogy from physiology: An animal deprived of sufficient dietary protein does poorly, but it does better than one fed an inadequate protein source such as gelatin. To oversimplify the situation, the latter animal still receives the branch-chain amino acids that stimulate protein synthesis and therefore squanders its energy in ineffectual efforts. In the case of cell death, one might suggest a similar scenario, with the added hypothesis that the cell actively monitors its resources as a self-evaluation mechanism. The more overall hypothesis would therefore be that of cell confusion: Cells are tuned to monitor the consistency of growth and stasis stimuli and their capacity to respond to the stimuli. In the event of a discordance of these parameters, the cells self-destruct either because a metabolic parameter has been drained or (for lymphocytes, thymocytes, and others on the front line of defense) because they carry specific instructions to do so ("Better dead than wrong" [J.J. Cohen, personal communication]). Such a hypothesis combines ideas from several sources, for instance, the "death set" model described by Tolkovsky et al. (this volume). Related scenarios are described by Pittman and DiBenedetto (this volume) and, for mitotic cells, by Lowe and Ruley (this volume). It is consistent with the concept that all cells contain the information for self-destruction [Ellis et al., 1991b]. It further would account for the frequent involvement of cell cycle genes in the death of even postmitotic cells and the often contradictory findings concerning growth factors. In metamorphosing animals, the inability of tissues to respond to growth hormones such as thyroxin (tadpoles) or ecdysterone (insects) would lead to their destruction in the same manner as that proposed for differentiating lymphocytes. The direction of research would therefore turn to the mechanism of intracellular monitoring.

VI. CALCIUM, CYTOSKELETON, PHOSPHORYLATION, FREE RADICALS

A rise in intracellular calcium is considered to be a major component of necrotic cell death [Trump and Berezesky, 1992] and is likewise considered by many to be an early, perhaps initiating or causal, component of apoptotic cell death [Schwartzman and Cidlowski, 1993]. There are several reasons why a rise in calcium could trigger apoptosis—activation of a Ca^{2+}, Mg^{2+} endonuclease; stimulation of Ca^{2+}-dependent proteases such as calpain [Squier et al., 1994]; activation of Ca^{2+}-mediated phospholipase [Orrenius et al., 1992] and many workers have detected a rise in cytosolic Ca^{2+} that lasts a few hours in many instances of apoptosis in lymphocytes or differentiating thymocytes [Dowd et al., 1991; Ojcius et al., 1991; Story et al., 1992; Baffy et al., 1993; Ning and Murphy, 1993; Schwartzman and Cidlowski, 1993; Jiang et al., 1994]. There are, however, many problems with this argument, including the identification of cells in which apoptosis is preceded by a drop in $[Ca^{2+}]_i$ [Magnelli et al., 1993; see also Whyte et al., 1993]; cells including immature thymocytes that are insensitive to $[Ca^{2+}]_i$ [Andjelic et al., 1993]; and situations in which a raised $[Ca^{2+}]_i$ can be dissociated from cell death [Barbieri et

al., 1992; Lennon et al., 1992; Iseki et al., 1993; Duke et al., 1994; Kluck et al., 1994; Matsubara et al., 1994]. Many of the researchers reporting these results attribute apoptosis to secondary processes that may in some instances be triggered by calcium. Among these secondary processes are partitioning of Ca^{2+} to the nucleus [Nicotera and Rossi, 1994], calmodulin [Dowd et al., 1991], calcineurin (a calcium- and calmodulin-dependent serine/threonine phosphatase [Fruman et al., 1992; Bonnefoy-Berard et al., 1994]), glutathione S-transferase [Flomerfelt et al., 1993], and calpain [Squier et al., 1994], in addition to calcium-dependent nucleases and proteases. One argument that is now fading into the background is the presumption that inhibition of the calcium-dependent nuclease by aurintricarboxylic acid (ATC) documents the centrality of the nuclease. ATC is now known to inhibit many reactions [Orrenius et al., 1992; Schwartzman and Cidlowski, 1993]. Many studies purporting to measure the nuclease use millimolar levels of Ca^{2+} [Schwartzman and Cidlowski, 1993]. While there may be valid reasons to defend such concentrations, they must remain suspect until the mechanism for this requirement is fully understood.

Many of the observations of changes invoked by calcium ionophores, such as elevated glutathione S-transferase [Flomerfelt et al., 1993], cytoskeletal changes [Squier et al., 1994] (confirmed by our laboratory in cell death during insect metamorphosis), correlation of calcineurin phosphatase with apoptosis [Fruman et al., 1992], inhibition of apoptosis by staurosporine, an inhibitor of protein kinase C [Koseki et al., 1992], and movement of lysosomes [Zakeri et al., 1995] suggest that the larger overview should be that cell death involves a profound conversion of intracellular signal transduction and trafficking, mediated by changes in the state of phosphorylation of critical enzymes or components of the cytoskeleton, as well as perhaps major rearrangement of the cytoskeleton or its ability to target materials to specific regions of a cell. These changes may be the physiological translation of the concept of cell confusion [Lockshin and Zakeri, 1994; Zakeri et al., 1994] and deserve further exploration.

VII. CONTROL OF CELL DEATH

The problem can be stated as follows:

1. Many different types of signals can lead cells into apoptosis. *In vitro*, withdrawal of growth factors or interference with growth factor receptors is commonly used to induce cell death. The patterned death of embryonic cells may reflect this situation. Among thymocytes and lymphocytes, several types of signals, including over-stimulation of T-cell receptors, treatment with glucocorticoids, viral gene activity, and irradiation, can independently lead to the death of the cell.
2. These several pathways lead, presumptively through different transductions, to similar morphological collapse of cells. Several instances have been reported in which no protein synthesis or other nuclear involvement is necessary, thus suggesting that the signal transduction may be direct.
3. In other instances there is a clear nuclear response, involving the swift up- and down-regulation of several genes, some of which are common to different instances of cell death.

The second step is perhaps the most interesting, but it is the most difficult to study. Presumptively the best way to approach it is to have a clearly delineated, natural form of cell death and, using a well-defined marker such as the up-regulation of a specific gene, to work backwards to trace the signal passage. Genes may change activity in dying cells for several reasons: they may actively generate the collapse of the cell; they may respond to changing intracellular conditions; they may be derepressed by loss of suppressor mechanisms; or they may be a participant in the process, such as a suppressor of complement activation

(Ahuja-Singh and Zakeri, personal communication). The race to identify cell death genes, "thanatogenes" or "killer genes," has focused on up-regulation *in vitro* and knockouts or transgenic up-regulation *in vitro*. To determine the actual role of any gene or gene product, we need to be able to put it under the regulation of a controllable promoter so that we can up-regulate and down-regulate the gene at will.

VIII. PROBABILITY THAT CELL DEATH IS A SIGNIFICANT ASPECT OF AGING

Virtually all of many efforts to find a unifying theory of aging have foundered when rigorous questions were asked, and there is no reason to assume that cell death as an explanation for aging would fare any better. No one hypothesizes that aging occurs because we run out of cells. Rather it is the process of dying that is so interesting, especially in the production of immunocompetent cells, the loss of neurons, and the generation of tumors, but also in all tissues that show functional deficit with age.

Cell death in differentiating lymphocytes and thymocytes creates many situations of concern, including decreased ability to mount an effective primary or secondary response, production of autoantibodies through failure to destroy potentially antiself cells, and failure of immunosurveillance to eradicate newly generated tumors. The discovery that at least one oncogene functions by interfering with cell death allows us to postulate many similar scenarios, similar to the sequence of *ced* mutants in *Caenorhabditis*. In these situations it is quite possible to suppose that pathology can derive from abnormal regulation of the machinery of cell death and to predict that that machinery can be manipulated for therapeutic purposes. Several biotechnology companies are already aggressively pursuing this aim.

Concerning the central nervous system and by extrapolation other tissues of low mitotic index, it is unlikely that these tissues are so directly regulated by cell death. Here our interest is different: We have no reason to assume that these tissues lack a mechanism of active cell death, and the deaths of these cells as well presumptively occur by an active mechanism. The concept of an active mechanism presupposes a series of intracellular and extracellular signals whereby the stress or agony of a cell is manifest. In situations as poignant as loss of cardiomyocytes or brain cells, through SDAT or other means, or the loss of cells as, in the face of a nonlytic virus, they self-destruct, we have a different responsibility: to identify the signals that these cells emit under duress and to identify the threshold at which they fail and the cause of that failure. If we can accomplish this, we can act in a proactive fashion to shore up or protect those cells and to help them through the crisis. If we can do this, then the study of programmed, active, or physiological cell death or apoptosis will have truly benefited gerontology.

ACKNOWLEDGMENTS

The research described in this paper was performed with the support of grants from the National Institute on Aging to R.A.L. and Z.Z. (R01-10101) and to Z.Z. (K04). We thank Dr. Daniela Quaglino (U. Modena, Italy) for the electron micrographs and Theresa Latham for excellent assistance.

REFERENCES

Amati B, Littlewood TD, Evan GI, Land H (1993): The c-Myc protein induces cell cycle progression and apoptosis through dimerization with Max. EMBO J 12:5083–5087.

Andjelic S, Jain N, Nikolic-Zugic J (1993): Immature thymocytes become sensitive to calcium-mediated apoptosis with the onset of CD8, CD4, and the T cell receptor expression: A role for *bcl-2*. J Exp Med 178:1745–1751.

Baffy G, Miyashita T, Williamson JR, Reed JC (1993): Apoptosis induced by withdrawal of interleukin-3 (IL-3) from an IL-3–dependent hematopoietic cell line is associated with repartitioning of intracellular calcium and is blocked by enforced Bcl-2 oncoprotein production. J Biol Chem 268:6511–6519.

Barbieri D, Troiano L, Grassilli E, Agnesini C, Cristofalo EA, Monti D, Capri M, Cossarizza A, Franceschi C

(1992): Inhibition of apoptosis by zinc: A reappraisal. Biochem Biophys Res Commun 187:1256–1261.

Beaulaton J, Lockshin RA (1977): Ultrastructural study of the normal degeneration of the intersegmental muscles of *Antheraea polyphemus* and *Manduca sexta* (Insecta, Lepidoptera) with particular reference to cellular autophagy. J Morphol 154:39–58.

Beaulaton J, Lockshin RA (1982): The relation of programmed cell death to development and reproduction: Comparative studies and an attempt at classification. Int Rev Cytol 29:215–235.

Bennett MR, Evan GI, Newby AC (1994): Deregulated expression of the *c-myc* oncogene abolishes inhibition of proliferation of rat vascular smooth muscle cells by serum reduction, interferon-gamma, heparin, and cyclic nucleotide analogues and induces apoptosis. Circ Res 74:525–536.

Bissonnette RP, Echeverri F, Mahboubi A, Green DR (1992): Apoptotic cell death induced by *c-myc* is inhibited by *bcl-2*. Nature 359:552–554.

Bonnefoy-Berard N, Genestier L, Flacher M, Revillard JP (1994): The phosphoprotein phosphatase calcineurin controls calcium-dependent apoptosis in B cell lines. Eur J Immunol 24:325–329.

Brown DG, Sun X-M, Cohen GM (1993): Dexamethasone-induced apoptosis involves cleavage of DNA to large fragments prior to internucleosomal fragmentation. J Biol Chem 268:3037–3039.

Bursch W, Kleine L, Tenniswood M (1990): The biochemistry of cell death by apoptosis. Biochem Cell Biol 68:1071–1074.

Buttyan R, Zakeri Z, Lockshin RA, Wolgemuth D (1988): Cascade induction of c-fos, c-myc, and heat shock 70 k transcripts during regression of the rat ventral prostate gland. Mol Endocrinol 2:650–657.

Carson DA, Carrera CJ, Wasson DB, Yamanaka H (1988): Programmed cell death and adenine deoxynucleotide metabolism in human lymphocytes. Adv Enzyme Regul 27:395–404.

Carson DA, Seto S, Wasson DB, Carrera CJ (1986): DNA strand breaks, NAD metabolism, and programmed cell death. Exp Cell Res 164:273–281.

Cheng Y, Wixom P, James-Kracke MR, Sun AY (1994): Effects of extracellular ATP on Fe^{2+}-induced cytotoxicity in PC-12 cells. J Neurochem 63:895–902.

Chiou S-K, Rao L, White E (1994): Bcl-2 blocks p53-dependent apoptosis. Mol Cell Biol 14:2556–2563.

Colombel M, Olsson CA, Ng P-Y, Buttyan R (1992): Hormone-regulated apoptosis results from reentry of differentiated prostate cells onto a defective cell cycle. Cancer Res 52:4313–4319.

Colotta F, Polentarutti N, Sironi M, Mantovani A (1992): Expression and involvement of *c-fos* and *c-jun* protooncogenes in programmed cell death induced by growth factor deprivation in lymphoid cell lines. J Biol Chem 267:18278–18283.

Davidoff AN, Mendelow BV (1993): Puromycin-elicited

c-myc mRNA superinduction precedes apoptosis in HL-60 leukaemic cells. Anticancer Res 13:2257–2260.

Dowd DR, MacDonald PN, Komm BS, Haussler MR, Miesfeld R (1991): Evidence for early induction of calmodulin gene expression in lymphocytes undergoing glucocorticoid-mediated apoptosis. J Biol Chem 266:18423–18426.

Driscoll M (1992): Molecular genetics of cell death in the nematode *Caenorhabditis elegans*. J Neurobiol 23:1327–1351.

Duke RC, Witter RZ, Nash PB, Young JD-E, Ojcius DM (1994): Cytolysis mediated by ionophores and pore-forming agents: Role of intracellular calcium in apoptosis. FASEB J 8:237–246.

Ellis HM, Horvitz HR (1986): Genetic control of programmed cell death in the nematode *C. elegans*. Cell 44:817–829.

Ellis RE, Jacobson DM, Horvitz HR (1991a): Genes required for the engulfment of cell corpses during programmed cell death in *Caenorhabditis elegans*. Genetics 129:79–94.

Ellis RE, Yuan J, Horvitz HR (1991b): Mechanisms and functions of cell death. Annu Rev Cell Biol 7:663–698.

Enright H, Hebbel RP, Nath KA (1994): Internucleosomal cleavage of DNA as the *sole* criterion for apoptosis may be artifactual. J Lab Clin Med 124:63–68.

Evan G, Harrington E, Fanidi A, Land H, Amati B, Bennett M (1994): Integrated control of cell proliferation and cell death by the *c-myc* oncogene. Philos Trans R Soc Lond B Biol Sci 345:269–275.

Evan GI, Wyllie AH, Gilbert CS, Littlewood TD, Land H, Brooks M, Waters CM, Penn LZ, Hancock DC (1992): Induction of apoptosis in fibroblasts by c-myc protein. Cell 69:119–128.

Fernandes RS, Cotter TG (1994): Apoptosis or necrosis: Intracellular levels of glutathione influence mode of cell death. Biochem Pharmacol 48:675–681.

Fesus L (1993): Biochemical events in naturally occurring forms of cell death. FEBS Lett 328:1–5.

Flomerfelt FA, Briehl MM, Dowd DR, Dieken ES, Miesfeld RL (1993): Elevated glutathione S-transferase gene expression is an early event during steroid-induced lymphocyte apoptosis. J Cell Physiol 154:573–581.

Freeman RS, Estus S, Johnson EM Jr (1994): Analysis of cell cycle-related gene expression in postmitotic neurons: Selective induction of *cyclin D1* during programmed cell death. Neuron 12:343–355.

Fruman DA, Mather PE, Burakoff SJ, Bierer BE (1992): Correlation of calcineurin phosphatase activity and programmed cell death in murine T cell hybridomas. Eur J Immunol 22:2513–2517.

Furuya Y, Isaacs JT (1993): Differential gene regulation during programmed death (apoptosis) *versus* proliferation of prostatic glandular cells induced by androgen manipulation. Endocrinology 133:2660–2666.

González-Martín C, De Diego I, Crespo D, Fairén A (1992): Transient c-*fos* expression accompanies naturally occurring cell death in the developing interhemispheric cortex of the rat. Dev Brain Res 68:83–95.

Grassilli E, Carcereri de Prati A, Monti D, Troiano L, Menegazzi M, Barbieri D, Franceschi C, Suzuki H (1992): Studies of the relationship between cell proliferation and cell death. II. Early gene expression during concanavalin A–induced proliferation or dexamethasone-induced apoptosis of rat thymocytes. Biochem Biophys Res Commun 188:1261–1266.

Guenette RS, Daehlin L, Mooibroek M, Wong K, Tenniswood M (1994): Thanatogen expression during involution of the rat ventral prostate after castration. J Androl 15:200–211.

Halaby R, Zakeri Z, Lockshin RA (1994): Metabolic events during programmed cell death in insect labial glands. Biochem Cell Biol 72:597–601.

Harrington EA, Fanidi A, Evan GI (1994): Oncogenes and cell death. Curr Opin Genet Dev 4:120–129.

Heintz N (1993): Cell death and the cell cycle: A relationship between transformation and neurodegeneration. Trends Biochem Sci 18:157–159.

Hengartner MO, Ellis RE, Horvitz HR (1992): *Caenorhabditis elegans* gene *ced-9* protects cells from programmed cell death. Nature 356:494–499.

Hengartner MO, Horvitz HR (1994a): Activation of *C. elegans* cell death protein CED-9 by an amino-acid substitution in a domain conserved in Bcl-2. Nature 369:318–320.

Hengartner MO, Horvitz HR (1994b): Programmed cell death in *Caenorhabditis elegans*. Curr Opin Genet Dev 4:581–586.

Hockenbery DM, Oltvai ZN, Yin X-M, Milliman CL, Korsmeyer SJ (1993): Bcl-2 functions in an antioxidant pathway to prevent apoptosis. Cell 75:241–251.

Horvitz HR, Sternberg PW, Greenwald IS, Fixsen W, Ellis HM (1983): Mutations that affect neural cell lineages and cell fates during the development of the nematode *Caenorhabditis elegans*. Cold Spring Harb Symp Quant Biol 48:453–463.

Iseki R, Kudo Y, Iwata M (1993): Early mobilization of Ca^{2+} is not required for glucocorticoid-induced apoptosis in thymocytes. J Immunol 151:5198–5207.

Jiang S, Chow SC, Nicotera P, Orrenius S (1994): Intracellular Ca^{2+} signals activate apoptosis in thymocytes: Studies using the Ca^{2+}-ATPase inhibitor thapsigargin. Exp Cell Res 212:84–92.

Kerr JFR, Winterford CM, Harmon BV (1994): Apoptosis: Its significance in cancer and cancer therapy. Cancer 73:2013–2026.

Kerr JFR, Wyllie AH, Currie AR (1972): Apoptosis: A basic biological phenomenon with wide-ranging implications in tissue kinetics. Br J Cancer 26:239–257.

Kluck RM, McDougall CA, Harmon BV, Halliday JW (1994): Calcium chelators induce apoptosis—Evidence that raised intracellular ionised calcium is not essential for apoptosis. Biochim Biophys Acta Mol Cell Res 1223:247–254.

Korsmeyer SJ (1992): Bcl-2 initiates a new category of oncogenes: Regulators of cell death. Blood 80: 879–886.

Koseki C, Herzlinger D, Al-Awqati Q (1992): Apoptosis in metanephric development. J Cell Biol 119:1327–1333.

Kumar S, Kinoshita M, Noda M, Copeland NG, Jenkins NA (1994): Induction of apoptosis by the mouse *Nedd2* gene, which encodes a protein similar to the product of the *Caenorhabditis elegans* cell death gene *ced-3* and the mammalian IL-1β-converting enzyme. Genes Dev 8:1613–1626.

Latour I, Pregaldien J-L, Buc-Calderon P (1992): Cell death and lipid peroxidation in isolated hepatocytes incubated in the presence of hydrogen peroxide and iron salts. Arch Toxicol 66:743–749.

Lennon SV, Kilfeather SA, Hallett MB, Campbell AK, Cotter TG (1992): Elevations in cytosolic free Ca^{2+} are not required to trigger apoptosis in human leukaemia cells. Clin Exp Immunol 87:465–471.

Levi-Montalcini R, Aloe L (1981): Mechanism(s) of action of nerve growth factor in intact and lethally injured sympathetic nerve cells in neonatal rodents. In Bowen ID, Lockshin RA (eds): Cell Death in Biology and Pathology. New York: Chapman and Hall, pp 295–327.

Lockshin RA (1969a): Programmed cell death. Activation of lysis of a mechanism involving the synthesis of protein. J Insect Physiol 15:1505–1516.

Lockshin RA (1969b): Lysosomes in insects. In Dingle JT, Fell HB (eds): Lysosomes in Biology and Pathology. Amsterdam: North Holland Publishing, pp 363–391.

Lockshin RA, Beaulaton J (1979): Cytological studies of dying muscle fibers of known physiological parameters. Tissue Cell 11:803–819.

Lockshin RA, Beaulaton J (1981): Cell death: Questions for histochemists concerning the causes of the various cytological changes. Histochem J 13:659–666.

Lockshin RA, Williams CM (1964): Programmed cell death. II. Endocrine potentiation of the breakdown of the intersegmental muscles of silkmoths. J Insect Physiol 10:643–649.

Lockshin RA, Williams CM (1965a): Programmed cell death. I. Cytology of the degeneration of the intersegmental muscles of the *Pernyi* silkmoth. J Insect Physiol 11:123–133.

Lockshin RA, Williams CM (1965b): Programmed cell death. IV. The influence of drugs on the breakdown of the intersegmental muscles of silkmoths. J Insect Physiol 11:803–809.

Lockshin RA, Zakeri Z (1994): Programmed cell death: Early changes in metamorphosing cells. Biochem Cell Biol 72:589–596.

Magnelli L, Cinelli M, Turchetti A, Chiarugi VP (1993): Apoptosis induction in 32D cells by IL-3 withdrawal is

preceded by a drop in the intracellular calcium level. Biochem Biophys Res Commun 194:1394–1397.

Manfredini R, Grande A, Tagliafico E, Barbieri D, Zucchini P, Citro G, Zupi G, Franceschi C, Torelli U, Ferrari S (1993): Inhibition of c-*fes* expression by an antisense oligomer causes apoptosis of HL60 cells induced to granulocytic differentiation. J Exp Med 178:381–389.

Matsubara K, Kubota M, Kuwakado K, Hirota H, Wakazono Y, Okuda A, Bessho R, Lin YW, Adachi S, Akiyama Y (1994): Variable susceptibility to apoptosis induced by calcium ionophore in hybridomas between HL-60 promyelocytic and CEM T-lymphoblastic leukemia cell lines: Relationship to constitutive Mg^{2+}-dependent endonuclease. Exp Cell Res 213:412–417.

Milner AE, Grand RJA, Waters CM, Gregory CD (1993): Apoptosis in Burkitt lymphoma cells is driven by c-*myc*. Oncogene 8:3385–3391.

Munck A (1971): Glucocorticoid inhibition of glucose uptake by peripheral tissues: Old and new evidence, molecular mechanisms, and physiological significance. Perspect Biol Med 14:265–289.

Nazareno MB, Horton MJ, Szego CM, Seeler BJ (1981): Antibodies against estradiol-binding lysosomal lipoproteins gain access to the nuclear compartment of preputial gland cells under estrogen influence. Endocrinology 108:1156–1163.

Nicotera P, Rossi AD (1994): Nuclear Ca^{2+}: Physiological regulation and role in apoptosis. Mol Cell Biochem 135:89–98.

Ning Z-Q, Murphy JJ (1993): Calcium ionophore–induced apoptosis of human B cells is preceded by the induced expression of early response genes. Eur J Immunol 23:3369–3372.

Oberhammer F, Fritsch G, Pavelka M, Froschl G, Tiefenbacher R, Purchio T, Schulte-Hermann R (1992): Induction of apoptosis in cultured hepatocytes and in the regressing liver by transforming growth factor-β1 occurs without activation of an endonuclease. Toxicol Lett 64-65:701–704.

Oberhammer F, Fritsch G, Schmied M, Pavelka M, Printz D, Purchio T, Lassmann H, Schulte-Hermann R (1993a): Condensation of the chromatin at the membrane of an apoptotic nucleus is not associated with activation of an endonuclease. J Cell Sci 104:317–326.

Oberhammer FA, Hochegger K, Fröschl G, Tiefenbacher R, Pavelka M (1994): Chromatin condensation during apoptosis is accompanied by degradation of lamin A+B, without enhanced activation of cdc2 kinase. J Cell Biol 126:827–838.

Oberhammer F, Wilson JW, Dive C, Morris ID, Hickman JA, Wakeling AE, Walker PR, Sikorska M (1993b): Apoptotic death in epithelial cells: Cleavage of DNA to 300 and/or 50 kb fragments prior to or in the absence of internucleosomal fragmentation. EMBO J 12:3679–3684.

O'Connor PM, Wassermann K, Sarang M, Magrath I, Bohr VA, Kohn KW (1991): Relationship between DNA cross-links, cell cycle, and apoptosis in Burkitt's lymphoma cell lines differing in sensitivity to nitrogen mustard. Cancer Res 51:6550–6557.

Ojcius DM, Zychlinsky A, Zheng LM, Young JD-E (1991): Ionophore-induced apoptosis: Role of DNA fragmentation and calcium fluxes. Exp Cell Res 197:43–49.

Orrenius S, McCabe MJ Jr, Nicotera P (1992): Ca^{2+}-dependent mechanisms of cytotoxicity and programmed cell death. Toxicol Lett 64-65:357–364.

Pan Z, Perez-Polo R (1993): Role of nerve growth factor in oxidant homeostasis: Glutathione metabolism. J Neurochem 61:1713–1721.

Quinlan MP (1993): E1A 12S in the absence of E1B or other cooperating oncogenes enables cells to overcome apoptosis. Oncogene 8:3289–3296.

Ratan RR, Murphy TH, Baraban JM (1994): Oxidative stress induces apoptosis in embryonic cortical neurons. J Neurochem 62:376–379.

Ryan JJ, Danish R, Gottlieb CA, Clarke MF (1993): Cell cycle analysis of p53-induced cell death in murine erythroleukemia cells. Mol Cell Biol 13:711–719.

Saunders JW Jr (1966): Death in embryonic systems. Science 154:604–612.

Saunders JW Jr, Fallon JF (1966): Cell death in morphogenesis. In Locke M (ed): Major Problems in Developmental Biology (25th Symposium of the Society for Developmental Biology). New York: Academic Press, p 289.

Schwartzman RA, Cidlowski JA (1993): Mechanism of tissue-specific induction of internucleosomal deoxyribonucleic acid cleavage activity and apoptosis by glucocorticoids. Endocrinol 133:591–599.

Schweichel JU, Merker HJ (1973): The morphology of various types of cell death in prenatal tissues. Teratology 7:253–266.

Shen W, Kamendulis LM, Ray SD, Corcoran GB (1992): Acetaminophen-induced cytotoxicity in cultured mouse hepatocytes: Effects of Ca^{2+}-endonuclease, DNA repair, and glutathione depletion inhibitors on DNA fragmentation and cell death. Toxicol Appl Pharmacol 112:32–40.

Smeyne RJ, Vendrell M, Hayward M, Baker SJ, Miao GG, Schilling K, Robertson LM, Curran T, Morgan JI (1993): Continuous c-fos expression precedes programmed cell death in vivo. Nature 363:166–169.

Squier MKT, Miller ACK, Malkinson AM, Cohen JJ (1994): Calpain activation in apoptosis. J Cell Physiol 159:229–237.

Steller H, Abrams JM, Grether ME, White K (1994): Programmed cell death in Drosophila. Philos Trans R Soc Lond B Biol Sci 345:247–250.

Story MD, Stephens LC, Tomasovic SP, Meyn RE (1992): A role for calcium in regulating apoptosis in rat thymocytes irradiated in vitro. Int J Radiat Biol 61:243–251.

Strange R, Li F, Saurer S, Burkhardt A, Friis RR (1992): Apoptotic cell death and tissue remodelling during mouse mammary gland involution. Development 115:49–58.

Szego CM (1974): The lysosome as a mediator of hormone action. Recent Prog Horm Res 30:171–233.

Szego CM, Seeler BJ (1973): Hormone-induced activation of target-specific lysosomes: Acute translocation to the nucleus after administration of gonadal hormones in vivo. J Endocrinol 56:347–360.

Szego CM, Seeler BJ, Steadman RA, Hill DF, Kimura AK, Roberts JA (1971): The lysosomal membrane complex. Focal point of primary steroid hormone action. Biochem J 123:523–538.

Tata JR (1966): Requirement for RNA and protein synthesis for induced regression of tadpole tail in organ culture. Dev Biol 13:77–94.

Toren A, Rechavi G (1994): The anti-apoptotic role of infectious agents in lymphoid malignancies characterized by c-myc deregulation. Br J Haematol 87:675–677.

Trump BF, Berezesky IK (1992): The role of cytosolic Ca^{2+} in cell injury, necrosis and apoptosis. Curr Opin Cell Biol 4:227–232.

Vaux DL, Weissman IL (1993): Neither macromolecular synthesis nor Myc is required for cell death via the mechanism that can be controlled by Bcl-2. Mol Cell Biol 13:7000–7005.

Vaux DL, Weissman IL, Kim SK (1992): Prevention of programmed cell death in Caenorhabditis elegans by human bcl-2. Science 258:1955–1957.

Walker PR, Kokileva L, Leblanc J, Sikorska M (1993): Detection of the initial stages of DNA fragmentation in apoptosis. BioTechnology 15:1032–1040.

Walker PR, Weaver VM, Lach B, Leblanc J, Sikorska M (1994): Endonuclease activities associated with high molecular weight and internucleosomal DNA fragmentation in apoptosis. Exp Cell Res 213:100–106.

Weaver VM, Lach B, Walker PR, Sikorska M (1993): Role of proteolysis in apoptosis: Involvement of serine proteases in internucleosomal DNA fragmentation in immature thymocytes. Biochem Cell Biol 71:488–500.

White K, Grether ME, Abrams JM, Young L, Farrell K, Steller H (1994): Genetic control of programmed cell death in Drosophila. Science 264:677–683.

Whittemore ER, Loo DT, Cotman CW (1994): Exposure to hydrogen peroxide induces cell death via apoptosis in cultured rat cortical neurons. Neuro Report 5:1485–1488.

Whyte MKB, Hardwick SJ, Meagher LC, Savill JS, Haslett C (1993): Transient elevations of cytosolic free calcium retard subsequent apoptosis in neutrophils in vitro. J Clin Invest 92:446–455.

Woo K, Sikorska M, Weaver VM, Lockshin RA, Zakeri

Z (1994): DNA fragmentation and DNA synthesis during insect metamorphosis. 10th Int Symp Cell Endocrinol p 3.

Wyllie AH (1993): Apoptosis (The 1992 Frank Rose Memorial Lecture). Br J Cancer 67:205–208.

Yonish-Rouach E, Grunwald D, Wilder S, Kimchi A, May E, Lawrence J-J, May P, Oren M (1993): p53-Mediated cell death: Relationship to cell cycle control. Mol Cell Biol 13:1415–1423.

Yuan J, Horvitz HR (1990): The Caenorhabditis elegans genes ced-3 and ced-4 act cell autonomously to cause programmed cell death. Dev Biol 138:33–41.

Yuan J, Horvitz HR (1992): The Caenorhabditis elegans cell death gene ced-4 encodes a novel protein and is expressed during the period of extensive programmed cell death. Development 116:309–320.

Zakeri Z, Ahuja HS (1995): Cell death in limb bud development. Biochem Cell Biol (in press).

Zakeri Z, Bursch W, Tenniswood M, Lockshin RA (1995): Cell death: Programmed, apoptosis, necrosis, or other. Cell Death Differ 2:87–96.

Zakeri Z, Quaglino D, Latham T, Woo K, Lockshin RA (1995): Programmed cell death in the tobacco hornworm, Manduca sexta: Alterations in protein synthesis. Microsc Res Technol (in press).

Zakeri ZF, Quaglino D, Latham T, Lockshin RA (1993): Delayed internucleosomal DNA fragmentation in programmed cell death. FASEB J 7:470–478.

ABOUT THE AUTHORS

RICHARD A. LOCKSHIN is Professor of Biology at St. John's University. He received his B.S. (honors) in Biological Sciences and his M.A. and Ph.D. (1963) in Biology from Harvard University. His doctoral thesis was on programmed cell death. He spent two years as a postdoctoral fellow in Embryology at the Institute of Animal Genetics, University of Edinburgh, after which he accepted a faculty position in the Department of Physiology, University of Rochester School of Medicine and Dentistry. In 1975 he moved to St. John's University, where he served as Chair of the Department from 1983 to 1992. He continues to pursue the analysis of metabolic changes in programmed cell death.

ZAHRA ZAKERI is Associate Professor of Biology at Queens College and the Graduate Center of the City University of New York. She received her B.S. (honors) from York College, her M.S. from C.W. Post, and her Ph.D. from St. John's University, where she studied the early response genes of adenovirus. She then spent four years in the Department of Human Genetics at Columbia University School of Medicine, where she moved from postdoctoral fellow to research associate, studying the activation of heat shock and other genes during differentiation of the testis and first began to explore the activation of genes during the involution of the prostate. After one year as an instructor in the Department of Physiology of Robert Wood Johnson School of Medicine, she moved to Queens College, where she has won several awards. Her research interests include cell death and other aspects of pattern formation in the mammalian limb as well as the development of techniques for the detection of cell death and the up-regulation of genes during cell death.

Cellular Aging and Cell Death: 181–208
© 1996 Wiley-Liss, Inc.

Programmed Cell Death During Development of Animals

Carolanne E. Milligan and Lawrence M. Schwartz

I. INTRODUCTION

At a cellular level, embryonic development includes the events of genesis, migration, differentiation, maturation, and death. Cell death occurs throughout the life span of the organism and represents the ultimate differentiative decision by a cell [for reviews, see Glucksmann, 1951; Saunders, 1966; Kerr et al., 1972; Wyllie et al., 1980; Moon, 1981; Hurle, 1988; Shier, 1988; Clarke, 1990; Oppenheim, 1991; Schwartz and Osborne, 1993]. Unlike necrosis that results from pathological conditions or injury, programmed cell death (PCD) is a fundamental component of normal development and homeostasis in all multicellular organisms. The term *programmed cell death* was coined by Lockshin and Williams [1965] to describe the developmentally regulated loss of specific larval muscle following the emergence of adult moths. With the current attraction of biologists from fields other than development, the term has been used in a wider context than originally intended. Many investigators, most notably immunologists, use the term PCD to refer to any cell death that is dependent on a gene-mediated process or "program," independent of the signal that initiates the process. Under this broader application of the term, the irradiation-induced death of immature thymocytes, which does depend on *de novo* gene expression, is an example of PCD. This is despite the fact that irradiation is not a normal physiological mediator of cellular developmental decisions. Therefore, in any discussion of PCD, it is important to bear in mind the kind of phenomenon being addressed [see Schwartz and Osborne, 1993, for discussion].

In this chapter we use the term *PCD* to refer to the temporally and spatially restricted loss of a specific population of cells during development. It should also be said that this chapter is not intended to be a comprehensive review of all known examples of PCD, since much of that information is presented in earlier reviews [Glucksmann, 1951; Saunders, 1966; Hinchliffe, 1981; Lockshin, 1981; Hurle, 1988; Clarke, 1990; Oppenheim, 1991] (see Table I). Instead, we try to provide insight into some of the basic principles that govern the cellular and molecular mechanisms responsible for PCD.

While PCD has recently attracted considerable interest, anecdotal observations regarding this process have been published for almost a century [Beard, 1896; Collin, 1906; Ernst, 1925; Glucksmann, 1930]. However, these earlier authors often regarded the presence of dying cells within an embryo as indicative of the culling of only a small number of aberrant cells. It was not until the careful quantitative studies of Viktor Hamburger and Rita Levi-Montalcini [1949] that the magnitude of PCD was appreciated. That landmark paper demonstrated that approximately 50% of the sympathetic neurons and motoneurons produced during embryogenesis in chicks died prior to birth.

The observations of Hamburger and Levi-Montalcini raised a fundamental question: Why would an organism waste precious resources to generate cells that will be ultimately induced to commit suicide? PCD appears to provide the organism with profound developmental plas-

TABLE I. Examples of Programmed Cell Death

Location	Time/rationale	Animal studied	References
Yolk and endoderm	Bastula	Chick	Gluckmann [1951]
Polar trophectoderm and inner cell mass	70–89 stage bastula	Mouse	Copp [1978]; El Shershaby and Hinchliffe [1974]
Endoderm	Gastrula	Frog	Glucksmann [1951]
Thickening of amnion	Folding of amnion	Chameleon	Glucksmann [1951]
Primitive node	Primitive streak	Chick	Glucksmann [1951]
Epidermis, mesenchyme and mesothelium of mid-ventral line	Union of body halves before formation of sternum	Chick	Glucksmann [1951]
Cranial and caudal somites	Embryo/loss of segmentation	Chick	Glucksmann [1951]
		Mouse	Jeffs and Osmond [1992]
Notochord	Detachment from endoderm and during partial regression	Fish, rabbit	Glucksmann [1951]
Midventral body wall	Before growth of dorsolateral tissue	Chick	Glucksmann [1951]
Nervous system			
Presumptive neural tissue	Primitive streak	Chick	Glucksmann [1951]
Neural plate	Before invagination	Chick	Glucksmann [1951]
Neural tube	Before and during invagination, during detachment from ectoderm	Chick	Glucksmann [1951]
Neurons	Every neuronal population undergoes a period of PCD (refer to review articles listed in text)	Nematode, flies, moth, birds, mammals	Cunningham [1981], Cowan et al. [1984], Oppenheim [1991]
Glia cells			
Astrocytes	Removal of glial slings	Rat	Hankin et al. [1988]
	Matching with axons	Cat	Silver et al. [1993]
Oligodendrocytes		Rat	Barres et al. [1992]
Eye			
Optic vesicle	Invagination to form optic cup	Vertebrates	Glucksmann [1951], Silver and Hughes [1973]
Lens	Detachment from ectoderm	Vertebrates	Glucksmann [1951], Ishizaki et al. [1993]
Conjunctival papilla	Regression and form change	Chick	Glucksmann [1951], Silver and Hughes [1973]
Hyaloid capillary network	Regression	Mouse	Szimai and Balazs [1958]
Ear			
Auditory vesicle	Invagination and detachment from ectoderm	Vertebrates	Glucksmann [1951]
Octocyst	Fusion with acusticus ganglion	Fish, birds, mammals	Glucksmann [1951]

Structure	Process	Organism	Reference
Nose			
Olfactory pit	During invagination and detachment from palate	Birds and mammals	Glucksmann [1951]
Nose plug	Regression	Birds and mammals	Glucksmann [1951]
Stratified cuboidal epithelium	Formation of ciliated epithelium	Birds and mammals	Glucksmann [1951]
Vasculature			
Vascular rudiments	Lumen opening	Humans	Glucksmann [1951]
Aorta	Bifurcation	Chick	Glucksmann [1951]
Ductus asteriosus	Regression	Chick	Glucksmann [1951]
Limbs			
Anterior necrotic (ANZ) and posterior necrotic (PNZ)	Embryo	Birds and mammals	Glucksmann [1951], Hinchliffe [1981]
Interdigital cells	Embryo	Birds and mammals	Glucksmann [1951], Ballard and Holt [1968], Garcia-Martinez et al. [1993], Rotello et al. [1994]
Pulvillar	Transition from pupa to adult	Flies	Whitten [1969]
Muscle, cartilage, and bone			
Myotome	Formation of second and permanent muscle	Teleosts, frog, birds, and mammals	Glucksmann [1951]
Dermatone	Dissolution of medial part	Birds and mammals	Glucksmann [1951]
Muscle cells	Early differentiation of myoblasts and remodeling of insertions	Humans	Glucksmann [1951], Webb [1974]
Sclerotome	Prior to cartilage formation	Birds and mammals	Glucksmann [1951]
Protochondral tissue	Prior to formation of matrix	Cyclostomes	Glucksmann [1951]
Hypertrophic cartilage	Prior to ossification	Birds and mammals	Glucksmann [1951]
Mandible, vertebrae, and long bones	Dense prechondral mesenchyme	Birds and mammals	Glucksmann [1951]
Mandible, midline	Union of bilateral analgen	Chicks	Glucksmann [1951]
Mesenchyme of mandible	Before ingrowth of dorsolateral tissue	Chicks	Glucksmann [1951]
Kidney			
Pronephros, mesonephros, metanephros	Regression during the formation of nephron and tubules	Mice, rats, humans	Glucksmann [1951], Koseki et al. [1991], Coles et al. [1993]
Palate			
Palate midline epithelium	Regression during formation of palate	Rat and mouse	Pratt and Greene [1976], Mori et al. [1994]
Reproductive organs			
Müllerian ducts (in males)	Regression	Hamster, human	Glucksmann [1951], Price et al. [1977]
Wolffian ducts (in females)	Regression	Hamster, mouse	Glucksmann [1951]
Germ cells	Prespermatogonia, oogonia and oocytes	Hamster, mouse	Miething [1992], Coucouvanis et al. [1993], Pesce et al. [1993]

(continued)

TABLE I. Examples of Programmed Cell Death (*Continued*)

Location	Time/rationale	Animal studied	References
Glands			
Adrenal cortex	Regresses in response to decrease in ACTH	Rat	Wyllie et al. [1973]
Acinar cells	Die following pancreatic duct ligation	Rat	Walker [1987]
Ovary (atretic follicles, corpora lutea)	Ovulatory (menstrual) cycle	Hamster, human	Sandow et al. [1979], Hopwood and Levison [1975]
Uterine endometrium	Ovulatory (menstrual) cycle	Hamster, human	Sandow et al. [1979], Hopwood and Levison [1975]
Mammary epithelium	Regression after lactation	Mammals	Wellings and DeOme [1963], Helimen and Ericsson [1968], Walker et al. [1989], Strange et al. [1992]
Hepatocytes	Die following liver hyperplasia	Rat	Columbano et al. [1985]
Skin			
Fetal periderm, intermediate epidermal cells, and appendages	Development and remodeling	Human	Polakowska et al. [1994]
Gut			
Epithelial cells of small intestine	Shortening during metamorphosis	Frog (tadpole)	Ishizuya-Oka and Shimozawa [1992],
Intestinal crypt cells	Fetal development	Rat	Harmon et al. [1984]
Hemopoietic cells	Homeostasis	Mammals	
Thymocytes	Negative selection	Birds and mammals	Claesson and Hartmann [1976], MacDonald and Lees [1990], Murphy et al. [1990], Surh and Sprent [1994]
Bursa of Fabricius cells	Regression	Birds	Bracci-Laudiero et al. [1993]
Transitory structures			
Tadpole tail	Metamorphosis	Frogs	Tata [1966]
Tail	Regression	Human	Fallon and Simandl [1978]
Intersegmental muscles	Metamorphosis	Moth, flies	Finlayson et al. [1956], Kimura and Truman [1990]
Ptelianal muscles	Metamorphosis	Flies	Kimura and Tanimura [1992]

ticity, since it can be used to resolve a wide range of problems in differentiation. For example, given the complexity of the nervous system (approximately 10 billion cells within the human brain), it seems impossible to pre-establish individually the phenotypic properties of each cell. A more efficient "strategy" might be to generate large populations of cells, all of which share the same general instructions regarding differentiation, migration, target specificity, and so forth. Those that fail to complete these instructions successfully die [Rager and Rager, 1978; Cunningham, 1982; Cowan et al., 1984; Oppenheim, 1991]. Many neurons appear to compete for limited amounts of trophic support from the target cells. For example, a decrease or increase in the number of surviving motoneurons can be engineered experimentally by altering the size of the target cell population. Limb extirpation results in the loss of almost all the lumbar motor neurons, while the grafting of a supernumerary limb bud rescues many of the otherwise condemned cells [Hamburger, 1958; Hollyday and Hamburger, 1976; Hamburger and Oppenheim, 1982]. Within the nervous system, such apparent competition is not restricted to motoneurons and their skeletal muscle targets. For example, in the optic nerve massive PCD is a normal component of development within oligodendrocyte populations, thereby allowing matching between optic nerve axons and the cells that myelinate them [Barres et al., 1992]. Extrapolating from these and other data, Raff [1992] has suggested that it is, in fact, the fate of all cells within a developing organism to commit suicide unless granted a reprieve from other cells with which they interact.

PCD can function to remove not only surplus cells but also those that have outlived their usefulness. For example, many structures are required transiently during development. This is seen most dramatically in animals that undergo metamorphosis, such as amphibians and insects. In these animals, structures required for larval life such as the tadpole tail are selectively lost with progression to the adult stage [Tata, 1966]. The transient use of highly dif-

ferentiated cells is not restricted to larval animals. Following lactation in mammals, massive regression of the mammary gland epithelium follows weaning [Strange et al., 1992].

PCD also plays an essential role in protecting organisms from deleterious cells. During T-cell maturation, large numbers of immature thymocytes are generated that bear self-reactive T-cell receptors. Should these autoreactive cells mature and leave the thymus, they have the potential to initiate an autoimmune response. This catastrophic outcome is prevented by inducing these self-reactive cells to die during the process of negative selection [Claesson and Hartmann, 1976; MacDonald and Lees, 1990; Murphy et al., 1990].

While the focus of this chapter, and the field in general, are cell death in animals, plants also display PCD during development [Greenberg et al., 1994]. As in animals, PCD in plants serves a variety of functions, such as allowing cells to die to hollow out the vasculature and to differentiate sexually dimorphic structures. Despite the current intense interest in the field, there are still large areas that have been inadequately addressed. For example, while PCD has been well characterized in several chordate species, this moderately diverse phylum is only 1 of over 29 recognized animal phyla. While the study of PCD has been devoted to a small number of the invertebrate phyla [Finalyson et al., 1956; Whinen, 1969; Sulston and Horvitz, 1977; Hedgecock et al., 1983; Lauzon et al., 1993; Abrams et al., 1994], the vast majority of animal species have not been examined for this process. What is clear, however, is that PCD is an evolutionarily conserved process, although the evolutionary conservation of the actual PCD pathways and components of those pathways are still poorly described. From the current investigations of PCD it appears that all organisms have three basic processes that must occur during programmed cell death: (1) the cell must receive a signal to die, (2) that cell must then activate its cell death machinery (i.e., commit suicide), and (3) the debris from the dying/dead cell must be removed. While there is considerable variability in the species-specific execution of these three steps, these general rules seem to apply.

II. SIGNALS THAT CAN TRIGGER PCD

Given that PCD occurs within a developmental context, it is not surprising that the signals used to trigger this process are physiological and can be used to initiate other differentiative decisions as well. The molecules that trigger PCD can act at a distance, such as hormones, or more intimately, via cell–cell contacts. In principle, all cells in the body can be exposed to the trigger, but only a subset are programmed by lineage-specific decisions to be capable of responding to the trigger by dying. For example, in the tobacco hawkmoth *Manduca sexta*, specific embryonically derived abdominal muscles and neurons die following the emergence of the adult [Finlayson, 1956; Lockshin and Williams, 1965; Truman and Schwartz, 1984]. The signal that initiates these deaths is a decline in the circulating titer of the insect molting hormone 20-hydroxyecdysone [Schwartz and Truman, 1983]. While all cells in the body presumably experience withdrawal of this hormone, only selected cells die in response to it. More interestingly, each muscle and neuron that dies following adult eclosion has its own unique threshold for which ecdysteroids can activate the cell death program. The intersegmental muscles (ISMs) become committed to die first and are then followed by different groups of uniquely identified neurons. This can be demonstrated by injecting 20-hydroxyecdysone into animals at various times and determining if this treatment rescues otherwise condemned cells. With judicious timing of steroid injection, it is possible to allow a subset of these cells to die while rescuing others.

One problem with relying on specific molecules to signal cell death is that their appearance during development may not be restricted to those times when cell loss is desired. For example, with every molt in *Manduca* there is an elevation and decline in the ecdysteroid titer [Riddiford, 1993]. If the ISMs and their motor neurons died in response to declines in

the ecdysteroid titer following a larval molt, animals would be unable to generate the behaviors required for subsequent molts and would remain trapped within the cuticle. Therefore, for these cells to respond to a decline in the ecdysteroid titer as a "death signal," they must have differentiated to the point where they are competent to activate their PCD program. This question has been specifically addressed in the giant silkmoth *Antheraea polyphemus* [Schwartz and Truman, 1982; 1984a,b]. The trigger for ISM death in this insect is the peptide eclosion hormone (EH), a factor that is released into the hemolymph just prior to adult emergence [Truman and Riddiford, 1970]. Interestingly, if animals are injected with EH on the day prior to adult eclosion, the ISMs do not die. Therefore, some mechanism must exist to ensure that the ISMs become competent to die only at the right time in development. This "fail-safe" mechanism is mediated by the natural decline in the ecdysteroid titer, which determines both the timing of EH release and the ability of the muscles to respond to it.

In addition to the actions of hormones secreted from distant sites, as mentioned above, cell death can also be triggered via cell–cell contact. The best characterized example of this process is negative selection in the mouse thymus [Smith et al., 1989; Swat et al., 1991]. Stromal cells within the thymus present "self"-peptide within the groove of the major histocompatability complex (MHC) to passing immature thymocytes. Those T cells that have matured properly, but fail to recognize the presented antigens, survive and exit to the periphery to await future stimulation by novel pathogens. T cells bearing receptors that recognize self-antigens are induced to die. In fact, it is estimated that upwards of 90% of the T cells that migrate into the thymus never exit [Scollay and Shortman, 1985].

Most of our insight into the regulation of cell death has come from studies of the vertebrate nervous system. While our understanding of the precise triggers for cell death is not complete, dependence on trophic support does appear to be a critical component of survival. The first hypothesis suggesting the existence of a "trophic factor" was made by Levi-Montalcini and Hamburger [1950], followed by the subsequent demonstration that the survival of sympathetic neurons was dependent on nerve growth factor (NGF) [Levi-Montalcini, 1952; Levi-Montalcini and Angeletti, 1966]. It is thought that the limited availability of such trophic factors is one of the mechanisms by which the size of the cell population within the nervous system is adjusted to allow proper matching between neurons and their targets [see Oppenheim, 1991, for review]. For example, approximately 50% of the lumbar spinal motoneurons produced during embryogenesis die before birth [Hamburger, 1958; Hamburger and Oppenheim, 1982]. Survival of the motoneuron is dependent on its interaction with muscle targets since removal of the limb induces greater than 90% motoneuron death [Hamburger, 1958; Hollyday and Hamburger, 1976; Hamburger and Oppenheim, 1982]. This then raises the question of how target cells regulate the production of trophic factors. One mechanism appears to be electrical activity within the target cells, which is regulated by synaptic transmission by the motoneurons. This has been demonstrated by a number of experimental manipulations. Administration of neuromuscular blocking agents, such as curare, rescues motoneurons from cell death [Pittman and Oppenheim, 1978; Oppenheim et al., 1978], while electrical stimulation of the hind limb prior to innervation reduces motoneurons survival [Oppenheim and Nunez, 1982]. One interpretation is that the muscle initially produces sufficient amounts of trophic factor to support all arriving motoneurons. When synaptogenesis occurs, the muscle becomes electrically active, which in turn signals a reduction in trophic factor production and/or availability. Neurons establishing appropriate synaptic contacts would presumably receive the signal to persist, while the others would undergo PCD. There are other cellular interactions, however, that also influence the survival of neurons. A neuron consists of both dendrites (afferent; inputs) and an axon (efferent; output). The influence of afferent

input on a neuron's decision to live or die needs also to be considered. For example, in the visual system, cells of the superior colliculus receive afferent input from the retinal ganglion cells. If this input is removed via enucleation at birth, then cell death in the superior colliculus is greatly enhanced [Delong and Sidman, 1962; Giordana et al., 1980, 1981]. This observation has led to the hypothesis that neuronal survival represents a balance between appropriate afferent input and target interaction, with imbalances resulting in death [reviewed by Cunningham, 1982; see also Furber et al., 1987]. One component of maintaining these balances is intracellular calcium levels. Prolonged elevations at specific concentrations can serve to regulate a neuron's survival [see Franklin and Johnson, 1992].

Such complex regulatory mechanisms are present for other systems as well. As mentioned above, engagement of the T-cell receptor on immature thymocytes results in the immediate loss of these cells [Smith et al., 1989; Surh and Sprent, 1994]. However, once the cell is positively selected and enters the periphery (maturation), receptor engagement is the signal for proliferation. Therefore, the same signal delivered in different developmental contexts has profoundly different consequences. Furthermore, while external signals can initiate cell death, so too can intracellular ones. In some cases, cell death may represent an aborted cell cycle in which mistakes or imbalances in regulatory molecules may shuttle the cell into the cell death cascade. Evidence for this can be found in experiments with mice carrying germline deficiencies in the p53 tumor suppressor gene [Clarke et al., 1993; Lowe et al., 1993]. While immature thymocytes from normal animals undergo cell death following irradiation, those from p53–/– mice do not [Clarke et al., 1993; Lowe et al., 1993]. It appears that p53 serves to arrest cells at the G_1 interface in order to allow for repair of radiation-damaged DNA. If the DNA is not correctly repaired then the cell undergoes cell death. Additional evidence for cell death resulting from an aborted cell cycle is found in the rat ventral prostate. Following castration, prostate cells begin to replicate their DNA prior to undergoing the DNA degradation that occurs with apoptosis [Colombel et al., 1992]. An imbalance of the cell cycle may also be responsible for peripheral neuronal death as a result of trophic factor withdrawal [Freeman et al., 1994]. In superior cervical ganglion cells, there is an increase of cyclin D1 message following withdrawal of NGF. Other cyclins that are required components of the cell cycle are neither present nor differentially regulated in these postmitotic cells. It has been hypothesized that the loss of trophic support signals the neuron to attempt to re-enter the cell cycle. Since all of the appropriate components are not available, the intracellular pathways are unbalanced and the cell cycle aborts, resulting in death. Therefore, it appears that some signals that initiate cell death initiate imbalances in the cell's extracellular and/or intracellular interactions.

III. CELLS UNDERGOING PCD COMMIT SUICIDE

Like other stages of differentiation, it has been demonstrated that PCD is an active process requiring specific gene products. As illustrated in Figure 1, cells differentiate down a pathway that makes them responsive to lineage-specific signals. In fact, recent evidence suggests that motoneurons need to achieve a certain level of maturity before they become dependent on muscle-derived trophic support for survival [Mettling et al., 1995]. Once the cell is signaled to die, specific genes within the condemned cell are then either activated or repressed, which in turn leads to the demise of the cell [Wyllie, 1984; Ellis and Horvitz, 1986; Fesus et al., 1987; Wielckens et al., 1987; Serwmaa and Rytomaa, 1988, Schwartz et al., 1990a]. The first support for this hypothesis came from studies in the 1960s, when investigators treated developing animals with actinomycin D and cycloheximide, inhibitors of RNA and protein synthesis, respectively [Tata, 1966; Lockshin, 1969; Pratt and Green, 1976; Martin et al., 1988; Oppenheim et al., 1990]. They found that, instead of being noxious, these metabolic inhibitors prevented cell death. Di-

rect evidence that specific genes were required for PCD was provided by studies from the Horvitz laboratory examining PCD in the free-living nematode *Caenorhabditis elegans* [reviewed by Ellis et al., 1991; and Lintz and Driscoll, this volume]. In these animals, 131 of the normal 1,090 cells die during development [Sulston and Horvitz, 1977]. Using a genetic screen, mutations in 14 different genes were identified that altered the normal pattern of cell loss and suggested a hierarchy of genes involved in cell death [Ellis et al., 1991]. One gene, *ced-9* (*ce*ll *d*eath abnormal), functions to protect cells from cell death: Gain of function mutations block most PCDs while loss of functional alleles are lethal due to massive ectopic cell loss [Hengartner et al., 1992]. This gene has recently been identified as a structural and functional homolog of the mammalian anti-apoptosis gene, *bcl-2* [Vaux et al., 1992; Hengartner and Horvitz, 1994]. *ced-9* appears to function upstream of two essential cell death genes, *ced-3* and *ced-4*. Loss of function mutations in either of these genes block all programmed cell deaths [Ellis et al., 1986, Yuan and Honitz, 1990]. While the identity of *ced-4* appears to be novel [Yuan and Horvitz, 1990; Yuan and Horvitz, 1992; Yuan et al., 1993], *ced-3* is the nematode homolog of a mammalian protease, interleukin-1β–converting enzyme (ICE) [Yuan et al., 1993]. ICE or an ICE-like protease has been shown to be involved in the death of several cell types *in vitro,* including transfected fibroblasts [Miura et al., 1993], APO-1/Fas-mediated apoptosis [Kuida et al., 1995; Enari et al., 1995; Los et al., 1995] and NGF-deprived sympathetic neurons [Galiardini et al., 1994]. In addition to having a regulatory role in the death of motoneurons deprived of trophic support *in vitro,* an ICE-like protease also appears to have a role in mediating the death of motoneurons and interdigital cells during the period of naturally occurring death *in ovo* [Milligan et al., 1995b]. Several members of the ICE protease family have been identified recently, including ced-3, nedd-2/ich-1 [Kumar et al., 1994; Wang et al., 1994] and a protease-like ICE (prICE)/cpp-32/YAMA/apopain [Lazebnik et al., 1994; Fernandes-Alnemri et al., 1994; Tewari et al., 1995;

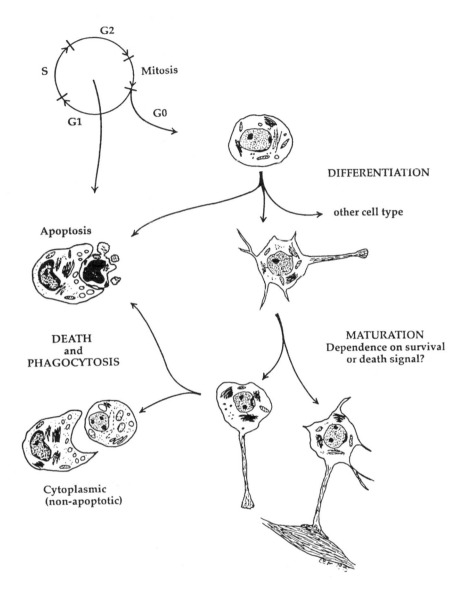

Fig. 1. *Developmental decisions of a cell. During its life span, a cell must exit the cell cycle, differentiate, and mature. At any point during this process cell death can occur, with the type of cell death (as well as the biochemical and molecular mechanisms) possibly depending on its state of differentiation and/or maturation. For example, an undifferentiated cell may give rise to an epidermal cell that further differentiates into a motoneuron. With maturation, the motoneuron develops a dependence on muscle-derived trophic support. Motoneurons that fail to receive this trophic support undergo cell death. The reference to cytoplasmic or non-apoptotic cell death refers to the type illustrated in Figures 2D and 3B.*

Nicholson et al., 1995]. PrICE/cpp-32/Yama/ apopain is distinct from ICE in that it does not have the ability to cleave the pro-IL1-β to generate the active cytokine, but does hydrolyze poly(ADP-ribose)polymerase (PARP) to generate the specific active cleavage products [Tewari et al., 1995; Nicholson et al., 1995]. Active PARP is thought to be involved in the cleavage of DNA that accompanies apoptosis.

In addition to *C. elegans*, the fruitfly *Drosophila melanogaster* represents a powerful model for the genetic analysis of development. Recently, the Steller laboratory has identified a gene, *reaper*, that appears to be required for cell death during embryogenesis [White et al., 1994]. Loss of *reaper* results in the abolition of all embryonic cell deaths, while ectopic expression induces the loss of cells. *Reaper* appears to have homology with the cell death domains of APO-1/Fas and the tumor necrosis receptor TNFR1, suggesting that it may be a phylogenetically conserved signal transduction molecule involved in the initiation of apoptosis [Golstein et al., 1995].

Unfortunately, most organisms are not as amenable for genetic analysis as nematodes and flies. Consequently, it is difficult to use genetic screens to identify genes essential for cell death or for other developmental processes. An alternate approach is to use molecular techniques to isolate mRNAs that are differentially expressed in cells prior to or during their death. However, in most systems dying cells tend to be intermingled with healthy ones, making the data obtained from such a study difficult to interpret. Nature has provided a few organ systems in which massive, relatively synchronous cell death takes place within a relatively homogeneous population of cells. In these model systems, it has been possible to clone differentially expressed genes whose expression is tightly correlated with death (see Table II). Furthermore, genetic manipulations and mutations in mice have also provided valuable tools for beginning to understand some of the genes that may be regulated during PCD [see Coucouvanis et al., 1995].

One model system that has proven amenable to a molecular analysis of PCD is the inter- segmental muscle of the moth *M. sexta* [reviewed by Schwartz, 1992]. As mentioned above, the ISMs are composed of giant cells (each syncytial cell being ~5 mm long and up to 1 mm in diameter) that undergo a rapid, synchronous death following adult eclosion (emergence). The loss of the ISMs is correlated with the repression and activation of a number of genes [Schwartz et al., 1990a]. Using a variety of techniques, the Schwartz laboratory has shown that three of the repressed genes encode the 28.1 kD subunit of the multicatalytic proteinase [Jones et al., 1995], as well as actin and myosin heavy chain [Schwartz et al., 1993a]. In addition, nine other genes have been cloned that are up-regulated with death, including polyubiquitin [Schwartz et al., 1990b] and apolipoprotein III [Sun et al., 1995].

Massive loss of specific cell populations is also seen in the mouse immune system, where a variety of techniques have been employed to identify essential cell death molecules. For example, two laboratories independently identified the APO-1/Fas molecule on the surface of many cells, most notably activated lymphocytes, that appears to signal cell death [Trauth et al., 1989; Itoh et al., 1991]. Antibodies that cross-link this APO-1/Fas molecule activate cell death [Trauth et al., 1989]. APO-1/Fas has been cloned and shown to be a member of the tumor necrosis receptor gene family [Itoh et al., 1991]. Recently, the ligand for APO-1/Fas, Fas ligand, has been shown to be related to tumor necrosis factor [Suda et al., 1993]. Spontaneous mutations in both APO-1/Fas and Fas ligand have been identified in mouse and are the previously described *lpr* (lymphoproliferative disease) and *gld* (global lymphoproliferative disorder) mice, respectively [Watanabe et al., 1992; Takahashi et al., 1994]. Loss of this signalling pathway in mice is associated with lymphoproliferation and autoimmune disorders, suggesting that these genes are involved in negative selection [Watanabe et al., 1992; Takahashi et al., 1994].

A spontaneous mutation in humans that has provided great insight into the regulation of PCD results in the development of Burkitt's lymphoma [Balchshi et al., 1985; Cleary and

TABLE II. Genes Associated With Specific Cell Deaths

Gene	Product and function	System	Expression	References
Genes that appear to be required for cell death				
ces-2	Inhibit cell death	Nematode pharynx		Ellis et al. [1991]
ced-3	Protease	Nematode		Yuan and Horvitz [1990]
		Rat fibroblasts		Miura et al. [1993]
Interleukin-1β Converting	Cysteine protease (functionally	Chick superior ganglion cells,		Gagliardini et al. [1994],
	homologous to ced-3)	motoneurons interdigital cells		Milligan et al. [1995b]
Enzyme (ICE)-like		Rat fibroblasts		Miura et al. [1993]
proteases				
ced-4		Nematode		Yuan and Horvitz [1990]
nur-77	Immediate early gene transcription factor	Mouse T cells		Liu et al. [1994], Woronicz et al. [1994]
reaper	Activator of PCD (?)	Drosophila embryo		White et al. [1994]
apt-4	(?)	Mouse T cells		Smith, Kallinich, McLaughlin, Schwartz, and Osborne (unpublished data)
p53	Tumor suppressor gene	Mouse T cells		Lowe et al. [1993], Clark et al. [1993]
	(transcription factor)	Regression mammary gland		Strange et al. [1992]
		Mouse choroid plexus epithelium		Symonds et al. [1994]
E1A	Viral protein	Mouse cell lines		Debbas and White [1992]
c-myc	Protooncogene transcription factor	Rat ventral prostate		Buttyan et al. [1988]
BAX	Heterodimerizes with and inactivates	Mouse T cells		Shi et al. [1992]
	Bcl-2	Mouse T cells		Oltvai et al. [1993]
Bcl-Xs	Inactivates Bcl-2	Rat cell lines		Yin et al. [1994]
				Boise et al. [1993]
c-rel	Transcription factor	Chick bone marrow cells		Abbadie et al. [1993]
APO-1/Fas	Cell surface receptor	Mouse nonlymphoid cells		Clarke et al. [1993], Trauth et al. [1993]
Fas ligand	Cell surface molecule	Mouse cytotoxic T cells		Suda et al. [1993]
Genes that appear to prevent cell death				
ced-9	(?)	Nematode		Ellis et al. [1991]
Bcl-2	Protooncogene	Nematode		Vaux et al. [1992]
	Membrane associated protein	Mouse lymphoid and myeloid cells		Hengartner and Horvitz [1994]
	Functional homolog of ced-9	Rat superior cervical ganglion cells		Garcia et al. [1992]
	Free radical scavenger (?)	Mouse neuronal cell lines		Kane et al. [1993]
		Human cell lines		Jacobson et al. [1993]
p35	Baclovirus protein	Moth cell lines		Clem et al. [1991]
	(behaves as Bcl-2)	Drosophila embryos		Hay et al. [1994]

(continued)

TABLE II. Genes Associated With Specific Cell Deaths (Continued)

Gene	Product and function	System	Expression	References
egl-1	(?)	Nematode neurons		Ellis et al. [1991]
ces-1, ces-2	(?)	Nematode pharynx		Ellis et al. [1991]
erythropoietin	Peptide hormone	Mouse erythroid progenitor cells		Koury and Bondurant [1990]
E1B	Adenoviral gene, blocks E1A	Mouse cell lines		Rao et al. [1992]
Rb	Tumor suppressor retinoblastoma gene (transcription factor)	Mouse lens		Morgenbesser et al. [1994]
IAP	Baculovirus protein	Moth cell lines		Crook et al. [1993]
EBV-LMP1	Epstein-Barr Virus (induces Bcl-2)	Human B cells		Henderson et al. [1991]
Genes whose expression is associated with cell death				
ced-1, -2, -5, -6, -7, -8, -10	Required for phagocytosis of corpse	Nematode neurons		Ellis et al. [1991]
polyubiquitin	Tags proteins for proteolysis	Moth ISMs and neurons	Increases	Schwartz et al. [1990a]
actin	Cytoskeletal component	Moth ISMs and neurons	Decreases	Schwartz et al. [1993a]
		Rat T cells	Decreases	Owens et al. [1991]
myosin heavy chain	Cytoskeletal component	Moth ISMs	Decreases	Schwartz et al. [1993a]
28.1 kD subunit of the multicatalytic proteinase	Proteolysis	Moth ISMs	Increases	Jones et al. [1995]
apolipoprotein III	Lipid transfer (?)	Moth ISMs and neurons	Increases	Sun et al. [1995]
Sgp-2, TRPM-2, clusterin	Lipid transfer (?) Complement inhibitor (?)	Rat ventral prostate	Increases	Buttyan et al. [1988]
c-fos	Transcription factor	Rat ventral prostate	Increases	Buttyan et al. [1988]
		Mouse interdigital cells, Motoneurons, palate		Smenye et al. [1993]
Hsp70	Heat shock protein	Rat ventral prostate	Increases	Buttyan et al. [1988]
TGF-β	Growth factor	Rat ventral prostate	Increases	Buttyan et al. [1988]
stromelysin	Protease	Regressing mouse mammary gland	Increases	Lefebure et al. [1992]
TIMP	Protease inhibitor	Regressing mouse mammary gland	Increases	Strange et al. [1992]

UPA	Protease	Regreeeing mouse mammary gland	Increases	Strange et al. [1992]
TPA	Protease	Regressing mouse mammary gland	Increases	Strange et al. [1992]
Rp-2	Integral membrane protein (?)	Rat T cells	Increases	Owens et al. [1991]
Rp-8	DNA-binding protein (?)	Rat T cells	Increases	Owens et al. [1991]
egr-1	Transcription factor	Mouse T cells	Increases	Liu et al. [1994]
apt-5	(?)	Mouse T cells	Increases	McLaughin, Smith, Schwartz, and Osborne (unpublished data)
transglutaminase	Protein cross-linking enzyme	Rat hepatocytes	Increases	Fesus et al. [1988]
low-affinity NGF receptor	Trophic factor receptor	Rat superior cervical ganglion cells	Decreases	Freeman et al. [1994]
neurofilament M	Cytoskeletal component	Rat superior cervical ganglion cells	Decreases	Freeman et al. [1994]
tyrosine hydroxylase	Enzyme (?)	Rat superior cervical ganglion cells	Decreases	Freeman et al. [1994]
cyclin D3	Cell cycle regulator	Rat superior cervical ganglion cells	Decreases	Freeman et al. [1994]
cdk-4	Cyclin-dependent protein kinase	Rat superior cervical ganglion cells	Decreases	Freeman et al. [1994]
cdk-5	Cycline-dependent protein kinase	Rat superior cervical ganglion cells	Decreases	Freeman et al. [1994]
PCNA	DNA replicating and repair protein	Rat superior cervical ganglion cells	Decreases	Freeman et al. [1994]
Rb	Tumor suppressor retinoblastoma gene (transcription factor?)	Rat superior cervical ganglion cells	Decreases	Freeman et al. [1994]
cyclin D1	Cell cycle regulator	Rat superior cervical ganglion cells	Increases	Freeman et al. [1994]

Sklar, 1985; Tsujimuto et al., 1985]. This t(14;18) translocation brings the *bcl-2* (B cell lymphoma) gene under the control of the immunoglobulin enhancer. In these individuals, B cells overproduce bcl-2, which in turn blocks cell death. Massive numbers of B cells accumulate, and, when specific secondary mutations arise, the cells become transformed [Balchshi et al., 1985; Cleary and Sklar, 1985; Tsujimuto et al., 1985]. As mentioned above, the bcl-2 protein shares 22% identity with the *C. elegans* ced-9 protein [Vaux et al., 1992]. Furthermore, introduction of *bcl-2* into *ced-9* loss-of-function mutant nematodes results in the partial rescue of the phenotype, suggesting that they share functional homology as well [Vaux et al., 1992; Hengartner and Horvitz, 1994]. While the specific mechanism by which bcl-2 prevents cell death is not known, it has been suggested that it may act as a free radical scavenger [Kane et al., 1993]. In fact, mice homozygous for targeted germline deficiencies in *bcl-2* survive to birth, although they do die approximately 2–3 weeks later [Veis et al., 1993]. Interestingly, these mice display defects consistent with an inability to block free radical damage, further suggesting that *bcl-2* may play an essential role in this pathway [Veis et al., 1993]. Furthermore, recent studies have demonstrated that the family of bcl-2–related proteins appears to interact as positive and negative regulators of cell death [Boise et al., 1993; Yin et al., 1994]. These data support the hypothesis that cell death is regulated by several factors and pathways.

Several transcription factors have been implicated in PCD including: *c-fos* [Buttyan et al., 1988; Smenye et al., 1993], *c-myc* [Evans et al., 1992; Shi et al., 1992], *nur-77* [Liu et al., 1994; Woronicz et al., 1994], *c-rel* [Abbadie et al., 1993], the tumor suppressor gene p53 [Lowe et al., 1993; Clarke et al., 1993], and the retinoblastoma gene (*Rb*) product [Morgenbesser et al., 1994; Slebos et al., 1994]. Several of these genes are regulators of the cell cycle and therefore may play a role in a cell's decision to undergo mitosis, differentiate, or die. In fact, this has recently been demonstrated to be true during the differentiation of the lens in Rb-deficient mice [Morgenbesser et al.,

1994]. In these animals loss of Rb function leads to unchecked proliferation, altered differentiation, and inappropriate apoptosis in lens fibers. In mice doubly deficient in Rb and p53, these effects are not seen, suggesting that in these Rb-deficient cells apoptosis is dependent on p53 [Morgenbesser et al., 1994].

Another biochemical event that may be involved in cell death is a decrease in glutathione [Pan and Perez-Polo, 1993; Jackson et al., 1994; Sampath et al., 1994; Mayer and Noble, 1994]. For example, there is a decrease in glutathione in rat pheochromocytoma cells (PC12 cells) following NGF withdrawal [Jackson et al., 1994; Sampath et al., 1994]. Furthermore, if the fall in glutathione is prevented, cell death is blocked in PC12 cells, as well as rat embryonic cortical cells, oligodendrocytes, and L929 fibroblasts [Pan and Perez-Polo, 1993; Jackson et al., 1994; Sampath et al., 1994; Mayer and Noble, 1994; Ratan et al., 1994]. The decrease in glutathione has been hypothesized to increase free radical generation, which in turn triggers death. This correlates with the apparent free radical squelching ability of bcl-2 [Kane et al., 1993]. The decrease in glutathione may play a role in the activation of specific proteases. For example, a glutathione disulfide has been demonstrated to inhibit the protease ICE [Thronberry et al., 1992]. A decrease in glutathione could then activate ICE or an ICE-like protease, which would then lead to the cell's death. Possible interactions with these types of molecules further suggest that there are several factors that contribute to the cell death cascade.

The results from initial studies employing inhibitors of RNA and protein synthesis made it tempting to speculate that there is a "deathase" that is a fundamental mediator of cell death. However, at present such a molecule has yet to be found. Instead, it appears that the interaction of specific sets of gene products is responsible for mediating specific cell deaths. Furthermore, while there is not an apparent universal mediator of cell death, the same gene products can have opposing effects in cells, depending on the developmental context. For example, both the protooncogene *c-myc* and the immediate early gene *nur-77* have been shown to be up-regulated in many cell types following mitogenic stimulation. However, under specific physiological conditions, expression of these genes has been shown to mediate cell death. Presumably what is different in these two situations is not the presence of *c-myc* or *nur-77* but rather the proteins with which they interact within the cell. Consequently, caution must be exercised when labeling specific molecules as "cell death" gene products.

It has been suggested by Raff [1992] that unless a cell appropriately interacts with its neighbors or targets it will die. It appears that during its lifetime a cell makes a series of decisions, each one further executing the "survival" program. Depending on the signals received by a cell, it may activate the cell death cascade at any stage of its life. A loss of balance between maintaining the survival program and integrating extra- and intracellular signals may result in death. For example, a neuroectodermal cell that has differentiated into a mature motoneuron develops a dependence on target-derived trophic support for survival. Loss of this support may shift the balance of intracellular regulatory pathways, resulting in cell death. The precise biochemical, molecular, and morphological characteristics of that cell death may differ depending on the cell's state of maturation.

III.A. All Cells Undergoing PCD Do Not Display the Same Morphological Features

Anatomical studies of PCD in different species or in different cellular lineages within a species have revealed that not all PCDs display the same morphological features. The best characterized pattern of cell death is apoptosis, a process that involves cellular condensation, genomic DNA fragmentation, the deposition of electron-dense chromatin along the inner margin of the nuclear envelope, the formation of membrane "blebs" that contain portions of the nucleus and intact organelles and ultimately the phagocytosis of these apoptotic bodies by neighboring cells [Kerr et al., 1972, Wyllie et al., 1980] (Fig. 2C). A characteristic biochemical marker for many, but not all [e.g., see Tomei et al., 1993], apoptotic deaths is DNA frag-

mentation induced by an endogenous endonuclease [Wyllie, 1984; Duke et al., 1983; Cohen and Duke, 1984]. This results in the cleavage of DNA into approximately 180 bp oligonucleotide fragments. When the DNA from cells undergoing apoptosis is size fractionated in an agarose gel, a characteristic "ladder" can be resolved (Fig. 3A). It is now possible to detect this DNA fragmentation anatomically in individual cells by performing an *in situ* nick translation (TUNEL) reaction [Gavrieli et al., 1992]. This has provided investigators with a very powerful tool since it is now possible to detect individual apoptotic cells within tissue sections as well as more easily determine the percentage of apoptotic cells within a given population.

While the majority of programmed cell deaths presented in the literature occur with this stereotypic morphological pattern, not all cells undergoing PCD die via apoptosis [Pilar and Landmesser, 1976; Clarke, 1990; Schwartz et al., 1993b]. In fact, it has been suggested that non-apoptotic deaths may represent a more phylogenetically primitive pattern of cell loss [Cornillon et al., 1994]. Many nonapoptotic cell deaths are seen in the vertebrate nervous system. When Pilar and Landmesser [1976] examined the ultrastructural morphology of dying ciliary ganglion cells they observed two distinct cell death associated morphologies that they termed *nuclear* and *cytoplasmic*. These morphologies were also subsequently observed in several other regions of the developing nervous system where they were referred to as "type I" and "type II," respectively [Chu-Wang and Oppenheim, 1978; Giordano et al., 1980; Cunningham et al., 1982]. Although the nuclear type of cell death corresponds to apoptosis, the cytoplasmic type of cell death shares few structural features with apoptosis. While these cells undergo cytoplasmic condensation, they do not display genomic DNA fragmentation, chromatin deposition, or membrane blebbing; instead the initial changes associated with death are found in the cytoplasm with swelling of the mitochondria and a breakdown of ribosomes [Pilar and Landmesser, 1976; Clarke, 1990; Schwartz et al., 1993b](Figs. 2D, 3B). In fact, in some cases they are not necessarily phago-

Apoptosis   Non-apoptotic PCD

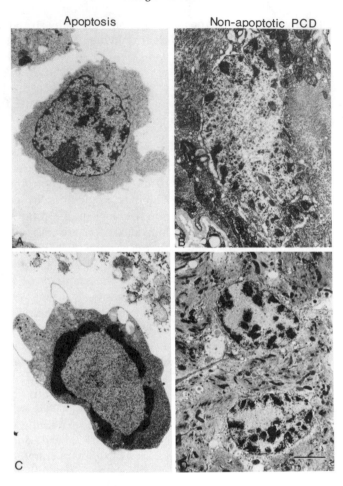

Fig. 2. *Morphological characteristics of cell death. Transmission electron micrographs of a non-stimulated thymocyte (A) and one treated with glucocorticoids (C). The thymocyte treated with glucocorticoids undergoes apoptosis and exhibits marginal condensation of the nuclear chromatin (C). ISMs from a moth M. sexta prior to (B; healthy) and after (D; dying) adult eclosion are also shown. The ISMs of the moth do not exhibit the characteristic features of apoptosis.*

cytosed by neighboring cells [Jones, Fahrbach, and Schwartz, in preparation].

While it would be convenient to assume that these different morphologies reflect differences in species or cellular lineages, preliminary data suggest that some cells can generate both morphologies depending on the signals they receive. For example, during chick embryogenesis, the majority of dying ciliary ganglion cells display a cytoplasmic pattern of death. However, when these cells are induced to die by removal of their target, essentially 100% of the dying cells display an apoptotic morphology [Pilar and Landmesser, 1976]. Therefore, it appears that the same cell can activate two different pathways for death depending on the nature of the trigger. Interestingly, this apoptotic death occurs in neurons that have not had the opportunity to interact with their target. Consequently, the execution of cytoplasmic versus apoptotic cell death pathways may reflect maturation-specific choices [see Cunningham et al., 1982]. At present it is not known if these two morphologies utilize overlapping molecular components or represent distinct differentiative decisions.

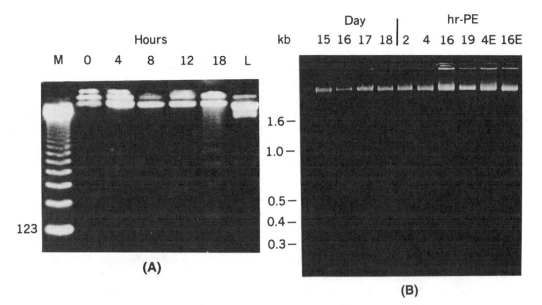

Fig. 3. *Genomic DNA was collected from thy-mocytes at selected hours after glucocorticoid treatment, size fractionated on an agarose gel and detected with ethidium bromide (A). The characteristic DNA ladder is observed by 4 hours following glucocorticoid treatment; 0 hour represents DNA collected from unstimulated cells (M, molecular weight standards; L, lambda molecular weight marker). DNA was also collected from the ISMs of M. sexta on pupal days 15–18 and several hours posteclosion (PE) (B). At no time examined was there evidence of DNA fragmentation. [Modified from Schwartz et al., 1990b, with permission of the publisher.]*

III.B. How Do Cells Die?

What actually occurs during PCD that results in the death of a cell? A decrease in ATP synthesis may be responsible for PCD, but a dramatic decrease in ATP would inhibit membrane ionic pumps, thereby allowing ions and water to flood the cell and initiate a necrotic death. Furthermore, while there is DNA degradation during apoptosis, several cell types, such as lens cells and erythrocytes, remain viable for extended periods of time without nuclei. In fact, enucleated cytoplasts have been shown to respond to specific stimuli by undergoing an apoptotic death [Lazebinik et al., 1993], and prior expression of bcl-2 can protect them from this death [Jacobson et al., 1994]. Macromolecular synthesis inhibitors have also been demonstrated to inhibit PCD, suggesting that the lack of RNA or protein synthesis is not immediately lethal. It is interesting to note that if indeed a cell can undergo cell death without a nucleus, this suggests that the machinery necessary for cell death must already be resident in the cell and can be activated in response to specific signals. The observed *de novo* gene expression associated with PCD in many systems may reflect transcription of regulatory molecules that access this resident machinery as seen in CTL-mediated cell death [Heusel et al., 1994]. Alternatively, protein synthesis may be required to initiate post-translational modification of death-associated molecules such as the proteolytic activation of ICE family members. For most systems, it is possible to determine when a cell is irreversibly committed to death; however, it is difficult to identify the cell's physiological "last breath."

IV. RESOLUTION OF CELL DEATH

While much attention has been devoted to understanding the signals and molecular pathways that mediate PCD, less attention has been focused on how the body deals with removing the cellular debris that follows death. This is a nontrivial consideration given the number of cells that undergo PCD and the inflammatory nature of intracellular constituents. Consequently, phagocytosis of the dead cell appears to be an essential component of almost all cell deaths [Ellis et al., 1991b]. Understanding the cellular, molecular, and biochemical mechanisms of how dead cells are disposed will provide powerful tools for understanding issues relevant to development, aging, and pathology.

The entire process of PCD, including its final resolution, is tightly controlled. In almost all cases, cellular debris is phagocytically removed by other cells. The genetics of this process is best understood in *C. elegans,* where seven genes have been identified that are involved in the process of phagocytosis of the corpse [Ellis et al., 1991b]. These genes appear to represent two distinct complementation groups; one group may encode cell surface ligands that identify the cell as a corpse and the other receptors for these signals on the phagocytosing cell. While mutations in these genes prevent corpse removal, they have no impact on the death process itself.

While inflammatory responses in invertebrates are poorly defined, much is known about its destructive potential in mammals. Consequently, if cells undergoing PCD were allowed to lyse, as occurs with necrosis, the cellular components (involving neutrophils, eosinophils, macrophages, and lymphocytes) and humoral components (such as histamine, prostaglandins, leukotrienes, lymphokines, and antibodies) of the inflammatory response could be disastrous to surrounding cells. Although neighboring cells have been found to phagocytose dying cells, the predominant professional scavenger is the macrophage [Kerr et al., 1972]. In fact, almost every morphological study examining tissues displaying PCD describes the presence of macrophages [reviewed by Milligan, 1992]. The CNS has been well characterized in this regard, where macrophages remove dead cells that arise during PCD, axon retraction, and glial scaffold elimination [Hamburger and Levi-Montalcini, 1949; Chu-Wang and Oppenheim, 1978; Sohal and

Weidman, 1978; Valentino and Jones, 1981; Hume et al., 1983; Perry et al., 1985; Innocenti et al., 1983a,b; Hankin et al., 1988; Milligan et al., l991a]. Macrophages are also reported to phagocytose the dead interdigital mesenchyme of the fetal rat foot [Ballard and Holt, 1968], acinar cells in the pancreas following experimental duct ligation [Walker, 1987], as well as endometrial cells coincident with menstruation [Hopwood and Levison, 1976]. Their role in disposing of cells in the immune system has also been documented, in both the spleen [Swartzendruber and Congdon, 1963] and the thymus [Bowen and Lewis, 1980]. The phagocytosis of cells undergoing PCD is so efficient that dying cells are often found within macrophages prior to the final death of the cell [Hume et al., 1983].

While the pattern of macrophage recruitment during PCD is well characterized, little is known about the molecular mechanisms responsible for targeting macrophages to these regions. Since macrophages are usually not resident in the tissues where PCD occurs, chemotactic signals must recruit them to the regions where their activity is required. In addition, some cell–cell interactions must be involved to ensure that only dying cells are engulfed. Given the specificity of this reaction in the CNS, some investigators have hypothesized that degenerating neuronal elements provide a chemotactic signal for macrophages [Hume et al., 1983; Perry et al., 1985, 1987, 1991; Milligan et al., 1991a,b]. Given the small number of cells that die at any given time during neural development, it is difficult to determine the nature of these chemotactic factors. To circumvent this limitation, experimental manipulations were performed to exaggerate the normal extent of cell death during development. For example, lesions to the visual cortex at birth result in massive and rapid invasion of macrophages into the degenerating dorsal lateral geniculate nucleus (dLGN) [Milligan et al, 1991b]. Since macrophages are not evident within the dLGN during the first few days of normal postnatal development, these cells must migrate in from distant sites. This suggests that a chemotactic signal must be produced by cells within the dLGN as a consequence of either the axotomy or the subsequent neuronal degeneration. One chemotactic molecule has recently been identified in this system [Milligan, 1992; Milligan, et al., 1995a]. This factor, brain-derived chemotactic factor (BDCF), exhibits opioid-like activity that initiates macrophage chemotaxis via a receptor with pharmacological characteristics of a δ-opioid receptor. Present experimental evidence suggests that this opioid works in association with other factors to initiate the migration of macrophages to sites of injury-induced cell death in the nervous system.

The mechanisms by which macrophages identify cells to be phagocytosed appear to result from alterations in the cell surface of dying cells. As suggested for *C. elegans*, it is thought that specific cell–cell interactions are required between degenerating and phagocytosing cells in the mammalian CNS [Cammermyer, 1965a,b; LaPushin and deHarven, 1971; Duvall et al., 1985; Savill et al., 1990; Rotello et al., 1994; also see Savill et al., 1993, for review]. These interactions appear to be mediated by a macrophage cell surface lectin that interacts with an altered vitronectin receptor on apoptotic cells [Duvall et al., 1985; Savill et al., 1990]. This interaction may be mediated via macrophage-secreted thrombosporin, which binds to both dying cells and the integrin receptors (and other receptors) on macrophages [see Savill et al., 1993, for review]. Another factor that appears to mediate macrophage recognition and phagocytosis of dying cells is phosphatidylserine [see Savill et al., 1993]. While this phospholipid is usually restricted to the inner monolayer of the plasma membrane, changes in membrane hydrophobicity or charge in dying cells appears to result in the exposure of phosphatidylserine on the external surface of the membrane [Fadok et al., 1992a, 1993]. Different macrophage populations then appear to target specifically either the vitronectin receptor or the exposed phosphotidylserine to mediate phagocytosis [Fadok et al., 1992b].

In both nematodes [Ellis et al., 1991b] and moths [Jones, Fahrbach and Schwartz, in preparation], cell death occurs independently of phagocytosis. The same appears to be true in mammals. For example, the initial morphological changes in neurons of the dLGN associated with cell death occur prior to massive macrophage invasion into the brain [Milligan et al., 1991b]. However, a recent report has presented evidence that macrophages may serve not just to remove debris, but also to kill target cells in the developing mouse eye [Lang and Bishop, 1993]. In this study, the authors used transgenic mice that targeted toxin to subsets of mature macrophages, including those found in the eye cavity. In contrast to wild-type mice, the hyaloid vasculature persisted in the transgenic animals, suggesting that in the absence of hyalocytes (macrophages) cell death of the cells associated with the hyloid vasculature was not initiated. However, there is always the possibility that the cell death still occurred, but, in the absence of phagocytosis, gross tissue morphology was maintained. Furthermore, in some systems it appears that the macrophage response following cell death may facilitate tissue remodeling during development [Milligan et al., 1991b].

V. CONCLUSION

After examining examples of PCD from many animal species, it is evident that no one animal serves as a model to provide information concerning all the details of this highly complex process. For example, much of our current understanding of the genetics of cell death has come from studies of select invertebrate systems. The vast majority of animal phyla have not been examined with regard to PCD, and additional insights could come from studies of other species. These studies may allow us to better dissect the biochemical pathways responsible for PCD as well as determine which components are phylogenetically conserved.

In the simplest terms, PCD requires that a cell receive a signal to die, subsequently activate the biochemical and molecular cascade that leads to death, and have its corpse removed. It is clearly not possible to provide a detailed analysis of this complex developmental process in a single chapter. Instead, we have attempted to provide an overview that ad-

SOME OF THE EVENTS THAT MEDIATE PROGRAMMED CELL DEATH

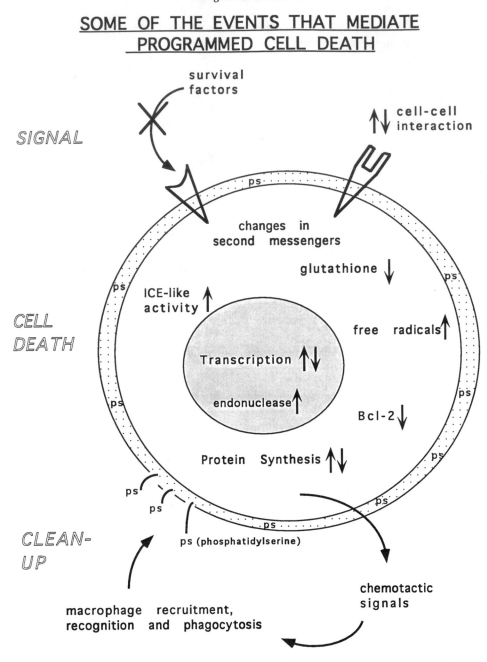

Fig. 4. *Schematic of some of the features of cell death, including signals to die, intracellular changes, and removal of the corpse.*

dresses some of the major issues in the field (Fig. 4). It is critical to note, however, that the details of the specific mechanisms involved in cell death differ between species, as well as between cell populations within a species. An understanding of PCD not only provides insight into a basic developmental process, but also has real potential for therapeutic applications. Since all nucleated cells have the molecular machinery to commit suicide, the value of understanding and consequently being able to regulate this process in a lineage-specific manner cannot be overestimated. The ability to induce the death of specific clones of cells, such as malignant cells, could provide a relatively benign but powerful way of targeting and specifically eliminating cancer cells. Conversely, the ability to rescue valuable but condemned cells, such as neurons in Alzheimer's disease or CD4[+] T cells in HIV infection, could potentially block disease symptoms. The opportunities offered by our increasingly detailed understanding of PCD in various systems will ultimately provide valuable tools for clinical applications.

ACKNOWLEDGMENTS

The authors thank Kathie Eagleson, Eric Findeis, Steve Robinson, and Margaret Jones for helpful discussions and critical reading of this manuscript. This work was supported in part by a grant from the Spinal Cord Research Foundation to C.E.M. and NIH grants GM40458 and AG000495 to L.M.S.

REFERENCES

Abbadie C, Kabrun N, Bovali I, Smardova J, Stehelin D, Vandenbunder B, Enrietto PJ (1993): High levels of c-rel expression are associated with programmed cell death in the developing avian embryo and in bone marrow cells in vitro. Cell 75:899–912.

Abrams JM, White K Fessler LI, Steller H. (1993): Programmed cell death during *Drosophila* embryogenesis. Development 117:29–43.

Bakhshi A, Jensen JP, Goldman P, Wright JJ, McBride OW, Epstein AL, Korsmeyer SJ (1985): Cloning the chromosomal breakpoint of t(14:18) human lympho-

mas: Clustering around J_H on chromosome 14 and near a transcriptional unit on 18. Cell 41:899–906.

Ballard KJ, Holt SJ (1968): Cytological and cytochemical studies on cell death and digestion in the fetal rat foot: The role of macrophage and hydrolytic enzymes. J Cell Sci 3:245–262.

Barres BA, Hart IK, Coles HS, Burne JF, Voyvodic JT, Richardson WD, Raff MC (1992): Cell death and control of cell survival in the oligodendrocyte lineage. Cell 70:31–46.

Beard J (1896): The history of a transient nervous apparatus in certain Ichthyopsida. An account of the development and degeneration of ganglion cells and nerve fibers. Zool Jahrbucher Abt Morphol 9:1–106.

Boise LH, Gonzalez-Garcia M, Postema CE, Ding D, Lindstein T, Turka LA, Mao X, Nunez G, Thompson CB (1993): *bcl-x*, a *bcl-2*-related gene that functions as a dominant regulation of apoptotic cell death. Cell 79:597–608.

Bowen ID, Lewis GHJ (1980): Acid phosphatase activity and cell death in the mouse thymus. Histochemistry 65:173–179.

Bracci-Laudiero L, Vigneti E, Iannicola C, Aloe L (1993): NGF retards apoptosis in the chick embryo bursal cells *in vitro*. Differentiation 53:61–66.

Buttyan R, Zakeri Z, Lockshin R, Wolgemuth D (1988): Cascade induction of c-fos, c-myc and heat shock 70k transcripts during repression of the rat ventral prostrate gland. Mol Endocrinol 2:650–657.

Buttyan R, Olsson CA, Pintar J, Chang C, Bandyk M, Ng P-Y, Sawczuk IS (1989): Induction of the *TRPM-2* gene in cells undergoing programmed cell death. Mol Cell Biol 9:3473–3481.

Cammermeyer J (1965a): I. Juxtavascular karyokinesis and microglia cell proliferation during retrograde reaction in the mouse facial nucleus. Ergebn Anat Entwiek-Gesch 38:1–22.

Cammermeyer J (1965b): VI. Histiocytes, juxtavascular mitotic cells and the microglia cells during retrograde changes in the facial nucleus of rabbits of varying age. Ergebn Anat Entwiek-Gesch 38:195–229.

Chu-Wang IW, Oppenheim RW (1978a): Cell death of motoneurons in the chick embryo spinal cord. I. A light and electron microscopic study of naturally occurring and induced cell loss during development. J Comp Neur 177:33–58.

Chu-Wang I, Oppenheim RW (1978b): Cell death of motoneurons in the chick embryo spinal cord. II. A quantitative and qualitative analysis of degeneration in the ventral root, including evidence for axon outgrowth and limb innervation prior to cell death. J Comp Neur 177:59–86.

Claesson MH, Hartmann NR (1976): Cytodynamics in the thymus of young adult mice: A quantitative study on the loss of thymic blast cells and nonproliferative small thymocytes. Cell Tissue Kinet 9:273–291.

Clarke PG (1990): Developmental cell death: Morphological diversity and multiple mechanisms. Anal Embryol 181:195–213.

Clarke AR, Purdie CA, Harrison DJ, Morris RG, Bird CC, Hooper ML, Wyllie AH (1993): Thymocyte apoptosis induced by p53-dependent and independent pathways. Nature 362:849–852.

Clarke AR, Gledhill S, Hooper MC, Bird CC, Wyllie AH (1994): p53 dependence of early apoptotic and proliferative responses with the mouse intestinal epithelium following gamma-irradiation. Oncogene 9:1767–1773.

Cleary ML, Sklar J (1985): Nucleotide sequence of a t(14;18) chromosomal breakpoint in follicular lymphoma and demonstration of a breakpoint cluster region near a transcriptionally active locus on chromosome 18. Proc Natl Acad Sci USA 82:7439–7443.

Clem RJ, Fechheimer M, Miller LK (1991): Prevention of apoptosis by a baculovirus gene during infection of insect cells. Science 254:1388–1390.

Cohen JJ, Duke RC (1984): Glucocorticoid activation of a calcium-dependent endonuclease in thymocyte nuclei leads to cell death. J Immunol 132:38–42.

Coles HSR, Burne JF, Raff MC (1993): Large-scale normal cell death in the developing rat kidney and its reduction by epidermal growth factor. Development 118:777–784.

Collin R (1906): Recherches cytologiques sur le développement de la cellule nerveuse. Nevraxie 8:181–308.

Columbano A, Ledda-Columbano GM, Coni PP, Faa G, Liguori C, Santa Cruz G, Pani P (1985): Occurrence of cell death (apoptosis) during the involution of liver hyperplasia. Lab Invest 52:670–675.

Colombel M, Ng CA, Ng P-Y, Buttyan R (1992): Hormone-regulated apoptosis results from re-entry of differentiated prostate cells into a defective cell cycle. Cancer Res 52:4313–4319.

Cornillon S, Foa C, Davoust J, Buonavista N, Gross JD, Golstein P (1994): Programmed cell death in Dictyostelium. J Cell Sci 107:2691–2704.

Coucouvanis EC, Sherwood SW, Carswell-Crumpton C, Spack EG, Jones PP (1993): Evidence that the mechanism of prenatal germ cell death in the mouse is apoptosis. Exp Cell Res 209:238–247.

Coucouvanis EC, Martin GR, Nadeau JH (1995): Genetic approaches for studying programmed cell death during development of the laboratory mouse. In Schwartz LM, Osborne BA (eds): Methods in Cell Biology Series: Cell Death. New York: Academic Press 46:388–440.

Cowan WM, Fawcett JW, O'Leary DDM, Stanfield BB (1984): Regressive events in neurogenesis. Science 225:1258–1265.

Crook NE, Clem RJ, Miller LK (1993): An apoptosis-inhibiting baculovirus gene with a zinc-finger-like motif. J Virol 67:2167–2174.

Cunningham TJ (1982): Naturally occurring death and its regulation by developing neural pathways. Int Rev Cytol 74:163–186.

Cunningham TJ, Mohler IM, Giordano DL (1982): Naturally occurring neuron death in the ganglion cell layer of the neonatal rat: Morphology and evidence for regional correspondence with neuron death in the superior colliculus. Dev Brain Res 2:203–215.

Debbas M, White E (1992): Wild-type p53 mediates apotosis by E1A, which is inhibited by E1B. Genes Dev 7:546–554.

Delong GR, Sidman RL (1962): Effects of eye removal at birth on histogenesis of the mouse superior colliculus: An autoradiographic analysis with tritiated thymidine. J Comp Neurol 118:205–224.

Duke RC, Chervenak R, Cohen JJ (1983): Endogenous endonuclease-induced DNA fragmentation: An early event in cell-mediated cytolysis. Proc Natl Acad Sci USA 80:6361–6365.

Duvall E, Wyllie AH, Morris RG (1985): Macrophage recognition of cells undergoing programmed cell death (apoptosis). Immunology 56:351–358.

El Shershaby AM, Hinchliffe JR (1974): Cell redundancy in the zona intact preimplantation mouse blastocyst: A light and electron microscope study of dead cells and their fate. J Embryol Exp Morphol 31:643–654.

Ellis HM, Horvitz HR (1986): Genetic control of programmed cell death in the nematode Caenorhabditis elegans. Cell 44:817–829.

Ellis R, Yuan J, Horvitz HR (1991a): Mechanisms and functions of cell death. Ann Rev Cell Biol 7:663–698.

Ellis RE, Jacobson DM, Horvitz HR (1991b): Genes required for the engulfment of cell corpses during programmed cell death in Caenorhabditis elegans. Genetics 129:79–94.

Enari M, Hug H, Nagata S (1995): Involvement of an ICE-like protease in Fas-mediated apoptosis. Nature 375:78–81.

Ernst M (1926): Ueber Untergang von Zellen warhrend der normalen Entwicklung bei Wirbeltieren. Z Anat Entwicklungsgesch 79:228–262.

Evan GI, Wyllie AH, Gilbert CS, Littlewood TD, Land H, Brooks M, Waters CM, Penn LZ, Hancock DC (1992): Induction of apoptosis in fibroblasts by c-myc protein. Cell 69:119–128.

Fadok VA, Voelker DR, Campbell PA, Cohen JJ, Bratton DL, Henson PM (1992a): Exposure of phosphatidylserine on the surface of apoptotic lymphocytes triggers specific recognition and removal by macrophages. J Immunol 148:2207–2216.

Fadok VA, Savill JS, Haslett C, Bratton DI, Doherty DE, Campbell PA, Henson PM (1992b): Different populations of macrophages use either the vitronectin receptor or the phosphatidylserine receptor to recognize and remove apoptotic cells. J Immunol 149:4029–4035.

Fadok VA, Laszio DJ, Noble PW, Weinstein L, Riches DWH, Henson PM (1993): Particle digestibility is required for induction of phosphatidylserine recognition mechanism used by mouse macrophages to phagocytose apoptotic cells. J Immunol 151:4274–4285.

Fallon JF, Simandl BK (1978): Evidence of a role for cell death in the disappearance of the embryonic human tail. Am J Anat 152:111–129.

Fernandes-Alnemri T, Litwack G, Alnemri ES (1994): CPP32, a novel human apoptotic protein with homology to *Caenorhabditis elegans* cell death protein Ced-3 and mammalian interleukin-1 beta-converting enzyme. J Biol Chem 269:30761–30764.

Fesus L, Thomazy V, Falus A (1987): Induction and activation of tissue transglutaminase during programmed cell death. FEBS Lett 224:104–108.

Finlayson LH (1956): Normal and induced degeneration of abdominal muscles during metamorphosis in the Lepidoptera. Q J Microsc Sci 97:215–233.

Franklin JL, Johnson EM Jr (1992): Suppression of programmed neuronal death by sustained elevation of cytoplasmic calcium. TINS 15:501–508.

Freeman RS, Estus S, Johnson EM Jr (1994): Analysis of cell cycle-related gene expression in postmitotic neurons: Selective induction of cyclin D1 during programmed cell death. Neuron 12:343–355.

Furber S, Oppenheim RW, Prevette D(1987): Naturally occurring neuron death in the ciliary ganglion of the chick embryo following removal of preganglionic input: Evidence for the role of afferents in ganglion cell survival. J Neurosci 7:1816–1832.

Gagliardini V, Fernandez P-A, Lee RKK, Drexler HCA, Rotello RJ, Fishman MC, Yuan J (1994): Prevention of vertebrate neuronal death by the *CrmA* gene. Science 263:826–828.

Garcia I, Martinou I, Tsujimoto Y, Martinou JC (1992): Prevention of programmed cell death of sympathetic neurons by the bcl-2 protoncogene. Science 258:302–304.

Garcia-Martinez V, Macias D, Ganan Y, Garcia-Lobo JM et al. (1993): Internucleosomal DNA fragmentation and programmed cell death (apoptosis) in the interdigital tissue of chick leg bud. J Cell Sci 106:201–208.

Giordano DL, Cunningham TJ (1981): Effects of differentiation on neuron death in the superior colliculus. Anat Rec 199:95A.

Giordano DL, Murray M, Cunningham TJ (1980): Naturally occurring neuron death in the optic layers of the superior colliculus of the postnatal rat. J Neurocytol 9:603–614.

Glucksmann A (1951): Cell death in normal vertebrate ontogeny. Biol Rev 26:59–86.

Golstein P, Darguet D, Depraetere V (1995): Homology between *Reaper* and the cell death domains of Fas and TNFR1. Cell 81:185–186.

Greenberg JT, Guo A, Klessig DF, Ausubel FM (1994): Programmed cell death in plants: A pathogen-triggered response activated coordinately with multiple defense functions. Cell 77:551–563.

Hamburger V (1958): Regression versus peripheral control of differentiation in motor hypoplasia. Am J Anat 102:365–410.

Hamburger V, Levi-Montalcini R (1949): Proliferation, differentiation and degeneration in the spinal ganglia of the chick embryo under normal and experimental conditions. J Exp Zool 111:457–502.

Hamburger V, Levi-Montalcini R (1950): Some aspects of neuroembryology. In Weiss P (ed): Genetic Neurobiology. Chicago: University of Chicago Press, pp 125–160.

Hamburger V, Oppenheim RW (1982): Naturally occurring neuronal death in vertebrates. Neurosci Comm 1:39–55

Hankin M, Schneider B, Silver J (1988): Death of the subcallosal glial sling is correlated with the formation of the cavum septi pellucidi. J Comp Neurol 272:191–202.

Harmon B, Bell L, Williams L (1984): An ultrastructural study on the "meconium corpuscles" in rat fetal intestinal epithelium with particular reference to apoptosis. Anat Embryol 169:119–124.

Hay BA, Wolff T, Rubin GM (1994): Expression of baculovirus p35 prevents cell death in *Drosophila*. Development 120:2121–2129.

Hedgecock EM, Sulston J, Thompson J (1983): Mutations affecting programmed cell death in the nematode *Caenorilabditis elegans*. Science. 220:1277–1279.

Helminen HJ, Ericsson JLE (1968): Studies on mammary gland involution II. Ultrastructural evidence for auto- and heterophagocytic pathways for cytoplasmic degradation. J Ultrastruct Res 25:214–227.

Henderson S, Rowe M, Gregory C, Croom-Carter D, Wald F, Longnecker R, Kieff E, Rickinson AB (1991): Induction of bcl-2 expression by Epstein-Barr virus latent membrane protein 1 protects infected B cells from programmed cell death. Cell 65:1107–1115.

Hengartner MO, Ellis R, Horvitz HR (1992): *Caenorhabditis elegans* gene ced-9 protects cells from programmed cell death. Nature 356:494–501.

Hengartner, MO, Horvitz HR (1994): *C. elegans* cell survival gene ced-9 encodes a functional homolog of the mammalian protooncogene bcl-2. Cell 76:665–670.

Hengartner MO, Horvitz HR (1994): Activation of *C. elegans* cell death protein ced-9 by an amino-acid substitution in a domain conserved in bcl-2. Nature 369:318–320.

Heusel JW, Wesselschmidt RL, Shresta S, Russell JH, Ley TJ (1994): Cytotoxic lymphocytes require granzyme B for the rapid induction of DNA fragmentation and apoptosis in allogeneic target cells. Cell 76:977–987.

Hinchliffe JR (1981): Cell death in embryogenesis. In Bowen ID, Lockshin RA (eds): Cell Death in Biology and Pathology. New York: Chapman and Hall, pp. 35–38.

Hollday M, Hamburger V (1976): Reduction of the naturally occurring motor neuron loss by enlargement of the periphery. J Comp Neurol 170:311–320.

Hopwood P, Levison PA (1976): Atrophy and apotosis in the cyclical human endometrium. J Path 119:159–166.

Hume DA, Perry VH, Gordon S (1983): Immunohistochemical localization of a macrophage-specific antigen in developing mouse retina: Phagocytosis of dying neurons and differentiation of microglia cells to form a regular array in the layers. J Cell Biol 97:253–257.

Hurle J M (1988): Cell death in developing systems. Methods Achiev Exp Pathol 13:55–86.

Innocenti GM, Koppel H, Clarke S (1983a): Transitory macrophages in the white matter of the developing visual cortex. I. Light and electron microscopic characteristics and distribution. Dev Brain Res 11:39–53.

Innocenti GM, Clarke S, Koppel H (1983b): Transitory macrophages in the white matter of the developing visual cortex. II. Development and relations with axonal pathways. Dev Brain Res 11:55–66.

Ishizaki Y, Voyvodic JT, Burne JF, Raff MC (1993): Control of lens epithelial cell survival. J Cell Biol 121:899–908.

Ishizuya-Oka A, Shimozawa A (1992): Programmed cell death and heterolysis of larval epithelial cells by macrophage-like cells in the auran small intestine in vivo and in vitro. J Morph 213:185–195.

Itoh N, Yonehara S, Ishii A, Sameshima M, Hase A, Seto Y, Nagata S (1991): The polypeptide encoded by the cDNA for human cell surface antigen FAS can mediate apoptosis. Cell 66:233–243.

Jackson GR, Sampath D, Werrbach-Perez K, Perez-Polo JR (1994): Effects of nerve growth factor on catalase and glutathione peroxidase in hydrogen peroxide-resistant pheochromocytoma subclone. Brain Res 634:69–76.

Jacobson MD, Burne JF, King MP, Miyashita T, Reed JC, Raff MC (1993): Bcl-2 blocks apoptosis in cells lacking mitochondrial DNA. Nature 361:365–369.

Jacobson MD, Burne JF, Raff MC (1994): Programmed cell death and bcl-2 protection in the absence of a nucleus. EMBO J 13:1899–1910.

Jeffs P, Osmond M (1992): A segmental pattern of cell death during development of the chick embryo. Anat Embryol 185:589–598.

Jones MEE, Haire MF, Kloetzel P-M, Mykles DL, Schwartz LM (1995): Changes in the structure and function of the multicatalytic proteinase (proteaseome) during programmed cell death in the intersegmental muscle of the hawkmoth Manduca sexta. Dev Biol 169:436–447.

Kane DJ, Sarafian TA, Anton R, Hahn H et al. (1993): Bcl-2 inhibition of neural death: Decreased generation of reactive oxygen species. Science 262:1274–1277.

Kerr JFR, Wyllie A, Currie AR (1972): Apoptosis: A basic biological phenomenon with wide-ranging implications in tissue kinetics. Br J Cancer 26:239–257.

Kimura KI, Truman JW (1990): Postmetamorphic cell death in the nervous and muscular systems of Drosophila melanogaster. J Neurosci 10:403–411.

Kimura K, Tanimura T (1992): Mutants with delayed cell death of the ptelianal head muscles in Drosophila. J Neurogenet 8:57–69.

Koseki C, Herzlinger D, Al-Awqati Q (1992): Apoptosis in metanephric development. J Cell Biol 119:1327–1333.

Koury MJ, Bondurant MC (1990): Erythroprotein retards DNA breakdown and prevents programmed death in erythroid progenitor cells. Science 248:378–381.

Kuida K, Lippke JA, Ku G, Harding MW, Livingston DJ, Su MS-S, Flavell RA (1995): Altered cytokine export and apoptosis in mice deficient in interleukin-1β converting enzyme. Science 267:2000–2003.

Kumar S, Kinoshita M, Noda M, Copeland NG, Jenkins NA (1994): Induction of apoptosis by the mouse Nedd2 gene, which encodes a protein similar to the product of the Caenorhabditis elegans cell death gene ced-3 and the mammalian IL-1β-converting enzyme. Genes Devel 8:1613–1626.

Lang RA, Bishop JM (1993): Macrophages are required for cell death and tissue remodelling in the developing mouse eye. Cell 74:453–462.

LaPushin RW, de Harven E (1971): A study of glucocorticosteroid-induced pyknosis in the thymus and lymph node of adrenalectomized rat. J Cell Biol 50:583–597.

Lauzon RJ, Patton CW, Weissman IL (1993): A morphological and immunohistochemical study of programmed cell death in Botryllus schlosseri (tunicata, Ascidiacea). Cell Tissue Res 272:115–127.

Lazebnik YA, Cole S, Cooke CA, Nelson WG, Earnshaw WC (1993): Nuclear events of apotosis in vitro in cell-free mitotic extracts: A model system for analysis of the active phase of apoptosis. J Cell Biol 123:7–22.

Lazebnik YA, Kaufmann SH, Desnoyers S, Polrler GG, Earnshaw WC (1994): Cleavage of poly(ADP-ribose) polymerase by a proteinase with properties like ICE. Nature 371:346–347.

Lefebvre O, Wolf C, Limacher J-M, Hutin P (1992): The breast cancer-associated stromelysin-3 gene is expressed during mouse mammary gland apoptosis. J Cell Biol 119:997–1002.

Levi-Montalcini R (1952): Effects of mouse tumor transplantation on the nervous system. Ann NY Acad Sci 55:330–344.

Levi-Montalcini R, Angeletti PV (1966): Nerve growth factor. Physiol Rev 48:534–569.

Liu Z-G, Smith S, McLaughlin KA, Schwartz LM, Osborne BA (1994). Apoptotic signals delivered through the T cell receptor require the immediate early gene nur77. Nature 367:281–284.

Lockshin RA, Williams CM (1965): Programmed cell death: Cytology of degeneration in the intersegmental muscles of the silkmoth. J Insect Physiol 11:123–33.

Lockshin RA (1969): Activation of lysis by a mechanism involving the synthesis of protein. J Insect Physiol 15:1505–1516.

Lockshin RA (1981): Cell death in metamorphosis. In

Bowen ID, Lockshin RA (eds): Cell Death in Biology and Pathology. London: Chapman and Hall, pp 79–122.

Los M, Van de Craen M, Penning LC, et al. (1995): Requirement of an ICE/CED-3 protease for Fas/APO-1-mediated apoptosis. Nature 375:81–83.

Lowe S, Schmidtt E, Smith SE, Osborne BA, Jacks T (1993): p53 is required for radiation-induced apoptosis in mouse thymocytes. Nature 262:847–849.

Martin DP, Schmidt RE, Distefano PS, Lowry OH, Clarke JG, Johnson EM Jr (1988): Inhibitors of protein synthesis and RNA synthesis prevent neuronal death caused by nerve growth factor deprivation. J Cell Biol 106:829–844.

Mayer M, Noble M (1994): N-acetyl-L-cysteine is a pluripotent protein against cell death and enhancer of trophic factor-mediated cell survival in vitro. Proc Natl Acad Sci USA 91:7496–7500.

Mettling C, Gouin A, Robinson M, el M'Hamdi H, Camu W, Bloch-Gallego E, Buisson B, Tanaka H, Davies AM, Henderson CE (1995): Survival of newly postmitotic motoneurons is transiently independent of exogenous trophic support. J Neurosci 15:3128–3137.

Miething A (1992): Germ-cell death during prespermatogenesis in the testis of the golden hamster. Cell Tissue Res 267:583–590.

MacDonald HR, Lees RK (1990): Programmed death of autoreactive thymocytes. Nature 343:642–644.

Matthews W, Tanaka K, Driscoll J, Ichihara A, Goldberg AL (1989): Involvement of the proteosome in various degradative processes in mammalian cells. Proc Natl Acad Sci USA 86:2597–2601.

McCall CA, Cohen JJ (1991): Programmed cell death in terminally differentiating keratinocytes: role of the endogenous nuclease. J Invest Dermatol 97:111–114.

Milligan CE, Cunningham TJ, Levitt P (1991a): Differential immunochemical markers reveal the normal distribution of brain macrophages and microglia in the developing rat brain. J Comp Neurol 314:125–135.

Milligan CE, Levitt P, Cunningham TJ (1991b): Brain macrophages and microglia respond differently to lesions of the developing and adult visual system. J Comp Neurol 314:136–146.

Milligan CE (1992): The response of brain macrophages and microglia to neuronal cell death. Thesis: The Medical College of Pennsylvania, Philadelphia, PA. University Microfilms.

Milligan CE, Webster L, Piros ET, Evans CJ, Cunningham TJ, Levitt P (1995a): Induction of opioid receptor-mediated macrophage chemotactic activities after neonatal brain injury. J Immunol 154:6571–6581.

Milligan CE, Prevette D, Yaginuma H, Homma S, Cardwell C, Fritz LC, Tomaselli KJ, Oppenheim RW, Schwartz LM (1995b): Peptide inhibitors of the ICE protease family arrest programmed cell death of motoneurons in vitro and in vivo. Neuron 15:1–10.

Miura M, Zhu H, Rotello R, Hartwieg E, Yuan J (1993): Induction of apoptosis in fibroblasts by IL-1 beta-converting enzyme, a mammalian homolog of the C. elegans cell death gene ced-3. Cell 75:653.

Moon RT (1981): Cell death: An integral aspect of development. Biologist 63:5–26.

Morgenbesser SD, Williams BO, Jacks T, DePinho RA (1994): p53 dependent apoptosis produced by Rb-deficiency in the developing mouse lens. Nature 371:72–74.

Mori C, Nakamura N, Okamoto Y, Osawa M, Shiota K (1994): Cytochemical identification of programmed cell death in the fusing fetal mouse palate by specific labelling of DNA fragmentation. Anat Embryol 190:21–28.

Murphy K, Heimberger A, Loh D (1990): Induction by antigen of intrathymic apoptosis of CD4+ CD8+ TCRlo thymocytes in vivo. Science 250:1720–1722.

Nicholson DW, Ali A, Thornberry NA, Vaillancourt JP, Ding CK, Gallant M, Gareau Y, Griffen PR, Labelle M, Lazebnik YA, Munday NA, Raju SM, Smulson ME, Yamin TT, Yu VL, Miller DK (1995): Identification and inhibition of the ICE/Ced-3 protease necessary for mammalian apoptosis. Nature 376:37–43.

Oltvai ZN, Milliman CL, Korsmeyer SL (1993): Bcl-2 heterodimerized in vivo with a conserved homolog. Bax, that accelerates programmed cell death. Cell 74:609–619.

Oppenheim RW (1991): Cell death during the development of the nervous system. Annu Rev Neurosci 14:453–501.

Oppenheim RW, Nunez R (1982): Electrical stimulation of hindlimb increases neuronal cell death in chick embryos. Nature 295:57–59.

Oppenheim RW, Chu-Wang I-W, Maderdrot JL (1978): Cell death of motoneurons in the chick embryo spinal cord. III. The differentiation of motoneurons prior to their induced degeneration following limb-bud removal. J Comp Neurol 177:87–112.

Oppenheim RW, Haverkamp LJ, Prevette D, McManaman JL, Appel SH (1988): Reduction of naturally occurring motoneuron death in vivo by a target derived neurotrophic factor. Science 240:919–922.

Owens GP, Hahn W, Cohen JJ (1991): Identification of mRNAs associated with programmed cell death in immature thymocytes. Mol Cell Biol 11:4177–4188.

Pan Z, Perez-Polo R (1993): Role of nerve growth factor in oxidant homeostasis: Glutathione metabolism. J Neurochem. 61:1713–1721.

Perry VH, Gordon S (1991): Macrophages and the nervous system. Intl Rev Cytol 125:203–244.

Perry VH, Brown MC, Gordon S (1987): The macrophage response to central and peripheral nerve injury. A possible role for macrophages in regeneration. J Exp Med 165:1218–1223.

Perry VH, Hume DA, Gordon S (1985): Immunohistochemical localization of macrophages and microglia in the adult and developing mouse brain. Neuroscience 15:313–326.

Pesce M, Farrace MG, Piacentini M, Dolci S, DeFelici

M (1993): Stem cell factor and leukemia inhibitory factor promote primordial germ cell survival by suppressing programmed cell death (apoptosis). Development 118:1089–1094.

Pilar G, Landmesser LJ (1976): Ultrastructural differences during embryonic cell death in normal and peripherally deprived ciliary ganglia. J Cell Biol 68:339–356.

Pittman RH, Oppenheim RW (1978): Neuromuscular blockade increases motoneurone survival during normal cell death in the chick embryo. Nature 271:364–366.

Polakowski RR, Piacentini M, Bartlett R, Goldsmith LA, Haaki HR (1994): Apoptosis in human skin development: Morphogenesis, periderm and stem cells. Dev Dynamics 199:176–188.

Pratt RM, Greene RM (1976): Inhibition of palatal epithelial cell death by altered protein synthesis. Dev Biol 54:135–145.

Price JM, Donahoe PK, Ito Y, Hendren WH (1977): Programmed cell death in the mullerian duct induced by mullerian inhibiting substance. Am J Anat 149:353–376.

Rabizadeh S, Lacount DJ, Friesen PD, Bredesen DE (1993): Expression of the baculovirus p35 gene inhibits mammalian neural cell death. J Neurochem 61:2318–2321.

Raff MC (1992): Societal controls on cell survival and cell death. Nature 356:397–400.

Rager G, Rager U (1978): Systems-matching by degeneration. I. A quantitative electron microscopic study of the generation and degeneration of retinal ganglion cells in the chicken. Exp Brain Res 33:65–78.

Rao L, Debbas M, Sabbatini P, Hockenberry D, Korsmeyer SJ, White E (1992): The adenovirus E1a proteins include apoptosis which is inhibited by the E1b 19 KDa and Bcl-2 proteins. Proc Natl Acad Sci USA 89:7742–7746.

Riddiford LM (1993): Hormone receptors and the regulation of insect metamorphosis. Receptor 3:203–209.

Rottello RJ, Fernandez P-A, Yuan J (1994): Anti-apogens and anti-engulfins: monoclonal antibodies reveal specific antigens on apoptotic and engulfment cells during chick embryonic development. Development 120:1421–1431.

Sampath D, Jackson GR, Werrback-Perez K, Perez-Polo JR (1994): Effects of nerve growth factor on glutathione peroxidase and catalase in PC12 cells. J Neurochem 62:2476–2479.

Sandow BA, West NB, Norman RL, Brenner RM (1976): Hormonal control of apoptosis in hamster uterine luminal epithelium. Am J Anat 156:15–36.

Saunders J (1966): Death in embryonic systems. Science 154:604–612.

Savill J, Dransfield I, Hogg N, Haslett C (1990): Vitronectin receptor-mediated phagocytosis of cells undergoing apoptosis. Nature 343:170–173.

Savill J, Fadok V, Henson P, Haslett C (1993): Phagocytic recognition of cells undergoing apoptosis. Immunol Today 14:131–136.

Schwartz LM (1992): Insect muscle as a model for programmed cell death. J Neurobiol 23:1312–1326.

Schwartz LM, Truman JW (1982): Peptide and steroid regulation of muscle degeneration in an insect. Science 215:1420–1421.

Schwartz LM, Truman JW (1983): Hormonal control of rates of metamorphic development in the tobacco hornworm Manduca sexta. Dev Biol 99:103–114.

Schwartz LM, Truman JW (1984a): Endocrine regulation of skeletal muscle atrophy and degeneration in the silkmoth Antheraea polyphemus. J Exp Biol 111:13–30.

Schwartz LM, Truman JW (1984b): Cyclic GMP may serve as a second messenger in peptide-induced muscle degeneration in an insect. Proc Natl Acad Sci USA 81:6718–6722.

Schwartz LM, Osborne BA (1993): Programmed cell death, apoptosis and killer genes. Immunol Today 14:582–590.

Schwartz LM, Kosz L, Kay BK (1990a): Gene activation is required for developmentally programmed cell death. Proc Natl Acad Sci 87:6594–6598.

Schwartz LM, Myer A, Kosz L, Engelstein M, Maier C (1990b): Activation of polyubiquitin gene expression during developmentally programmed cell death. Neuron 5:411–419.

Schwartz LM, Jones MEE, Kosz L, Kuah K (1993): Selective repression of actin and myosin heavy chain expression during the programmed death of insect skeletal muscle. Dev Biol 158:448–455.

Scollay R, Shortman K (1985): Cell traffic in the adult thymus: Cell entry and exit, cell birth and death. In Watson JD, Marbrook J (eds): Recognition and Regulation in Cell-Mediated Immunity. New York: Marcel Dekker, pp 3–30.

Serwmaa K, Rytomaa T (1988): Suicide death of rat chloroleukaemia cells by activation of the long interspersed repetitive DNA element (LIRN). Cell Tissue Kinet 21:33–43.

Shi Y, Szaly M, Paskar L, Boyer L, Singh B, Green DR (1990): Activation-induced cell death in T cell hybridomas is due to apoptosis. J Immunol 144:3326–3333.

Shier WT (1988): Why study mechanisms of cell death? Methods Achiev Exp Pathol 13:1–17.

Silver J, Hughes AFW (1973): The role of cell death during morphogenesis of the mammalian eye. J Morphol 140:159–170.

Silver J, Edwards MA, Levitt P (1993): Immunocytochemical demonstration of early appearing astroglial structures that form boundaries and pathways along axon tracts in the fetal brain. J Comp Neurol 328:415–436.

Slebos RJ, Lee MH, Plunkett BS, Kessis TD et al. (1994): p53 dependent G1 arrest involves pRB-related proteins and is disrupted by the human papillomavirus 11.E7 oncoprotein. Proc Natl Acad Sci USA 91:5320–5324.

Smenye RJ, Vendrell M, Hayward M, Baker SJ, Miao GG, Schilling K, Robertson LM, Curran T, Morgan JI (1993): Continuous *c-fos* expression precedes programmed cell death *in vivo*. Nature 363:166–169.

Smith CA, Williams G, Kingston R, Jenkinson EJ, Owen JT (1989): Antibodies to CD3/T-cell receptor complex induces death by apoptosis in immature T cells in thymic culture. Nature 337:181–184.

Sohal GS, Weidman TA (1978): Ultrastructural and sequence of embryonic cell death in normal and peripherally deprived trochlear nucleus. Exp Neurology 61:53–64.

Stanfield BB, O'Leary DDM, Fricks C (1982): Selective collateral elimination in early postnatal development restricts cortical distribution of rat pyramidal tract neurons. Nature 298:371–373.

Strange R, Li F, Sauer S, Burkhardt A, Friis RR (1992): Apoptotic cell death and tissue remodeling during mouse mammary gland involution. Development 115:49–58.

Suda T, Takahashi T, Goldstein P, Nagata S (1993): Molecular cloning and expression of the Fas ligand novel member of the tumor necrosis factor family. Cell 75:1169–1178.

Sulston JE, Horvitz HR (1977): Postembryonic cell lineages of the nematode *Caenorhabditis elegans*. Dev Biol 56:110–156.

Sun D, Ziegler R, Milligan CE, Fahrbach S, Schwartz LM (1995): Apolipophorin III is dramatically upregulated during the programmed death of insect skeletal muscle and neurons. J Neurobiol 26:119–129.

Surh CD, Sprent J (1994): T-cell apoptosis detected *in situ* during positive and negative selection in the thymus. Nature 372:100–103.

Swartzendruber DC, Congdon CC (1963): Electron microscope observations on tangible body macrophages in mouse spleen. J Cell Biol 19:641–646.

Swat W, Leszek I, von Boehmer H, Kisielow P (1991): Clonal deletion of immature CD4$^+$CD8$^+$ thymocytes in suspension culture by extrathymic antigen-presenting cells. Nature 351:150–153.

Symonds H, Krall L, Remington L, Saenz-Robles M, Lowe S, Jacks T, Van Dyke T (1994): p53 dependent apoptosis suppresses tumor growth and progression *in vivo*. Cell 78:703–711.

Szirmai JA, Balazs EA (1958): Studies on the structure of the vitreous. III. Cells in the cortical layer. Arch Ophthalmol 59:34–48.

Takahashi T, Tanaka M, Branna CI, Jenkins NA, Copeland NG, Suda T, Nagata S (1994): Generalized lymphoproliferative disease in mice caused by a point mutation in the Fas ligand. Cell 76:969–976.

Tata JR (1966): Requirement for RNA and protein synthesis for induced regression of the tadpole tail in organ culture. Dev Biol 13:77–94.

Tewari M, Quan LT, O'Rourke K, Desnoyers S, Zeng Z, Beidler DR, Poirier GG, Dixit VM (1995): Yama/ CPP32B, a mammalian homolog of Ced-3, is a crmA-inhibitable protease that cleaves the death substrate poly(ADP-ribose) polymerase. Cell 81:801–809.

Thornberry NA, Bull HG, Calayery JR, Chapman KT et al. (1992): A novel heterodimeric cysteine protease is required for interleukin-1β processing in monocytes. Nature 356:768–774.

Tomei LD, Shapiro JP, Cope FO (1993): Apoptosis in C3H/10T1/2 mouse embryonic cells: Evidence for internucleosomal DNA modification in the absence of double-strand cleavage. Proc Natl Acad Sci USA 90:853–857.

Trauth B, Klas C, Peters AMJ, Matzku S, Moller P, Falk W, Debatin KM, Krammer PH (1989): Monoclonal antibody mediated tumor regression by induction of apoptosis. Science 245:301–305.

Truman JW, Riddiford LM (1970): Neuroendocrine control of ecdysis in silkmoths. Science 167:1624–1626.

Tsujimoto Y, Gorham J, Cossman J, Jaffe E, Croce C (1985): The T(14;18) chromosome translocations involved in B cell neoplasms result from mistakes in VDJ joining. Science 229:1390–1393.

Valentino KL, Jones EG (1981): Morphological and immunocytochemical identification of macrophages in the developing corpus callosum. Anat Embryol 163:157–172.

Vaux DL, Weissman IL, Kim SK (1992): Prevention of programmed cell death in *Caenorhabditis elegans* by human bcl-2. Science 258:1955–1957.

Veis DJ, Sorenson CM, Shutter JR, Korsmeyer SJ (1993): Bcl-2-deficient mice demonstrate fulminant lymphoid apoptosis, polycystic kidneys, and hypopigmented hair. Cell 75:229–240.

Walker NI (1987): Ultrastructure of the rat pancreas after experimental duct ligation. I. The role of apoptosis and intraepithelial macrophages in acinar cell deletion. Am J Pathol 126:439–451.

Walker NI, Bennett RE, Kerr JFR (1989): Cell death by apoptosis during involution of the lactating breast in mice and rats. Am J Anat 185:19–32.

Watanabe-Fukunga R, Brannan CJ, Copeland NG, Jenkins NA, Nagata S (1992): Lymphoproliferation disorder in mice explained by defects in FAS antigen that mediates apoptosis. Nature 356:314–317.

Watson ML, Rao JK, Gilkeson GS, Ruiz P, Eicher EM, Pisetsky DS, Matsuzawa A, Rochelle JM, Seldin MF (1992): Genetic analysis of MRL-lpr mice: Relationship of the Fas apoptosis gene to disease manifestations and renal disease–modifying loci. J Exp Med 176:1645–1656.

Webb JN (1974): Muscular dystrophy and muscle cell death in the normal fetal development. Nature 252:233–234.

Wellings SR, DeOre KB (1963): Electron microscopy of milk secretion in the mammary gland of the C3H mouse. III. Cytomorphology of the involuting gland. JNCI 30:241–261.

White K, Gether ME, Abrams JM, Young L, Farrell K,

Steller H (1994): Genetic control of programmed cell death in *Drosophila*. Science 264:677–683.

Whitten JM (1969): Cell death during early morphogenesis: Parallels between insect limb and vertebrate limb development. Science 163:1456–1457.

Wielckens K, Delfs T, Moth A, Freese V, Kleeberg HJ (1987): Glucocorticoid-induced lymphoma cell death: The good and the evil. J Steroid Biochem 27:413–419.

Woronicz JD, Calnan B, Ngo V, Winoto A (1994): Requirement for the orphan steroid receptor Nur77 in apoptosis of T-cell hybridomas. Nature 367:277–281.

Wyllie AH (1984): Glucocorticoid-induced thymocyte apoptosis is associated with endogenous endonuclease activation. Nature 284:555–556.

Wyllie AH, Kerr JFR, Currie AR (1973): Cell death in the normal neonatal rat adrenal cortex. J Pathol 111:255–261.

Wyllie AH, Kerr JFR, Currie AR (1980): Cell death: Significance of apoptosis. Intl Rev Cytol 68:251–306.

Wyllie AH, Morris R, Smith A, Dunlop D (1984): Chromatin cleavage in apoptosis: association with condensed chromatin morphology and dependence upon macromolecular synthesis. J Pathol 142:66–77.

Yin X-M, Oltval ZN, Korsmeyer SJ (1994): BH1 and BH2 domains of bcl-2 are required for inhibition of apoptosis and heterodimerization with Bax. Nature 369:321–323.

Yonish-Rouach E, Resnitzky D, Lotem L, Sachs L, Kimechi A, Oren M (1991): Wildtype p53 induces apoptosis in myloid leukemic cells that is inhibited by interleukin-6. Nature 352:345–347.

Yuan J, Horvitz HR (1990): The *Caenorhabditis elegans* genes *ced-3* and *ced-4* act cell-autonomously to cause programmed cell death. Dev Biol 138:33–41.

Yuan JY, Horvitz HR (1992): The *Caenorhabditis elegans* cell death gene *ced-4* encodes a novel protein and is expressed during the period of extensive programmed cell death. Development 116:309–320.

Yuan J, Shaham S, Ledoux S, Ellis HM, Horvitz HR (1993): The *C. elegans* cell death gene ced-3 encodes a protein similar to mammalian interleukin-1 betaconverting enzyme. Cell 75:641–652.

ABOUT THE AUTHORS

CAROLANNE E. MILLIGAN is an Assistant Professor in the Department of Neurobiology and Anatomy at the Bowman Gray School of Medicine of Wake Forest University, Winston-Salem, North Carolina. She received a B.S. in Biology in 1985 from Saint Joseph's University in Philadelphia. In 1992 she earned a Ph.D. for work conducted in the laboratories of Pat Levitt and Timothy Cunningham at the Medical College of Pennsylvania in Philadelphia, investigating macrophage and microglia responses to neuronal cell death. This work included identification of a brain-derived macrophage chemotactic factor induced following neonatal brain injury. As a postdoctoral fellow in the Department of Biology at the University of Massachusetts, she identified and characterized genes that may be differentially expressed during naturally occurring motoneuron cell death in the embryonic chick spinal cord and was awarded fellowships from the National Institutes of Health and the Spinal Cord Research Foundation for this work. The work of her laboratory continues to focus on investigating the cellular and molecular changes involved in neuronal death.

LAWRENCE M. SCHWARTZ is an Associate Professor in the Department of Biology at the University of Massachusetts in Amherst. He teaches in the areas of Neurobiology, Animal Development, and Molecular Biology. He receives his B.A. from Northwestern University and his Ph.D. from the University of Washington in Seattle, where he worked with James W. Truman on the endocrine control of programmed cell death in insects. He then engaged in postdoctoral training in Membrane Biophysics at the University of Washington with Wolfhard Almers and Molecular Biology and Development at the University of North Carolina at Chapel Hill with Brian Kay. Dr. Schwartz joined the faculty at the University of Massachussets in 1988, where he was promoted to Associate Professor with tenure in 1992. His work focuses on the molecular mechanisms that mediate programmed cell death during development, focusing primarily on three distinct model systems: metamorphosis in insects, negative selection in the mouse immune system, and motor neuron loss during embryogenesis in chicks. Dr. Schwartz has been the recipient of a Research Career Development Award from the National Institutes of Health, serves on the editorial board of *Cell Death and Differentiation*, and has served on several federal advisory boards.

Cellular Aging and Cell Death: 209–234
© 1996 Wiley-Liss, Inc.

p53-Dependent Apoptosis in Tumor Progression and in Cancer Therapy

Scott W. Lowe and H. Earl Ruley

I. APOPTOSIS IN TUMOR BIOLOGY

Tissue homeostasis is a dynamic process that requires a balance between cell proliferation and cell loss. Neoplasia occurs when proliferation exceeds cell loss, producing net tissue growth. Much of cancer research over the last two decades has focused on genes that, when mutated, act in either a dominant or recessive manner to enhance proliferation. However, emerging evidence indicates that factors that reduce cell loss also contribute to tumorigenesis.

Tumor biologists have long noted a discrepancy between the observed rate of tumor growth compared with that expected from direct measurements of proliferating cells. In many instances, the observed growth rate is less than 5% of that predicted from proliferation estimates alone [reviewed by Wyllie, 1985], implying that continuous cell loss occurs from malignant tumors. It follows, therefore, that factors that influence cell loss can have a substantial effect on tumor growth. Decreases in cell loss from tumors may promote tumor progression to more malignant stages, whereas increases in cell loss cause tumor regression. Since the purpose of cancer therapy is to reverse the imbalance in homeostasis from cell accumulation to cell loss, factors that modulate cell loss from tumors may also influence the effectiveness of radiation or chemotherapy.

Cell death within tumors can occur by necrosis or apoptosis. Although necrosis contributes substantially to tumor cell loss, it cannot explain the magnitude of cell death observed in many tumors [Kerr et al., 1972]. Until recently, the contribution of apoptosis to cell death in tumors has been largely overlooked, perhaps because dying cells shrink and are rapidly phagocytosed by macrophages or neighboring cells. However, apoptosis can account for a large proportion of cell loss from both developing and regressing tumors [reviewed by Wyllie, 1985; Kerr et al., 1994]. In this chapter, we discuss recent evidence indicating that the p53 tumor suppressor can promote cell loss from tumors by apoptosis and the implications of this view for understanding the aggressive nature of malignancies acquiring p53 mutations.

II. THE p53 TUMOR SUPPRESSOR GENE

II.A. The Impact of p53 Mutation on Carcinogenesis

Mutations in the p53 tumor suppressor gene are among the most frequent genetic lesions observed in human cancer, occurring in greater than 50% of human malignancies. Moreover, p53 mutations occur in many tumor types, including tumors of the colon, lung, breast, and bone, as well as various leukemias and lymphomas [Greenblatt et al., 1994]. The frequency and spectrum of p53 mutations in human cancer implies that normal p53 possesses antineoplastic activities in many tissues.

Germline p53 mutations predispose individuals with Li-Fraumeni syndrome to the development of diverse tumor types [Malkin et al., 1990; Srivastava et al., 1990]. In sporadic cancers, however, p53 mutations are often a late event in tumorigenesis and occur during progression to more advanced stages of disease [e.g., Fearon and Vogelstein, 1990]. Indeed, p53 mutations are frequently associated with clinically aggressive cancers and poor patient prognosis [Harris and Hollstein, 1993]. Gene-targeting studies demonstrate that mice homozygous for p53 deletions develop normally but are highly predisposed to the occurrence of malignant tumors. Thus, p53 has no essential role in regulating cell growth or survival during embryogenesis but instead functions primarily to suppress neoplastic growth [Donehower et al., 1992].

These observations raise several fundamental questions concerning p53 function. How can inactivation of a protein with no obvious role in normal proliferation have such an impact on carcinogenesis? Which biological processes that limit neoplastic growth are affected by p53? What selective advantage is conferred upon cells acquiring p53 mutations? Why is selection for p53 inactivation a late event in the progression of certain tumor types?

II.B. Biochemical and Biological Activities of p53

Considerable progress has been made in understanding p53 at the biochemical level [Prives, 1994]. p53 is a sequence-specific DNA-binding protein [e.g., El Deiry et al., 1992] that functions as a transcription factor [Farmer et al., 1992; Kern et al., 1992; Seto et al., 1992; Zambetti et al., 1992a; Mack et al., 1993] and interacts directly with proteins involved in both DNA replication [Dutta et al., 1993] and DNA repair [Wang et al., 1994]. Although it is not yet certain which biochemical activity of p53 participates in tumor suppression, the ability of p53 to associate specifically with DNA may be essential. In support of this view, all tumor-derived p53 mutants encode proteins with reduced affinity for DNA [Vogelstein and Kinzler, 1992], and many of the most common mutations occur at residues that are essential for the p53–DNA interaction [Cho et al., 1994].

Efforts to understand the biological functions of p53 initially relied on gene transfer to examine the consequences of overexpressing

wild-type or dominant-negative p53 alleles on cellular processes. Such studies implicate p53 in negative growth control [Michalovitz et al., 1990; Martinez et al., 1991; Mercer et al., 1991], apoptosis [Yonish-Rouach et al., 1991; Shaw et al., 1992], senescence [Jenkins et al., 1984; Harvey and Levine, 1991; Harvey et al., 1993], and differentiation [Shaulsky et al., 1991]. Nevertheless, gene transfer experiments alone do not provide a satisfactory understanding of p53 activities. For example, the levels of p53 achieved by gene transfer may be substantially higher than the levels expressed under normal physiological circumstances. Because p53 expression is forced, overexpression studies cannot identify upstream signals that recruit p53 to participate in these processes. Moreover, mutant p53s may have dominant effects on cell growth and transformation and are not simply dominant-negative inhibitors of wild-type p53 [Dittmer et al., 1993]. Thus, the biological effects of mutant p53 do not necessarily reveal the effects of inactivating wild-type p53.

In genetically tractable organisms such as *Drosophila melanogaster* and *Saccharomyces cerevisiae*, gene function is typically studied using null mutants. Because the only genetic differences between normal and null animals are in the gene of interest, phenotypic differences between strains provide insight into *endogenous* gene function. Until recently, this approach has generally not been feasible in mammalian systems. However, gene-targeting technology has facilitated the generation of null mutants in mice. Because cells are readily derived from both normal and mutant mice, this powerful genetic approach can be extended to the analysis of gene function in cell culture.

Using embryonic fibroblasts derived from p53-deficient mice, endogenous p53 has been implicated in the cellular response to DNA damage [Kastan et al., 1992]. Normal embryonic fibroblasts possess cell cycle checkpoints that limit proliferation following genomic damage, such that γ-irradiation produces cell cycle arrest at the G_1/S and G_2/M boundaries. Fibroblasts derived from p53-deficient mice fail to arrest in G_1 following γ-irradiation [Kastan et al., 1992], indicating that p53 is required for the G_1 checkpoint. p53 protein levels and stability increase in normal cells following irradiation [Maltzman and Czyzyk, 1984; Kastan et al., 1991], suggesting a mechanism whereby p53 is activated to arrest cell growth. In addition, p53-deficient fibroblasts more readily become aneuploid and are more prone to gene amplification events compared with normal cells, suggesting that defects in this G_1 checkpoint lead to genomic instability [Livingstone et al., 1992; Yin et al., 1992].

p53 can also influence the survival of cells following DNA damage. Low doses of γ-radiation induce apoptosis in normal thymocytes but not those derived from p53-deficient animals [Clarke et al., 1993; Lotem and Sachs, 1993a; Lowe et al., 1993b]. As precedes radiation-induced growth arrest in fibroblasts, p53 protein levels increase prior to the onset of apoptosis [Lowe et al., 1993b]. Similarly, p53 is required for efficient apoptosis following treatment with DNA-damaging agents in myeloid progenitor cells and in the epithelial stem cells of the intestinal crypts [Lotem and Sachs, 1993a; Clarke et al., 1994; Merritt et al., 1994]. Tissue-specific factors therefore can determine whether p53 functions in growth arrest or apoptosis. However, this decision is not simply a cell type–specific phenomenon (see below).

II.C. p53 as a Guardian of the Genome

The ability of p53 to promote cell cycle arrest following DNA damage suggests a mechanism in which p53 acts to suppress neoplastic growth. In this model, p53 is an essential component of a cell cycle checkpoint that facilitates DNA repair following genomic damage, thereby reducing the likelihood that the damaged cell incorporates mutations [Lane, 1992]. Cells acquiring defects in p53 function have no immediate growth or survival advantage, but more readily sustain additional mutations—some of which are oncogenic. The driving force for p53 mutation is therefore *indirect*; subsequent lesions actually provide developing tumor cells with a growth or survival advantage.

In tissues that normally respond to DNA damage by activating p53-dependent apoptosis, p53 mutations also may indirectly promote tumorigenesis. Thus, p53 mutations enhance the survival of cells that sustain DNA damage, thereby increasing their probability of acquiring oncogenic mutations.[1] Consistent with this view, defects in thymocyte apoptosis are associated with a high incidence of thymic lymphoma in p53-deficient mice [Donehower et al., 1992; Jacks et al., 1994; Purdie et al., 1994].

This indirect mechanism of tumor suppression is consistent with the observation that p53 mutations are associated with genomic instability—a form of mutation—in human and murine tumors [Kihana et al., 1992; Blount et al., 1994; Purdie et al., 1994]. Moreover, because p53 is activated to suppress cell growth under circumstances that might be uncommon during embryogenesis (i.e., DNA damage), this model accounts for the fact that p53 is not required for normal growth and development.

The increase in mutation rate that may accompany p53 loss helps explain a paradox in cancer etiology. This paradox arises from the observation that at least five genetic lesions may be required for the full malignant phenotype [Fearon and Vogelstein, 1990], a number too large to achieve in a single cell at the normal rate of spontaneous mutation [Loeb, 1991]. Thus, tumor progression may necessitate that neoplastic cells acquire a "mutator" phenotype during tumorigenesis. Because p53 may facilitate DNA repair, it has been argued that p53 mutations, which occur frequently in human tumors, may account for the increased mutagenic potential of developing tumor cells.

---

[1]Note that in tissues that normally respond to DNA damage by p53-dependent apoptosis it is predicted that cells have an immediate survival advantage, in addition to having a higher incidence of subsequent mutation.

Implicit in this notion is that early loss of p53 function should have the greatest impact on tumorigenesis. Indeed, p53 is mutated early in the progression of certain tumor types [Blount et al., 1993] and mice and humans harboring germline p53 mutations are predisposed to spontaneous tumors [Malkin et al., 1990; Srivastava et al., 1990; Donehower et al., 1992; Jacks et al., 1994; Purdie et al., 1994].

Although this model is attractive, there are compelling reasons to doubt that this indirect mechanism is the *only* means by which p53 mutations contribute to cancer. First, dominant-negative p53 mutants collaborate with either activated *ras* oncogenes or the adenovirus E1A gene to transform primary cells to a tumorigenic state [Hinds et al., 1989; Debbas and White, 1993]. Transformation occurs within days of gene transfer and requires continuous expression of the mutant p53 allele [Zambetti et al., 1992b; Debbas and White, 1993]. This outcome would be unlikely if p53 mutations *solely* predisposed cells to additional oncogenic alterations. Second, p53 mutations occur late in the progression of many human malignancies [Fearon and Vogelstein, 1990; Yamada et al., 1991; Ichikawa et al., 1992; Tsuda and Hirohashi, 1992; Bookstein et al., 1993]. Finally, in carcinogen-induced murine skin cancers, p53 mutations typically occur *after* the acquisition of point mutations in *ras* protooncogenes, during tumor progression from benign papillomas to malignant carcinomas [Burns et al., 1991]. Studies using p53-deficient mice demonstrate that p53 loss does not influence the emergence of papillomas; instead, the absence of p53 accelerates the progression of papillomas to carcinomas [Kemp et al., 1993]. Taken together, these observations suggest that, in many developing tumors, p53 mutations provide a selective advantage only after other oncogenic alterations have occurred.

III. APOPTOSIS AND ONCOGENIC TRANSFORMATION

The role of p53 in apoptosis during oncogenic transformation illustrates a conceptually different way in which p53 acts as a tumor suppressor. This alternative view of p53 function stems from the observation that oncogenes that deregulate normal growth control also can promote apoptosis. Although the probability that oncogene-expressing cells engage the apoptotic program is modulated by environmental factors, the increased susceptibility of these cells to apoptosis has the potential consequence of suppressing oncogenic transformation. In many instances, apoptosis requires p53, even in cells where p53 normally produces only growth arrest. Consequently, p53 mutations can enhance the viability of oncogene-expressing cells without necessarily affecting proliferation, thereby promoting oncogenic transformation and tumor growth.

III.A. p53 and the Transforming Interactions Between Adenovirus E1A and E1B

Since oncogenic transformation of primary cells typically requires two or more oncogenes acting in concert, the transforming interactions between viral and/or cellular oncogenes provide useful systems for studying multistep carcinogenesis. For example, the adenovirus early region 1A (E1A) oncogene is unable to transform primary cells alone, but collaborates with either the adenovirus early region 1B (E1B) oncogene or activated *ras* oncogenes to transform primary cells to a tumorigenic state [Ruley, 1990]. E1A promotes proliferation and S phase entry, probably by associating with cellular proteins involved in negative growth control [e.g., Whyte et al., 1988]. By contrast, the E1B-encoded proteins p19^{E1B} and p55^{E1B} have no obvious effect on cell proliferation in the absence of E1A [van den Elsen et al., 1983] and do not substitute for *ras* in cotransformation with other oncogenes (e.g., *myc*) [Ruley, 1990]. Nevertheless, E1B expression allows the sustained proliferation of E1A-expressing cells [Rao et al., 1992].

The differential effect of E1A and E1B on apoptosis accounts for their collaborative interactions in adenovirus transformation. While E1A causes forced proliferation, it also promotes apoptosis, particularly under conditions of serum deprivation or high cell density [Rao et al., 1992]. Consequently, E1A-expressing cells cannot continue to proliferate. Regions of E1A required for apoptosis coincide with those involved in the induction of DNA synthesis [White et al., 1991], implying that these processes are tightly linked. E1B suppresses apoptosis in E1A-expressing cells, thereby allowing sustained proliferation without directly influencing cell growth [Rao et al., 1992]. Although these studies do not rule out the possibility that E1B has additional activities, they suggest that inhibition of apoptosis enhances oncogenic transformation.

The molecular mechanism underlying these observations involves modulation of apoptosis by p53. Although it has long been known that p53 levels and stability are increased in adenovirus-transformed cells (expressing both the E1A and E1B proteins), this effect was thought to be an inconsequential outcome of the physical interaction between p53 and $p55^{E1B}$ [Sarnow et al., 1982]. However, in cells that lack E1B and are susceptible to apoptosis, E1A metabolically stabilizes p53 protein, raising the possibility that p53 might participate in E1A-induced apoptosis [Lowe and Ruley, 1993]. Consistent with this notion, forced overexpression of p53 induces apoptosis in E1A-expressing cells, an effect that is blocked by co-expression of $p19^{E1B}$ [Debbas and White, 1993].

Efficient E1A-induced apoptosis requires p53 function, since p53-deficient fibroblasts expressing E1A are resistant to apoptosis following serum depletion [Lowe et al., 1994c] (Fig. 1). E1B has no additional effect on p53 stabilization [Lowe and Ruley, 1993] but inhibits apoptosis in p53-expressing cells [Lowe et al., 1994c]. The ability of E1B to block apoptosis involves two independent yet redundant activities. First, $p55^{E1B}$ directly interacts with p53 and interferes with p53-mediated transcriptional regulation [Yew and Berk, 1992]. Second, $p19^{E1B}$ acts downstream of p53 to inhibit apoptosis [White et al., 1991, 1992; Rao et al., 1992; Debbas and White, 1993]. It

appears that inactivation of p53 is the primary function of the E1B proteins,[2] because p53-deficiency substitutes for E1B in promoting the growth, survival, and transformation of E1A-expressing cells (Fig. 1) [Lowe et al., 1994c]. Taken together, these observations suggest that p53-dependent apoptosis is a cellular response to forced proliferation that ultimately limits the oncogenic potential of E1A-expressing cells.

III.B. Oncogenes That Modulate p53-Dependent Apoptosis

Table I summarizes a rapidly growing list of the viral and cellular oncogenes that can modulate p53-dependent apoptosis. All oncogenic alterations capable of promoting p53-dependent apoptosis share with E1A the ability to deregulate normal growth control (Table IA). Like E1A, the regions of c-myc that promote apoptosis coincide with those involved in proliferation, although myc-associated apoptosis can occur in noncycling cells [Evan et al., 1992]. Several other oncogenes that promote apoptosis have not been tested for the dependence of apoptosis on p53 function (e.g., E2A–PBX1 fusion protein [Dedera et al., 1993]). However, the fact that diverse oncogenic stimuli induce a similar cellular program suggests that p53-dependent apoptosis occurs as a general response to forced proliferation.

Other viral and cellular oncogenes inhibit p53-dependent apoptosis (Table IB). In many instances, oncogenes that repress p53-dependent apoptosis collaborate with those that promote apoptosis in oncogenic transformation. Some examples include transforming collaborations between the papilloma virus E6 and E7 oncoproteins [Munger et al., 1989], activated ras and mutant p53 oncogenes [Hinds et al., 1989], and c-myc and bcl-2 [Bissonnette et al., 1992; Fanidi et al., 1992]. Thus, modulation of apoptosis by p53 may be a critical event in oncogenic transformation.

---

[2]It should be noted, however, that the E1B-encoded proteins presumably have evolved to promote adenovirus replication and not oncogenic transformation.

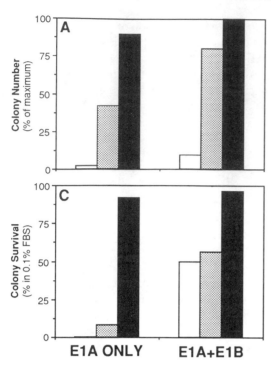
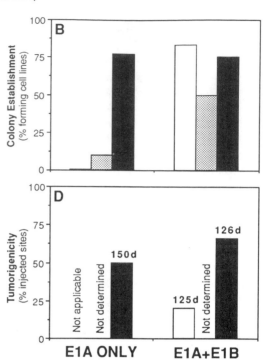

Fig. 1. *The effect of endogenous p53 and the adenovirus E1B proteins on the growth, survival, and transformation of E1A-expressing cells. An E1A expression vector (coexpressing hygromycin phosphotransferase) is introduced into p53$^{+/+}$ (open bars), p53$^{+/-}$ (shaded bars), or p53$^{-/-}$ (closed bars) mouse embryonic fibroblasts alone or in combination with a vector expressing both E1B-encoded proteins. E1A-expressing colonies were selected in hygromycin B [Lowe et al., 1994c]. Growth potential is assessed by the ability of transfected cells to form E1A-expressing colonies (A) and the frequency with which these* colonies are established into permanent cell lines (B). The susceptibility of cells to apoptosis is measured by the viability of individual E1A-expressing colonies 3 days after transfer to medium containing 0.1% fetal bovine serum, a treatment known to induce apoptosis in wild-type cells expressing E1A without E1B (C). Oncogenic transformation is determined by the frequency with which cell lines form tumors following subcutaneous injection into immunocompromised mice (D). The average latency of tumors is listed above each column. [Compiled from Lowe et al., 1994c, and S.W. Lowe, unpublished data.]

Although important, inhibition of p53-dependent apoptosis is not a prerequisite for oncogenic transformation. For example, *ras* cooperates with E1A in oncogenic transformation but does not block p53-dependent apoptosis [Lowe et al., 1993a, 1994c]. Similarly, coexpression of *c-myc* and *ras* also transforms primary cells to a malignant state, but *ras* does not block apoptosis associated with *c-myc* [Harrington et al., 1994]. This suggests that cells can acquire tumorigenic phenotypes through routes that alter the balance of growth and survival in different ways. Oncogenes such as p19^{E1B} and *bcl-2* appear to block apoptosis

directly, whereas the enhanced growth rate of *ras*-transformed cells may simply compensate for cell losses due to apoptosis. Thus, tumor growth can occur while cells remain genotypically susceptible to apoptosis, a suggestion consistent with the high rate of apoptosis observed in many malignant tumors.

III.C. Direct Mechanism of Tumor Suppression by p53

The ability of p53 to suppress oncogenic transformation by promoting apoptosis suggests a fundamentally different mechanism by which p53 limits tumor growth (Fig. 2) [Lowe

TABLE I. Oncogenic Alterations That Modulate p53-Dependent Apoptosis

	Gene	Origin	Mode of action	References
A*	Adenovirus E1A	Viral	Dominant	Debbas and White [1993], Lowe et al. [1993a, 1994c]
	Papillomavirus E7	Viral	Dominant	Howes et al. [1994], Pan and Gripe [1994]
	SV40 Antigen†	Viral	Dominant	Symonds et al. [1994]
	c-myc overexpression	Cellular	Dominant	Hermeking and Eick [1994]
	Activated ras	Cellular	Dominant	Tanaka et al. [1994]
	E2F1 overexpression	Cellular	Dominant	Wu and Levine [1994]
	Rb-deletions	Cellular	Recessive	Morgenbesser et al. [1994]
B‡	Adenovirus E1B	Viral	Dominant	Debbas and White [1993], Lowe et al. [1994a–c]
	Papillomavirus E6	Viral	Dominant	Howes et al. [1994], Pan and Griep [1994]
	SV40 Antigen‖	Viral	Dominant	Symonds et al. [1994]
	missence p53 mutants	Cellular	Dominant-negative	Debbas and White [1993], Wu and Levine [1994]
	p53-deletions	Cellular	Recessive	Lowe et al. [1993a; 1994c]; Tanaka et al. [1994]
	bcl-2 overexpression	Cellular	Dominant	Chiou et al. [1994]

*Oncogenic alterations that promote apoptosis in a p53-dependent manner.
†T-antigen mutant that expresses amino-terminal regions that bind the Rb family of proteins but not p53.
‡Oncogenic alterations that interfere or suppresses p53-dependent apoptosis.
‖Full-length or mutant T-antigen expressing the carboxy-terminal region containing the p53-binding domain.

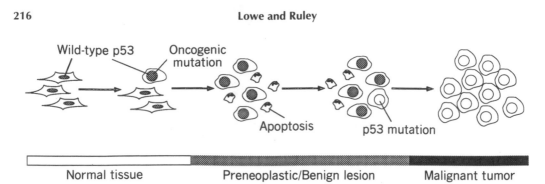

Fig. 2. *Model illustrating how p53 mutations may promote the progression of malignant tumors by attenuating apoptosis. Early oncogenic mutations enhance cell proliferation but also increase apoptosis. Tumor growth can occur, but is limited by environ-mental conditions that trigger apoptosis. p53 mutations dramatically reduce the ability of cells to engage the apoptotic program, allowing proliferation to continue unabated.*

et al., 1994c]. In this view, oncogenic mutations that initiate tumor growth enhance proliferation, but also induce a cellular response that compensates by promoting apoptosis. Tumor growth can occur, but is limited by the increased susceptibility of developing tumor cells to physiological conditions that trigger cell death. These conditions may include limiting concentrations of growth or survival factors [Evan et al., 1992; Lowe et al., 1994c], hypoxia [Graeber et al., 1994], physical constraints on tumor growth [Rao et al., 1992], abnormal differentiation [Howes et al., 1994; Morgenbesser et al., 1994], and treatment with cytotoxic agents used in cancer therapy (see Section IV). Only under these circumstances does p53 inactivation provide a selective advantage—thereby releasing cells from apoptosis susceptibility and allowing proliferation to continue unabated. This implies that p53 *directly* suppresses tumor progression by destroying aberrantly growing cells.

This direct mechanism of tumor suppression can account for several observations concerning p53. First, the need to suppress p53-dependent apoptosis can explain why continuous expression of dominant-negative p53 mutants is required to maintain the transformed phenotype. Second, the involvement of p53 in eliminating aberrantly growing cells is a strictly antineoplastic activity and is consistent with the observation that p53-deficient mice develop normally. Finally, this view accounts for

how p53 mutations promote the progression (but not the initiation) of some malignant tumors, because normal p53 is activated to suppress viability in cells previously altered by the inappropriate activities of certain oncogenes. This may provide the selective pressure for p53 inactivation in tumors where p53 mutations arise late, after other oncogenic mutations have already occurred.

III.C.1. p53 loss and attenuated apoptosis promote tumor progression in mice. Recent studies have strongly supported the notion that p53-dependent apoptosis occurs in response to aberrant proliferation and limits the progression of developing tumors. For example, loss of *Rb* function increases both proliferation and apoptosis in the mouse embryonic lens [Morgenbesser et al., 1994]. Cell death is dependent on p53, because lenses from mice homozygous for both *Rb* and p53 deletions contain few apoptotic cells. In transgenic mice, lens-specific expression of the papilloma virus E7 or E6 oncoproteins inactivates the *Rb*-related proteins or p53, respectively [Pan and Griep, 1994]. Expression of E7 alone promotes both proliferation and apoptosis, whereas the coexpression of E6 with E7 causes proliferation without apoptosis. In this system, inhibition of p53-dependent apoptosis by E6 correlates with the appearance of lens cell tumors.

In another transgenic strain, the absence or presence of p53 function determines whether E7-expressing retinal cells will produce retinoblastoma or undergo apoptosis [Howes et al., 1994]. Furthermore, in experimental tumors of the choroid plexus, a weakly oncogenic form of T-antigen (which inactivates the *Rb*-related proteins but not p53) disrupts proliferation, but also increases apoptosis [Symonds et al., 1994]. In the presence of functional p53, growth deregulation produces only hyperplasia. Inactivation of p53—by viral oncoprotein binding, by deletion, or by spontaneous mutation—rapidly leads to the development of highly malignant tumors. p53 loss has no significant effect on the proliferation, but dramatically reduces the occurrence of apoptosis.

III.C.2. p53 mutation, apoptosis, and anaplasia in Wilms' tumor.
The potential role of p53-dependent apoptosis in limiting tumor growth can explain the selective pressure for p53 mutation late in the progression of certain malignancies. Although this hypothesis has yet to be tested extensively, molecular and physiological changes occurring during the progression of Wilms' tumor support this view. Wilms' tumor is a pediatric malignancy of the kidney affecting 1 in 10,000 children. p53 mutations are rare in Wilms' tumor and are strictly associated with a more aggressive subtype characterized by an anaplastic morphology [Bardeesy et al., 1994].

In tumors of mixed morphology, p53 mutations are typically detected in focal regions of anaplasia but not in the surrounding nonanaplastic tumor. This demonstrates that anaplastic variants arise from nonanaplastic tumors and implies that p53 inactivation is an important event in this progression. As shown in Figure 3, focal regions of anaplasia also display a striking decrease in spontaneous apoptosis compared with the surrounding nonanaplastic tumor [Bardeesy et al., 1995]. These studies therefore associate inactivation of p53 with attenuated apoptosis and tumor progression. As discussed in Section V, the reduced susceptibility of anaplastic cells to apoptosis may explain the aggressive nature of these tumors.

IV. p53 AND THE EFFICACY OF CANCER THERAPY

IV.A. Factors That Determine the Efficacy of Anticancer Agents

The use of radiation and chemotherapy in the treatment of human malignancy often prolongs the disease-free interval without substantially increasing long-term patient survival. One reason for this failure is that many tumors respond poorly to therapy or become nonresponsive on tumor relapse. Since the identification of therapeutic agents has been empirical, the molecular mechanisms that determine treatment efficacy remain largely unknown.

Radiation and many chemotherapeutic agents induce DNA damage or cause disruptions in DNA metabolism, so the tumor-specific cytotoxicity of these agents has been attributed to their debilitating effects on actively proliferating cells. The interaction of an agent with its primary intercellular target (e.g., etoposide with topoisomerase II) presumably disrupts an essential metabolic function, ultimately leading to cell death. Therefore, efforts to understand the molecular basis of drug resistance have focused on how chemotherapeutic agents reach their primary intracellular targets or on the molecular nature of the drug–target interaction [Chabner and Myers, 1989; Chin et al., 1993].

The view that anticancer agents preferentially kill more rapidly proliferating cells predicts that p53 mutations should enhance the cytotoxicity of radiation and chemotherapy. This suggestion stems from the fact that many anticancer agents produce DNA damage either directly or indirectly [Chabner and Myers, 1989]. Since p53 may limit proliferation and promote repair following genomic damage, tumors lacking this checkpoint should carry a greater load of damaged DNA into S phase compared with tumors that arrest in G_1. However, neither p53-deficient mouse embryonic fibroblasts nor head and neck tumor lines harboring p53 mutations displayed increased radiosensitivity when compared to their p53-expressing counterparts [Brachman et al.,

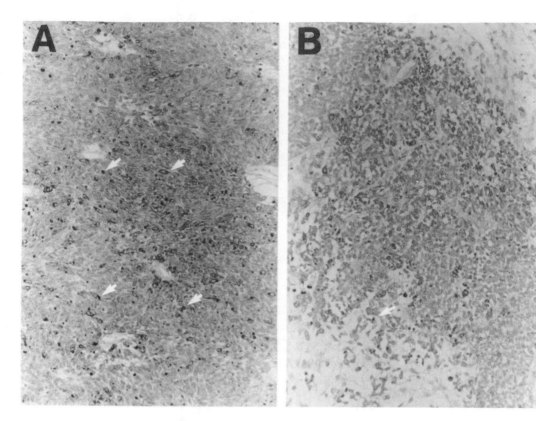

Fig. 3. *Association between p53 mutations and attenuated apoptosis in Wilms' tumor. The incidence of apoptosis occurring in a Wilms' tumor with mixed morphology is determined by the TUNEL assay, a method that detects the DNA degradation characteristic of cells undergoing apoptosis [Gavrieli et al., 1992]. The nonanaplastic region of the tumor contains a considerable number of apoptotic cells (A), whereas the frequency of apoptosis is dramatically reduced in an anaplastic region arising within the same tumor (B). Arrows identify some TUNEL-positive cells. [Courtesy of J. Pelletier, McGill University.]*

1993; Slichenmyer et al., 1993]. In fact, in many instances inactivation of p53 leads to radio resistance (see below).

An increasing amount of evidence indicates that anticancer agents do not generally target proliferating cells but instead induce apoptosis [Dive and Hickman, 1991; Kerr et al., 1994]. This implies that the cytotoxicity of many anticancer agents may be determined, in part, by events *subsequent* to the drug–target interaction. Furthermore, divergent types of cellular or metabolic damage may produce signals that converge on a common cell death program. Because apoptosis is a genetically driven process, defects in apoptosis could produce cross-resistance to many anticancer agents. Of

considerable importance, therefore, is a better understanding of factors that influence the threshold at which anticancer agents can trigger apoptosis.

IV.B. p53 Can Modulate the Cytotoxicity of Anticancer Agents

The induction of apoptosis by anticancer agents in embryonic fibroblasts and their oncogenically transformed derivatives illustrates the potential importance of apoptosis as a determinant of anticancer agent cytotoxicity. The usefulness of these systems stems largely from the observation that oncogenes that promote apoptosis in response to physiological stimuli also decrease the threshold at which anticancer agents trigger apoptosis. In many instances, efficient apoptosis requires endogenous p53 function.

For example, normal fibroblasts expressing E1A and endogenous p53 are readily killed by several chemotherapeutic agents, but E1A-expressing cells derived from p53-deficient mice are resistant (Fig. 4). Similarly, p53-expressing fibroblasts transformed by the combination of E1A and *ras* rapidly lose viability following treatment with relatively low doses of several anticancer agents (Fig. 5). The dying cells have both morphological and physiological features of apoptosis, including condensed chromatin, fragmented nuclei, and intranucleosomal DNA degradation [Lowe et al., 1993a]. p53 enhances the cytotoxicity of these agents, because p53-deficient cells remain viable and display few signs of apoptosis (Fig. 5; not shown). The differential toxicity

of these agents is not a consequence of differential proliferation, because resistant cells coexpressing E1A and *ras* that lack p53 grow *more rapidly* than sensitive cells that express p53 [Lowe et al., 1993a]. Importantly, non-transformed fibroblasts (both $p53^{+/+}$ and $p53^{-/-}$) are largely resistant to doses of anticancer agents that efficiently kill transformed cells by p53-dependent apoptosis.

Several other oncogenes appear to potentiate p53-dependent apoptosis following treatment with anticancer agents. For example, expression of either *c-myc* or activated *ras* oncogenes dramatically enhances apoptosis following DNA damage in embryonic fibroblasts, and efficient apoptosis requires normal p53 function [Tanaka et al., 1994; S.W. Lowe, unpublished data]. Similarly, overexpression of *c-myc* promotes apoptosis following treatment with cytotoxic stimuli in myeloid leuke-

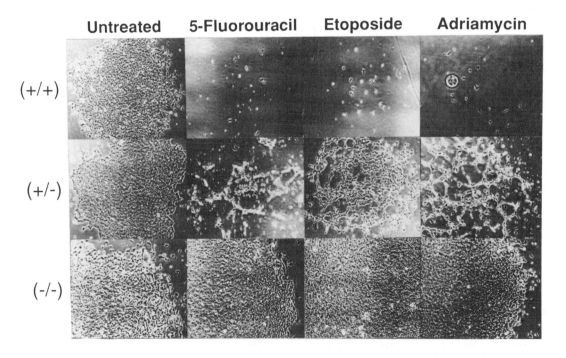

Fig. 4. *The effect of p53 on the cytotoxicity of several chemotherapeutic agents. E1A-expressing colonies derived from $p53^{+/+}$ (+/+), $p53^{+/-}$ (+/−), and $p53^{-/-}$ (−/−) mouse embryonic fibroblasts are left untreated or exposed to 1 μM 5-fluorouracil, 0.2 μM etoposide, or 0.2 μg/ml adriamycin. Representative colonies were photographed 72 (5-fluorouracil or etoposide) or 24 (adriamycin) hours after treatment. The untransfected fibroblasts are largely resistant to these concentrations of chemotherapeutic agents (not shown). [Reproduced from Lowe et al., 1993a, with permission of the publisher.]*

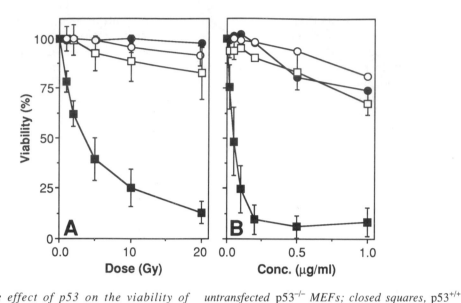

Fig. 5. *The effect of p53 on the viability of nontransformed and transformed fibroblasts treated with anticancer agents. The viability of untransfected mouse embryonic fibroblasts (MEFs) and fibroblasts transformed by E1A and ras was estimated by dye exclusion 36 hours after treatment with the indicated dose of γ-radiation or 24 hours after incubating in the indicated concentration of adriamycin. Closed circles, untransfected p53$^{+/+}$ MEFs; open circles,* untransfected p53$^{-/-}$ MEFs; closed squares, p53$^{+/+}$ cells transformed by E1A and T24 H-ras; open squares, p53$^{-/-}$ cells transformed by E1A and T24 H-ras. Radiation and adriamycin efficiently kill oncogenically transformed cells containing p53 but not the oncogenically transformed p53-deficient cells or the normal fibroblasts. [Data compiled from Lowe et al., 1993a, and reproduced from Lowe et al., 1994a, with permission of the publisher.]*

mia cell lines [Lotem and Sachs, 1993b]. p53 may be required for the *myc*-induced sensitivity, because coexpression of dominant-negative p53 mutants abolishes this effect. Rat embryonic fibroblasts transformed by the combination of papilloma virus E7 and *ras* oncogenes are much more radiosensitive then similar lines coexpressing dominant-negative p53 alleles, suggesting inactivation of p53 promotes radioresistance in these cells [Bristow et al., 1994]. It seems likely that other oncogenic stimuli will increase cellular susceptibility to apoptosis induced by cytotoxic agents. Because this sensitivity often involves p53 function, inactivation of p53 can promote cross-resistance to diverse cytotoxic agents.

IV.C. Effects of p53 on the Efficacy of Anticancer Therapy *In Vivo*

The data described above provide a provocative link between defects in an apoptotic pro-

gram and treatment resistance to anticancer agents. Nevertheless, while in vitro cell culture models provide simple systems for studying anticancer agent cytotoxicity and apoptosis, their relevance to tumor response may be complicated by *in vivo* factors such as hypoxia, nutrient supply, or drug metabolism. However, a systematic analysis of human tumor response is limited by the genetic variability that exists between tumors or tumor-derived cell lines. For example, comparisons among tumor lines that differ in their p53 status may be confounded by unknown mutations in other relevant genes.

Oncogenically transformed fibroblasts derived from normal and p53-deficient mice have been used to evaluate the role of p53 systematically in tumor response to anticancer therapies [Lowe et al., 1994b]. Specifically, coexpression of E1A and *ras* is sufficient to transform primary cells to a malignant state irrespective of their p53 status, but cells ex-

pressing wild-type p53 remain genetically susceptible to apoptosis [Lowe et al., 1994c].
Upon inoculation of these cells into immunocompromised mice, tumor growth should occur without strong selection for additional
mutations, allowing comparison of tumors that
differ primarily in their p53 status.

The response of these tumors displays a
striking dependence on p53: Tumors derived
from p53-expressing cells typically regress
following radiation or chemotherapy,
whereas p53-deficient tumors continue to
grow (Fig. 6). The regression of p53-express-
ing tumors apparently results from massive
apoptosis (Fig. 7). In contrast, p53-deficient
tumors contain only a modest increase in the
number of apoptotic cells following γ-irradiation. Several nonresponsive or recurrent
tumors derived from p53-expressing cells
acquire missense mutations in p53 at codons
mutated in human cancer. This observation
provides further evidence for the importance
of p53 in determining the therapeutic responsiveness of these tumors [Lowe et al.,
1994b].

IV.D. The Implications of Model Systems
for Human Cancer

The studies described above provide a rationale for understanding the response of human tumors to radiation and chemotherapy.
Certain oncogenic alterations, by increasing
cellular susceptibility to apoptosis (Table IA),
can provide the "therapeutic index" whereby
anticancer agents specifically target tumor
cells. Other lesions—for example, p53 mutation—can have the opposite effect, producing a general resistance to cancer therapy.
These experimental systems establish several important principles: (1) tumor sensitivity and resistance to anticancer agents can
be determined by the threshold in which anticancer agents trigger apoptosis, (2) defects
in apoptotic programs can cause tumor resistance to a variety of agents, and (3) genetic alterations that accompany malignant
transformation can determine the effectiveness of cancer therapy.

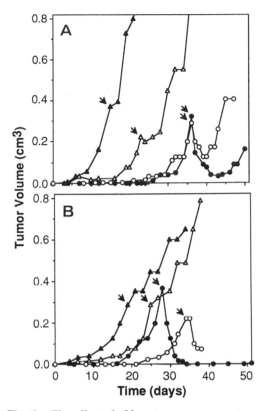

Fig. 6. *The effect of p53 on tumor response to radiation and adriamycin. Tumor volumes were estimated at various times after subcutaneous injection
of p53$^{+/+}$ (circles) or p53$^{-/-}$ (triangles) fibroblasts
transformed by E1A and ras into immunocompromised
mice. When tumors reached an appropriate volume
(indicated by the arrows), the mice were irradiated
with 7 Gy (**A**) or treated with adriamycin (10 mg/kg
total body weight) (**B**) and the tumors were monitored
for growth or regression. p53-Deficient tumors do not
respond to either form of treatment. [Data are compiled
from Lowe et al., 1994b, and reproduced with permission of the publisher.]*

V. p53 AND THE THERAPEUTIC
RESPONSE OF HUMAN TUMORS

The knowledge that p53 mutations promote
resistance to apoptosis and anticancer therapy
may ultimately lead to improvements in human oncology. However, the potential involvement of p53 and apoptosis in anticancer agent
cytotoxicity has arisen from basic research
using model systems, and the extent to which
similar mechanisms operate in human cancer

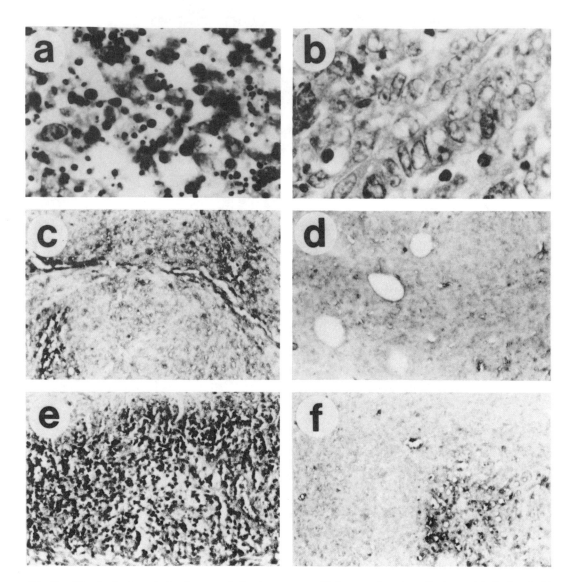

Fig. 7. *Histological analysis of tumors following γ-irradiation. Tumors derived from p53-expressing cells undergo massive apoptosis following irradiation, whereas p53-deficient tumors display only a modest increase in the number of apoptotic cells. Tumors are derived from either* p53$^{+/+}$ *(a,c,e) or* p53$^{-/-}$ *(b,d,f) transformed fibroblasts and were recovered untreated or 48 hours after irradiation with 7 Gy. (a,b): Hemotoxylin and eosin staining of γ-irradiated tumors; TUNEL staining of untreated (c,d) and irradiated (e,f) tumors. TUNEL-positive cells appear black. [Reproduced from Lowe et al., 1994a, with permission of the publisher.]*

has yet to be determined. Although the efficacy of cancer therapy in human tumors is influenced by many factors, emerging evidence is consistent with the notion that p53 is an important determinant of tumor responsiveness.

V.A. p53 Mutations and Patient Prognosis

The hypothesis that p53 participates in the therapeutic response predicts that cancer therapy would be less effective in patients harboring tumors with p53 mutations. Indeed, p53 mutations are associated with poor patient prognosis in a variety of tumor types. These include certain lymphomas [Rodriguez et al., 1991] and leukemias [Elrouby et al., 1993; Hsiao et al., 1994; Nakai et al., 1994], carcinomas of the breast [Thompson et al., 1992; Andersen et al., 1993; Thorlacius et al., 1993] and lung [Horio et al., 1993; Mitsudomi et al., 1993], soft tissue sarcomas [Cordoncardo et al., 1994], and Wilms' tumor [Bardeesy et al., 1994].

p53 proteins encoded by mutant alleles are often more stable than wild-type p53, leading to dramatic increases in p53 protein expression. In many cancers, histological detection of p53 overexpression is associated with poor patient prognosis [e.g., Harris, 1992]. However, p53 expression studies must be interpreted with caution, since it is now clear that p53 overexpression does not always correlate with p53 mutations [Hall and Lane, 1994].

Tumor types that typically suggest a favorable prognosis rarely contain p53 mutations. For example, testicular cancer is a highly curable tumor and rarely, if ever, contains p53 mutations [Heimdal et al., 1993; Pang et al., 1993; Fleischhacker et al., 1994]. p53 mutations are uncommon in acute lymphoblastic leukemia on presentation [Gaidano et al., 1991; Jonveaux and Berger, 1991], but are observed in many recurrent tumors [Hsaio et al., 1994]. Since relapsed tumor cells must have survived the original therapeutic regimen, p53 mutations may confer a survival advantage to cells undergoing cancer therapy.

Wilms' tumor is a highly curable malignancy where chemotherapeutic regimens in combination with surgical resection produce survival rates approaching 90%. However, patients harboring an aggressive subtype of Wilms' tumor (which represents 5%–10% of the total cases) often fail chemotherapy [Zuppan et al., 1988]. As discussed in Section III, these more aggressive tumors are characterized by their anaplastic morphology, which has been identified as the single most important prognostic indicator for Wilms' tumor [Zuppan et al., 1988]. In this regard, it is interesting that anaplastic tumor cells harboring p53 mutations are less prone to apoptosis compared with non-anaplastic cells containing normal p53 (Fig. 3). It seems plausible that p53 mutations reduce the susceptibility of tumor cells to apoptosis, thereby contributing to the poor response of anaplastic tumors to chemotherapy.

A direct correlation between p53 mutation and the probability of patient remission is noted in B-cell chronic lymphocytic leukemia (B-CLL). In one study, 27 of 29 patients with leukemias containing wild-type p53 entered remission following chemotherapy, whereas only 1 of 7 patients with tumors harboring p53 mutations responded [Elrouby et al., 1993]. This pattern of drug resistance is often attributed to the action of the multidrug resistance genes (mdr), which encode proteins that prevent intracellular accumulation of many anticancer agents [Chin et al., 1993]. However, drug resistance in B-CLL is not associated with increased expression of mdr-1 or mdr-3 [Elrouby et al., 1993], raising the possibility that the failure of chemotherapy in patients harboring leukemias with p53 mutations is not due to diminished drug accessibility to intracellular targets. It remains to be determined whether these leukemias are resistant to apoptosis.

V.B. Human Tumor Lines

p53 mutations are associated with resistance of some human tumor lines to anticancer agents. For example, defects in the p53-induced G_1 checkpoint are associated with increased radioresistance of Burkitt's lymphoma and ovarian carcinoma lines [O'Connor et al., 1993; Mcilwrath et al., 1994], and introduction of a dominant-negative p53 into a radio-sensitive ovarian carcinoma line makes it radioresistant [Mcilwrath et al., 1994]. More-over, adenovirus-mediated transfer of wild-type p53 into a p53-deficient lung carcinoma line enhances apoptosis following cisplatin treatment both *in vitro* and *in vivo* [Fujiwara et al., 1993]. Thus, while the analysis of human tumor lines is limited by the uncertainty of their genetic makeup, additional correlations be-tween p53 status, apoptosis, and drug cytotox-icity should prove informative.

It is important to note that not all studies associate p53 mutations with resistance to an-ticancer agents. For example, no correlation between p53 mutations and radiosensitivity is observed in head and neck squamous cell carci-noma lines [Brachman et al., 1993], and p53 mutations are associated with increased radiosensitivity in carcinogen-induced lung carcinomas in rats [Biard et al., 1994]. Stud-ies investigating the cytotoxicity of radiation or chemotherapy have traditionally used clonigenic assays to measure the long-term proliferative capacity of treated cells. How-ever, the growth of cells at clonal densities may provide additional constraints on pro-liferation that are unrelated to the induction of apoptosis or cell survival in the tumor environment. Nevertheless, these contradic-tions underscore the complexity of the thera-peutic response.

V.C. Complexities of Human Cancer

Although the analysis of null mutants allows p53 function to be studied directly, the route to p53 inactivation in human cancer is more complex. The integrity of the p53 pathway is determined by many factors, including the type of p53 mutation, the expression of modifying genes, and the tissue of tumor origin. Not sur-prisingly, the type of agent used in treatment may also have a major impact on the role of p53 in tumor responsiveness.

V.C.1. Effects of p53 mutations.
The most frequent route to p53 inactivation in hu-man tumors results from point mutations that produce altered proteins with single amino acid substitutions. p53 mutations have been identi-fied in more than 100 different codons [Greenblatt et al., 1994], implying that diverse structural alterations in p53 promote car-cinogenesis. Although these mutants can en-code proteins that inactivate wild-type p53 by a dominant-negative mechanism, missense mu-tations are often accompanied by deletion of the normal p53 allele [Fearon and Vogelstein, 1990]. This implies that many mutant proteins do not completely abolish the activity of wild-type p53.

The threshold for p53-dependent apoptosis is highly dosage dependent, and deletion of even a single p53 allele can provide a substan-tial survival advantage following an apoptotic stimulus. For example, thymocytes and bone marrow cells harboring a single p53 allele show an intermediate resistance to radiation-induced apoptosis compared with their wild-type and p53-deficient counterparts [Clarke et al., 1993; Lotem and Sachs, 1993a; Lowe et al., 1993b]. Similarly, oncogene-expressing cells derived from p53$^{+/-}$ fibroblasts are more resistant to apoptosis than isogenic cells ex-pressing two p53 alleles [Lowe et al., 1993a, 1994c]. This dosage dependence implies that the intrinsic ability of individual p53 mutants to inhibit wild-type p53 function, as well as the status of the remaining allele, may contrib-ute to heterogeneity in cellular response to an-ticancer agents.

V.C.2. Modifying genes.
Activities that suppress p53-dependent apoptosis may reduce the cytotoxicity of anticancer agents. In prin-ciple, this could result from the effects of other proteins on p53 expression or activity, from defects in the upstream signaling pathway, or from mutations in downstream targets or regu-lators. For example, the product of the *mdm-2* oncogene binds p53 and suppresses p53-me-diated transcriptional transactivation [Momand et al., 1992] and transformation suppression [Finlay, 1993]. *mdm-2* amplification occurs in many human sarcomas [Oliner et al., 1992], but whether *mdm-2* interferes with p53-depen-dent apoptosis or contributes to treatment re-sistance has not been examined. Second, the papilloma virus E6 oncoprotein contributes to

cervical cancer and inactivates p53 by promoting its degradation [Scheffner et al., 1990]. In animal models, E6 suppresses p53-dependent apoptosis [Howes et al., 1994; Pan and Griep, 1994].

Certain genes mutated in patients with ataxia telangiectasia genes may act upstream of p53, because fibroblasts from some patients are defective in p53 induction and growth arrest following γ-irradiation [Kastan et al., 1992]. However, the involvement of these genes in apoptosis has not been examined. The *bcl-2* oncogene inhibits apoptosis in p53-expressing cells and blocks apoptosis induced by p53 overexpression [Chiou et al., 1994]. Thus, *bcl-2* acts downstream of p53. Like p53 mutations, *bcl-2* activation inhibits apoptosis induced by multiple anticancer agents [Lotem and Sachs, 1993b; Miyashita and Reed, 1993; Dole et al., 1994; Maung et al., 1994]. Because many lesions can modulate p53 activities, the absence of the p53 mutation does not ensure that the p53 pathway is intact. This issue clearly complicates correlative studies investigating the association between p53 mutations and drug resistance.

V.C.3. Tissue specificity. The contribution of p53-dependent apoptosis to the efficacy of anticancer agents may be dependent on the tissue of tumor origin. This is certainly true in normal tissues, where DNA damage promotes p53-dependent apoptosis in only a few cell types (e.g., thymocytes, myeloid progenitor cells). Similarly, the ability of oncogenic mutations to "prime" normal cells for p53-dependent apoptosis may be determined by tissue-specific factors. The extent to which different tissues become sensitized to p53-dependent apoptosis during tumorigenesis may ultimately determine the clinical importance of p53 mutations in treatment resistance.

V.C.4. p53-dependent and -independent apoptotic programs. Presently, only a limited number of anticancer agents have been used to investigate the involvement of p53 in modulating the efficacy of cancer therapy.

Many of these agents are thought to damage DNA directly or indirectly, an event that may be required to elicit a p53 response. However, the presence of p53-independent apoptotic programs is well established, and at least some anticancer agents may promote apoptosis in the absence of functional p53. Some of the anticancer agents that induce p53-dependent apoptosis at low concentrations can induce apoptosis in p53-deficient cells if provided in sufficient quantities [Lowe et al., 1993a]. Presumably, the extent to which elevated doses of "p53-dependent" drugs are tolerated by cancer patients determines their effectiveness in treating tumors harboring p53 mutations.

Most therapeutic regimens utilize combinations of different anticancer agents. The possibility that these agents promote apoptosis through different pathways may complicate retrospective studies investigating the role of p53 mutations in drug resistance. At the same time, this complexity provides one of our greatest hopes in improving cancer therapy, since p53-dependent and -independent agents should prove effective when used in combination.

V.D. Side-Effects of Chemotherapy

If anticancer agents simply trigger apoptosis in susceptible cells, the threshold of apoptosis may provide the physiological basis for the therapeutic index differentiating normal cells from tumor cells. It is intriguing that certain oncogenes can alter this threshold, making preneoplastic or neoplastic cells more susceptible to apoptosis. Several cell types, including many hematopoietic cells and the epithelial stem cells of the intestinal crypts, normally have a low threshold for apoptosis. In these tissues, apoptosis is readily induced by therapeutic doses of radiation or chemotherapy, an effect that also requires functional p53 (see Section II). Ironically, the presence of normal p53 in these tissues may be deleterious for patients undergoing cancer therapy.

VI. MECHANISM OF APOPTOSIS IN ONCOGENE-EXPRESSING CELLS

Although much is understood about the mechanism whereby p53 facilitates growth arrest [Marx, 1993], considerably less is known about how p53 promotes apoptosis. p53 protein levels and transcriptional activity increase following DNA damage [Kastan et al., 1992], and the subsequent cell cycle arrest may result from p53-mediated activation of p21$^{CiP1/WAF1}$ expression, a cyclin-dependent kinase inhibitor [Harper et al., 1993; Xiong et al., 1993; El Deiry et al., 1994]. In contrast, transcriptional activation of genes by p53 may be unnecessary for apoptosis, because actinomycin D blocks transcription but not apoptosis following p53 overexpression [Caelles et al., 1994]. This conclusion is supported by the observation that p19^{E1B} prevents apoptosis following p53 overexpression but does not inhibit p53-mediated transcriptional activation of reporter genes [Shen and Shenk, 1994]. In contrast, p19^{E1B} blocks p53-mediated transcriptional repression, suggesting that apoptosis involves a fundamentally distinct mechanism from that involved in p53-dependent growth arrest [Shen and Shenk, 1994]. Nevertheless, it remains possible that apoptosis induced by p53 overexpression is distinct from that associated with endogenous p53.

VI.A. The Role of Oncogenes

Oncogenic stimuli can alter the function of p53 from inducing growth arrest to promoting apoptosis. For example, primary mouse embryonic fibroblasts respond to low levels of γ-radiation or other DNA-damaging agents by activating a p53-dependent checkpoint that leads to cell cycle arrest. However, the same doses trigger p53-dependent apoptosis in fibroblasts expressing oncogenes such as c-myc or E1A (Fig. 5). Of fundamental importance, therefore, may be an understanding of how certain oncogenes activate the apoptotic program in otherwise resistant cells.

Although oncogenes such as c-myc, E1A, and ras dramatically increase the susceptibility of fibroblasts to apoptosis, the expression of these oncogenes is not sufficient for apoptosis. It has been suggested that these cells are simply "primed" for apoptosis; they have access to the cellular machinery required for the execution of apoptosis, whereas their normal counterparts do not [Wyllie, 1993]. The ultimate decision to proliferate or die, then, is determined by other factors. Cells primed for apoptosis can continue to proliferate under favorable circumstances, but initiate apoptosis following DNA damage or by encountering physiological conditions that normally limit proliferation (e.g., growth factor deprivation).

The priming of cells for apoptosis is tightly linked to deregulated proliferation. For instance, c-myc, activated ras oncogenes, papilloma virus E7, and adenovirus E1A each promote S phase entry and apoptosis (Table I). Moreover, the regions of E1A and c-myc that promote apoptosis are indistinguishable from those that activate proliferation [White et al., 1991; Evan et al., 1992]. It is perhaps significant that the E1A and E7 oncogenes interfere with the p53-dependent G$_1$ cell cycle checkpoint [Lowe et al., 1993a; Demers et al., 1994; Hickman et al., 1994; Slebos et al., 1994]. For example, p53-expressing fibroblasts arrest in G$_1$ following γ-irradiation, but similar cells expressing E1A continue to proliferate into S phase, even as they die by apoptosis [Lowe et al., 1993a]. c-myc may also interfere with p53-mediated cell cycle arrest, because cells overexpressing c-myc proliferate in the presence of elevated p53 [Hermeking and Eick, 1994]. ras oncogenes promote genomic instability in cells that express wild-type p53, suggesting that constitutive ras activation, like p53 mutation, may also interfere with cell cycle checkpoints [Stark, 1993; Denko et al., 1994].

It is not clear whether p53 participates in priming cells for apoptosis. E1A expression leads to increases in p53 protein stability, even under circumstances where cells proliferate without significant apoptosis [Lowe and Ruley, 1993]. Overexpression of c-myc may also stabilize p53 [Hermeking and Eick, 1994]. Since normal cells express low levels of p53 and are resistant to apoptosis, increases in p53 stability may help to sensitize cells to apoptotic stimuli. Nevertheless, it remains to be determined if other oncogenes that promote p53-dependent apoptosis also stabilize p53.

VI.B. p53 Triggers Apoptosis

Apoptosis is *triggered* by stimuli that cause susceptible cells to *execute* the apoptotic program [Wyllie, 1993]. Therefore, the trigger (directly or indirectly) activates the apoptotic machinery to cause cell shrinkage, chromatin condensation, DNA fragmentation, and ultimately cell death. In the absence of priming by oncogenes, the trigger produces only growth arrest. For instance, low serum concentrations, high cell density, and DNA-damaging agents induce growth arrest in normal cells but trigger apoptosis in cells expressing E1A or *c-myc* [Bissonnette et al., 1992; Rao et al., 1992; Lotem and Sachs, 1993b; Lowe et al., 1993a, 1994c].

In oncogene-expressing cells, p53 functions primarily as a trigger for apoptosis. This is most apparent in p53-deficient tumor lines, where re-introduction of high levels of wild-type p53 typically induces apoptosis directly [e.g., Yonish-Rouach et al., 1991]. In contrast, most normal cells arrest following p53 overexpression [e.g., Michalovitz et al., 1990], implying that the tumor lines are primed for apoptosis prior to the introduction of p53. p53 does not participate in the execution phase of apoptosis, because apoptosis can occur in the absence of p53. For example, glucocorticoids readily induce apoptosis in p53-deficient thymocytes [Clarke et al., 1993; Lotem and Sachs, 1993a; Lowe et al., 1993b], and etoposide induces apoptosis in the p53-deficient HL-60 myeloid leukemia line [Kaufmann, 1989]. It appears that the machinery involved in the execution of apoptosis is unaffected by p53 loss, but is less likely to be engaged in the absence of biologically active p53.

VI.C. Upstream Events and Downstream Effectors

The physiological stimulus thought to signal engagement of the p53 pathway is DNA damage [Lu and Lane, 1993; Nelson and Kastan, 1994]. Clearly, DNA damage can trigger apoptosis in certain normal and oncogenically transformed cells, but it is not clear whether this is the only stimulus capable of activating the program. In several instances, p53-dependent apoptosis occurs under circumstances not known to damage DNA. These situations include apoptosis induced by serum deprivation in cells constitutively expressing E1A, *ras* oncogenes, or *c-myc* [Hermeking and Eick, 1994; Lowe et al., 1994c; Tanaka et al., 1994]; in aberrantly proliferating lens cells of *Rb*-deficient mice [Morgenbesser et al., 1994]; and in SV40 T-antigen–induced tumors of the choroid plexus [Symonds et al., 1994]. However, forced cell proliferation and unscheduled entry into S phase entry may in itself produce DNA damage. Alternatively, since the proliferative life span of cells may be determined in part by shortening of terminal chromosome sequences known as telomeres [Greider, 1994], telomere loss resulting from excessive cell division may also produce events similar to DNA damage and thereby trigger a p53 response.

The downstream effectors of p53 in the apoptotic program are presently unknown. In principle, p53 may interact with the apoptotic machinery directly, transcriptionally regulate apoptotic genes, or indirectly produce a cellular environment that facilitates apoptosis. Intriguing candidates for regulation by p53 during apoptosis are the products of the *bcl-2* gene family. As discussed in Section V, *bcl-2* acts downstream of p53 to inhibit apoptosis. *bax* is structurally related to *bcl-2*, but functionally promotes apoptosis [Oltvai et al., 1993]. *bcl-2* and *bax* form homodimeric complexes or can heterodimerize with each other, and the ratio of *bcl-2* to *bax* expression may ultimately determine cell survival following an apoptotic stimulus [Oltvai et al., 1993]. Decreases in *bcl-2* and increases in *bax* mRNA levels are associated with apoptosis induced by p53 overexpression in M1 leukemia cells, suggesting that p53 regulates apoptosis by influencing the Bcl-2/Bax ratio [Miyashita et al., 1994a,b; Selvakumaran et al., 1994]. However, p53-deficient cells transformed by E1A and *ras* do not overexpress *bcl-2*, and the *bcl-2* message does not decrease during radiation-induced apoptosis of mouse thymocytes [S.W. Lowe, unpublished data; E. Schmitt and T. Jacks, personal communication]. This raises the possibility that the p53-induced changes in *bcl-2* and *bax* expression are a consequence of nonphysiological levels of p53.

VII. FUTURE PERSPECTIVES

The participation of p53 in apoptosis has important implications for the development of neoplastic disease. In tissues that are normally prone to p53-dependent apoptosis, inactivation of p53 may promote tumor growth by allowing inappropriate cell survival following DNA damage. In other tissues, p53 mutation may enhance the progression of neoplastic or preneoplastic lesions by eliminating an apoptotic program that keeps tumor growth in check. Since p53 also participates in a cell cycle checkpoint that arrests cell growth in response to DNA damage [Kastan et al., 1992], several mechanisms may account for the high frequency of p53 mutations occurring in human tumors. Clearly, a more detailed understanding of these processes will be essential for unraveling the complexities of human cancer.

The knowledge that p53 can influence the cytotoxicity of anticancer agents may ultimately lead to improvements in cancer therapy. First, the identification of p53 mutations in certain tumor types may become an important factor in treatment decisions. This information may allow oncologists to determine the appropriateness of cancer therapy or to identify the most effective treatment regimen. Second, the re-introduction of normal p53 function into tumors harboring p53 mutations may sensitize these cells to apoptosis induced by radiation and chemotherapy, an approach that has proved successful for cisplatin in a lung carcinoma line [Fujiwara et al., 1994]. Since many tumors express mutant p53 proteins, drugs that restore the wild-type conformation to these mutants

may provide an alternative to gene therapy. Nevertheless, our knowledge of p53 and apoptosis is in its infancy, and much more needs to be understood before this information can contribute to advances in human oncology.

Most anticancer agents presently in use have been identified by empirical screens based on their ability specifically to kill tumor cells but are ineffective against many human cancers. Indeed, our limited understanding of why these agents work or fail has been a major impediment to the design of more effective therapies. Regardless of whether p53 is identified as a significant determinant of treatment efficacy in human cancer, the studies of p53 in model systems establish the importance of apoptosis in determining the cytotoxicity of anticancer agents. A more detailed understanding of this process should allow known drugs to be used more effectively and facilitate the rational design of better anticancer agents.

ACKNOWLEDGMENTS

The authors thank M. Maxwell for editorial advice and M. McCurrach, T. Jacks, and D.E. Housman for helpful and enlightening discussions.

REFERENCES

Andersen TI, Holm R, Nesland JM, Heimdal KR, Ottestad L, Borresen AL (1993): Prognostic significance of TP53 alterations in breast carcinoma. Br J Cancer 68:540–548.
Bardeesy N, Beckwith JB, Pelletier J (1995): Clonal expansion and attenuated apoptosis in Wilms' tumor are associated with p53 gene mutations. Cancer Res 55:215–219.
Bardeesy N, Falkoff D, Petruzzi MJ, Nowak N, Zabel B, Adam M, Aguiar MC, Grundy P, Shows T, Pelletier J (1994): Anaplastic Wilms' tumour, a subtype displaying poor prognosis, harbours p53 gene mutations. Nature Genet 7:91–97.
Biard DS, Martin M, Rhun YL, Duthu A, Lefaix JL, May E, May P (1994): Concomitant p53 gene mutation and increased radiosensitivity in rat lung embryo epithelial cells during neoplastic development. Cancer Res 54:3361–3364.
Bissonnette RP, Echeverri F, Mahboubi A, Green DR (1992): Apoptotic cell death induced by c-myc is inhibited by bcl-2. Nature 359:552–554.

Blount PL, Galipeau PC, Sanchez CA, Neshat K, Levine DS, Yin J, Suzuki H, Abraham JM, Meltzer SJ, Reid BJ (1994): 17p allelic losses in diploid cells of patients with Barrett's esophagus who develop aneuploidy. Cancer Res 54:2292–2295.

Blount PL, Meltzer SJ, Yin J, Huang Y, Krasna MJ, Reid BJ (1993): Clonal ordering of 17p and 5q allelic losses in Barrett dysplasia and adenocarcinoma. Proc Natl Acad Sci USA 90:3221–3225.

Bookstein R, MacGrogan D, Hilsenbeck SG, Sharkey F, Allred DC (1993): p53 is mutated in a subset of advanced-stage prostate cancers. Cancer Res 53: 3369–3373.

Brachman DG, Beckett M, Graves D, Haraf D, Vokes E, Weichselbaum RR (1993): p53 mutation does not correlate with radiosensitivity in 24 head and neck cancer cell lines. Cancer Res 53:3667–3369.

Bristow RG, Jang A, Peacock J, Chung S, Benchimol S, Hill RP (1994): Mutant p53 increases radioresistance in rat embryo fibroblasts simultaneously transfected with HPV16-E7 and/or activated H-ras. Oncogene 9:1527–1536.

Burns PA, Kemp CJ, Gannon JV, Lane DP, Bremner R, Balmain A (1991): Loss of heterozygosity and mutational alterations of the p53 gene in skin tumours of interspecific hybrid mice. Oncogene 6:2363–2369.

Caelles C, Heimberg A, Karin M (1994): p53-dependent apoptosis in the absence of transcriptional activation of p53-target genes. Nature 370:220–223.

Chabner BA, Myers CE (1989): Clinical pharmacology of cancer chemotherapy. In DeVita VTJ, Hellman S, Rosenberg SA (eds): Cancer: Principles and Practice of Oncology. Philadelphia: J.B. Lippencott, pp 349–395.

Chin KV, Pastan I, Gottesman MM (1993): Function and regulation of the human multidrug resistance gene. Adv Cancer Res 60:157–80.

Chiou SK, Rao L, White E (1994): Bcl-2 blocks p53-dependent apoptosis. Mol Cell Biol 14:2556–2563.

Cho Y, Gorina S, Jeffrey PD, Pavletich NP (1994): Crystal structure of a p53 tumor suppressor–DNA complex: Understanding tumorigenic mutations. Science 265:346–355.

Clarke AR, Gledhill S, Hooper ML, Bird CC, Wyllie AH (1994): p53 dependence or early apoptotic and proliferative responses within the mouse intestinal epithelium following gamma-irradiation. Oncogene 9:1767–1773.

Clarke AR, Purdie CA, Harrison DJ, Morris RG, Bird CC, Hooper ML, Wyllie AH (1993): Thymocyte apoptosis induced by p53-dependent and independent pathways. Nature 362:849–852.

Cordoncardo C, Latres E, Drobnjak M, Oliva MR, Pollack D, Woodruff JM, Marechal V, Chen JD, Brennan MF, Levine AJ (1994): Molecular abnormalities of mdm2 and p53 genes in adult soft tissue sarcomas. Cancer Res 54:794–799.

Debbas M, White E (1993): Wild-type p53 mediates apoptosis by E1A, which is inhibited by E1B. Genes Dev 7:546–554.

Dedera DA, Walter EK, LeBrun DP, Sen-Majumdar A, Stevens ME, Barsh GS, Cleary ML (1993): Chimeric homeobox gene E2A-PBX1 induces proliferation, apoptosis, and malignant lymphomas in transgenic mice. Cell 74:833–843.

Demers GW, Foster SA, Halbert CL, Galloway DA (1994): Growth arrest by induction of p53 in DNA damaged keratinocytes is bypassed by human papillomavirus 16 E7. Proc Natl Acad Sci USA 91:4382–4386.

Denko NC, Giaccia AJ, Stringer JR, Stambrook PJ (1994): The human Ha-ras oncogene induces genomic instability in murine fibroblasts within one cell cycle. Proc Natl Acad Sci USA 91:5124–5128.

Dittmer D, Pati S, Zambetti G, Chu S, Teresky AK, Moore M, Finlay C, Levine AJ (1993): Gain of function mutations in p53. Nature Genet 4:42–46.

Dive C, Hickman JA (1991): Drug-target interactions: Only the first step in the commitment to a programmed cell death? Br J Cancer 64:192–196.

Dole M, Nunez G, Merchant AK, Maybaum J, Rode CK, Bloch CA, Castle VP (1994): Bcl-2 inhibits chemotherapy-induced apoptosis in neuroblastoma. Cancer Res 54:3253–3259.

Donehower LA, Harvey M, Slagle BL, McArthur MJ, Montgomery CA, Butel JA, Bradley A (1992): Mice deficient for p53 are developmentally normal but susceptible to spontaneous tumours. Nature 356:215–220.

Dutta A, Ruppert JM, Aster JC, Winchester E (1993): Inhibition of DNA replication factor RPA by p53. Nature 365:79–82.

El Deiry WS, Harper JW, Oconnor PM, Velculescu VE, Canman CE, Jackman J, Pietenpol JA, Burrell M, Hill DE, Wang YS, Wiman KG, Mercer WE, Kastan MB, Kohn KW, Elledge SJ, Kinzler KW, Vogelstein B (1994): Waf1/Cip1 is induced in p53-mediated G1 arrest and apoptosis. Cancer Res 54:1169–1174.

El Deiry WS, Kern SE, Pietenpol JA, Kinzler KW, Vogelstein B (1992): Definition of a consensus binding site for p53. Nature Genet 1:45–49.

Elrouby S, Thomas A, Costin D, Rosenberg CR, Potmesil M, Silber R, Newcomb EW (1993): p53 gene mutation in B-cell chronic lymphocytic leukemia is associated with drug resistance and is independent of MDR1/MDR3 gene expression. Blood 82:3452–3459.

Evan GI, Wyllie AH, Gilbert CS, Littlewood TD, Land H, Brooks M, Waters C, Penn LZ, Hancock DC (1992): Induction of apoptosis in fibroblasts by c-myc protein. Cell 69:119–128.

Fanidi A, Harrington EA, Evan GI (1992): Cooperative interaction between c-myc and bcl-2 proto-oncogenes. Nature 359:554–556.

Farmer G, Bargonetti J, Zhu H, Friedman P, Prywes R, Prives C (1992): Wild-type p53 activates transcription in vitro. Nature 358:83–86.

Fearon ER, Vogelstein B (1990): A genetic model for colorectal tumorigeneisis. Cell 61:759–767.

Finlay CA (1993): The mdm-2 oncogene can overcome wild-type p53 suppression of transformed cell growth. Mol Cell Biol 13:301–306.

Fleischhacker M, Strohmeyer T, Imai Y, Slamon DJ, Koeffler HP (1994): Mutations of the p53 gene are not detectable in human testicular tumors. Modern Pathol 7:435–439.

Fujiwara T, Grimm EA, Mukhopadhyay T, Cai DW, Owenschaub LB, Roth JA (1993): A retroviral wild-type p53 expression vector penetrates human lung cancer spheroids and inhibits growth by inducing apoptosis. Cancer Res 53:4129–4133.

Fujiwara T, Grimm EA, Mukhopadhyay T, Zhang WW, Owenschaub LB, Roth JA (1994): Induction of chemosensitivity in human lung cancer cells in vivo by adenovirus-mediated transfer of the wild-type p53 gene. Cancer Res 54:2287–2291.

Gaidano G, Ballerini P, Gong JZ, Inghirami G, Neri A, Newcomb EW, Magrath IT, Knowles DM, Dalla-Favera R (1991): p53 mutations in human lymphoid malignancies: Association with Burkitt lymphoma and chronic lymphocytic leukemia. Proc Natl Acad Sci USA 88:5413–5417.

Gavrieli Y, Sherman Y, BenSasson SA (1992): Identification of programmed cell death in situ via specific labeling of nuclear DNA fragmentation. J Cell Biol 119:493–501.

Graeber TG, Peterson JF, Tsai M, Monica K, Fornace AJ, Giaccia AJ (1994): Hypoxia induces accumulation of p53 protein, but activation of a G_1-phase checkpoint by low-oxygen conditions is independent of p53 status. Mol Cell Biol 14:6264–6277.

Greenblatt MS, Bennett WP, Hollstein M, Harris CC (1994): Mutations in the p53 tumor suppressor gene: clues to cancer etiology and molecular pathogenesis. Cancer Res 54:4855–4878.

Greider CW (1994): Mammalian telomere dynamics: Healing, fragmentation, shortening and stabilization. Curr Opin Genet Dev 4:203–211.

Hall PA, Lane DP (1994): P53 in tumour pathology—can we trust immunohistochemistry—revisited. J Pathol 172:1–4.

Harper JW, Adami GR, Wei N, Khandan K, Elledge SJ (1993): The p21 cdk-interacting protein Cip1 is a potent inhibitor of G_1 cyclin-dependent kinases. Cell 75:805–816.

Harrington EA, Fanidi A, Evan GI (1994): Oncogenes and cell death. Curr Opin Genet Dev 4:120–129.

Harris AL (1992): p53 expression in human breast cancer. Adv Cancer Res 59:69-88.

Harris CC, Hollstein M (1993): Clinical Implications of the p53 tumor suppressor gene. N Eng J Med 329:1318–1327.

Harvey DM, Levine AJ (1991): p53 alteration is a common event in the spontaneous immortalization of primary BALB/c murine embryo fibroblasts. Genes Dev 5:2375–2385.

Harvey M, Sands AT, Weiss RS, Hegi ME, Wiseman RW, Pantiazis P, Giovanella BC, Tainsky MA, Bradley A, Donehower LA (1993): In vitro growth characteristics of embryo fibroblasts isolated from p53-deficient mice. Oncogene 1993:2457–2467.

Heimdal K, Lothe RA, Lystad S, Holm R, Fossa SD, Borresen AL (1993): No germline TP53 mutations detected in familial and bilateral testicular cancer. Genes Chromosom Cancer 6:92–97.

Hermeking H, Eick D (1994): Mediation of c-myc induced apoptosis by p53. Science 265:2091–2093.

Hickman ES, Picksley SM, Vousden KH (1994): Cells expressing HPV16 E7 continue cell cycle progression following DNA damage induced p53 activation. Oncogene 9:2177–2181.

Hinds P, Finlay C, Levine AJ (1989): Mutation is required to activate the p53 gene for cooperation with the ras oncogene and transformation. J Virol 63:739–746.

Horio Y, Takahashi T, Kuroishi T, Hibi K, Suyama M, Niimi T, Shimokata K, Yamakawa K, Nakamura Y, Ueda R, Takahashi T (1993): Prognostic significance of p53 mutations and 3p deletions in primary resected non-small cell lung cancer. Cancer Res 53:1–4.

Howes KA, Ransom LN, Papermaster DS, Lasudry JGH, Albert DM, Windle JJ (1994): Apoptosis or retinoblastoma: Alternative fates of photoreceptors expressing the HPV-16 E7 gene in the presence or absence of p53. Gene Dev 8:1300–1310.

Hsiao MH, Yu AL, Yeargin J, Ku D, Haas M (1994): Nonhereditary p53 mutations in T-cell acute lymphoblastic leukemia are associated with the relapse phase. Blood 83:2922–2930.

Ichikawa A, Hotta T, Takagi N, Tsushita K, Kinoshita T, Nagai H, Murakami Y, Hayashi K, Saito H (1992): Mutations of p53 gene and their relation to disease progression in B-cell lymphoma. Blood 79:2701–2707.

Jacks T, Remington L, Williams BO, Schmitt EM, Halachmi S, Bronson RT, Weinberg RA (1994): Tumor spectrum analysis in p53-mutant mice. Curr Biol 4:1–7.

Jenkins JR, Rudge K, Currie GA (1984): Cellular immortalization by a cDNA clone encoding the transformation-associated phosphoprotein p53. Nature 312:651–654.

Jonveaux P, Berger R (1991) Infrequent mutations in the p53 gene in primary human T-cell acute lymphoblastic leukemia. Leukemia 5:839–840.

Kastan MB, Onyekwere O, Sidransky D, Vogelstein B, Craig RW (1991): Participation of p53 protein in the cellular response to DNA damage. Cancer Res 51:6304–6311.

Kastan MB, Zhan Q, el-Deiry WS, Carrier F, Jacks T, Walsh WV, Plunkett BS, Vogelstein B, Fornace A Jr (1992): A mammalian cell cycle checkpoint pathway utilizing p53 and GADD45 is defective in ataxia-telangiectasia. Cell 71:587–597.

Kaufmann SH (1989): Induction of endonucleolytic DNA cleavage in human acute myelogenous leukemia cells by etoposide, camptothecin, and other cytotoxic anticancer drugs: A cautionary note. Cancer Res 49:5870–5878.

Kemp CJ, Donehower LA, Bradley A, Balmain A (1993): Reduction of p53 gene dosage does not increase initiation or promotion but enhances malignant progression of chemically induced skin tumors. Cell 74:813–822.

Kern SE, Pietenpol JA, Thiagalingam S, Seymour A, Kinzler KW, Vogelstein B (1992): Oncogenic forms of p53 inhibit p53-regulated gene expression. Science 256:827–830.

Kerr JF, Wyllie AH, Currie AR (1972): Apoptosis: a basic biological phenomenon with wide-ranging implications in tissue kinetics. Br J Cancer 26:239–257.

Kerr JFR, Winterford CM, Harmon BV (1994): Apoptosis—Its significance in cancer and cancer therapy. Cancer 73:2013–2026.

Kihana T, Tsuda H, Teshima S, Okada S, Matsuura S, Hirohashi S (1992): High incidence of p53 gene mutation in human ovarian cancer and its association with nuclear accumulation of p53 protein and tumor DNA aneuploidy. Jpn J Cancer Res 83:978–984.

Lane DP (1992): p53, guardian of the genome. Nature 358:15–16.

Livingstone LR, White A, Sprouse J, Livanos E, Jacks T, Tlsty TD (1992): Altered cell cycle arrest and gene amplification potential accompany loss of wild-type p53. Cell 70:923–935.

Loeb LA (1991): Mutator phenotype may be required for multistage carcinogenesis. Cancer Res 51:3075–3079.

Lotem J, Sachs L (1993a): Hematopoietic cells from mice deficient in wild-type p53 are more resistant to induction of apoptosis by some agents. Blood 82:1092–1096.

Lotem J, Sachs L (1993b): Regulation by bcl-2, c-myc, and p53 of susceptibility to induction of apoptosis by heat shock and cancer chemotherapy compounds in differentiation-competent and -defective myeloid leukemic cells. Cell Growth Differ 4:41–47.

Lowe SW, Bodis S, Bardeesy N, McClatchey A, Remington L, Ruley HE, Fisher DE, Jacks T, Pelletier J, Housman DE (1994a): Apoptosis and the prognostic significance of p53 mutation. Cold Spring Harbor Symp Quant Biol 59:419–436.

Lowe SW, Bodis S, McClatchey A, Remington L, Ruley HE, Fisher D, Housman DE, Jacks T (1994b): p53 status and the efficacy of cancer therapy in vivo. Science 266:807–810.

Lowe SW, Jacks T, Housman DE, Ruley HE (1994c): Abrogation of oncogene-associated apoptosis allows transformation of p53-deficient cells. Proc Natl Acad Sci USA 91:2026–2030.

Lowe SW, Ruley HE (1993): Stabilization of the p53 tumor suppressor is induced by adenovirus E1A and accompanies apoptosis. Genes Dev 7:535–545.

Lowe SW, Ruley HE, Jacks T, Housman DE (1993a): p53-dependent apoptosis modulates the cytotoxicity of anticancer agents. Cell 74:954–967.

Lowe SW, Schmitt EM, Smith SW, Osborne BA, Jacks T (1993b): p53 is required for radiation-induced apoptosis in mouse thymocytes. Nature 362:847–849.

Lu X, Lane DP (1993): Differential induction of transcriptionally active p53 following UV or ionizing radiation: defects in chromosome instability syndromes? Cell 75:765–778.

Mack DH, Vartikar J, Pipas JM, Laimins LA (1993): Specific repression of TATA-mediated but not initiator-mediated transcription by wild-type p53. Nature 363:281–283.

Malkin D, Li FP, Strong LC, Fraumeni J J. F., Nelson CE, Kim DH, Kassel J, Gryka MA, Bischoff FZ, Tainsky MA, Friend SH (1990): Germ line p53 mutations in a familial syndrome of breast cancer, sarcomas, and other neoplasms. Science 250:1233–1238.

Maltzman W, Czyzyk L (1984): UV irradiation stimulates levels of p53 cellular tumor antigen in nontransformed mouse cells. Mol Cell Biol 4:1689–1694.

Martinez J, Georgoff I, Martinez J, Levine AJ (1991): Cellular localization and cell cycle regulation by a temperature-sensitive p53 protein. Genes Dev 5:151–159.

Marx J (1993): How p53 suppresses cell growth. Science 262:1644–1645.

Maung ZT, Madean FR, Reid MM, Pearson ADJ, Proctor SJ, Hamilton PJ, Hall AG (1994): The relationship between bcl-2 expression and response to chemotherapy in acute leukaemia. Br J Haematol 88:105–109.

Mcilwrath AJ, Vasey PA, Ross GM, Brown R (1994): Cell cycle arrests and radiosensitivity of human tumor cell lines: dependence on wild-type p53 for radiosensitivity. Cancer Res 54:3718–3722.

Mercer WE, Shields MT, Lin D, Appella E, Ullrich SJ (1991): Growth suppression induced by wild-type p53 protein is accompanied by selective down-regulation of proliferating-cell nuclear antigen expression. Proc Natl Acad Sci USA 88:1958–1962.

Merritt AJ, Potten CS, Kemp CJ, Hickman JA, Balmain A, Lane DP, Hall PA (1994): The role of p53 in spontaneous and radiation-induced apoptosis in the gastrointestinal tract of normal and p53-deficient mice. Cancer Res 54:614–617.

Michalovitz D, Halevy O, Oren M (1990): Conditional inhibition of transformation and of cell proliferation by a temperature-sensitive mutant of p53. Cell 62:671–680.

Mitsudomi T, Oyama T, Kusano T, Osaki T, Nakanishi R, Shirakusa T (1993): Mutations of the p53 gene as a predictor of poor prognosis in patients with non-small-cell lung cancer. J Nat Cancer Inst 85:2018–2023.

Miyashita T, Harigai M, Hanada M, Reed JC (1994a): Identification of a p53-dependent negative response element in the bcl-2 gene. Cancer Res 54:3131–3135.

Miyashita T, Krajewski S, Krajewska M, Wang HG, Lin HK, Liebermann DA, Hoffman B, Reed JC (1994b): Tumor suppressor p53 is a regulator of bcl-2 and bax gene expression *in vitro* and *in vivo*. Oncogene 9:1799–1805.

Miyashita T, Reed JC (1993): Bcl-2 oncoprotein blocks chemotherapy-induced apoptosis in a human leukemia cell line. Blood 81:151–157.

Momand J, Zambetti GP, Olson DC, George D, Levine AJ (1992): The mdm-2 oncogene product forms a complex with the p53 protein and inhibits p53-mediated transactivation. Cell 69:1237–1245.

Morgenbesser SD, Williams BO, Jacks T, DePinho RA (1994): p53-dependent apoptosis produced by Rb-deficiency in the developing mouse lens. Nature 371:72–74.

Munger K, Werness BA, Dyson N, Phelps WC, Howley PM (1989): The E6 and E7 genes of the papillomavirus type 16 together are necessary and sufficient for transformation of primary human keratinocytes. J Virol 63:4417–4421.

Nakai H, Misawa S, Taniwaki M, Horiike S, Takashima T, Seriu T, Nakagawa H, Fujii H, Shimazaki C, Maruo N, Akaogi T, Uike N, Abe T, Kashima K (1994): Prognostic significance of loss of a chromosome 17p and p53 gene mutations in blast crisis of chronic myelogenous leukaemia. Br J Haematol 87:425–427.

Nelson WG, Kastan MB (1994): DNA strand breaks: The DNA template alterations that trigger p53-dependent DNA damage response pathways. Mol Cell Biol 14:1815–1823.

O'Connor PM, Jackman J, Jondle D, Bhatia K, Magrath I, Kohn KW (1993): Role of the p53 tumor suppressor gene in cell cycle arrest and radiosensitivity of Burkitt's lymphoma cell lines. Cancer Res 53:4776–4780.

Oliner JD, Kinzler KW, Meltzer PS, George DL, Vogelstein B (1992): Amplification of a gene encoding a p53-associated protein in human sarcomas. Nature 358:80–83.

Oltvai ZN, Milliman CL, Korsmeyer SJ (1993): Bcl-2 heterodimerizes *in vivo* with a conserved homolog, Bax, that accelerates programed cell death. Cell 74:609–619.

Pan HC, Griep AE (1994): Altered cell cyde regulation in the lens of HPV-16 E6 or E7 transgenic mice: Implications for tumor suppressor gene function in development. Gene Dev 8:1285–1299.

Peng HQ, Hogg D, Malkin D, Bailey D, Gallie BL, Bulbul M, Jewett M, Buchanan J, Goss PE (1993): Mutations of the p53 gene do not occur in testis cancer. Cancer Res 53:3574–3578.

Prives C (1994): How loops, beta sheets, and alpha helices help us to understand p53. Cell 78:543–546.

Purdie CA, Harrison DJ, Peter A, Dobbie L, White S, Howie SEM, Salter DM, Bird CC, Wyllie AH, Hooper ML, Clarke AR (1994): Tumour incidence, spectrum and ploidy in mice with a large deletion in the p53 gene. Oncogene 9:603–609.

Rao L, Debbas M, Sabbatini P, Hockenbery D, Korsmeyer S, White E (1992): The adenovirus E1A proteins induce apoptosis which is inhibited by the E1B 19K and Bcl-2 proteins. Proc Natl Acad Sci USA 89:7742–7746.

Rodriguez MA, Ford RJ, Goodacre A, Selvanayagam P, Cabanillas F, Deisseroth AB (1991): Chromosome 17- and p53 changes in lymphoma. Br J Haematol 79:575–582.

Ruley HE (1990): Transforming collaborations between *ras* and nuclear oncogenes. Cancer Cells 2:258–268.

Sarnow P, Ho YS, Williams J, Levine AJ (1982): Adenovirus E1b-58kd tumor antigen and SV40 large tumor antigen are physically associated with the same 54 kd cellular protein in transformed cells. Cell 28:387–394.

Scheffner M, Werness BA, Huibregtse JM, Levine AJ, Howley PM (1990): The E6 oncoprotein encoded by human papillomavirus types 16 and 18 promotes the degradation of p53. Cell 63:1129–1136.

Selvakumaran M, Lin HK, Miyashita T, Wang HG, Krajewski S, Reed JC, Hoffman B, Liebermann D (1994): Immediate early up-regulation of bax expression by p53 but not TGF beta 1: A paradigm for distinct apoptotic pathways. Oncogene 9:1791–1798.

Seto E, Usheva A, Zambetti GP, Momand J, Horikoshi N, Weinmann R, Levine AJ, Shenk T (1992): Wild-type p53 binds to the TATA-binding protein and represses transcription. Proc Natl Acad Sci USA 89:12028–12032.

Shaulsky G, Goldfinger N, Peled A, Rotter V (1991): Involvement of wild-type p53 in pre-B-cell differentiation in vitro. Proc Natl Acad Sci USA 88:8982–8986.

Shaw P, Bovey R, Tardy S, Sahli R, Sordat B, Costa J (1992): Induction of apoptosis by wild-type p53 in a human colon tumor-derived cell line. Proc Natl Acad Sci USA 89:4495–4499.

Shen YQ, Shenk T (1994): Relief of p53-mediated transcriptional repression by the adenovirus E1B 19-kDa protein or the cellular bcl-2 protein. Proc Natl Acad Sci USA 91:8940–8944.

Slebos RJC, Lee MH, Plunkett BS, Kessis TD, Williams BO, Jacks T, Hedrick L, Kastan MB, Cho KR (1994): p53-dependent G(1) arrest involves pRB-related proteins and is disrupted by the human papillomavirus 16 E7 oncoprotein. Proc Natl Acad Sci USA 91:5320–5324.

Slichenmyer WJ, Nelson WG, Slebos RJ, Kastan MB (1993): Loss of a p53-associated G_1 checkpoint does not decrease cell survival following DNA damage. Cancer Res 53:4164–4168.

Srivastava S, Zou Z, Pirollo K, Blattner W, Chang E

(1990): Germ-line transmission of a mutated p53 gene in a cancer-prone family with Li-Fraumeni syndrome. Nature 348:747–749.

Stark GR (1993): Regulation and mechanism of mammalian gene amplification. Cancer Res 61:87–113.

Symonds H, Krall L, Remington L, Saenz-Robles M, Lowe SW, Jacks T, Van Dyke T (1994): p53-dependent apoptosis suppresses tumor growth and progression *in vivo*. Cell 78:703–711.

Tanaka N, Ishihara M, Kitagawa M, Hisashi H, Kimura T, Matsuyama T, Lamphier MS, Alzawa S, Mak TW, Taniguchi T (1994): Cellular commitment to oncogene-induced transformation or apoptosis is dependent on the transcription factor IRF-1. Cell 77:829–839.

Thompson AM, Anderson TJ, Condie A, Prosser J, Chetty U, Carter DC, Evans HJ, Steel CM (1992): p53 allele losses, mutations and expression in breast cancer and their relationship to clinico-pathological parameters. Int J Cancer 50:528–532.

Thorlacius S, Borresen AL, Eyfjord JE (1993): Somatic p53 mutations in human breast carcinomas in an Icelandic population: A prognostic factor. Cancer Res 53:1637–1641.

Tsuda H, Hirohashi S (1992): Frequent occurrence of p53 gene mutations in uterine cancers at advanced clinical stage and with aggressive histological phenotypes. Jpn J Cancer Res 83:1184–1191.

van den Elsen P, Houweling A, van der Eb AJ (1983): Morphological transformation of human adenoviruses is determined to a large extent by gene products of region E1A. Virology 131:362–368.

Vogelstein B, Kinzler KW (1992): p53 function and dysfunction. Cell 70:523–526.

Wang XW, Forrester K, Yeh H, Feitelson MA, Gu JR, Harris CC (1994): Hepatitis B virus X protein inhibits p53 sequence-specific DNA binding, transcriptional activity, and association with transcription factor ERCC3. Proc Natl Acad Sci USA 91:230–2234.

White E, Cipriani R, Sabbatini P, Denton A (1991): Adenovirus E1B 19-kilodalton protein overcomes the cytoxicity of E1A proteins. J Virol 65:2968–2978.

White E, Sabbatini P, Debbas M, Wold WM, Kusher DI, Gooding LR (1992): The 19-kilodalton adenovirus E1B transforming protein inhibits programmed cell death and prevents cytolysis by tumor necrosis factor alpha. Mol Cell Biol 12:2570–2580.

Whyte P, Buchkovich KJ, Horowitz JM, Friend SH, Raybuck M, Weinberg RA, Harlow E (1988): Association between an oncogene and an antioncogene: The adenovirus E1A proteins bind to the retinoblastoma gene product. Nature 334:124–129.

Wu XW, Levine AJ (1994): p53 and E2F-1 cooperate to mediate apoptosis. Proc Natl Acad Sci USA 91:3602–3606.

Wyllie AH (1985): The biology of cell death in tumours. Anticancer Res 5:131–136.

Wyllie AH (1993): Apoptosis (The 1992 Frank Rose Memorial Lecture). Br J Cancer 67:205–208.

Xiong Y, Hannon GJ, Zhang H, Casso D, Kobayashi R, Beach D (1993): p21 is a universal inhibitor of cyclin kinases. Nature 366:701–705.

Yamada Y, Yoshida T, Hayashi K, Sekiya T, Yokota J, Hirohashi S, Nakatani K, Nakano H, Sugimura T, Terada M (1991): p53 gene mutations in gastric cancer metastases and in gastric cancer cell lines derived from metastases. Cancer Res 51:5800–5805.

Yew PR, Berk AJ (1992): Inhibition of p53 transactivation required for transformation by adenovirus early 1B protein. Nature 357:82–85.

Yin Y, Tainsky MA, Bischoff FZ, Strong LC, Wahl GM (1992): Wild-type p53 restores cell cycle control and inhibits gene amplification in cells with mutant p53 alleles. Cell 70:937–948.

Yonish-Rouach E, Resnitzky D, Lotem J, Sachs L, Kimchi A, Oren M (1991): Wild-type p53 induces apoptosis of myeloid leukaemic cells that is inhibited by interleukin-6. Nature 352:345–347.

Zambetti GP, Bargonneti J, Walker K, Prives C, Levine AJ (1992a): Wild-type p53 mediates positive regulation of gene expression through a specific DNA sequence element. Genes Dev 6:1443–1152.

Zambetti GP, Olson D, Labow M, Levine AJ (1992b): A mutant p53 protein is required for maintenance of the transformed phenotype in cells transformed with p53 plus ras cDNAs. Proc Natl Acad Sci USA 89:3952–3956.

Zuppan CW, Beckwith JB, Luckey DW (1988): Anaplasia in unilateral Wilm's tumor: A report from the national Wilm's tumor study pathology center. Hum Pathol 19:1199–1209.

ABOUT THE AUTHORS

SCOTT W. LOWE is presently a Research Fellow at Cold Spring Harbor Laboratory. Dr. Lowe received a Bachelor of Science Degree in Biochemistry from the University of Wisconsin—Madison in 1986, where he worked with Dr. Allen D. Attie on the molecular and physiological basis of hypercholesterolemia. In 1988, he initiated graduate studies in the Department of Biology at the Massachusetts Institute of Technology under the supervision of Dr. H. Earl Ruley. During this period, he collaborated extensively with Drs. David Housman and Tyler Jacks, also at M.I.T. After receiving his Ph.D. in January 1994, Dr. Lowe remained at M.I.T. as an Anna Fuller Post-Doctoral Fellow in Dr. Housman's laboratory. While at M.I.T., his research focused on the role of apoptosis in oncogenic transformation and as a determinant of anticancer agent cytotoxicity. He continues to investigate the molecular biology of apoptosis in neoplasia.

H. EARL RULEY is a Professor of Microbiology and Immunology and the Director of the Cancer Etiology Program at Vanderbilt University School of Medicine. Dr. Ruley received his Ph.D. degree in 1980 from the University of North Carolina, Chapel Hill, for work on the molecular biology of mitochondrial DNA. He received postdoctoral training at the Imperial Cancer Research Fund in the laboratory of Dr. Michael Fried, where he worked on mechanisms of cell transformation by polyoma viruses. He later held positions of Staff Investigator at the Cold Spring Harbor Laboratory and Assistant Professor at the Massachusetts Institute of Technology. Since 1982, Dr. Ruley's research has focused on the molecular biology of growth control and development.

Cellular Aging and Cell Death: 235–253
© 1996 Wiley-Liss, Inc.

Programmed and Pathological Cell Death in *Caenorhabditis elegans*

Robyn Lints and Monica Driscoll

I. INTRODUCTION

In recent years it has come to be appreciated that cells do not die only as a consequence of injury or disease—they can also be genetically directed to undergo death during the normal course of development and homeostasis. Disease states can result if cell death is inappropriately timed, as observed in degenerative disorders, or if cells do not die when they should, as in the case of various malignancies and cancers. Without question, an understanding of the mechanisms of cell death is of key significance to human health.

Studies of cell death in both invertebrate and vertebrate systems have revealed that the genetic instructions for the regulation and execution of normal programmed cell death, also referred to as *apoptotic death*, have been remarkably conserved. Analyses in the simple nematode *Caenorhabditis elegans* have provided significant insight into the mechanism of programmed cell death. Less is clear about the mechanisms of inappropriate or pathological cell death, but a detailed molecular model of one inherited neurodegenerative condition identified in the nematode is being elaborated that may provide a means of identifying the genetic requirements for pathological cell death.

Here we review molecular and genetic characterization of programmed and pathological cell death in *C. elegans* and consider how similar mechanisms of cell death may influence health and aging of higher organisms.

II. OVERVIEW OF CELL DEATH: TWO MORPHOLOGICALLY DISTINCT PATTERNS OF CELL DEATH HAVE BEEN IDENTIFIED

Two types of cell death have been distinguished on the basis of morphological changes observed in the dying cell [Kerr et al., 1972; Wyllie et al., 1980]. One, termed *apoptosis*, is characterized by shrinkage and fragmentation of cytoplasm, compaction of chromatin and eventual destruction of cellular organelles. Frequently DNA is degraded by intranucleosomal cleavage that generates a characteristic DNA ladder of fragments that differ in size by one nucleosome repeat length [Wyllie, 1980]. Cellular remains are usually removed by phagocytosis and do not invoke an inflammatory response. In several cases it has been demonstrated that cell death by apoptosis is an active process, requiring RNA and protein synthesis, although this is not a universal feature of this type of cell death [reviewed by Ellis et al., 1991b]. Death by apoptosis often occurs as part of normal development or homeostasis.

A second type of cell death, termed *necrosis*, contrasts with apoptosis in several respects. First, necrotic cell death does not appear to be part of normal development or homeostasis. Rather, this type of death generally occurs as a consequence of cellular injury or in response to extreme changes in physiological conditions. Second, the morphological changes observed during necrosis differ greatly from those observed in apoptotic cell death. Necrotic cell death is characterized by gross cellular swelling and distention of subcellular organelles such as mitochondria and endoplasmic reticulum. Clumping of chromatin is observed, and DNA degradation occurs by cleavage at random sites. Necrosis is believed to occur independently of *de novo* protein synthesis and is generally thought to reflect the chaotic breakdown of the cell. However, given that many cells of diverse origins exhibit stereotyped responses to cellular injury, it is conceivable that a conserved "execution" program, activated in response to injury, may exist. It should be noted that some have argued that more than just these two patterns of cell death can be distinguished [Clarke, 1990].

III. *CAENORHABDITIS ELEGANS* AS A MODEL SYSTEM FOR THE STUDY OF CELL DEATH

C. elegans is well suited for the study of both normal and aberrant cell death at the cellular, genetic, and molecular levels. There is no model system in which development is better understood. The animal is essentially transparent throughout its life cycle, and individual nuclei can be readily visualized using differential interference contrast optics. These attributes have enabled the complete sequence of somatic cell divisions, from the fertilized egg to the 959-cell adult hermaphrodite, to be determined [Sulston and Horvitz 1977; Sulston et al., 1983; Sulston, 1988]. Elucidation of the lineage map has revealed that in certain lineages particular divisions generate cells that die at specific times and locations and that the identities of these ill-fated cells are invariant from one animal to another. This detailed knowledge of normal development has allowed easy identification of mutants with aberrant patterns of cell death [reviewed by Ellis et al., 1991b; Hengartner and Horvitz, 1994a].

It has been possible to identify many of the genes that cause normal and aberrant cell death in *C. elegans*, progress that stems from the amenability of this organism to genetic and molecular analyses. The nematode is readily cultured in the laboratory and has a very short life cycle, reaching maturity within 2.5 days

at 25°C. There are two sexes, hermaphrodites and males. A hermaphrodite is capable of producing approximately 300 progeny via self-fertilization. Thus, mutagenized animals are capable of generating a vast number of F_2 progeny that are homozygous for any induced mutation. Males can mate with hermaphrodites and are utilized in genetic crosses to transfer genes between animals. Genetic analysis has reached a sophisticated level with the development of straightforward approaches for generating transgenic animals [Fire, 1986; Mello et al., 1991]. To date, thousands of mutations that affect development and behavior have been identified, and hundreds of genes have been positioned on the *C. elegans* genetic map. In addition, a near complete physical map of the genome now exists, consisting of overlapping cosmid and YAC DNA clones that are aligned with the genetic map [Coulson et al., 1986, 1988]. For *C. elegans* researchers this means that once the position of a gene of interest has been determined by genetic mapping, clones corresponding to that region of the genome can be readily obtained. Cloning of *C. elegans* genes has been made easier still due to the considerable progress made in sequence analysis of the genome [Wilson et al., 1994], which has a target completion date of 1998 [R.H. Waterston, Washington University, personal communication].

IV. PROGRAMMED CELL DEATH IN *C. ELEGANS*
IV.A. Programmed Cell Death During Development

The nematode life cycle consists of a period of embryonic development within the eggshell, four larval stages, and, finally, adulthood. Of the 1,090 cells born in the developing hermaphrodite, 131 undergo cell death. The pattern of cell death observed and the identity of cells affected is invariant [Sulston and Horvitz 1977; Kimble and Hirsh, 1979; Sulston et al., 1983; Sulston, 1988] and is thus termed *programmed cell death* (PCD).

The majority of cell deaths occur in the embryonic phase of development, during which 113 of the 671 cells generated during this period die. In no instance do both daughter cells arising from a division undergo cell death—one cell is always viable. It appears that most of the cells fated to undergo PCD have neuronal potential, although some cells of putative hypodermal and muscular fate also die. Their potential fate has been inferred from the fate adopted by these cells in mutant animals where PCD is blocked [White et al., 1991; Horvitz et al., 1982; Avery and Horvitz, 1987].

It is noteworthy that elimination of cells during *C. elegans* development occurs prior to the expression of any differentiated characteristics by these cells. This contrasts with the process of neuronal selection observed during the establishment of the vertebrate nervous system, in which neuronal cells differentiate and send axons toward their target tissue where they compete to form contacts [reviewed by Oppenheim, 1991].

IV.B. Morphology of Programmed Cell Deaths

The morphological changes that occur in dying cells have been documented using light and electron microscopy (EM). Although EM studies have focused on the death of two postembryonic cells, called P11.aap and P11.aaap[1] [Robertson and Thompson, 1982], the changes observed during the demise of these cells are thought to be similar to those seen in the PCDs of other cell types in the animal [Sulston and Horvitz, 1977; Sulston et al., 1983; Ellis and Horvitz, 1986]. These studies divide the death process into five distinguish-

able stages; however, it should be noted that the morphological changes observed are continuous and progressive.

The cell fated to undergo cell death can be distinguished almost immediately after its generation from the parent cell division. The doomed cell is generally smaller than its sister cell and is surrounded in part by a cytoplasmic process extending from the neighboring cell. During this first "stage" of cell death the nucleus and the cytoplasm of the dying cell do not differ in appearance from those of its sister cell. In the second stage, the refractility of the dying cell, as observed through the light microscope, increases significantly, although internal structures differ little in appearance from those of the sister cell. The viable sister cell, which in the case of P11.aap or P11.aaap will become a motor neuron, begins to differentiate, initiating axon extension. In the dying cell this differentiation does not occur. However, if cell death is blocked by mutation, this cell is spared and differentiates to extend a process [White et al., 1991]. In the third stage, the first signs of intracellular disruption in the ill-fated cell become evident, and phagocytosis by the neighboring cell commences. At the ultrastructural level, the cytoplasm becomes condensed and the nuclear envelope becomes dilated. Within the nucleus, chromatin forms aggregates, and electron-dense particles (possibly the altered nucleolus) are present. Portions of the cell are split into membrane-bound fragments by the phagocytotic arms of the neighboring cell. During the fourth stage of cell death phagocytosis continues. Internal and plasma membranes adopt a whorled appearance, and mitochondria often appear distorted and are frequently observed within autophagic vacuoles. The nucleus appears convoluted, and dense chromatin-like material is evident. In the final stage, the cell shrinks and disappears. The nuclear membrane breaks up, dispersing chromatin-like fragments within the cytoplasm. Whorls of membrane and electron-dense material can be detected within vacuoles of the neighboring hypodermis, which engulfs all detectable remains of the dead cell.

---

[1]In the nomenclature of the *C. elegans* cell lineage, a blast cell is given a name that begins with an uppercase letter and may include a number. Each daughter cell is named by adding to the name of the mother cell a letter that indicates the position of the cell relative to its sister; a for anterior, p for posterior, l for left, r for right, d for dorsal, and v for ventral. The name of the blast cell and the subsequent progeny are separated by a period. Thus, P11.aap is the posterior daughter of the anterior daughter of the anterior daughter of the P11 blast cell.

The full time course of cell death is rapid, taking as little as 1 hour. Interestingly, the morphological changes associated with PCD in *C. elegans* are commonly observed in both invertebrate and vertebrate apoptotic cell death. Common features include initial condensation of the cytoplasm, nuclear chromatin aggregation, fragmentation of the dying cell into membrane-bound fragments, and phagocytosis by neighboring cells. The morphological similarities of cell death in very divergent organisms suggested that a conserved mechanism for removal of cells during development might exist. As described below, genetic and molecular analyses of cell death have revealed that functional homologs of the *C. elegans* PCD pathway operate in higher organisms.

IV.C. Genetic Analysis of Programmed Cell Death in *C. elegans*

Genetic analysis of cell death in *C. elegans* has identified mutations that affect four aspects of the cell death process: (1) regulation of cell death activation, (2) execution of cell death, (3) engulfment of the dying cell, and (4) degradation of the DNA released from the dying cell (Fig. 1). Control of cell death activation appears to involve both cell-specific and global regulatory factors. Once activated, cell death proceeds in much the same way in different cell types, progressing through the "stages" outlined above.

Below we describe genes identified in *C. elegans* that are essential for the execution of PCD, genes required for disposal of the cell corpse and its DNA, and, finally, the genes that regulate the activation of the cell death pathway.

IV.C.1. Genes required for cell death.
ced-3 and *ced-4* (*cell death* abnormal) are essential for the execution of PCD in the nematode [Ellis and Horvitz, 1986]. In *ced-3* and *ced-4* mutants all 131 of the cells that normally die in wild-type animals instead survive. These surviving cells differentiate, taking on the fate of lineally related cells. Interestingly, none of these surviving cells undergo further cell divisions, suggesting that PCD does not serve to eliminate proliferating blast cells. At first glance, *ced-3* and *ced-4* mutants (which bear 131 extra cells) appear normal in general morphology, locomotion, fertility, and mating behavior. However, closer examination of these mutants has revealed some behavioral and developmental differences from wild-type animals, including slower development and defective chemotaxis [Ellis et al., 1991b]. It appears then that PCD may improve the fitness of the animal.

The nucleotide sequences of both *ced-3* and *ced-4* have been determined. The predicted protein sequence for CED-3 is related to a family of novel cytoplasmic cysteine proteases that includes the human ICE protein (*Interleukin-*

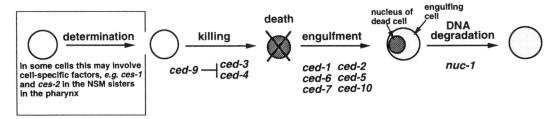

Fig. 1. *The genetic pathway for developmentally programmed cell death in* Caenorhabditis elegans. *Cell-specific genes, such as* ces-1 *and* ces-2 *of the NSM sisters, may act to determine which cells die. These determination genes may circumvent the action of CED-9, a protein that acts globally to inhibit programmed cell death. CED-9 blocks cell death by preventing CED-3 and CED-4 from activating or executing the cell death program. Two*

groups of genes, ced-1, ced-6, ced-7, *and* ced-2, ced-5, ced-10, *encode products that participate in corpse removal via phagocytosis. Finally, the nuclease encoded by the* nuc-1 *gene degrades the DNA from the dead cell. Vertebrate genes (open-faced type)* bcl-2 *and* Ice *have been shown to be functional homologs of* ced-9 *and* ced-3, *respectively [modified from Ellis et al., 1991b, with permission of the publisher.]*

1β-*C*onverting *E*nzyme), human Ich-1, and its murine counterpart Nedd2 (*n*eural precursor cell-*e*xpressed, *d*evelopmentally *d*own-regulated) [Cerretti et al., 1992; Kumar et al. 1992; Thornberry et al., 1992; Yuan et al., 1993; Wang et al., 1994]. Although the overall sequence identity shared between CED-3 and these proteins is relatively low (27%), amino acids that constitute the active site of ICE are strictly conserved in CED-3, indicating that this *C. elegans* homolog is also likely to function as a protease. Expression of ICE and Nedd2 in certain mammalian culture cells can induce cell death by apoptosis [Miura et al., 1993; Gagliardini et al., 1994; Kumar et al., 1994]. Significantly, expression of *ced-3* in Rat-1 cells can also induce death with remarkable efficiency [Miura et al., 1993]. Thus ICE and CED-3 are considered functional homologs, capable of inducing the cell death process.

How then is cell destruction achieved by the action of these proteases? Clearly a knowledge of their targets will provide significant insight. At present the substrate of the *C. elegans* protease is unknown. The human ICE protein was initially identified by virtue of its ability to process the precursor of interleukin-1β to generate an active cytokine [Black et al., 1989; Kostura et al., 1989]. While the involvement of interleukin-1β in several immune processes such as inflammation, septic shock, and certain aspects of hemopoiesis is well documented [reviewed by Dinarello and Wolff, 1993], its function (if any) in apoptosis is not clear. It seems likely that ICE or a related protease may have additional substrates and that these may be more directly involved in cell death. One important implication of the heterologous expression experiments that revealed that CED-3 can induce apoptosis in vertebrate cells is that the targets of this protease family must possess some common structural features. Thus identification of additional substrates, in either vertebrate or nematode systems, will be potentially informative to our understanding of events that ensue downstream of CED-3 and ICE activity.

The *ced-4* gene encodes a novel hydrophilic protein that contains two regions with some similarity to the Ca^{2+}-binding EF-hand motif [Yuan and Horvitz, 1992]. The EF-hand motif is a 29 amino acid helix–loop–helix domain that includes precisely spaced amino acids that coordinate calcium binding via side chain oxygens within the loop [reviewed by Kretsinger, 1987]. Although it has not been demonstrated that CED-4 binds Ca^{2+}, the potential involvement of Ca^{2+} binding in cell death is intriguing given that perturbation of intracellular Ca^{2+} levels has been implicated in triggering cell death elsewhere [Cheung et al., 1986; Choi, 1987; 1988a,b; Choi and Rothman, 1990]. To date no mammalian homologs for this gene have been identified.

Genetic mosaic studies indicate that CED-3 and CED-4 are most likely to act within the cells that die [Yuan and Horvitz, 1990]. Not surprisingly, transcription of these genes is highest during embryogenesis, coinciding with the timing of substantial cell death [Yuan and Horvitz, 1992; Yuan et al., 1993]. The relative abundance of *ced-3* and *ced-4* message during this time suggests that many cells, not only those programmed to die, express these genes. The decision to activate the cell death program may therefore depend on some translational or post-translation regulatory mechanism that controls availability or activity of CED-3 and CED-4.

IV.C.2. Genes required for the engulfment of the dying cells. Mutations in six genes (*ced-1, ced-2, ced-5, ced-6, ced-7, ced-10*) prevent phagocytosis of the dying cell [Hedgecock et al., 1983; Ellis et al., 1991a]. In the absence of functional engulfment, the dead cell adopts a refractile appearance and persists in this highly refractile stage for hours. Since cells programmed to die can proceed through the initial stages of cell death, it appears that engulfment is not the cause of death. Thus, phagocytosis serves to remove cell corpses rather than create them.

None of engulfment *ced* mutations listed above completely block phagocytosis. This low expressivity suggested that more than a single pathway for engulfment exists so that cell corpses can still be removed by one pathway should the other pathway be blocked. Examination of engulfment in animals carrying double and triple *ced* mutations suggests that one pathway requires the genes *ced-2, ced-5,* and *ced-10* and the other *ced-1, ced-6,* and *ced-7* [Ellis et al., 1991a].

At present the precise functions encoded by the engulfment genes are unclear. Further molecular and biochemical analyses should provide some insight concerning their respective roles in cell corpse disposal.

IV.C.3. A nuclease is required for DNA degradation.

In a *nuc-1* genetic background, cell death and engulfment proceed normally; however, the DNA from dead cells persists as a compact mass of Feuglen reactive material [Hedgecock et al., 1983]. Biochemical studies indicate that *nuc-1* encodes or controls the activity of a Ca^{2+}-, Mg^{2+}-independent deoxyribonuclease [Hevelone and Hartman, 1988]. This enzyme is not required exclusively for DNA degradation in PCD, as it is also necessary for DNA degradation in the gut [Sulston, 1976].

The process of DNA degradation in *C. elegans* PCD appears to differ from that observed in vertebrate apoptotic cells in two respects. First, it has been suggested that thymocyte nucleases may induce apoptosis [Arends et al., 1990]. This contrasts with the role of *nuc-1* in *C. elegans* PCD, which is needed for efficient clean up of the cell corpses and not for cell death itself. However, it remains possible that other genes may encode an endonuclease activity that participates in the earliest phases of the nematode cell death. Second, the biochemical properties of nucleases involved in *C. elegans* and mammalian cell death may differ. For example, thymocyte and prostate nuclease activities are dependent on Ca^{2+} [Cohen and Duke, 1984; Wyllie et al., 1984; Kyprianou et al., 1988], whereas NUC-1 acts independently of this cation [Hevelone and Hartman, 1988].

IV.D. Regulation of Programmed Cell Death

The picture emerging from genetic characterization of cell death in *C. elegans* is that PCD in this organism is actively repressed in many

cells and that reversal of this negative regulation is necessary for cell death to occur. Genetic screens have identified genes that regulate the cell death program in select groups of cells as well as a gene that functions globally. The following section describes the genetic properties of these cell death regulators and our current understanding of their molecular nature.

ces-1 and *ces-2* are two genes that encode cell-specific factors that regulate the death of a few cells in the developing pharynx [Ellis and Horvitz, 1991]. Mutations in these genes do not disrupt PCD elsewhere in the animal. Genetic interactions between *ces-1* and *ces-2* mutant alleles suggest that CES-1 is required to protect against PCD and that CES-2 interferes with this protection in cells developmentally programmed to die.

ced-9 was first identified by a rare gain-of-function (*gf*) mutation that prevents most programmed cell deaths during nematode development [Hengartner et al., 1992]. Loss-of-function (*lf*) mutations in *ced-9* cause ectopic cell deaths in many sublineages. Since the absence of the *ced-9* product results in the death of cells that would otherwise survive in wild-type development, it seems likely that the CED-9 protein functions in many (if not all) cells to inhibit cell death. Genetic analysis has established that *ced-9* functions in the same pathway as *ced-3* and *ced-4*, acting upstream of these genes. This is evidenced by the fact that the affects of *ced-9(lf)* mutations are completely blocked by *lf* mutations in *ced-3* and *ced-4*.

ced-9 mRNA levels, as determined by Northern hybridization analysis, are highest during embryogenesis, then decrease to a low constitutive level in later larval stages. This high level of *ced-9* expression during early development may ensure that cells programmed to survive generate sufficient levels of the CED-9 protein to suppress cell death.

ced-9 is predicted to encode a 280 amino acid protein that shares significant, though limited, sequence and structural homology with the product of the mammalian protooncogene *bcl-2*. [Hengartner and Horvitz, 1994b]. The mammalian gene was discovered and molecularly cloned based on its involvement in a t(14,18) chromosomal translocation in follicular lymphomas [Fukuhara et al., 1979; Yunis et al., 1982]. This rearrangement causes *bcl-2* expression to become elevated and deregulated and, as a consequence, delays the PCD of affected B cells [Tsujimoto et al., 1984; Bakhshi et al., 1985; Cleary et al., 1986; Tsujimoto and Croce, 1986; Seto et al., 1988]. Overexpression of Bcl-2 can also have life-prolonging effects on many nonlymphoid cell types [Bissonnette et al., 1992; Fanidi et al., 1992; Garcia et al., 1992; Allsopp et al., 1993; Batistatou et al., 1993].

The ability of CED-9 and Bcl-2 to protect cells against programmed cell death together with their structural similarities suggested that these proteins could be functional homologs. Indeed, expression of *bcl-2* in *C. elegans* can suppress some of the ectopic cell deaths observed in *ced-9(lf)* mutants [Vaux et al., 1992; Hengartner and Horvitz, 1994b]. However, some noteworthy differences in the activities of Bcl-2 and CED-9 are apparent. For example, the *ced-9(gf)* mutation that suppresses all developmentally programmed cell deaths encodes an amino acid substitution for a conserved residue in the CED-9 protein. The corresponding substitution in Bcl-2 renders this mammalian protein inactive, suggesting that the role of this conserved residue may differ in the two proteins [Hengartner and Horvitz, 1994c]. In addition, CED-9 and Bcl-2 function during different periods of development and therefore may be responsive to different cellular signals. CED-9 is highly expressed in early development and is essential for survival through the embryonic period of development [Hengartner et al., 1992]. In contrast, *bcl-2* appears to be to some degree dispensable in early development since *bcl-2* $^{/-}$ mice survive postnatally for 2–5 weeks before succumbing to fulminant lyphomenia and polycystic renal disease [Vies et al., 1993].

ced-9 and *bcl-2* are members of a rapidly growing gene family of factors that are involved in determining cell survival and cell viability [reviewed by Williams and Smith, 1993; Oltvai and Korsmeyer, 1994]. It has been demonstrated that this family consists of factors that both promote cell death (for example Bcl-x_s and Bax) and repress it (such as Bcl-2 and BcI-x_L). Biochemical analysis of the vertebrate regulators reveals that these proteins function in homo- and heteromeric complexes formed between family members [Boise et al., 1993; Oltvai et al., 1993]. Whether such a multimeric complex acts to suppress apoptosis or to induce it depends on its composition. In the current model for cell death regulation the life and death fate of a cell is determined by the balance of positively- and negatively-acting family members within the cell. In *C. elegans*, it appears that CED-9 activity may also be modulated by as yet unidentified factors [Hengartner and Horvitz, 1994c]. It will be of great interest to determine whether these factors are also members of the *ced-9/bcl-2* family.

At present the mechanism by which this family regulates cell death is poorly understood. In vertebrates, overexpression of *bcl-2* in many and varied cell types can protect not only against the effects of serum or growth factor deprivation but also against exposure to a wide range of toxic agents and physiological insults [Henderson et al., 1991; Sentman et al., 1991; Strasser et al., 1991; Alnemri et al., 1992; Miyashita et al., 1992; Hennet et al., 1993; Zhong et al., 1993; Martinou et al., 1994]. How such broadly acting protection might be achieved is unclear. Identification of CED-9/Bcl-2 targets may provide some clues.

In summary, genetic analysis of cell death in the nematode has identified several key players required for the regulation and execution of the cell death pathway. Significantly, structurally related vertebrate genes that appear to perform similar functions have been identified, indicating that the general framework of the cell death mechanism has been conserved over the course of evolution.

V. RELEVANCE OF CELL DEATH TO AGING

The study of cellular aging in cell culture systems has established that cells exhibit a limited life span and have an inherent capacity to divide a specific number of times before undergoing cell death [Hayflick, 1965]. These observations suggest that the longevity of a cell is genetically determined.

It is logical to wonder whether the genetic programs for cell death contribute in any way to the aging process of an organism. The study of cell death and aging in *C. elegans*, however, suggest that these two processes are independent. Mutations in cell death genes, which block developmental PCD, do not affect the life span of the animal: *ced-3* and *ced-4* mutant animals live no longer than wild type [T. Johnson, University of Colorado, personal communication]. This provides a strong *"in vivo"* argument that PCD is separable from aging in this organism.

There are a few genes that do dramatically influence the life span of *C. elegans*. These are described briefly here but are dealt with more extensively in the chapter by Tom Johnson (this volume).

Mutant alleles of the *age-1* gene can increase the mean life span the nematode by as much as 60%, an extension that is due to an increase in the postreproductive life rather than an alteration in the length of development, the length of reproduction, or efficiency of feeding [Friedman and Johnson, 1988a,b]. Analysis of *age-1* mutants has revealed that these animals have elevated levels of catalase and Cu/Zn-superoxide dismutase (SOD) at old age, providing evidence to support the idea that increased tolerance to oxidative stress is correlated with longer life span [Larsen, 1993; Vanfleteren, 1993].

In *C. elegans* increased sperm production has been implicated in the shortening of the male life span [Van Voorhies, 1992]. In *spe-26* mutants primary spermatocytes fail to develop into spermatids, and life span is 65% longer than that of the average wild-type nematode. How inhibition of sperm maturation or other affects of *spe-26* influence life span remains to be determined.

Mutations in genes affecting life cycle progression have also been found to alter life span. In the absence of food, *C. elegans* enters an alternative third larval form called the dauer larva. Dauer larva are mobile but do not feed

and can survive for more than 60 days. In the presence of food, the dauer larva de-differentiate and proceed through the fourth larval stage to become fertile adults. Life spans of adult animals that have recovered from the dauer larva stage are similar in length to those of animals that have never entered the dauer larva stage (approximately 18 days) [Klass and Hirsh, 1976]. Thus, in the dauer stage the "aging clock" (if there is one) appears to be suspended. Alternatively, it is possible that development must be completed before aging can begin [Johnson and Lithgow, 1992].

Mutations in two genes that regulate formation of the dauer larva, *daf-2* and *daf-16*, influence life span [Kenyon et al., 1993]. Mutations in *daf-2* cause hermaphrodites to live more than 200% longer than wild-type, and mutations in *daf-16* eliminate the life extension conferred by the *daf-2* mutation. Genetic studies indicate that *daf-16* functions downstream of *daf-2* to promote dauer formation [Riddle et al., 1981; Vowels and Thomas, 1992]. Thus, the model is that *daf-2* mutations extend life span by inappropriately activating *daf-16*. What role *daf-16* performs in the dauer larva development is unclear, but it has been suggested that this gene could control a life span extension program utilized by dauers. Alternatively, as nutritional limitation is known to extend nematode life span, *daf-2* mutations may induce an altered metabolism in adult hermaphrodites, thereby affecting their longevity.

Clearly, a pressing issue in *C. elegans* aging is to identify the proteins that alter life span and to determine their biochemical activities. Given the numerous examples of functional conservation of proteins among eukaryotic species, the study of aging mutants in invertebrates may well serve to identify vertebrate counterparts that influence senescence in higher organisms.

VI. PATHOLOGICAL CELL DEATH IN *C. ELEGANS*

The second type of cell death observed in *C. elegans* is referred to as inappropriate or pathological cell death. These cell deaths do not occur in the normal development of the animal and result from mutations that (1) cause misregulation of the PCD pathway, (2) evoke inappropriate cell phagocytosis, or (3) induce neuronal swelling and subsequent degeneration by a mechanism distinct from that of programmed cell death.

VI.A. Pathological Cell Death Can Occur as a Consequence of Misregulation of Programmed Cell Death

Aberrant cell death can occur as a consequence of inappropriate activation of PCD in cells that are usually destined to survive in development. Misregulation of the PCD pathway can have severe consequences for the survival of the organism or may only affect a select subgroup of cells in the animal. The effect of loss-of-function mutations in *ced-9* provides a good example whereby misregulation adversely affects the viability of the organism. As described above, *ced-9* functions as a global negative regulator of the programmed cell death pathway. Loss-of-function mutations in the gene induce extensive ectopic cell death, resulting in lethality [Hengartner et al., 1992]. Less severe affects are observed when mutations disrupt regulation of normal cell death in only a small subset of cells. For example, mutations in the *egl-1* gene (*egg-l*aying defective) cause inappropriate cell death of only the HSN neurons, which innervate the vulva and control egg expulsion in the hermaphrodite [Trent et al., 1983; Ellis and Horvitz, 1986]. In this case, the life span of the animal is dramatically shortened after it reaches reproductive maturity as progeny mature and hatch within the parent.

VI.B. Cell Death Caused by Inappropriate Engulfment

The Pn.p cells are a group of postembryonically derived precursors that give rise to hypodermal and vulval cells in later stages of development. Semidominant mutations in two genes, *lin-24* and *lin-33* (*lin*eage abnormal), cause Pn.p cells to adopt an abnormal morphology and degenerate [Kim, 1994]. This cell death does not occur via the normal pro-

grammed cell death pathway as mutations in *ced-3* and *ced-4* do not block degeneration. Moreover, *lin-24–* and *lin-33–*induced deaths are dependent on the activity of the engulfment genes *ced-2, ced-5*, and *ced-10*. This contrasts with normal PCD in which death *per se* can occur independently of engulfment gene activity. Precisely why Pn.p cells are targeted for cell phagocytosis in the *lin-24* and *lin-33* backgrounds is unknown. It is possible that the *lin-24* and *lin-33* mutations lead to alterations in cell morphology or cell surface proteins, which label them as targets for phagocytosis by their neighbors.

VI.C. Abnormal Cell Deaths Characterized by Cellular Swelling and Cell Lysis

Rare dominant mutations (*d*) in two genes, *mec-4* and *deg-1*, cause cellular swelling and subsequent degeneration of specific neurons in the animal [Chalfie and Sulston, 1981; Chalfie and Au, 1989; Chalfie and Wolinsky, 1990]. Dominant mutations in the *mec-4* gene (*mec*hanosensory abnormal) cause death of the six touch receptor neurons required for the sensation of gentle touch, while a dominant mutation in *deg-1* (*deg*eneration) induces degeneration of a group of neurons that includes the posterior interneurons of the touch sensory circuit (the PVC cells) [Chalfie et al., 1985].

Death-inducing mutations in *mec-4* and *deg-1* do not disrupt the development of affected neurons. For example, in *deg-1(d)* mutants the PVC neurons are born early in embryonic development, differentiate, and even function briefly before they die in the second and third larval stages [Chalfie and Wolinsky, 1990]. The touch receptor neurons are also observed to express terminally differentiated properties before they degenerate [Mitani et al., 1993]. *mec-4(d)-* and *deg-1(d)*–induced cell deaths are therefore considered "late onset," analogous to many human neurodegenerative disorders that do not manifest themselves until late in life. Unlike many degenerative disorders of the human nervous system, however, *mec4(d)-* and *deg-1(d)*–induced cell deaths do not affect the viability of the organism. Mutant animals sur-

vive long after affected neurons have degen-
erated, exhibiting little consequence of the loss,
other than an inability to sense gentle touch.

VI.C.1. The morphology of *mec-4(d)-* and *deg-1(d)*–induced cell deaths. Although *mec-4(d)* and *deg-1(d)* mutations induce cell death in different groups of cells, the morphological changes observed during cell death appear identical. In the initial stages of degeneration, the cell body of the neuron swells, and in some instances enlargement of the nucleus is also observed. The cell body continues to swell to several times its original diameter and then finally disappears. This process occurs over several hours. Superficially, the morphological changes that transpire during these cell deaths resemble the pattern of events that occur during necrotic cell death [Kerr et al., 1972; Wyllie et al., 1980]. Future detailed ultrastructural characterization of the degeneration process should reveal whether the dying neurons exhibit changes in common with the necrotic response.

VI.C.2. *mec-4(d)-* and *deg-1(d)*–induced cell death occurs by a mechanism distinct from programmed cell death. Morphological profiles and genetic studies suggest that *mec4(d)-* and *deg-1(d)*–induced cell death occurs by a mechanism distinct from that employed in developmentally programmed cell death. In PCD cells become compact and refractile, whereas *mec-4* and *deg-1* dominant mutations induce cellular swelling (Fig. 2). The time frame within which these two types of death occur is also significantly different. PCD is complete within 1 hour while *mec-4(d)-* and *deg-1(d)*–induced cell deaths span several hours. Moreover, it has been demonstrated that *ced-3(lf)* and *ced-4(lf)* mutations cannot prevent *mec-4(d)-* and *deg-1(d)*–induced cell degeneration [Chalfie and Wolinsky, 1990]. Conversely, the presence of *mec-4* and *deg-1* dominant alleles does not disrupt PCD in the animal [M. Chalfie, Columbia University, personal communication]. These observations demonstrate that *mec-4(d)* and *deg-1(d)* kill cells by a mechanism distinct from that employed in PCD. It does appear, however, that

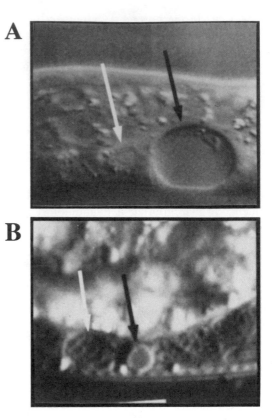

Fig. 2. *Morphological changes observed in degenerin-induced pathological cell death and developmentally programmed cell death (viewed with Nomarski differential interference contrast optics). (**A**): Pathological cell death in a* mec-4(d) *animal. Depicted is the central region of an animal showing an AVM touch receptor neuron at an advanced stage of swelling (black arrow). Adjacent, an unaffected cell is shown (white arrow). (**B**): Programmed cell death of a P11.aap cell. The distinctive "button-like" appearance of the dying cell (black arrow) is readily apparent during the period when the cell exhibits maximum refractility.*

other aspects of the death process may be shared by both programmed and degenerative cell death. For example, some of the engulfment *ced* genes involved in the disposal of cell corpses produced by PCD also seem to be required for removal of *mec-4(d)*– and *deg-1(d)*–generated corpses [S. Chung and M. Driscoll, unpublished data].

VI.C.3. *mec-4* and *deg-1* genetics. The three dominant *mec-4* mutations described above represent a relatively rare class of *mec-*

4 mutant alleles. The vast majority of mutations in this gene are recessive (*r*) and do not induce cell degeneration. Instead, recessive mutations in *mec-4* render the animal insensitive to gentle touch [Chalfie and Sulston, 1981; Chalfie and Au, 1989]. Electron microscopic examination of the touch receptor neurons in animals carrying these recessive, loss-of-function mutations reveals no morphological or developmental abnormalities associated with the absence of functional MEC-4. Examination of *mec-4/lacZ* expressed in transgenic animals indicates that the gene is expressed only within the six touch receptor neurons [Mitani et al., 1993]. The gene is expressed soon after each cell is born and the product appears to persist throughout the life of the animal [L. Gong and M. Driscoll, unpublished data].

The *deg-1* gene was originally defined by the single dominant mutant allele described above. Loss-of-function mutations in *deg-1* have been isolated by screening for suppression of the mechanosensory defect conferred by the death of the PVC interneurons in *deg-1(d)* animals. Interestingly, *deg-1* null mutants exhibit a wild-type touch response, suggesting that the *deg-1* product may be redundant in these cells [Chalfie and Wolinsky, 1990].

VI.C.4. *mec-4* and *deg-1* are members of a gene family called *degenerins*. The predicted primary sequences for MEC-4 and DEG-1 share approximately 51% amino acid identity [Driscoll and Chalfie, 1991]. These related genes have defined a new gene family termed *degenerins* because of their ability to mutate to forms that can induce cell degeneration. Since the initial discovery of *deg-1* and *mec-4* additional family members in *C. elegans* have been identified. These are *mec-10* [Huang and Chalfie, 1994] and *unc-105* [B. Schrank and R. H. Waterston, Washington University, personal communication].

Like *mec-4*, *mec-10* is required for sensitivity to gentle touch sensation. Loss-of-function mutations in *mec-10* are recessive and render the touch receptor neurons nonfunctional but do not affect their development or cell morphology [Chalfie and Au, 1989]. The *mec-10*

gene was isolated using a PCR-based approach designed specifically to isolate additional degenerin family members from the nematode [Huang and Chalfie, 1994]. This gene is expressed exclusively in the six touch receptors and four other neurons. It has been suggested that the *mec-6* gene may encode another degenerin family member [Huang and Chalfie, 1994]. This assertion is based on the fact that *mec-6* (*1*) is required for *mec-4(d)*– and *deg-1(d)*–induced cell degeneration and, (2) is essential for touch sensitivity [Chalfie and Wolinsky, 1990], and (3) the mammalian degenerin-like channels include three homologous subunits (described below).

unc-105 was defined by semidominant mutations that cause hypercontraction of body wall muscle in the animal [Park and Horvitz, 1986a]. Examination of mutant animals with polarized optics reveals that muscle cells are highly irregular and disorganized and exhibit reduced birefringency. These morphological defects suggest that the *unc-105* product is a component of muscle. However, it is possible that the UNC-105 protein is localized in the neurons that innervate this tissue, effecting muscle response to neuronal stimulation. Interestingly, *unc-105* null mutants, like those of *deg-1*, are phenotypically wild type, suggesting that *unc-105* may be functionally redundant [Park and Horvitz, 1986b].

Members of the degenerin gene family are not highly related, sharing at most 64% amino acid identity. However, conservation is sufficiently high to demonstrate that the general structure and likely topology of these molecules is conserved (Fig. 3A). Degenerin proteins are predicted to possess two hydrophobic stretches that span the membrane (designated MSDI and MSDII). A short hydrophobic H-5–like sequence is situated adjacent to MSDII and may also be positioned in the membrane [Jan and Jan, 1994]. Since degenerins lack a signal sequence, it is predicted that both the amino- and carboxy-terminal ends of the protein would lie on the intracellular side of the cell membrane. Consequently, the bulk of the protein would be displaced extracellularly (Fig.

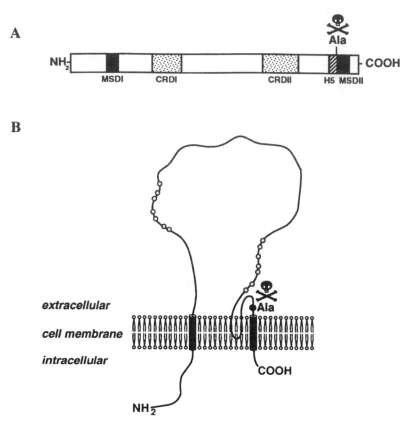

Fig. 3.   *Features of the primary sequence of the predicted* mec-4 *product. (A): Schematic representation of MEC-4. The deduced maximum length of the protein is 768 amino acids [M. Driscoll and M. Chalfie, unpublished data]. The protein is predicted to contain two hydrophobic regions, MSDI and MSDII (black boxes), each theoretically capable of spanning the cell membrane. The two cysteine-rich domains, CRDI and CRDII (stippled boxes), are located in the predicted extracellular portion of the protein. MSDII is immediately preceded by an H-5–like domain (hatched box) (see Jan and Jan, 1994), a sequence that may contribute to the formation of the channel pore. The site of the dominant, degeneration-induc-* *ing amino acid substitution (Ala 442 in the published sequence) lies close to the second predicted membrane spanning domain. (B): Theoretical topology of the predicted MEC-4 protein. The absence of an N-terminal signal sequence infers that the amino and carboxy termini of the protein are located intracellularly and that the bulk of the protein is situated on the extracellular side of the membrane. Cysteine residues of CRDI and CRDII are represented by open circles, MSDI and MSDII are represented by the black cylinders, and the position of the conserved Ala affected by all* mec-4(d) *dominant mutations is shown (black circle).*

3B). Interestingly, the predicted extracellular domain contains two cysteine-rich stretches in which the spacing of cysteine residues is similar to that observed in Cys-rich motifs of the low-density lipoprotein (LDL), EGF, and related receptor proteins. Potentially these Cys-rich regions may participate in some receptor function performed by degenerin proteins.

The likely association of degenerin proteins with the plasma membrane, together with cellular swelling induced by the dominant mutations in *mec-4* and *deg-1*, has lead to the speculation that degenerin proteins may function as part of a channel required for some aspect of membrane permeability or cell volume control. The structure of MSDII supports this possibility. This region is predicted to form an amphipathic α-helix in which polar and charged residues are aligned along one helical face [Hong and Driscoll, 1994]. On the basis of this structural similarity to characterized ion channels it is proposed that the polar face of MSDII lines the ion channel pore.

VI.C.5. Mammalian family members encode subunits of a novel Na+ channel.

A family of vertebrate genes that exhibit sequence similarity to the *C. elegans* degenerin family has been identified [Canessa et al., 1993; Canessa et al., 1994]. The rat genes α-, β-, and γ-rENaC (*rat epithelial Na+ channel*) encode three subunits of an amiloride-sensitive Na+ channel responsible for Na+ absorption in epithelia lining the distal part of the kidney tubule, the urinary bladder, the distal colon, and the lung [Canessa et al., 1993; O'Brodovich et al., 1993; Duc et al., 1994]. Interestingly, all three proteins are required for full channel activity, suggesting that they function as a heteromeric complex. These degenerin-related proteins encode two potential membrane-spanning domains and the second Cys-rich region identified the *C. elegans* family members. In fact, MSDII of the α-rENaC can specifically replace the equivalent segment of MEC-4 without destroying the ability of the *C. elegans* protein to mediate touch sensitivity *in vivo* [Hong and Driscoll, 1994].

VI.C.6. *C. elegans* degenerins may associate to form a heteromeric channel.

Characterization of the rat degenerin family members suggests that functional channels are formed by association of different family members. Several observations in *C. elegans* argue that heteromeric channels are formed in the worm. Specifically, genetic interaction between family members expressed in cells of the touch circuit have been noted [Chalfie and Wolinsky, 1990; Huang and Chalfie, 1994]. For example, functional *mec-6* is essential for touch sensitivity but is also required for *mec-4(d)*– and *mec-10(d)*–induced death of the touch receptor neurons [Chalfie and Wolinsky, 1990; Huang and Chalfie, 1994]. A similar situation also appears to exist in the PVC interneurons where functional *mec-6* is necessary for *deg-1(d)*–induced cell death [Chalfie and Wolinsky, 1990]. This interdependence suggests that, by analogy with the rat degenerin channel, a functional channel may be formed in the touch receptors by association of MEC-4, MEC-10, and MEC-6 and in other neurons by DEG-1, MEC-6, and another subunit.

VI.C.7. The molecular nature of the death-inducing mutations in *mec-4(d)* and *deg-1(d)*.

All death-inducing mutations in MEC-4 and DEG-1 involve substitution for a single conserved alanine residue predicted to lie immediately N-terminal the predicted pore-lining domain, MSD II [Driscoll and Chalfie, 1991]. In two of the three *mec-4(d)* alleles this conserved Ala residue is replaced by Thr, while in the third *mec-4(d)* allele and in the single *deg-1(d)* allele Val is substituted [García-Añoveros et al., 1995; Shreffler et al., 1995]. *In vivo* analysis of engineered substitutions at this conserved position in MEC-4 has revealed that the size of the substituting amino acid is the critical feature of the death-inducing proteins. Replacement of Ala with large side chain amino acids induces touch cell degeneration in transgenic animals. In contrast, substitution of Ala with amino acids of smaller or comparable size has no deleterious affect. Based on these observations, a model for degenerin-induced cell degeneration has been proposed in which steric hindrance near the pore-lining domain, MSDII, plays a causative role in inducing cellular swelling (Fig. 4). In this model

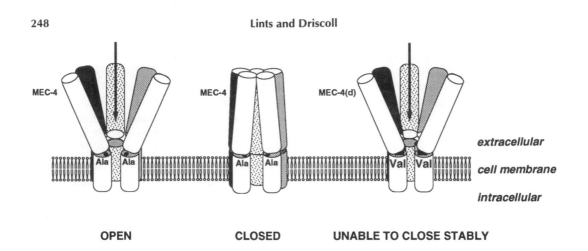

extracellular

cell membrane

intracellular

OPEN   **CLOSED**   **UNABLE TO CLOSE STABLY**

Fig. 4. *A hypothetical model for* mec-4(d)–*induced cell death. In the touch receptor neurons the heteromeric degenerin channel is composed of MEC-4, MEC-10, and possibly MEC-6 protein subunits. In wild-type animals, channel opening and closing is achieved by confirmation changes centered around a "hinge" region located near the cell membrane.*

The presence of Ala in this region allows free movement of the hinge so the channel can be open and closed as required. In mec-4(d) *animals, a Val substitution at this position prevents hinge movement. Consequently the channel is unable to close properly, allowing unregulated flow of ions and solutes into the cell that leads to cellular swelling.*

the presence of a large side chain amino acid adjacent to the channel pore prevents channel closure, perhaps by preventing a conformational change required to close the channel. The channel, rendered constitutively open, allows unregulated flow of ions and solutes across the membrane. This increase in membrane permeability may directly cause the cell to swell.

Evidence from other systems supports the idea that sustained opening of channels can lead to cell death. For example, the toxic affects of glutamate on the mammalian central nervous system are mediated through excitatory ion channels [Choi, 1988a, b; Choi and Rothman, 1990]. Disruption of channel activity may, in fact, prove to be a recurrent theme in neurodegenerative and muscular disorders. In humans, acetylcholine receptor defects underlie congenital myasthenic syndromes [Engel, 1990] and in the nematode mutations in an acetylcholine-related protein, DEG-3, cause degeneration of specific subsets of neurons in the animal [Treinin and Chalfie, 1995]. Based on biochemical analysis of vertebrate counterparts it is thought that this mutation in DEG-3 causes sustained activity of the channel, thereby allowing unregulated flow of Ca^{2+}

into affected neurons. Mutations that alter the α-subunit of the sodium channel of skeletal muscle and disrupt channel activation are responsible for hyperkalemic periodic paralysis, a human disorder characterized by cold-induced muscle stiffness [Ptacek et al., 1991] and paramyotonia congenita, a dominant disorder characterized by periodic muscle weakness [McClatchey et al., 1992; Ptacek et al., 1993]. Both autosomal recessive generalized myotonia (Becker's disease) and autosomal dominant myotonia congenita (Thomsen's disease), which are characterized by skeletal muscle stiffness, result from mutations in the skeletal muscle chloride channel (CLC-1) [Koch et al., 1992]. Cystic fibrosis, a recessive disorder, is also caused by a defective Cl^- channel [Bear et al., 1992]. Interestingly, a mutation in the human homolog of β-rENaC has been detected in suffers of Liddle's syndrome, an autosomal dominant condition in which defective Na^+ transport, particularly in the kidney, induces severe hypertension [Liddle et al., 1963]. The mutation identified affects amino acids of the intracellular carboxy-terminus of the protein thought to be required for protein–protein interaction with cytoskeleton [Shimkets et al., 1994].

The model for degenerin-induced cell death suggests that the activity of the aberrant proteins kills the cells directly by altering membrane permeability. While this represents the simplest model for degenerin-induced cell death, other explanations are also possible. For example, the necrotic cell death observed may reflect the activation of a cell death pathway triggered in response to cellular injury. In *C. elegans*, mutations in nonchannel genes have been isolated that elicit a necrotic-like cell death response. For example, mutations in the transcription factor, LIN-26, alter cell fate causing hypodermal cells to become neuroblasts that subsequently swell and die [Labouesse et al., 1994]. In many organisms and cell types necrotic cell death appears to be a common response to extreme changes in physiological conditions induced by wide variety of chemical and physical insults. Potentially this common response may reflect the activation of a conserved necrotic cell death pathway that serves to eliminate physiologically compromised cells. Degenerin-induced cell death in *C. elegans* provides a very good model system from which to explore the possible existence of such a necrotic cell death pathway.

VII. CONCLUDING REMARKS

Genetic dissection of both programmed and pathological cell death in *C. elegans* has paved the way for the identification of molecules essential for the initiation and implementation of cell death. Genes involved in these processes have proven to be remarkably well conserved between nematodes and mammals: *ced-3, ced-9,* and the *C. elegans* degenerins all have mammalian counterparts that function in fundamentally similar ways. Such findings underscore the importance of the nematode model system in deciphering cell death mechanisms that have common elements in invertebrates and vertebrates. The ability to move back and forth experimentally between the nematode and mammalian systems, exploiting the strengths of both, has (and not doubt will continue to) rapidly advanced our understanding of death mechanisms.

The nematode has provided concrete examples of how inappropriate cell death results in pathological states, ranging from developmental death of the organism to loss of specific sensory neurons. In humans, inappropriate cell death has been clearly linked to numerous degenerative conditions, including Alzheimer's disease and ALS. Ischemic cell death in kidney, heart, and brain has been touted as the major cause of morbidity and mortality in industrialized nations [Cheung et al., 1986]. Moreover, the failure of cells to die when they normally should, as in the case of certain B-cell lymphomas, results in oncogenesis. Unquestionably, understanding cell death will be a key to extending life, and we can look to studies in the elegant worm to provide clues for altering cell viability in higher organisms.

REFERENCES

Allsopp TE, Wyatt S, Paterson HF, Davies AM (1993): The proto-oncogene *bcl-2* can selectively rescue neurotrophic factor-dependent neurons from apoptosis. Cell 73:295–307.

Alnemri ES, Fernandes TF, Halder S, Croce CM, Litwack G (1992): Involvement of *BCL-2* in glucocorticoid-induced apoptosis of human pre-B-leukemias. Cancer Res 52:491–495.

Arends MJ, Morris RG, Wyllie AH (1990): Apoptosis. The role of the endonuclease. Am J Pathol 136:593–608.

Avery L, Horvitz HR (1987): A cell that dies during wild-type *C. elegans* development can function as a neuron in a *ced-3* mutant. Cell 51:1071–1078.

Bakhshi A, Jensen JP, Goldman P, Wright JJ, McBride OW, Epstein AL, Korsmeyer SJ (1985): Cloning the chromosomal breakpoint of t(14;18) human lymphomas: Clustering around J_H on chromosome 14 near a transcriptional unit on 18. Cell 41:899–906.

Batistatou A, Merry DE, Korsmeyer SJ, Green LA (1993): *Bcl-2* affects survival but not neuronal differentiation of PC12 cells. J Neurosci 13:4422–4428.

Bear CE, Li C, Kartner N, Bridges RJ, Jensen TJ, Ramjeesingh M, Riodan JR (1992): Purification and functional reconstitution of the cystic fibrosis transmembrane conductance regulator (CFTR). Cell 68:809–818.

Bissonnette RP, Echeverri F, Mahboubi A, Green DR (1992): Apoptotic cell death induced by *c-myc* is inhibited by *bcl-2*. Nature 359:552–554.

Black RA, Kronheim SR, Sleath PR (1989): Activation of interleukin-1β by a co-induced protease. FEBS Lett 247:386–390.

Boise LH, González-Garcia M, Postema CE, Ding L, Lindsten T, Turka LA, Mao X, Nuñez G, Thompson CB (1993): *bcl-x*, a *bcl-2*-related gene that functions as a dominant regulator of apoptotic cell death. Cell 74:597–608.

Canessa CM, Horisberger J-D, Rossier BC (1993): Epithelial sodium channel related to proteins involved in neurodegeneration. Nature 361:467–470.

Canessa CM, Schild L, Buell G, Thorens B, Gautschl I, Horisberger J-D, Rossier BC (1994): Amiloride-sensitive epithelial Na$^+$ channel is made up of three homologous subunits. Nature 367:463–467.

Cerretti DP, Kozlosky CJ, Mosley B, Nelson N, Van Ness K, Greenstreet TA, March CJ, Kronheim SR, Druck T, Cannizzaro LA, Huebner K, Black RA (1992): Molecular cloning of the interleukin-1β converting enzyme. Science 256:97–100.

Chalfie M, Au M (1989): Genetic control of differentiation of the *Caenorhabditis elegans* touch receptor neurons. Science 243:1027–1033.

Chalfie M, Sulston J (1981): Developmental genetics of the mechanosensory neurons of *C. elegans*. Dev Biol 82:358–370.

Chalfie M, Sulston JE, White JG, Southgate E, Thomson JN, Brenner S (1985): The neural circuit for touch sensitivity in *Caenorhabditis elegans*. J Neurosci 5:956–964.

Chalfie M, Wolinsky E (1990): The identification and suppression of inherited neurodegeneration in *Caenorhabditis elegans*. Nature 345:410–416.

Cheung JY, Bonventre JV, Malis CD, Leaf A (1986): Calcium and ischemic injury. N Engl J Med 314:1670–1677.

Choi DW (1987): Ionic dependence of glutamate neurotoxicity. J Neurosci 7:369–379.

Choi DW (1988a): Calcium-mediated neurotoxicity: Relationship to specific channel types and role in ischemic damage. Trends Neurosci 11:465–469.

Choi DW (1988b): Glutamate neurotoxicity and diseases of the nervous system. Neuron 1:623–634.

Choi DW, Rothman SM (1990): The role of glutamate neurotoxicity in hypoxic-ischemic neuronal death. Annu Rev Neurosci 13:171–182.

Clarke PGH (1990): Developmental cell death: Morphological diversity and multiple mechanism. Anat Embryol 181:195–213.

Cleary ML, Smith SD, Sklar J (1986): Cloning and structural analysis of cDNAs for *bcl-2* and a hybrid *bcl 2/immunoglobulin* transcript resulting from the t(14;18) translocation. Cell 47:19–28.

Cohen JJ, Duke RC (1984): Glucocorticoid activation of a calcium-dependent endonuclease in thymocyte nuclei leads to cell death. J Immunol 132:38–42.

Coulson A, Sulston J, Brenner S, Karn J (1986): Toward a physical map of the genome of the nematode *Caenorhabditis elegans*. Proc Natl Acad Sci USA 83:7821–7825.

Coulson A, Waterston R, Kiff J, Sulston J, Kohora Y (1988): Genome linking with artificial yeast chromosomes. Nature 335:184–186.

Dinarello CA, Wolff SM (1993): The role of interleukin-1 in disease. N Engl J Med 328:106–113.

Driscoll M, Chalfie M (1991): The *mec-4* gene is a member of a family of *Caenorhabditis elegans* genes that can mutate to induce neuronal degeneration. Nature 349:588–593.

Duc C, Farman N, Canessa CM, Bonvalet J-P, Rossier BC (1994): Cell-specific expression of the epithelial sodium channel α, β, and γ subunits in aldosterone-responsive epithelia from the rat: Localization by in situ hybridization and immunocytochemistry. J Cell Biol 127:1907–1921.

Ellis HM, Horvitz HR (1986): Genetic control of programmed cell death in the nematode *C. elegans*. Cell 44:817–829.

Ellis RE, Horvitz HR (1991): Two *C. elegans* genes control the programmed deaths of specific cells in the pharynx. Development 112:591–603.

Ellis RE, Jacobson DM, Horvitz HR (199la): Genes required for the engulfment of cell corpses during programmed cell death in *Caenorhabditis elegans*. Genetics 129:79–94.

Ellis RE, Yuan J, Horvitz HR (199lb): Mechanisms and functions of cell death. Annu Rev Cell Biol 7:663–698.

Engel AG (1990): Congenital disorders of neuromuscular transmission. Sem Neurol 10:12–26.

Fanidi A, Harrington EA, Evan GI (1992): Cooperative interaction between *c-myc* and *bcl-2* proto-oncogenes. Nature 359:554–556.

Fire A (1986): Integrative transformation of *Caenorhabditis elegans*. EMBO J 5:2673–2680.

Friedman DB, Johnson TE (1988a): A mutation in the *age-1* gene in *Caenorhabditis elegans* lengthens life and reduces hermaphrodite fertility. Genetics 118:75–86.

Friedman DB, Johnson TE (1988b): Three mutants that extend both mean and maximum lifespan of the worm, *Caenorhabditis elegans*, define the *age-1* gene. J Gerontol Bio Sci 43:B102–109.

Fukuhara S, Rowley JD, Variakojis D, Golomb HM (1979): Chromosome abnormalities in poorly differentiated lymphocytic lymphomas. Cancer Res 39:3119–3128.

Gagliardini V, Fernandez P-A, Lee RKK, Drexler HCA, Rotcllo RJ, Fishman MC, Yuan J (1994): Prevention of vertebrate neuronal death by the *crmA* gene. Science 263:826–828.

Garcia I, Martinou I, Tsujimoto Y, Martinou J-C (1992): Prevention of prograrnmed cell death of sympathetic neurons by the *bcl-2* proto-oncogene. Science 258:302–304.

García-Anoveros J, Ma C, Chalfie M (1995): Regulation of *Caenorhabditis elegans* degenerin proteins by a putative extracellular domain. Curr Biol 5:441–448.

Hayflick L (1965): The limited *in vitro* lifetime of human diploid cell strains. Exp Cell Res 37:614–636.

Hedgecock EM, Sulston JE, Thomson JN (1983): Mutations affecting programmed cell deaths in the nematode *Caenorhabditis elegans*. Science 220:1277–1279.

Henderson S, Rowe M, Gregory C, Croom-Carter D, Wang F, Longnecker R, Kieff E, Rickinson A (1991): Induction of *bcl-2* expression by Epstein-Barr virus latent membrane protein 1 protects infected B cells from programmed cell death. Cell 65: 1107–1115.

Hengartner MO, Ellis RE, Horvitz HR (1992): *Caenorhabditis elegans* gene *ced-9* protects cells from programmed cell death. Nature 356:494–499.

Hengartner MO, Horvitz HR (1994a): Programmed cell death in *Caenorhabditis elegans*. Curr Opin Gen Dev 4:581–586.

Hengartner MO, Horvitz HR (1994b): *C. elegans* cell survival gene *ced-9* encodes a functional homolog of the mammalian proto-oncogene *bcl-2*. Cell 76:665–676.

Hengartner MO, Horvitz HR (1994c): Activation of *C. elegans* cell death protein CED-9 by an amino-acid substitution in a domain conserved in Bcl-2. Nature 369:318–320.

Hennet T, Bertoni G, Richter C, Peterhans E (1993): Expression of BCL-2 protein enhances the survival of mouse fibrosarcoid cells in tumor necrosis factor-mediated cytotoxicity. Cancer Res 53:1456–1460.

Hevelone J, Hartman PS (1988): An endonuclease from *Caenorhabditis elegans*: partial purification and characterization. Biochem Genet 26:447–461.

Hong K, Driscoll M (1994): A transmembrane domain of the putative channel subunit MEC-4 influences mechanotransduction and neurodegeneration in *C. elegans*. Nature 367:470–473.

Horvitz HR, Ellis HM, Sternberg PW (1982): Programmed cell death in nematode development. Neurosci Comment 1:56–65.

Huang M, Chalfie M (1994): Gene interactions affecting mechanosensory transduction in *Caenorhabditis elegans*. Nature 367:467–470.

Jan LY, Jan YN (1994): Potassium channels and their evolving gates. Nature 371:119–122.

Johnson TE, Lithgow GJ (1992): The search for the genetic basis of aging: The identification of geronto-genes in the nematode *Caenorhabditis elegans*. J Am Geriatr Soc 40:936–945.

Kenyon C, Chang J, Gensch E, Rudner A, Tabtiang R (1993): A *C. elegans* mutant that lives twice as long as wild type. Nature 366:461–464.

Kerr JFR, Wyllie AH, Currie AR (1972): Apoptosis: A basic biological phenomenon with wide-ranging implications in tissue kinetics. Br J Cancer 26:239–257.

Kim S-C (1994): Two *C. elegans* Genes That Can Mutate to Cause Degenerative Cell Death. Ph.D. Thesis, MIT.

Kimble J, Hirsh D (1979): The postembryonic cell lineages of the hermaphrodite and male gonads in *Caenorhabditis elegans*. Dev Biol 70:396–417.

Klass MR, Hirsh D (1976): Non-aging developmental variants of *Caenorhabditis elegans*. Nature 260:523–525.

Koch MC, Steinmeyer K, Lorenz C, Ricker K, Wolf F, Otto M, Zoll B, Lehmann-Horn F, Grzeschik K-H, Jentsch TJ (1992): The skeletal muscle chloride channel in dominant and recessive human myotonia. Science 257:797–800.

Kostura MJ, Tocci MJ, Limjuco G, Chin J, Cameron P, Hillman AG, Chartrain NA, Schmidt JA (1989): Identification of a monocyte specific pre-interleukin 1β convertase activity. Proc Natl Acad Sci USA 86:5227–5231.

Kretsinger RH (1987): Calcium coordination and the calcium fold: Divergent versus convergent evolution. Cold Spring Harbor Symp Quant Biol 52:499–510.

Kumar S, Kinoshita M, Noda M, Copeland NG, Jenkins NA (1994): Induction of apoptosis by the mouse *Nedd2* gene, which encodes a protein similar to the product of the *Caenorhabditis elegans* cell death gene *ced-3* and the mammalian IL-1β-converting enzyme. Genes Dev 8:1613–1626.

Kumar S, Tomooka Y, Noda M (1992): Identification of a set of genes with developmentally down-regulated expression in the mouse brain. Biochem Biophys Res Commun 185:1155–1161.

Kyprianou N, English HF, Isaacs JT (1988): Activation of a Ca^{2+}-Mg^{2+}–dependent endonuclease as an early event in castration-induced prostatic cell death. Prostate 13:103–117.

Labouesse M, Sookhareea S, Horvitz HR (1994): The *Caenorhabditis elegans* gene *lin-26* is required to specify the fates of hypodermal cells and encodes a presumptive zinc-finger transcription factor. Development 120:2359–2368.

Larsen PL (1993): Aging and resistance to oxidative damage in *Caenorhabditis elegans*. Proc Natl Acad Sci USA 90:8905–8909.

Liddle GW, Bledsoe T, Coppage WS (1963): A familial renal disorder simulating primary aldosteronism but with negligible aldosterone secretion. Trans Assoc Am Physicians 76:199–213.

Martinou J-C, Dubois-Dauphin M, Staple JK, Rodriguez I, Frankowski H, Missotten M, Albertini P, Talabot D, Catsicas S, Pietra C, Huarte J (1994): Over-expression of BCL-2 in transgenic mice protects neurons from naturally occurring cell death and experimental ischemia. Neuron 13:1017–1030.

McClatchey AI, Van Der Bergh P, Pericak-Vance MA, Raskind W, Verellen C, McKenna-Yasek D, Rao K, Haines JL, Bird T, Brown RH Jr, Gusella JF (1992): Temperature-sensitive mutations in the III–IV cytoplasmic loop region of the skeletal muscle sodium channel gene in paramyotonia congenita. Cell 68:769–774.

Mello CC, Kramer JM, Stinchcomb D, Ambros V (1991): Efficient gene transfer in *C. elegans*: Extrachromosomal maintenance and integration of transforming sequences. EMBO J 10:3959–3970.

Mitani S, Du H, Hall DH, Driscoll M, Chalfie M (1993): Combinatorial control of touch receptor neuron expression in *Caenorhabditis elegans*. Development 119:773–783.

Miura M, Zhu H, Rotello R, Hartwieg EA, Yuan J (1993): Induction of apoptosis in fibroblasts by IL-1β-converting enzyme, a mammalian homolog of the *C. elegans* gene, *ced-3*. Cell 75:653–660

Miyashita T, Reed JC (1992): *bcl-2* gene transfer increases relative resistance of S49.1 and WEHI7.2 lymphoid cells to cell death and DNA fragmentation induced by glucocorticoids and multiple chemotherapeutic drugs. Cancer Res 52:5407–5411.

O'Brodovich H, Canessa C, Ueda J, Rafii B, Rossier BC, Edelson J (1993): Expression of the epithelial Na⁺ channel in the developing rat lung. Am J Physiol 265:C491–C497.

Oltvai ZN, Korsmeyer SJ (1994): Checkpoints of dueling dimers foil death wishes. Cell 79:189–192.

Oltvai ZN, Milliman CL, Korsmeyer SL (1993): Bcl-2 heterodimerizes *in vivo* with a conserved homolog, Bax, that accelerates programmed cell death. Cell 74:609–619.

Oppenheim RW (1991): Cell death during the development of the nervous system. Annu Rev Neurosci 14:453–501.

Park E-C, Horvitz HR (1986a): Mutations with dominant effects on the behavior and morphology of the nematode *Caenorhabditis elegans*. Genetics 113:821–852.

Park E-C, Horvitz HR (1986b): *C. elegans unc-105* mutations affect muscle and are suppressed by other mutations that affect muscle. Genetics 113:853–867.

Ptacek LJ, George AL Jr, Griggs RC, Tawill R, Kallen RG, Barchi RL, Robertson M, Leppert MF. (1991): Identification of a mutation in the gene causing hyperkalemic periodic paralysis. Cell 67:1021–1027.

Ptacek LJ, Gouw L, Kwieciński H, McManis P, Mendell JR, Barohn RJ, George AL Jr, Barchi RL, Robertson M, Leppert MF (1993): Sodium channel mutations in paramyotonia congenita and hyperkalemic periodic paralysis. Ann Neurol 33:300–307.

Riddle DL, Swanson MM, Albert PS (1981): Interacting genes in the nematode dauer larva formation. Nature 290:668–671.

Robertson AMG, Thomson JN (1982): Morphology of programmed cell death in the ventral nerve cord of *Caenorhabditis elegans* larvae. J Embryol Exp Morphol 67:89–100.

Sentman CL, Shutter JR, Hockenberry D, Kanagawa O, Korsmeyer SJ (1991): *bcl-2* inhibits multiple forms of apoptosis but not negative selection in thymocytes. Cell 67:879–888.

Seto M, Jaeger U, Hockett RD, Graninger W, Bennett S, Goldman P, Korsmeyer SJ (1988): Alternative promoters and exons, somatic mutations and deregulation of the Bcl-2–Ig fusion gene in lymphoma. EMBO J 7:123–131.

Shimkets RA, Warnock DG, Bositis CM, Nelson-Williams C, Hansson JH, Schambelan M, Gill JR Jr, Ulick S, Milora RV, Findling JW, Canessa CM, Rossier BC, Lifton RP (1994): Liddle's syndrome: Heritable human hypertension caused by mutations in the β subunit of the epithelial sodium channel. Cell 79:407–414.

Shreffler W, Magardino T, Shekdar K, Wolinsky E (1995): The *unc-8* and *sup-40* genes regulate ion channel function in *Caenorhabditis elegans* motorneurons. Genetics 139:1261–1272.

Strasser A, Harris AW, Cory S (1991): *bcl-2* transgene inhibits T cell death and perturbs thymic self-censorship. Cell 67:889–899.

Sulston JE (1976): Post-embryonic development in the ventral cord of *Caenorhabditis elegans*. Philos Trans R Soc Lond B 275:287–298.

Sulston JE (1988): Cell lineage. In Wood WB (ed): The Nematode *Caenorhabditis elegans*. Cold Spring Harbor, NY: Cold Spring Harbor Laboratory Press, pp 123–155.

Sulston JE, Horvitz HR (1977): Post-embryonic cell lineages of the nematode *Caenorhabditis elegans*. Dev Biol 56:110–156.

Sulston JE, Schierenberg E, White JG, Thomson JN (1983): The embryonic cell lineage of the nematode *Caenorhabditis elegans*. Dev Biol 100:64–119.

Thornberry NA, Bull HG, Calaycay JR, Chapman KT, Howard AD, Kostura MJ, Miller DK, Molineaux SM, Weidner JR, Aunins J, Elliston KO, Ayala JM, Casano FJ, Chin J, Ding GJ-F, Egger LA, Gaffney EP, Limjuco G, Palyha OC, Raju SM, Rolando AM, Salley JP, Yamin T-T, Lee TD, Shively JE, MacCross MM, Mumford RA, Schmidt JA, Tocci MJ (1992): A novel heterodimeric cysteine protease is required for interleukin-1β processing in monocytes. Nature 356:768–774.

Treinin M, Chalfie M (1995): A mutated acetylcholine receptor subunit causes neuronal degeneration in *C. elegans*. Neuron 14:871–877.

Trent C, Tsung N, Horvitz HR (1983): Egg-laying defective mutants of the nematode *Caenorhabditis elegans*. Genetics 104:619–647.

Tsujimoto Y, Croce CM (1986): Analysis of the structure, transcripts, and protein products of *bcl-2*, the gene involved in human follicular lymphoma. Proc Natl Acad Sci USA 83:5214–5218.

Tsujimoto Y, Finger LR, Tunis J, Nowell PC, Croce CM (1984): Cloning of the chromosomic breakpoint of neoplastic B cells with the t(14;18) chromosome translocation. Science 226:1097–1099.

Van Voorhies WA (1992): Production of sperm reduces nematode life span. Nature 360:456–458.

Vanfleteren JR (1993): Oxidative stress and aging in *Caenorhabditis elegans*. Biochem J 292:605–608.

Vaux DL, Weissman IL, Kim SK (1992): Prevention of programmed cell death in *Caenorhabditis elegans* by human *bcl-2*. Science 258:1955–1957.

Vies DJ, Sorenson CM, Shutter JR, Korsmeyer SJ (1993): Bcl-2-deficient mice demonstrate fulminant lymphoid apoptosis, polycystic kidneys and hypopigmented hair. Cell 75:229–240.

Vowels JJ, Thomas JH (1992): Genetic analysis of chemosensory control of dauer formation in *Caenorhabditis elegans*. Genetics 130:105–123.

Wang L, Masayuki M, Bergeron L, Zhu H, Yuan J (1994): *Ich-1*, an *Ice/ced-3*–related gene, encodes both positive and negative regulators of programmed cell death. Cell 78:739–750.

White JG, Southgate E, Thomson JN (1991): On the nature of undead cells in the nematode *Caenorhabditis elegans*. Philos Trans R Soc Lond B 331:263–271.

Williams GT, Smith CA (1993): Molecular regulation of apoptosis: Genetic controls on cell death. Cell 74:777–779.

Wilson R, Ainscough R, Anderson K, Baynes C, Berks M, Bonfield J, Burton J, Connell M, Copsey T, Cooper J, Coulson A, Craxton M, Dear S, Du Z, Durbin R, Favello A, Fraser A, Fulton L, Gardner A, Green P, Hawkins T, Hillier L, Jier M, Jones M, Johnston L, Kershaw J, Kirsten J, Laisster N, Latreille P, Lightning J, Lloyd C, Mortimer B, O'Callaghan M, Parsons J, Percy C, Rifken L, Roopra A, Saunders D, Shownkeen R, Sims M, Smaldon N, Smith A, Smith M, Sonnhammer E, Staden R, Sulston J, Thierry-Mieg J, Thomas K, Vaudin M, Vaughan K, Waterston R, Watson A, Weinstock L, Wilkinson-Sproat J, Wohldman P (1994): 2.2 Mb of contiguous nucleotide sequence from chromosome III of *C. elegans*. Nature 368:32–38.

Wyllie AH (1980): Glucocorticoid-induced thymocyte apoptosis is associated with endogenous endonuclease activation. Nature 284:555–556.

Wyllie AH, Kerr JFR, Currie AR (1980): Cell death: The significance of apoptosis. Int Rev Cytol 68:251–306.

Wyllie AH, Morris RG, Smith AL, Dunlop D (1984): Chromatin cleavage in apoptosis: Association with condensed morphology and dependence on macromolecular synthesis. J Pathol 142:67–77.

Yuan J, Horvitz HR (1990): The *Caenorhabditis elegans* genes *ced-3* and *ced-4* act cell autonomously to cause programmed cell death. Dev Biol 138:33–41.

Yuan J, Horvitz HR (1992): The *Caenorhabditis elegans* cell death gene *ced-4* encodes a novel protein and is expressed during the period of extensive programmed cell death. Development 116:309–320.

Yuan J, Shaham S, Ledoux S, Ellis HM, Horvitz HR (1993): The *C. elegans* cell death gene *ced-3* encodes a protein similar to mammalian Interleukin-1β–Converting Enzyme. Cell 75:641–652.

Yunis JJ, Oken MM, Kaplan ME, Ensrud KM, Howe RR, Theologides A (1982): Distinctive chromosomal abnormalities in histologic subtypes of non-Hodgkin's lymphomas. N Eng J Med 307:1231–1236.

Zhong L-T, Sarafian T, Kane DJ, Charles AC, Mah SP, Edwards RH, Breseden DE (1993): *bcl-2* inhibits death of central neural cells induced by multiple agents. Proc Natl Acad Sci USA 90:4533–4537.

ABOUT THE AUTHORS

ROBYN LINTS is a postdoctoral fellow in Dr. Monica Driscoll's laboratory at Rutgers University, New Jersey. Her research in this laboratory has focused on mechanisms of aberrant cell death in human degenerative disease through genetic and molecular analyses of animal model systems. She received her Master of Science degree with first class honors from the University of Auckland, New Zealand. For her doctoral studies she examined transcriptional regulation of metabolism in *Aspergillus nidulans* from a genetic and molecular perspective, under the supervision of Professor Michael J. Hynes in the Department of Genetics, University of Melbourne, Australia. After obtaining her Ph.D. in 1993 she joined Dr. Driscoll's laboratory. Her postdoctoral research is funded by the Muscular Dystrophy Association of America and by a grant from the Amyotrophic Lateral Sclerosis Association.

MONICA DRISCOLL is an Assistant Professor at Rutgers, the State University of New Jersey, where she teaches undergraduate and graduate courses in molecular biology and advanced genetics. After receiving an BA in Chemistry at Douglass College of Rutgers University in 1979, she pursued graduate studies at Harvard University in the laboratory of Dr. Helen Greer. In 1985 she earned a Ph.D. for her molecular genetic studies of gene regulation in yeast. She did postdoctoral work in the laboratory of Dr. Martin Chalfie at Columbia University in New York, investigating mutant genes that induce inappropriate neuronal degeneration in the nematode *Caenorhabditis elegans*. As an Assistant Professor she has continued to research the molecular genetics of late onset inherited neurodegeneration in simple model systems. She is currently a Fellow of the Alfred P. Sloan Foundation.

Cellular Aging and Cell Death: 255–265
© 1996 Wiley-Liss, Inc.

Apoptosis of Undifferentiated and Terminally Differentiated PC12 Cells

Randall N. Pittman and Angela J. DiBenedetto

I. INTRODUCTION

A massive number of neurons and glia die in the developing nervous system. In all cases, death appears to be "programmed" (defined as being either part of a developmental program or requiring new RNA and protein synthesis), and in many, but not all, cases cells exhibit features of apoptosis [for reviews, see Clarke, 1990; Oppenheim, 1991; Ellis et al., 1991]. Although neurons die during all stages of differentiation, in vertebrates approximately half the neurons in many areas of the nervous system die during the period when synapses are being formed between neurons and their targets [Oppenheim, 1991]. Developing neurons appear to compete for a limited supply of target-derived trophic (growth) factors. Trophic factors could have their survival-promoting effects by either supporting life (e.g., by transcriptional or translational control of proteins involved in cellular homeostasis) or by suppressing death (e.g. by blocking a developmental program that would kill the neurons). Although long-term effects of trophic factors include increasing regulatory enzymes and proteins, the immediate survival-enhancing effect of trophic factors during development appears to be due to their blocking an endogenous death program [Martin et al., 1988; Oppenheim et al., 1990; Scott and Davies, 1990; Raff, 1992].

There is considerable interest in identifying specific cellular and molecular steps in programmed cell death of neurons, and several *in vitro* model systems have been established [Martin et al., 1988; Scott and Davies, 1990; Rukenstein et al., 1991; Edwards et al., 1991, Koike, 1992; Mesner et al., 1992; Pittman et al., 1993] to define the sequence of events in neuronal cell death induced by removing trophic factors such as nerve growth factor (NGF). Although an ideal model system for studying programmed cell death of neurons does not exist, desirable properties for such a system would include: *(1)* a homogeneous population of neurons available in large quantities, *(2)* a paradigm whereby neurons undergo cell death with a precise time course either as part of a developmental program or following removal of a trophic factor, and *(3)* cell death that accurately reflects events *in vivo*. It is not clear if any of the *in vitro* models established thus far accurately reflect events *in vivo*; however, the model system established in our laboratory exhibits the first two properties. We chose the pheochromocytoma cell line PC12 as a model system because of its well-characterized response to NGF [Greene and Tischler, 1976; for review, Levi and Alema, 1991; Halegoua et al., 1991] and because, upon differentiation with NGF, cells are similar to sympathetic neurons, which undergo programmed cell death, requiring new RNA and protein synthesis following removal of NGF [Martin et al., 1988; Edwards et al., 1991]. Advantages of this model system are that a homogenous population of cells is vailable in large quantics that can be maintained in either a differentiated (sympathetic neuron-like) or undifferentiated (chromaffin cell-like) state, and apoptosis can be induced in either population. As with other cell lines, a major concern is the degree of similarity between cellular, biochemical, and molecular events in PC12 cells and those of neurons *in vivo*. Except for morphological changes accompanying death [Oppenheim, 1991] and the important observation that RNA and protein synthesis inhibitors block cell death [Oppenheim et al., 1990], few details are available for neuronal death in the developing nervous system *in vivo*. Therefore, a detailed comparison between death in our system and those *in vivo* is presently not possible. Thus far, data obtained with differentiated PC12 cells indicate that cellular and biochemical changes occurring after removal of NGF are very similar to those occurring in primary cultures of sympathetic neurons [Martin et al., 1988; Edwards et al., 1991; Pittman et al., 1993; Deckwerth and Johnson, 1993].

II. DEATH OF UNDIFFERENTIATED PC12 CELLS

As with other cell types, necrosis occurs in undifferentiated PC12 cells following insults that increase intracellular calcium (e.g., calcium ionophores), deplete ATP (e.g., rotenone + 2-deoxyglucose), or increase reactive oxygen species (e.g., hydrogen peroxide). In each of these cases, necrotic death of PC12 cells is similar to that seen in other cell types [Hawkins et al., 1972; Trump et al., 1984] and at a morphological level is characterized by formation of unidirectional membrane blebs that are devoid of organelles. Moderate increases in intracellular calcium and reactive oxygen species in PC12 cells, however, appear to induce apoptosis rather than necrosis (unpublished observations).

Removing serum-containing culture medium (a source of growth factors) and replacing it with serum-free medium induces apoptosis in undifferentiated PC12 cells [Batistatou and Greene, 1991; Rukenstein et al., 1991; Pittman et al., 1993] Following serum removal, 95% of the cells die over a 3–4 day period and exhibit classic features of apoptosis, including condensation of chromatin, fragmentation of DNA into nucleosomal

"ladders," and extensive formation of apoptotic bodies. Cell death can be blocked by a variety of growth factors such as NGF, FGF, insulin, IGF-I, and IGF-II, or with agents that increase intracellular cAMP (forskolin), or with cell-permeable cAMP analogs (Fig. 1) [Rukenstein et al., 1991; Pittman et al., 1993]. Of particular interest is the observation that aurintricarboxylic acid [ATA], an agent that inhibits endonucleases as well as a number of other enzymes, blocks death of PC12 cells induced by removing serum and growth factors [Batistatou and Greene, 1991]. Additional pharmacological studies and temporal analysis of endonuclease activation support the idea that endonuclease activation and subsequent DNA fragmentation is an important and possibly controlling event in death of undifferentiated PC12 cells in serum-free medium [Batistatou and Greene, 1993]. Inhibitors of protein and RNA synthesis do not block activation of endonucleases or inhibit death following removal of serum [Rukenstein et al., 1991; Batistatou and Greene, 1993], and, in fact, inhibitors of RNA synthesis appear to increase the rate of cell death following removal of serum from PC12 cells [Pittman et al., 1993; Pittmann, unpublished observations]. RNA and protein synthesis inhibitors do, however, block death of differentiated PC12 cells [Pittman et al., 1993] (see below), as well as sympathetic neurons following removal of NGF from cultures [Martin et al., 1988].

Recent studies investigating biochemical pathways responsible for death of PC12 cells in serum-free medium have suggested that the key cell cycle kinase $p34^{cdc2}$, is activated during death of PC12 cells [Brooks et al., 1993] and that the Ras signaling pathway probably plays an important role in death [Farinelli et al., 1993]. The Ras signaling pathway is involved in a number of cellular functions, including coupling actions of growth factors and mitogens to the cell cycle and proliferation. PC12 cells transfected with a dominant negative Ras do not die following removal of serum [Farinelli et al., 1993], which suggests that the Ras signaling pathway plays an important

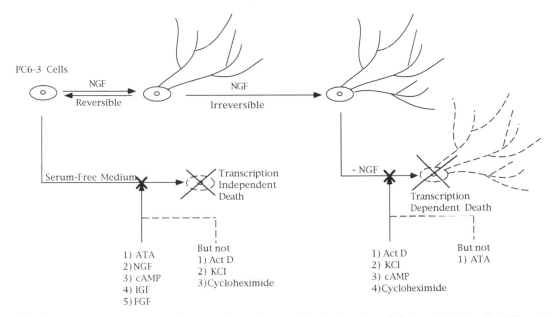

Fig. 1. *Transcription-independent and -dependent paradigms for apoptosis of PC6-3 cells. The PC6-3 subline of PC12 cells undergoes transcription-independent cell death following removal of serum and growth factors from the medium. Cell death can be blocked by a number of agents, including growth factors (NGF, FGF, IGF), cAMP analogs, and the endonuclease inhibitor ATA, but is not blocked by RNA and protein synthesis inhibitors or agents such as KCl that depolarize cells. When grown in the presence of*

NGF for less than 5–7 days, PC6-3 cells differentiate, but, upon removal of NGF, cells de-differentiate and begin proliferating (see also Fig. 3). In the continued presence of NGF, PC6-3 cells eventually become terminally differentiated, and, following removal of NGF, cells die. Death of terminally differentiated PC6-3 cells can be blocked by inhibitors of RNA and protein synthesis, by cAMP analogs, and by agents such as KCl that depolarize cells, but cannot be blocked by the endonuclease inhibitor ATA.

role in death of PC12 cells and is consistent with a "mitotic catastrophe" occurring in PC12 cells shifted into serum-free medium [Farinelli et al., 1993]. It also supports the idea that altered kinase activity in the Ras signaling pathway is a potential key component in the endogenous death program of undifferentiated PC12 cells. Therefore, characterizing patterns of phosphorylation with particular emphasis on the Ras signaling pathway should help identify cellular changes involved in cell death.

To summarize, data obtained by Greene's group as well as in several other labs suggest that apoptosis of undifferentiated PC12 cells is independent of transcription and translation and involves an endonuclease. Death may represent a "mitotic catastrophe" brought on by cells attempting to divide in the presence of a block in the cell cycle (fragmented DNA may

actually represent such a block). Testing this hypothesis and analyzing biochemical pathways (e.g., the Ras pathway) activated during this process are likely to be the focus of research efforts by several groups in the near future [see Tolkovsky et al., this volume].

III. VARIANT PC12 CELLS THAT DO NOT UNDERGO APOPTOSIS IN SERUM-FREE MEDIUM

A subline of PC12 cells, designated PC6-4, has been isolated that does not die when serum is removed from cultures [Pittman et al., 1993]. These cells do not fragment or grossly alter the state of their DNA and do not form apoptotic bodies or exhibit other motile events characteristic of apoptotic PC12 cells. Thus far, morphological and biochemical characteristics of this cell line are consistent with cells either aborting an endogenous death sequence early or never initiating steps in a cell death pathway. Based on the various mechanisms of action of agents that block death of PC12 cells in serum-free medium (discussed above; Fig. 1), properties of cells that would allow them to survive in serum-free medium would include increased basal levels of cAMP and/or increased levels of molecules involved in the response to cAMP such as protein kinase A. Autocrine production of a growth factor such as NGF, FGF, or IGF would also be an effective means for PC6-4 cells to survive following removal of serum. Preliminary data from our laboratory indicate that none of these mechanisms appear to be responsible for survival of PC6-4 cells in serum-free medium. Further biochemical characterization of these cells as well as testing them in other paradigms of apoptotic death should provide insights into mechanisms responsible for neural death and possibly those involved in neural survival. Isolation of additional variant or mutant PC12 cell lines capable of surviving in the absence of growth factors, or surviving other insults that "trigger" apoptosis, should be useful for analyzing biochemical pathways involved in cell death and for providing information on the number of different pathways that can lead to apoptosis.

IV. DEATH OF DIFFERENTIATED PC12 CELLS

Neurons *in vitro* and *in vivo* undergo programmed and/or apoptotic cell death following loss of trophic support. In most cases, it appears that new RNA and protein synthesis are required for cell death. Although *in vivo* model systems are critically important for understanding the biological context and relevance of cell death, the inability to manipulate or study isolated cellular events easily makes it difficult to use *in vivo* systems for identifying and characterizing biochemical and molecular steps involved in cell death.

Primary cultures of neurons overcome these problems to some extent by allowing environ-mental and cellular parameters to be specifically manipulated, and in many cases primary cultures of neurons provide a fairly homogeneous population of cells. The major drawback of primary cultures of neurons is that limited numbers of cells are available, which makes it difficult to characterize molecular events and biochemical pathways responsible for cell death. We started with the basic approach of identifying conditions in which a neuronal cell line became dependent on the trophic factor, NGF, for survival with the expectation that these cells would then provide an easily available homogeneous population of neural cells that could be used to identify events and genes involved in cell death.

IV.A. Characteristics of Terminally Differentiated PC12 Cells

Typically, PC12 cells are differentiated with NGF [Greene and Tischler, 1976], although FGF can also induce differentiation [Togari et al., 1985; Rydel and Greene, 1987]. If NGF (or FGF) is removed from cultures, then cells dedifferentiate and resume proliferation [Greene and Tischler, 1976]. Therefore, the initial step in establishing a model for a terminally differentiated neuronal state was to identify culture conditions that allowed PC12 cells to remain differentiated or to die following removal of NGF. Such a system might be useful not only for studying events in neuronal apoptosis of cells in either an undifferentiated or a differentiated state but also for investigating molecular events involved in irreversible commitment to a postmitotic phenotype.

Tissue culture conditions were identified that induced dependence on NGF for survival, and, following removal of NGF, PC12 cells required new RNA and protein synthesis for death [Pittman et al., 1993]. Although a number of parameters were found to be important, the factor that appeared most critical for converting PC12 cells to a terminally differentiated state dependent on NGF for survival was selecting an appropriate subline of PC12 cells (the PC6-3 subline was chosen from 19 sublines tested). It was also found that growing cells initially in a typical PC12 medium (RPMI containing 10% horse serum, 5% fetal calf serum, and 100 ng/ml NGF) for 7–10 days followed by subculturing cells and growing them in a medium more appropriate for primary cultures of neurons (DME/F12 containing 100 ng/ml NGF and 5%–10% fetal calf serum or N2 serum-free components) increased the number of cells that became dependent on NGF for survival.

After 2 weeks in the presence of NGF, differentiated PC12 cells resemble primary cultures of sympathetic or sensory neurons with rounded cell bodies and an extensive network of neurites. If NGF is removed from these cultures approximately 90% of the cells die over a 48 hour period. About 80%–90% of the cell death can be blocked with actinomycin D or cycloheximide, which is consistent with the idea that, upon removal of NGF, one or more genes are activated that are required for cell death. Besides actinomycin D and cycloheximide, other agents that block death include 35 mM KCl (which depolarizes cells and increases basal levels of intracellular free calcium) [Koike et al., 1989], NGF, bFGF, and cAMP analogs (Fig.1). Agents are routinely added to cultures at the time NGF is removed and have little effect on survival if added 12–14 hours after removing NGF.

IV.B. Morphological Changes Accompanying Cell Death

The sequence of morphological changes occurring after removing NGF have been characterized using time lapse videomicroscopy (major events are depicted above the time line in Fig. 2). Cell death occurs asynchronously in the population such that some cells begin dying as early as 6–8 hours after removing NGF, while others do not begin dying until 30 hours after removing NGF. Although the time when cell death begins in the population as a whole is quite heterogeneous, the basic sequence a cell goes through is fairly reproducible and consistent from cell to cell (a small number of cells, however, do exhibit a different mode or time course of death). For the majority of cells, the initial morphological change observed at the light microscopic level

is a thinning and "beading" of neurites. Over the next 30–60 minutes, the cell body begins to lose its smooth shape and begins to appear somewhat "lumpy." During this period some neurites begin to fragment, and initial changes are observed in chromatin when stained with a dye such as Hoescht 33342. Small condensed areas of chromatin begin to appear, and initial DNA laddering into nucleosomal fragments can be detected at this time. It is also about this time that a cell becomes irreversibly committed to die (addition of agents such as NGF are no longer able to block death). Soon after fragmentation of neurites, changes in chromatin, and commitment to death, the cell body becomes highly motile. Large membrane "blebs" are actively formed and resorbed over a period of several hours. During this highly motile period the cell often resembles a mitotically active cell undergoing cytokinesis. Eventually the cell becomes quiescent for 30 minutes or more and exhibits very little motile activity. Routinely, the final stage of death is initiated when a single transparent/translucent unidirectional bleb (resembling a bleb formed during necrosis) forms and ruptures (the cell now takes up trypan blue), leaving the cell as a compacted mass of debris.

The most striking morphological change during cell death is the highly motile, almost choreiform-like movements of the cell body, which often last several hours. This is contrasted with the smaller apoptotic bodies that are extended and resorbed from undifferentiated PC12 cells (as well as a minority of differentiated PC12 cells). Because differentiated and undifferentiated cells have very different cell morphologies, the difference in "bleb" size and formation may simply reflect differences in cytoskeletal arrangement. It is likely that the extensive movements of the cell body result from dynamic changes of the cytoskeleton. Because movements are initiated at about the time the cell becomes committed to die, it may suggest that the same biochemical/cellular changes responsible for committing the cell to die may also initiate cytoskeletal changes (as well as nuclear changes that also occur at about

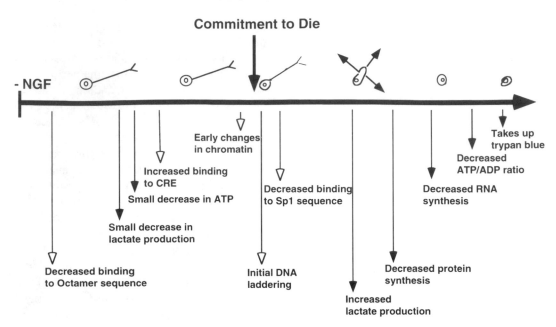

Fig. 2. *Cellular changes that occur after removing NGF from differentiated PC6-3 cells. Morphological changes are shown above the time line and include thinning and eventual loss of neurites, a period of active motility of the cell body involving formation and retraction of membrane "blebs," and finally a period of cellular quiescence followed by loss of viability. Biochemical and molecular changes occur-* *ring in the cytoplasm (solid arrows) and nucleus (open arrows) are shown below the time line in relationship to overall morphological changes. The approximate time when cells become committed to die (addition of NGF does not block cell death) relative to cellular changes is shown by the heavy arrow above the time line.*

the same time). Although the final stage of apoptotic death in these cells consists of a unidirectional bleb similar to ones that form during necrotic death, necrotic blebs and the "terminal" bleb in apoptosis are different in several ways: *(1)* whereas only a single bleb usually forms during the final stage of apoptosis, multiple blebs normally form during necrosis; *(2)* necrotic blebs often pinch off from the cell, while apoptotic blebs do not detach from the cell; and *(3)* apoptotic blebs are most often transparent and do not contain particles moving by Brownian motion inside them, whereas necrotic blebs are often translucent or phase dark and often have particles moving about inside them.

IV.C. Biochemical Changes Accompanying Cell Death

Although genetic and molecular studies have provided insights into molecules involved in

apoptosis, an understanding of the cellular mechanisms actually responsible for cell death will require identifying biochemical events and pathways controlled by or responsive to these various molecular changes. Nuclear changes including DNA fragmentation and microcondensation of the chromatin occur at about the time cells commit to die (Fig. 2); therefore, as with many other cells undergoing apoptosis [McConkey et al., 1989], changes in DNA may represent an important event in cell death of terminally differentiated PC12 cells. Endonuclease activation with subsequent DNA fragmentation may be a critical event in death of undifferentiated PC12 cells (see above); however, it does not appear to be required for death of differentiated PC12 cells. Inhibition of endonuclease activity has very little effect on the time course of cell death following removal of NGF from differentiated PC12 cells

[Pittman et al., 1993], indicating that, although it is temporally correlated with irreversible commitment to death, either a redundant event performs the same function as DNA fragmentation or another mechanism is responsible for cell death.

A number of metabolic and energetic parameters have been measured following removal of NGF from terminally differentiated PC12 cells (Fig. 2) [Mills et al., 1995]. Although some parameters such as lactate production and ATP levels show small decreases soon after removing NGF, overall metabolic parameters change little until very late in the death process. Cells maintain glucose uptake, RNA and protein synthesis, and ATP/ADP ratios until just before they die, which is consistent with cells actively killing themselves rather than dying by atrophy. Whereas few metabolic parameters change during early stages of death, lactate accumulation is biphasic during the course of cell death. Lactate levels provide information on glycolysis (increasing glycolysis increases lactate) as well as the TCA cycle and oxidative phosphorylation (decreasing activity in either increases lactate). During the first 12 hours after removing NGF (a period when little death occurs) a small but significant 5%–10% decrease in lactate production occurs, whereas at later times dying cells produce about twofold more lactate than control cells. This large increase in lactate production in later stages of cell death is inconsistent with free radicals being a major causative factor in death of differentiated cells, because the glycolytic enzyme glyceraldehyde-3-phosphate dehydrogenase is inactivated by very low levels of free radicals [Cochrane, 1991].

IV.D. Early Molecular Events Following Removal of NGF

Work by Wilcox et al. [1990] has shown that, 1 hour after removing NGF from cultured sympathetic neurons, latent herpes simplex virus-1 (HSV-1) begins reactivating. This indicates that biochemical changes are occurring in the nucleus as early as 1 hour after removing NGF-

from NGF-dependent neurons. Activation of HSV-1 in nonneuronal cells requires binding of an Oct-1 complex to an octamer sequence in the promoters of HSV-1 immediate early genes [O'Hare and Goding, 1988; Preston et al., 1988]. Several observations are consistent with a neuronal Oct-2 acting as a repressor for HSV-1 immediate early genes [Wheatley et al., 1991; Lillycrop et al., 1991]. A decrease in binding of the neuronal Oct-2 at promoters in immediate early genes is hypothesized to be responsible for reactivation of latent HSV-1. It seems likely that the same biochemical changes resulting in HSV-1 immediate early gene activation following removal of NGF (such as changes in the phosphorylation state of neuronal Oct-2 or a protein that binds to Oct-2) may also occur at host genes and activate (or inactivate) them and initiate a cascade of molecular events resulting in cell death. An important survival mechanism for viruses is to respond quickly when a cell is likely to die. For viruses present in host cells that undergo apoptosis, it seems reasonable that the virus would evolve a system of activation and cellular exit that was intimately related to early molecular events in cell death. Therefore, if current theories concerning reactivation of HSV-1 are correct, then an early event in neuronal cell death may be changes in DNA binding of an Oct-2–like transcription factor.

To define more clearly the early molecular events occurring soon after removing NGF, experiments were performed to characterize changes in nuclear protein binding to consensus DNA sequences for selected transcription factors [Wang and Pittman, 1993]. Nuclear proteins were isolated from terminally differentiated PC12 cells at different times after removing NGF and electrophoretic mobility shift assays used to characterize the binding of nuclear proteins to consensus sequences for Sp1 (a constitutive transcription factor that binds to GC-rich sequences present in many promoters), E2F (a transcription factor that activates a number of genes involved in the cell cycle), CREB (transcription factor that binds the cAMP response enhancer element,

CRE), and octamer (transcription factors in the POU family having a number of different functions including modulating differentiation). Based on DNA–protein binding experiments, molecular events occurring prior to neuronal cell death can be divided into three phases: *(1)* within 1–2 hours of removing NGF, binding to the octamer motif decreases and continues to decrease steadily until cells are committed to die; *(2)* after 5–7 hours an increase in binding to CRE occurs and is maintained through the period, when over half the cells are committed to die; and *(3)* after 14 hours (the point at which 50%–60% of the cells are committed to die and fragmentation of DNA is prominent) decreases in binding to the Sp1 motif occur.

Electrophoretic mobility shift assays were also performed with extracts from sympathetic ganglia taken before, during, and after the period of developmental cell death. These experiments indicate that changes in binding to CRE and octamer motifs also occur during the period of developmental cell death *in vivo* [Wang and Pittman, 1993]. These results along with those for terminally differentiated PC12 cells described above are consistent with alterations in protein binding to CRE and/or octamer motifs being involved in death of neurons both *in vitro* and *in vivo*. However, changes in DNA–protein binding could reflect other events occurring in these cells and have little or nothing to do with cell death. To provide support for changes in protein binding to CRE and/or octamer motifs being involved in neuronal cell death, primary cultures of sympathetic neurons were treated with double-stranded oligonucleotides and effects on cell survival determined. In theory, double-stranded oligonucleotides should act as dominant negative "promoters" unable to couple to transcriptional events, but capable of binding and sequestering transcription factors. Double-stranded octamer oligonucleotides but not other oligonucleotides increase cell death of primary cultures of sympathetic neurons [Wang and Pittman, 1993]. Single-stranded octamer oligonucleotides or double-stranded mutant octamer oligonucleotides that do not bind nuclear proteins do not increase cell death.

Gel shift assays and oligonucleotide delivery experiments are consistent with the hypothesis that cell death results from a cascade of cellular and molecular events and that an early event in programmed neuronal cell death is a decrease in binding of transcription factor(s) to octamer motif sequences. Based on these data, a working hypothesis for initial events in neuronal cell death would be that following removal of NGF the phosphorylation state of a number of proteins including an Oct-2–like (O2L) transcription factor or protein associated with an O2L complex is altered. Changes in phosphorylation would decrease DNA binding of O2L, which would activate genes where O2L acts as a repressor. One or more of the genes activated by loss of O2L repressor activity would initiate a cascade of molecular and biochemical events eventually resulting in cell death.

V. TRANSITION FROM UNDIFFERENTIATED TO TERMINALLY DIFFERENTIATED CELLS

In the presence of NGF, PC12 cells stop proliferating and begin differentiating [Greene and Tischler, 1976; Levi and Alema, 1991]. Differentiation involves a large number of morphological, biochemical, and molecular changes and is readily reversed following removal of NGF. For typical PC12 cells (but not the PC6-3 subline isolated in our laboratory) removing NGF initiates a sequence of events, including retraction of neurites, rounding of the cell body, and cell division [Greene and Tischler, 1976]. Similar events occur in PC6-3 cells after exposure to NGF for less than 5 days; however, in the continued presence of NGF, an increasing number of cells begins dying rather than de-differentiating following removal of NGF. Upon gross inspection of cultured cells, no obvious morphological change correlates with the transition from de-differentiation to death. Time-lapse videomicroscopy studies spanning the period of time when dependence on NGF for survival is acquired are currently being performed and should provide insights into events associated with this transition.

Undifferentiated PC12 cells have an endogenous pathway for apoptosis that can be triggered by serum withdrawal [Rukenstein et al., 1991], peroxide [Pittman and Chandra, unpublished observations], or calcium load [Pittman, unpublished observations]. Apoptosis does not require new RNA or protein synthesis in undifferentiated cells; therefore, a preexisting death pathway that can be activated by a variety of insults must be present in these cells. Whether these different triggers of cell death induce a similar post-translational change that initiates a common pathway for apoptosis or whether each represents a different cellular stress that activates the death pathway at distinctly different points is unknown. As NGF differentiates PC6-3 cells, it appears to access directly or indirectly and then control this preexisting death pathway, or, alternatively, NGF may induce and then control components of an additional death pathway (Fig. 3). A fairly trivial mechanism that would allow NGF to control a cell death pathway indirectly would be for NGF to down-regulate receptors for growth factors in serum that induce proliferation (e.g., receptors for EGF) and for growth

factors that generate second messengers that normally block the endogenous apoptotic pathway (e.g., receptors for IGF-I). Therefore, following removal of NGF, cells could no longer respond to proliferative or survival signals present in serum and would respond as if serum or growth factors are no longer present, that is, they die.

Morphological and biochemical characteristics of death associated with undifferentiated PC12 cells appear very different from those associated with death of differentiated PC12 cells. Particularly noteworthy are the different morphological features (small apoptotic bodies vs. large apoptotic bodies/blebs), and differential effects of inhibitors of RNA or protein synthesis (block death of differentiated cells), and inhibitors of endonucleases (block death of undifferentiated cells). Differences in apoptotic body shape and size may simply reflect major differences in cytoskeletal arrangement of differentiated compared with undifferentiated cells, while differences in the requirement for RNA and protein synthesis may indicate that one of the molecules (e.g., a regulatory component of endonuclease activation) in the apoptotic pathway must be induced in differentiated cells. At present, it is not clear whether undifferentiated

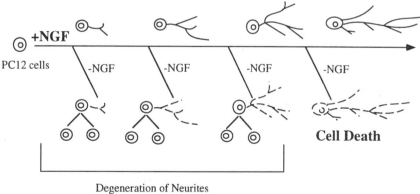

Degeneration of Neurites
Followed by Cell Proliferation

Fig. 3. *As part of differentiation, NGF may induce and/or gain control of an endogenous death program in PC12 cells. In the continuous presence of NGF, PC12 cells stop dividing and begin differentiating. If NGF is removed, neurites degenerate and cells pro-* *liferate in the presence of serum. In the PC6-3 subline of PC12 cells, however, cells eventually reach a stage of differentiation where cells die rather than de-differentiate and proliferate following removal of NGF.*

and differentiated PC12 cells share a similar apoptotic pathway or, upon differentiation, these cells acquire a partially or completely different apoptotic pathway. It would be surprising, but very interesting (and possibly quite informative), if intracellular pathways for apoptosis are different in differentiated and undifferentiated PC12 cells.

In summary, the PC6-3 subline of PC12 cells provides a powerful model system for characterizing events in neuronal cell death as well as biochemical and molecular changes associated with transition from an undifferentiated to a terminally differentiated state. Apoptosis of terminally differentiated PC6-3 cells induced by removing NGF is characterized by sequential morphological, biochemical, and molecular changes that appear quite different from those occurring in undifferentiated cells. Characterizing and comparing cell death in these different states of differentiation should provide insights into how widely apoptotic mechanisms can range within the same cell type.

REFERENCES

Batistatou A, Greene L (1991): Aurintricarboxylic acid rescues PC12 cells and sympathetic neurons from cell death caused by nerve growth factor deprivation: Correlation with suppression of endonuclease activity. J Cell Biol 115:461–471.

Batistatou A, Greene L (1993): Internucleosomal DNA cleavage and neuronal cell survival/death. J Cell Biol 122:523–532.

Brooks SF, Gibson A, Rubin LL (1993): Apoptosis induced by NGF-withdrawal from differentiated PC12 cells involves activation of p34cdc2 kinase. Soc Neurosci Abstr 19:368.12.

Clarke P (1990): Developmental cell death: Morphological diversity and multiple mechanisms. Anat Embryol 181:195–213.

Cochrane CG (1991): Cellular injury by oxidants. Am J Med 91 (3C):23S–30S.

Deckwerth TL, Johnson EM (1993): Temporal analysis of events associated with programmed cell death (apoptosis) of sympathetic neurons deprived of nerve growth factor. J Cell Biol 123:1207–1222.

Edwards SN, Buckmaster AE, Tolkovsky AM (1991): The death programme in cultured sympathetic neurons can be suppressed at the posttranslational level by nerve growth factor, cyclic AMP, and depolarization. J Neurochem 57:2140–2143.

Ellis RE, Yuan J, Horvitz HR (1991): Mechanisms and functions of cell death. Annu Rev Cell Biol 7:663–698.

Farinelli SE, Ferrari G, Greene LA (1993): A possible link between the cell cycle and apoptosis of PC12 cells. Soc Neurosci Abstr 19:368.11.

Greene L, Tischler A (1976): Establishment of a noradrenergic clonal line of rat adrenal pheochromocytoma cells which respond to nerve growth factor. Proc Natl Acad Sci USA 73:2424–2428.

Halegoua S, Armstrong RC, Kremer NE (1991): Dissecting the mode of action of neuronal growth factor. Cur Top Microbiol Immunol 165:119–170.

Hawkins HK, Rricsson JLE, Biberfeld P, Trump BF (1972): Lysosome and phagosome stability in lethal cell injury: Morphologic tracer studies in cell injury due to inhibition of energy metabolism, immune cytolysis and photosensitivity. Am J Pathol 68:255–288.

Koike T (1992): Molecular and cellular mechanism of neuronal degeneration caused by nerve growth factor deprivation approached through PC12 cell culture. Prog Neurophychopharmacol Biol Psychiatry 16:95–106.

Koike T, Martin D, Johnson E (1989): Role of Ca^{2+} channels in the ability of membrane depolarization to prevent neuronal death induced by trophic-factor deprivation: Evidence that levels of internal Ca^{2+} determine nerve growth factor dependence of sympathetic ganglion cells. Proc Natl Acad Sci USA 86:6421–6425.

Levi A, Alema, S (1991): The mechanism of action of nerve growth factor. Annu Rev Neurosci 31:205–228.

Lillycrop K, Dent C, Wheatley S, Beech M, Ninkina N, Wood J, Latchman D (1991): The octamer-binding protein oct-2 repress HSV immediate-early genes in cell lines derived from latently infectable sensory neurons. Neuron 7:381–390.

Martin D, Schmidt R, DiStefano P, Lowry O, Carter J, Johnson E (1988): Inhibitors of protein synthesis and RNA synthesis prevent neuronal death caused by nerve growth factor deprivation. J Cell Biol 106:829–844.

McConkey D, Hartzell P, Nicotera P, Orrenius S (1989): Calcium-activated DNA fragmentation kills immature thymocytes. FASEB J 3:1843–1849.

Mesner P, Winters T, Green S (1992): Nerve growth factor withdrawal-induced cell death in neuronal PC12 cells resembles that in sympathetic neurons. J Cell Biol 119:1669–1680.

Mills JC, Nelson D, Erecinska M, Pittman RN (1995): Metabolic and energetic changes during apoptosis in neural cells. J Neurochem 65 (in press).

O'Hare P, Goding CR (1988): Herpes simplex virus regulatory elements and the immunoglobulin octamer domain bind a common factor and are both targets for virion transactivation. Cell 52:435–445.

Oppenheim R (1991): Cell death during development of the nervous system. Annu Rev Neurosci 14:453–501.

Oppenheim R, Prevette D, Tytell M, Homma S (1990): Naturally occurring and induced neuronal death in the chick embryo *in vivo* requires protein and RNA synthesis: Evidence for the role of cell death genes. Dev Biol 138:104–113.

Pittman RN, Wang S, DiBenedetto AJ, Mills JC (1993): A system for characterizing cellular and molecular events in programmed neuronal cell death. J Neurosci 13:3669–3680.

Preston CM, Frame MC, Campbell MEM (1988): A complex formed between cell components and an HSV structural polypeptide binds to a viral immediate early gene regulatory DNA sequence. Cell 52:425–434.

Raff M (1992): Social controls on cell survival and cell death. Nature 356:397–400.

Rukenstein A, Rydel R, Greene L (1991): Multiple agents rescue PC12 cells from scrum-free cell death by translation- and transcription-independent mechanisms. J Neurosci 11:2552–2563.

Rydel R, Greene L (1987): Acidic and basic fibroblast growth factors promote stable neurite outgrowth and neuronal differentiation in cultures of PC12 cells. J Neurosci 7:3639–3653.

Scott SA, Davies AM (1990): Inhibition of protein synthesis prevents cell death in sensory and parasympa-

thetic neurons deprived of neurotrophic factor *in vitro*. J Neurobiol 21:630–638.

Togari A, Dickens G, Kuzuya H, Guroff G (1985): The effect of fibroblast growth factor on PC12 cells. J Neurosci 5:307–316.

Trump BF, Berezesky IK, T. Sato T, Laiho KU, Phelops PC, De Claries N (1984): Cell calcium, cell injury and cell death. Environ Health Prespect 57:281–287.

Wang S, Pittman RN (1993): Altered protein binding to the octamer motif appears to be an early event in programmed neuronal cell death. Proc Natl Acad Sci USA 90:10385–10389.

Wheatley S, Dent C, Wood J, and Latchman D (1991): A cellular factor binding to the TAATGARAT DNA sequence prevents the expression of the HSV immediate-early genes following infection of nonpermissive cell lines derived from dorsal root ganglion neurons. Exp Cell Res 194:78–82.

Wilcox CL, Smith RL, Freed CR, Johnson EM (1990): Nerve growth factor–dependence of herpes simplex virus latency in peripheral sympathetic and sensory neurons *in vitro*. J Neurosci 10:1268–1275.

ABOUT THE AUTHORS

RANDALL N. PITTMAN is Associate Professor of Pharmacology at the University of Pennsylvania School of Medicine in Philadelphia, where he teaches medical and graduate school classes in pharmacology and developmental neurobiology. He was a Morehead Scholar at the University of North Carolina, Chapel Hill, and received his B.S. in Chemistry in 1972. His interest in cell death began in 1974 while working with Ron Oppenheim on the effects of neuromuscular activity on cell death of developing motoneurons. He received his Ph.D. in 1981 from the University of Colorado, where he worked on functional properties of β-adrenergic receptors in the laboratory of Perry Molinoff. Postdoctoral work with Paul Patterson at Harvard Medical School and California Institute of Technology involved mechanisms of neurite outgrowth in the developing nervous system. Since 1984, Dr. Pittman has been on the faculty of the University of Pennsylvania. Early work in his laboratory involved the role of proteases in axon outgrowth and regeneration for which he received the Mike Taff Memorial Award for Scientific Research from the American Paralysis Association. More recently, the focus of research in his laboratory has been on cellular and molecular events in programmed neuronal cell death.

ANGELA J. DiBENEDETTO is a postdoctoral fellow in the laboratory of Dr. Randall N. Pittman at the University of Pennsylvania School of Medicine, where she conducts research directed at the identification of genes involved in neuronal cell death. After receiving a B.S. degree from the State University of New York at Binghamton, she was awarded an NSF predoctoral fellowship and pursued doctoral research at Cornell University in the laboratory of Dr. Mariana Wolfner. There she earned a Ph.D. in 1989 for work on sex-specific gene expression in *Drosophila*. Before her present work on cell death, her postdoctoral research with Dr. Pittman involved the study of tissue plasminogen activator and its role in neurite outgrowth.

Cellular Aging and Cell Death: 267–282
© 1996 Wiley-Liss, Inc.

Signaling for Survival: The Biochemistry of NGF Dependence in Primary Sympathetic Neurons

Aviva M. Tolkovsky, E. Anne Buckmaster, Susan N. Edwards,
Catherine D. Nobes, and Kanwar Virdee

I. INTRODUCTION

Nineteen ninety-four will undoubtedly be remembered as the year when apoptosis and its discoverers [Kerr et al., 1972; Wyllie et al., 1980] finally attained the eminence that they deserve. The realization that induction of apoptosis might be exploited to kill neoplastic cells and clear cells in inflammatory disorders [see Kerr et al., 1994; Fisher, 1994; Savill, 1993 for reviews] has generated immense interest because any cell that enters apoptosis is thereby committed to die, with the added bonus that neighboring cells remain unaffected. In the nervous system, however, neurons become largely postmitotic by early childhood, and therefore each cell that dies cannot be replaced. A prime concern of therapies aimed at treatment of neurodegenerative disorders, and of neurodegeneration due to injury, is therefore to prevent cell death rather than implement it. Moreover, although diseased neurons may die by apoptosis, it appears that cells displaying the earliest detectable signs of apoptosis have already passed the death commitment point [Martin et al., 1988; Edwards and Tolkovsky, 1994], and it is therefore impossible to rescue these cells. Until it is clear that the molecular processes that designate a cell to undergo apoptosis, as opposed to other forms of cell death like necrosis or autophagy [Clarke, 1990], begin prior to the death commitment point, the increasing demand for proof of apoptosis in cases of neurodegeneration may not be helpful in defining a strategy for survival. It is thus of vital importance to concentrate on defining the molecular basis of the death commitment point, whatever the mechanisms that execute the cell death program after this point turn out to be.

Like proliferating cells, postmitotic neurons, and perhaps all cells [Raff, 1992], are prevented from dying by multiple survival factors [Barres et al., 1993]. These factors are supplied by autocrine or paracrine mechanisms, but survival may also be supported from within the cell by oncogenic kinases such as bsr-abl [McGahon et al., 1994] and by members of the *bcl-2* gene family, which include the death-suppressing genes *ced-9* [Hengartner et al., 1992; Hengartner and Horvitz, 1994], *bcl-2* [Vaux et al., 1988; Nunez et al., 1990], and *bcl-XL* [Boise et al., 1993], and their homologous inhibitors *bax* [Oltvai et al., 1993] and *bcl-XS*, [Boise et al., 1993; for reviews, see Korsmeyer et al., 1993, Vaux et al., 1994, Reed, 1994]. Overexpression of *bcl-2* also supports survival of cultured neurons [Garcia et al., 1992; Allsopp et al., 1993]. The existence of survival genes may help to explain why apoptosis in cancer may not be easy to achieve. Cells expressing *bcl-2*, especially in combination with mutated p53 and oncogenes, are highly resistant to chemotherapeutic agents [Arends et al., 1994] and evolve new mutations at alarming rates. Since, on the other hand, neurons are longlived postmitotic cells (neuronal cancers are virtually nonexistent in adults), the selective expression of antiapoptotic genes such as *bcl-2* in neurons might be an appropriate aim for survival therapy. If so, how might their expression in particular neurons be induced or delivered and, at the same time, the action of their cellular antagonists annulled?

The realization that diverse survival factors suppress death by activating post-translational mechanisms [Edwards et al., 1991; Ruckenstein et al., 1991; Harrington et al., 1994] suggests that finding antagonists (in cancer) or agonists (in the nervous system) to these survival signaling pathways might be another useful therapeutic strategy for implementing death in cancer or neuronal survival. However, at present little is known about the signal transduction pathways that enable survival, nor have their cellular targets been defined. This chapter summarizes some of our studies on the mechanisms used by NGF-dependent superior cervical ganglion (SCG) neurons to prepare for the death commitment point [see also Pilar and Bruses, and Pittman in this volume and Martin et al., 1992] and outlines possible mechanisms that may be utilized to prevent their death due to survival factor deprivation (for a concise review on NGF and other neurotrophins see Snider, 1994, and for a general overview of physiological cell death in the nervous system see Clarke, 1990, and Oppenheim, 1991).

II. DEFINING THE DEATH COMMITMENT POINT

Based partly on the pioneering work by Levi-Montalcini [1965] and colleagues, Barde [1989] summarized the physiological mechanisms by which neurotrophic factors such as NGF control cell death in the nervous system as follows: NGF-dependent neurons die because they compete for a limited supply of NGF provided by their target tissues, a model recently borne out by the demonstration of total decimation of sympathetic neurons in transgenic mice in which NGF [Crowley et al., 1994] or the NGF receptor trkA [Smeyne et al., 1994] were deleted by homologous recombination. Furthermore, death due to NGF-deprivation occurs by a stochastic process in the sense that none of the neurons are specifically

predetermined to die, unlike neurons in the nematode nervous system where the same set of cells is always designated to die [Ellis et al., 1991]. In rats, 33%–40% of all SCG neurons die [Hendry and Campbell, 1976; Wright et al., 1983] between 3–6 days postnatally, soon after which the surviving neurons begin to show decreased dependence on NGF. This resistance to NGF deprivation increases with time, but the dependence on NGF for survival is never totally abrogated. SCG neurons removed from neonate rat pups are thus poised to die, and the properties of neurons at peak vulnerability in relation to NGF withdrawal can consequently be studied. Moreover, because the neurons can be further cultured to stages marked by ever decreasing NGF dependence, it is also possible to study the survival strategies that these neurons develop after their natural period of cell death is completed.

II.A. The Death Program in SCG Neurons May Not Require De Novo Gene Expression

Figure 1 summarizes the events that have been measured during the period approaching the death commitment point in SCG neurons *in vitro* after NGF withdrawal [Martin et al., 1988; Edwards et al., 1991; Martin et al., 1992; Deckwerth and Johnson, 1993; Edwards and Tolkovsky, 1994]. It shows that, after NGF deprivation, death can be prevented by activating two different mechanisms, as follows.

II.A.1. Inhibition of RNA and protein synthesis. During the first 12–20 hours after NGF withdrawal (depending on neuron age in culture), SCG neurons can be rescued by inhibitors of protein synthesis like cycloheximide (CHX) [Martin et al., 1988] or RNA synthesis (actinomycin D), as well as by NGF, cyclic AMP analogs (first demonstrated to be survival factors by Rydel and Greene, 1987), and, in older neurons, by depolarization with high potassium [Koike et al., 1989; Franklin and Johnson, 1992]. Complete rescue by CHX is achieved despite there having been a marked increase in catabolic metabolism, which is manifested early on by a rapid net conversion

After NGF is withdrawn

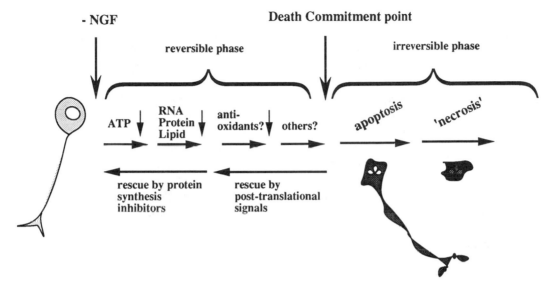

Fig. 1. *After NGF is withdrawn, neurons activate an increasing number of catabolic processes beginning with net loss of ATP. All of these steps appear to be reversible, reversibility being achieved through two mechanisms: (1) rescue by protein/ RNA synthesis inhibitors; and (2) rescue by activation of post-translational processes. Coincidentally with the death commitment point (when neurons are irreversibly committed to die) neurons display all the hallmarks of apoptosis, a particularly useful sign being DNA condensation into "blobs" that can be visualized using acridine orange or Hoechst 33342 stains. Older SCG neurons rapidly progress into "secondary necrosis," where autophagic processes appear to dominate. For further details, see description in text.*

of ATP into purine nucleosides that are transported out of the cells [Tolkovsky and Buck-master, 1989]. Following this phase one finds a decrease in integrity of RNA, proteins, and lipids and reduced uptake of nutrients [Deckwerth and Johnson, 1993]. The simplest explanation proposed by Martin et al. [1988] to account for how CHX prevents death after NGF-deprivation was that these inhibitors are suppressing the expression of a cell death program consisting of "suicide" genes. It follows that NGF must therefore be supporting survival by inhibiting the expression of these "suicide" genes and not by inducing the expression of an inhibitor, because, in the latter case, CHX would activate rather than inhibit cell death.

Other explanations for the effects of CHX, however, are also possible. For example, Ratan

et al. [1994] showed that death due to chemical depletion of glutathione in cortical neurons can also be prevented by inhibition of protein synthesis by CHX or anisomycin, but the survival-promoting actions of CHX were most strongly correlated with shunting the increased amounts of cysteine that became available due to inhibition of protein synthesis into glutathione synthesis. However, it is unlikely that the survival effects of CHX in SCG neurons is due solely to a similar sparing of cysteine, because the protective effects of CHX after profound glutathione depletion were achieved by blocking only 50% of protein synthesis, whereas in SCG neurons near-complete (over 95%) inhibition of protein synthesis is required to prevent cell death. We have found that persistent (>15 hours) exposure of SCG neurons *in vitro* to CHX or Act D at concentrations that block protein synthesis results in a threefold increase in peak influx of calcium through voltage-dependent calcium channels, thus indicating the presence of a labile inhibitor of voltage-dependent calcium channels [Murrell and Tolkovsky, 1993]. By analogy, it is possible that rapidly turning over, labile proteins such as proteases orchestrate the actions of more stable inhibitor(s) of cell death. Removal of NGF would allow the labile proteins to remove these inhibitors, but inhibition of protein synthesis after NGF withdrawal would cause the disappearance of labile proteins before they had the chance to degrade the inhibitors of cell death. In this scenario, NGF-maintained neurons could already be expressing a competent cell death program but one that is kept under control by death inhibitors.

Proteolytic degradation is widely used for control of key cellular functions. For example, degradation of cyclins defines the ultimate time during which entry into S and M phases of the cell cycle is permitted [Minshull, 1993]. Recently, specific proteases that cleave substrates after aspartate residues have been implicated in programmed cell death. The nematode gene *ced-3* was found to be homologous to the interleukin-1β–converting enzyme (ICE), which promotes apoptosis in fibroblasts [Yuan et al., 1993, Miura et al., 1993]. Another ICE-like gene, nedd2, has been found to be widely expressed in apoptotic tissues during mouse development [Kumar et al., 1994] and data from Kaufmann [1989] and Lazebnik et al. [1994] suggests that apoptosis induced by chemotherapeutic agents is also accompanied by activation of ICE-like proteases. Killing by cytotoxic lymphocytes is also mediated by the release of granules containing asp-specific serine proteases [Shi et al., 1992; Heusel et al., 1994]. Kumar et al. [1994] suggest that the expression of nedd2 is elevated in tissues undergoing apoptosis but is downregulated in tissues where programmed cell death is completed.

No candidate genes that are required to be expressed de novo in order to commit SCG neurons to die have been identified [see Pittman, this volume] but there is some evidence supporting the notion of pre-existence of an apoptotic machinery in young SCG neurons in that inhibitors of topoisomerase II (teniposide or mitoxanthrone) [Tomkins et al., 1994] or arabinose nucleosides [Wallace and Johnson, 1989] promote the activation of apoptosis in the presence of NGF. The question of how these inhibitors activate the execution of apoptosis is of great interest because it may help to pinpoint a crucial participant in the death program that is a target for the survival signal transduction pathway. The identity of this target is especially intriguing since we have found that cytosine arabinoside does not block the better characterized portion of the NGF-signal transduction pathway that lies between the NGF receptor trkA and cFos protein expression [Tomkins et al., 1994], which includes the activation of p21ras, a key mediator of survival (see below).

II.A.2. Inhibition by post-translational signals.
Although CHX can rescue neurons for several hours after NGF-deprivation, this period ends prior to the attainment of the absolute death commitment point because NGF, high K, and CPTcAMP can still rescue neurons after the ability of CHX to inhibit cell death has ceased [Edwards et al., 1991; Deckwerth and Johnson, 1993], and this activity is not dependent on protein synthesis. The simplest explanation is that completion of the CHX-inhibitory phase defines a point at which all the proteins that are essential for apoptosis to proceed are present but that the activation of these proteins can still be controlled by post-translational mechanisms. In several cell types CHX promotes rather than inhibits apoptosis [for example Vaux and Weissman, 1993]. It might therefore be argued that the inability of CHX to support survival as late as NGF or CPTcAMP is due to the development of a similar kind of "toxic effect" and not to the loss of ability to inhibit synthesis of crucial components of the suicide program. Nevertheless, this "toxic effect" too can be overcome by NGF and the other survival factors under conditions where protein synthesis is also inhibited, sug-

gesting that the actions of targets mediating CHX toxicity are also inhibitable by survival factors. The observation that NGF and other survival factors can rescue neurons after the CHX-inhibitory phase is completed regardless of whether protein synthesis is inhibited argues further against a mechanism of glutathione depletion underlying the ultimate point of death commitment.

Taken together, these data suggest that post-translational events are sufficient for suppressing apoptosis, and, furthermore, they indicate that the components of signaling that activate gene expression may not be involved in this short-term (up to 3 days) survival, although in the long-term protein synthesis is of course absolutely required. These findings may be significant for studies of neurodegeneration in that they suggest that survival of growth factor–dependent neurons may be implemented by directly activating cytoplasmic signaling pathways.

II.B. Older Neurons May Develop Cooperative Mechanisms to Prevent Accidental Cell Death

After death commitment, but almost in synchrony with it, cultured SCG neurons display the morphological and biochemical changes typical of apoptosis: Membrane blebbing occurs, access to the Hoechst dye 33342 is accelerated, and DNA becomes condensed into blobs that rapidly fragment into oligonucleosomal segments [Edwards and Tolkovsky, 1994]. If one were applying the trypan blue exclusion parameter widely used in thymocyte research as a test of viability, neurons at early stages of apoptosis would still be considered "alive," although they are clearly committed to die. In culture, the neurons flux through the apoptotic phase into a phase we have termed "ghost" because the DNA disappears; these cells are shriveled, indicating that they may have undergone autophagy, and eventually the plasma membrane ruptures. *In vivo*, when about 15,000 SCG neurons die over 72 hours [Wright et al., 1983], apoptotic neurons are phagocytosed partly by resident macrophages that become extensively activated by 3 days postnatally. Macrophages remain activated up to 10 days, the latest time followed [A. Tolkovsky and H. Perry, unpublished observations].

By comparing the kinetics of cell death after NGF deprivation in populations of younger and older cultured neurons, it has become clear that SCG neurons can change the relative timing of the four phases of death outlined in Figure 1, illustrated schematically in Figure 2. Neurons newly cultured from neonate pups show a lag phase of about 3–5 hours before onset of death, but neurons cultured for 7 days show an extended lag time of 16–20 hours. In contrast, an analysis of the rates of entry into apoptosis (from the end of the lag period) and exit from apoptosis to the "ghost" phase suggests that older neurons make these transitions twice as fast as their younger counterparts [Edwards and Tolkovsky, 1994]. The rapid flux through apoptosis into the later phases of death is a probable cause for the difficulties in showing clean DNA fragmentation in older neurons using conventional ethidium bromide staining [Edwards et al., 1991; Batistatou and Greene, 1991].

One mechanism by which the older neurons may develop more resistance to NGF deprivation is to build more rate-limiting steps that intervene between NGF withdrawal and death commitment. A detailed mathematical analysis of the ~-like shape of the rescue curves excludes, however, the possibility that cell death is driven by a sequential series of (≤4) steps and instead supports the idea that cooperativity plays a role in the control of death commitment. This idea is encapsulated in the concept of a "death set," outlined in Figure 3. If the "death set" is defined as a group of proteins/reactions that must be active synchronously in order to effect cell death, then, clearly, the larger the number of conditions that need to be fulfilled to activate the effector processes, the more resistant the cell becomes to cell death by accident. To account for the greater apparent rate at which the older cells flux through apoptosis, one might also postulate that older

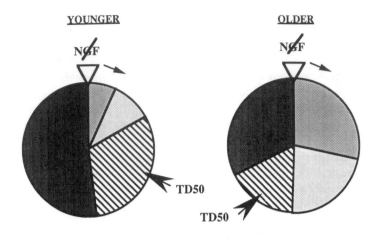

Fig. 2. *Susceptibility of neurons to NGF depriva-tion changes during their maturation. When NGF is withdrawn from neonatal neurons cultured briefly with NGF, neurons display a lag time of 3–5 hours before they plunge into apoptosis, the 50% death com-mitment point (TD50) occurring about 9 hours later, i.e. at about 12–14 hours. In neurons cultured for about 8 days, the lag period increases to about 16 hours, before neurons begin to plunge into apoptosis with a TD50 occurring about 7 hours later, i.e. at about 23 hours. Older neurons also appear to de-grade themselves faster than the young neurons once apoptosis has begun (the transitions healthy → Apo and Apo → Ghost are 2× the rates of those measured in young neurons). For further details, see descrip-tion in text.*

neurons contain higher concentrations of, or more active death-effector proteins. Our ob-servation that neurons are protected from death after NGF withdrawal when they are incubated at or below 23–24°C but not above 25–26°C (Fig. 4) suggests that one of the components may be an integral membrane protein. In keep-ing with the concept [Jacobson and Evan, 1994] that a central controller might coordi-nate and unleash the active phase of cell death, it is also possible to postulate that only one protein (or reaction) that serves as the central controller is the barrier to cell death in both young and older neurons but that higher con-centrations of this protein in the older neurons release more active ligand that autoregulates controller activity by binding to it at multiple sites. Thus, the more ligand available, the more rapidly the controller will be activated and with greater cooperativity. Cooperativity by synchrony is found in many processes where

tight control coupled to rapid onset is critical; for example, cytostatic factor has been postu-lated to consist of four reactions that must be activated concomitantly to arrest frog oocytes after first meiosis [Maller, 1993]. That mul-tiple steps may be used to provide a barrier against cell death was also suggested by Cohen [1993], although the notion of cooperativity is not implicitly stated.

Several other differences between younger and older neurons lend support to the sugges-tion that regulatory mechanisms have been developed in older neurons to prevent acciden-tal death (the resistance to NGF deprivation that develops in older cultures established from young neurons is also found in SCG neurons removed from older animals, suggesting that this phenomenon is not a culture artifact). For example, younger neurons are rescued much less efficiently with CHX: Only about 15% of all the newly isolated neonatal neurons that are

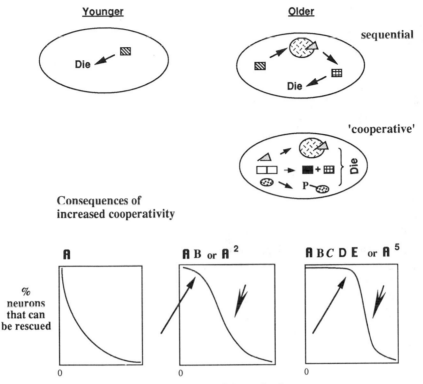

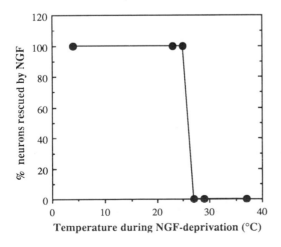

Time after NGF-deprivation

Fig. 3. *A barrier to cell death can be acquired by adding more conditions that need to be fulfilled (a "death set") in order to commit neurons to die. In this example, younger neurons contain one step to death commitment (here designated DIE), whereas older neurons contain three steps to death commitment. Theoretically, the three steps can be linked sequentially, but an analysis of our data suggests that the steps are arranged as if they were occurring concomitantly, thereby yielding a "cooperative" approach to death commitment. The graphs depict the shapes of the death commitment curves if one, two, or five components were required to be activated simultaneously in order to commit death: Note the longer lag phases (long arrows) and the increasing slopes (short arrows).*

Fig. 4. *SCG neurons incubated for 15 hours in the absence of NGF become committed to die if they were held at temperatures that are higher than the solid-to-fluid transition temperatures typical of plasma membranes but not if they are held below these temperatures. Single cell suspensions of SCG neurons from newborn rat pups were purified [Edwards and Tolkovsky, 1994] and incubated for 15 hours in L15 plating medium containing 3% rat serum at the indicated temperatures in the absence of NGF after which they were transferred to L15-CO₂ medium containing 3% rat serum and 20 ng/ml NGF and left to grow for 36 hours.*

maintained by NGF are rescued if neurons are plated on collagen in the absence of NGF, and about 50% are rescued if CHX is added to neurons plated on laminin. All of the neurons, however, are rescued at this stage by NGF when it is added together with CHX. Furthermore, newly plated neonatal neurons cannot be rescued by depolarization with high K. Since the "young" and older neurons appear to contain the same types of voltage-dependent calcium channels, it may be that either the sensory mechanisms for transducing the calcium signals are unavailable in the younger neurons or that the calcium-induced survival signals are overridden by inappropriately localized elevated calcium due to impaired calcium homeostasis. That high K is not nonspecifically toxic to the neurons is shown by their dying by apoptosis with normal kinetics [S.N. Edwards, unpublished data]. A similar change in the responsiveness to high K as a survival factor was described for chick sensory neurons [Acheson et al., 1987].

In summary, although some progress has been made in describing how SCG neurons die upon NGF-deprivation, our understanding of the molecular details of cell death mechanisms prior to the death commitment point in SCG neurons is still severely limited. It is not known which (or whether there are) components of the death program that are required to be synthesized *de novo* after NGF withdrawal. It is also not clear whether death in SCG neurons is mediated directly by these proteins or by their activation of signal transduction pathways, akin to those activated by TNF or Fas/Apo-l [Wong and Goeddel, 1994]. It is of interest in this regard to note that Rabizadeh et al. [1993] have shown that, in cells expressing the low-affinity NGF receptor p75, which is a member of the TNF receptor superfamily, apoptosis may be enhanced when p75 is unoccupied by NGF. Clearly, one of the most urgent tasks is to define the molecular identity of the processes underlying the death commitment point in these, and other, neurons.

III. MECHANISMS THAT SUPPRESS IMPLEMENTATION OF THE CELL DEATH PROGRAM

Although SCG neurons *in vivo* are supported solely by NGF at the time when they undergo physiological cell death, *in vitro* they can be supported by other survival factors with varying efficacies, thus potentially providing the opportunity to uncover several components of the survival signaling pathway. It is theoretically possible that all survival signals converge on a central controller, thus identifying a "hub" for all the survival signal transduction pathways. Alternatively, if death commitment is driven by a "death set," survival factors may operate on nonconvergent parallel pathways (Fig. 5) since inactivation of any one member of the set would be sufficient to disable the entire set from activating apoptosis.

III.A. p21Ras Sits at a Node of Several Survival Signal Transducing Pathways

In SCG neurons, NGF activates the tyrosine phosphorylation of trkA, which in turn mediates a series of phosphorylation events [Nobes and Tolkovsky, 1994]. Activation of trkA causes the activation of p21ras in SCG and other primary neurons, demonstrated by enhanced exchange of GTP for GDP [C. Nobes, unpublished data; Ng and Shooter, 1993]. In SCG neurons, the activation of p21ras appears to be both necessary for the survival response to NGF and sufficient for survival *per se* because we have found that (1) injection of oncogenic or cellular p21ras by pressure trituration, a kind of scrape loading [Borasio et al., 1989], promotes survival and neurite outgrowth [Nobes et al., submitted], and (2) injection of Fab fragments of the blocking antibody Y13-259 in the presence of NGF causes the neurons to die. It is clear that the ability of NGF to promote survival through p21ras is dependent on the kinase activity of trkA but is not dependent on PKC, since the inhibitor staurosporine (which inhibits trkA and PKC as well as other kinases) blocks all phosphotyrosine phosphorylation events induced by

Convergence model

'Death set' model

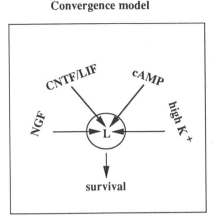
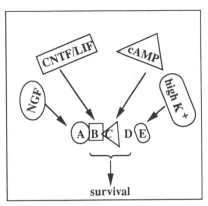

Fig. 5. *Two potential models for how survival signals induced by different survival factors might inactivate cell death. The convergence model postulates that signals induced by the four survival factors for SCG neurons converge on a common intermediate that, when modified either directly or indirectly, prevents cell death. In the death set model, it is postulated that if five conditions were needed to be fulfilled simultaneously in order for the neurons to become committed to die (as might be the case for older neurons), then inactivation of any one of these components will be sufficient to prevent cell death. Thus survival signals need not converge on a single step or molecule and so it may appear that there are multiple survival signalling pathways in such cells.*

NGF and inhibits survival, but the block in survival in the presence of staurosporine is overcome by the concomitant expression of p21ras(val12) protein [Nobes et al., 1991, and Nobes et al., submitted]. By contrast, no effects on TrkA signaling or survival are found with the staurosporine analog Ro31-8220.

The chlorophenylthio derivative of cyclic AMP (CPTcAMP) is an excellent long-term survival factor for SCG neurons *in vitro*, especially when the neurons are plated on laminin [Nobes and Tolkovsky, 1995]. Despite substantial neurite regeneration, the cell bodies remain small compared with the hypertrophy found with NGF [Rydel and Greene, 1987]. The survival effect of CPTcAMP is not dependent on neurite regeneration, however, since neurons in suspension maintained by CPTcAMP for 2 days are all rescued when plated on laminin in the presence of NGF or CPTcAMP [K. Virdee, unpublished data]. The survival activity of CPTcAMP is directly mediated by PKA, shown by the ability of the inactive Rp stereoisomer of thio-cyclic AMP (Rp-cAMPS) to block exclusively the survival effects of

CPTcAMP, but not those induced by NGF or the cytokines CNTF or LIF. The intracellular activation of PKA is further corroborated by the ability of Rp-cAMPS to block a histone kinase activity induced specifically in SCG neurons by CPTcAMP [Nobes and Tolkovsky, 1995].

CNTF (a survival factor first discovered for its potent survival actions on ciliary ganglion parasympathetic neurons [Barbin et al., 1984] and LIF (whose β-receptor subunit is also part of the CNTF signaling–receptor complex [Ip et al., 1992] activate the IL-6 receptor signaling subunit gp130. gp130 has been shown to mobilize soluble tyrosine kinases of the Jak/Tyk family [Boulton et al., 1994] and to activate transcription via p91 proteins [Daly and Reich, 1994]. CNTF and LIF are only short-term (<3 days) survival factors for purified SCG neurons [Burnham et al., 1994]. Like CPTcAMP, survival and regeneration in neonatal neurons in response to CNTF or LIF are also enhanced when the neurons are plated on laminin, although it is clear that the receptors for CNTF and LIF are capable of transmitting signals even when the neurons are plated on collagen

p21Ras as a hub for survival signalling pathways

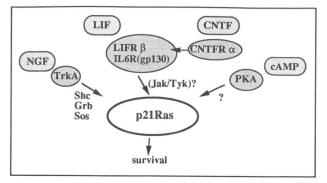

Bifurcation of survival signalling down stream from p21Ras

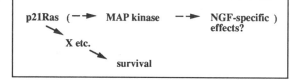

Fig. 6. *Top: P21ras appears to sit in a hub of three survival signaling pathways: that activated by NGF via trkA and tyrosine phosphorylation; that activated by CNTF (through its α-binding receptors) and LIF through mechanisms as yet unknown, possibly involving tyrosine phosphorylation; and that activated by cyclic AMP analogs through PKA and not through tyrosine phosphorylation events. Bottom: Although it has been postulated that MAP kinase is activated downstream from p21ras, in SCG neurons only NGF-induced survival is correlated to MAP kinase activation. The other survival factors CNTF/LIF and cAMP either activate MAP kinase only briefly or not at all, respectively, yet cAMP in particular is a long-term survival factor for these neurons. These MAP kinases are not obligate components of the survival signaling pathway in SCG neurons.*

(for example, over 85% of the neurons plated on collagen express c-Fos protein in response to LIF [C. Nobes, unpublished data]. After 3 days, CNTF and LIF induce SCG neurons to die, killing nearly 75% of older cultured SCG neurons even in the presence of NGF [Kessler et al., 1993; Burnham et al., 1994]. Why the cytokines induce cell death in the older neurons in the presence of NGF is obviously of prime interest because it may pinpoint some aspect of the NGF-survival signal transduction machinery downstream from trkA that is only found in older neurons.

To explore whether, in addition to NGF, the other survival factors utilize p21ras to mediate survival, we examined whether the Fab fragments of Y13–259, the neutralizing antibody to p21ras, affected the ability of these factors to promote survival. Surprisingly, we found that Fab fragments of Y13–259 blocked survival by all of these factors, suggesting that p21ras may indeed form a hub for several survival signal transduction pathways in SCG neurons. However, using the same preparation of Fab fragment to block p21ras activation, it has been shown that although p21ras activity is also required for survival of chick dorsal root ganglion sensory neurons it is not part of the NGF-dependent survival pathway in chick sympathetic neurons [Borasio et al., 1989, 1993]. These surprising results suggest that there may be multiple pathways linked to survival that lie downstream from trkA and that these can be utilized differentially depending on the cell type, or signaling context [see Heumann, 1994, for extensive review on neurotrophin signaling].

In other cell systems, it has been suggested

that bcr-abl binds to activators of p21ras [Puil et al., 1994], and systems overexpressing oncogenic p21ras may display reduced apoptosis by reducing nuclease availability [Arends et al., 1993]. However, the observation that tyrosine kinase activation by Fas ligand is required for apoptosis [Eischen et al., 1994], suggests that it may be possible that in some cells p21ras may actually be utilized to promote cell death.

III.B. Multiple Nonconvergent Kinase Cascades May Be Utilized for Survival of SCG Neurons

Since in many cell systems p21ras activation activates a kinase cascade that leads to an increase in MAP kinase activity, and MAP kinase has been suggested to be crucial in mediating proliferation and differentiation [see Marshall, 1994, for review], we examined whether MAP kinases are activated in SCG neurons in response to each of the survival factors discussed above [Virdee and Tolkovsky, 1995]. However, we found no correlation between MAP kinase activity and survival. Thus, although NGF activated the MAP kinases p44erkl and p42erk2 robustly and persistently for as long as NGF remained present in the medium (measured up to 48 hours), we found that CPTcAMP showed no interaction with MAP kinases, being neither activatory in the absence of NGF nor inhibitory in the presence of NGF. Furthermore, although the cytokines briefly activated both of the erks, this effect subsided within 30 minutes but survival was extended for 3 days. Yet, when NGF was added briefly to mimic the brief activation of the erks induced by the cytokines, no survival extension was obtained. These results, together with those implicating p21ras in the survival pathways of each of the factors, suggest that there exists a bifurcation in the signaling pathway downstream from p21ras. The branch leading to MAP kinase activation may have important cellular functions; for example, it may be involved in mediating the long-term, hypertrophic effects of NGF, but its activation is clearly not necessary for survival. Therefore,

another signaling pathway that branches out from p21ras appears to mediate the survival signals, perhaps in conjunction with coactivation of other kinases. Thus, we propose that SCG neurons, and by extension other types of neurons, may contain multiple signaling pathways that can be utilized to promote survival. The challenge now is to define the identity of these kinases and, most importantly, to define their intracellular targets. The therapeutic implications are that multiple signaling pathways may be available for survival in any one cell, and therefore inactivation of any one pathway will not necessarily cause these neurons to become irreversibly committed to die. Thus, even if a supply of NGF became limiting, it might still be possible to extend survival of these neurons by utilizing alternative signaling mechanisms. How these findings extend to other mammalian systems is of prime interest.

IV. A GENOMIC HSV VECTOR CAN BE USED TO EXTEND NEURONAL SURVIVAL

A major question is how to go about repairing neuronal dysfunction *in vivo*. There are at least three strategies that might be useful for preventing the loss of neurons *in vivo* or for obtaining some functional recovery (overlooking for the moment the immense problem of getting living but axotomized neurons to regenerate appropriate connections to their targets): (1) To arouse the expression of cell autonomous survival mechanisms, for example, that of bcl-2, or to eliminate the expression of their inhibitors—for this we must learn more about the regulatory mechanisms that underlie the expression of these genes in specific cell contexts; (2) to arouse a limited amount of mitosis—this raises the important questions of why neurons become postmitotic and whether the mechanisms are similar to those that underlie cellular senescence [see Lee et al., this volume]; (3) to deliver survival genes such as bcl-2 using appropriate vectors—the definition of an appropriate vector being the subject of considerable debate [Breakefield,

1993]. The observation that activation of PKA leads to long-term survival of SCG neurons led us to attempt to deliver the catalytic subunit of this kinase as a survival gene in the development of a first generation of viral vectors based on genomic HSV [Buckmaster and Tolkovsky, 1994]. The enzyme expressed under the control of a metallothionein promoter was cloned into the thymidine kinase gene of a wild-type HSV-1 (strain 17), which should not reactivate from latency, or into tsK, a mutated HSV containing a temperature-sensitive mutation in the immediate early gene ICP4 that does not allow the virus to replicate >38°C. As expected, the HSV systems were found to be highly neurotropic, infectivity following a predictable Poisson distribution. At a multiplicity of infection (moi) of about 3 pfu/cell, over 95% of 5–8-day cultured SCG neurons were found to be infected. We found that the viruses were excellent vehicles for transient expression due to their high infectivity and that successful induction of PKA activity could be achieved with this promoter. Thus, for short-term biochemical studies this system offers great advantages over microinjection or transient transfection using direct DNA delivery. Moreover, PKA activity remained elevated for about 18 hours and expression of PKA induced by zinc extended the survival of neurons deprived of NGF for about 3 days.

However, the many problems encountered revealed how complex the problem of achieving long-term expression of cognate mammalian genes might be: (1) Elevated activities of PKA (about 2× basal) only lasted for 18 hours, and expression required preincubation with NGF to be effectively induced. This might be due to the fact that in the absence of NGF the levels of protein synthesis decline, and the promoter used was not robust enough to overcome this down-regulation. This might be remedied by using a stronger promoter, preferably one not induced by potentially toxic metals. It may be of interest to note that in BHK cells infected with the same viral construct the induced levels of PKA activity measured *in vitro* were 10× basal, nearly fivefold higher than those found

in neurons, suggesting that SCG neurons *a priori* contain high levels of the inhibitory, regulatory subunits. This was confirmed by the near 12-fold increase in PKA activity revealed when CPTcAMP was added to the kinase assay medium to relieve the inhibition by the regulatory subunits; (2) The neurons responded to PKA elevation by synthesis of more RI, one of the isoforms of the inhibitory regulatory subunits of PKA. This counter-regulatory mechanism might be prevalent when one is attempting to express normal cellular genes, although in this specific case this problem might be overcome by using a noninhibitable form of PKA; (3) HSV contains cyclic AMP response elements that could potentially be activated by PKA [Smith et al., 1992], although we found no evidence for reactivation of the virus when CPTcAMP was deliberately added; (4) Although we had a mutant virus that was potentially replicadon defective, it appears that the block in ICP4 causes a huge accumulation of immediate early gene products that are toxic to the neurons, a conclusion supported by Johnson et al. [1992]. This could be clearly seen by the huge accumulation of the products in the trans-Golgi network, which was dispersed by addition of brefeldin A. Considering the early promise of amplicon [Spaete and Frenkel, 1982] and genomic HSV vectors, there is much work that remains to be done to overcome the toxicity displayed by these viral preparations to their host cells [see Friedmann, 1994, for review], and to the immune system after injection into the CNS *in vivo* [Wood et al., 1994], a toxicity that may extend to adenovirus vectors. The challenge now is to develop useful vectors for delivery of genes to postmitotic neurons both *in vitro* and *in vivo*.

V. CONCLUSIONS AND FUTURE PROSPECTS

Three major conclusions may be usefully reiterated here:
1. Apoptosis may be the point of least interest for studying the problem of neuronal cell death because once apoptosis

has begun, neurons cannot be rescued. We clearly need to discover more about the processes that lead up to and control the death commitment point and in the nervous system, in particular, not get hooked on proving apoptosis unless it can be proven to begin before death commitment. Perhaps it might be more useful to redefine cell death mechanisms in molecular terms; for example, we might define heriotz (*death* in the Basque language, [B Atoxaga, *Obabakaok,* Hutchinson, 1992]) as the kind of programmed cell death that is inhibitable by bcl-2.

2. It is clear that apoptosis in some neurons can be suppressed by activation of well-characterized signaling pathways, and we suggest that day-to-day death suppression most likely involves post-translational mechanisms. For strengthening neuronal survival in the CNS in particular, it may thus be possible to exploit traditional veins in pharmacological research that target known second-messenger systems, without the need to develop strategies for introducing proteins as large as NGF (26 kD) into the brain.

3. For long-term survival, it is likely that some form of gene expression will be required. New gene delivery systems must be explored, as none has yet shown unequivocal promise. Two other approaches, the activation of dormant survival genes, and the reactivation of neuronal mitosis, might be of some use against current problems facing gene therapy and transplantation research.

ACKNOWLEDGMENTS

We thank The Wellcome Trust, the MRC, and Action Research for supporting our work, none of which would have been possible without the help of Viv Wilkins and Molly Sheldon. We are indebted to our colleagues Bob Amess, Earl Clarke, Chris Tomkins, and Ruth Murrell for invaluable suggestions and thank Gian Borasio, Annette Markus, and Rolf Heumann at Ruhr Universitat, Bochum, Germany, for generous gifts of val12Ha-ras and the Fab fragments of Y13–259 and for lengthy discussions.

REFERENCES

Acheson A, Barde Y-A, Thoenen H (1987): High K-mediated survival of spinal sensory neurons depends on developmental age. Exp Cell Res 170:56–63.

Allsopp TE, Wyatt S, Patterson HF, Davies AM (1993): The proto-oncogene bcl-2 can selectively rescue neurotrophic factor-dependent neurons from apoptosis. Cell 73:295–307

Arends MJ, Mcgregor AH, Toft NJ, Brown EIH, Wyllie AH (1993): Susceptibility to apoptosis is differentially regulated by c-myc and mutated Ha-ras oncogenes and is associated with endonuclease availability. Br J Cancer 68:1127–1133.

Arends MJ, Mcgregor AH, Wyllie AH (1994): Apoptosis is inversely related to necrosis and determines net growth in tumors bearing constitutively expressed myc, ras, and HPV oncogenes. Amer J Pathol 144:1045–1057.

Barbin G, Manthorpe M, Varon S (1984): Purification of the eye ciliary neurotrophic factor. J Neurochem 43:1468–1478.

Barres BA, Schmid R, Sendnter M, Raff MC (1993): Multiple extracellular signals are required for long-term oligodendrocyte survival. Development 118: 283–295.

Batistatou A, Greene LA (1991): Aurintricarboxylic acid rescues PC12 cells and sympathetic neurons from cell death caused by NGF-deprivation. Correlation with suppression of endonuclease activity. J Cell Biol 115:461–471.

Boise LH, Gonzalez-Garcia M, Postema CE, Ding L, Lindsten T, Turka LA, Mao X, Nunez G Thompson CB (1993): Bcl-X, a bcl-2-related gene that functions as a dominant regulator of apoptotic cell death. Cell 74:597–608.

Borasio GD, John J, Wittinghofer A, Barde Y-A, Sendtner M, Heumann R (1989): Ras p21 protein promotes survival and fibre outgrowth of cultured embryonic neurons. Neuron 2:1087–1096.

Borasio GD, Markus A, Wittinghofer A, Barde Y-A, Heumann R (1993): Involvement of ras p21 in neurotrophin-induced response of sensory but not sympathetic neurons. J Cell Biol 121:665–672.

Boulton TG, Stahl N, Yancopoulos GD (1994): Ciliary neurotrophic factor/leukemia inhibitory factor/interleukin 6/oncostatin M family of cytokines induces tyrosine phosphorylation of a common set of proteins overlapping those induced by other cytokines and growth factors. J Biol Chem 269:11648–11655.

Breakefield XO (1993): Gene delivery into the brain using virus vectors. Nature Genet 3:187–189.

Buckmaster AE, Tolkovsky AM (1994): Expression of the cyclic AMP-dependent protein kinase (PKA) catalytic subunit from a herpes simplex vector extends the survival of sympathetic neurons in the absence of NGF. Eur J Neurosci 6:1316–1327.

Burnham P, Louis J-C, Magal E, Varon S (1994). Effects of ciliary neurotrophic factor on the survival and response to nerve growth factor of cultured rat sympathetic neurons. Dev Biol 161:96–106.

Barde Y-A (1989): Trophic factors and neuronal survival. Neuron 2:1525–1534.

Clarke PDH (1990): Developmental cell death: morphological diversity and multiple mechanisms. Anat Embryol 181:195–213.

Cohen JJ (1993): Apoptosis. Immunol Today 14:126–130.

Crowley C, Spencer SD, Nishimura MC, Chen KS, Pittsmeek S, Armanini MP, Ling LH, McMahon SB, Shelton DL, Levinson AD Phillips HS (1994): Mice lacking nerve growth factor display perinatal loss of sensory and sympathetic neurons yet develop basal forebrain cholinergic neurons. Cell 76:1001–1011.

Daly C, Reich NC (1994): Receptor to nucleus signalling via tyrosine phosphorylation of the p91 transcription factor. Trends Endocrinol Metab 5:159–164.

Deckwerth TL, Johnson EM (1993) Temporal analysis of events associated with programmed cell-death (apoptosis) of sympathetic neurons deprived of nerve growth-factor. J Cell Biol 123:1207–1222.

Edwards SN (1992): Regulation of Cell Death in Sympathetic Neurons. Ph.D. Thesis. University of Oxford.

Edwards SN, Tolkovsky AM (1994): Characterization of apoptosis in cultured rat sympathetic neurons after nerve growth factor (NGF) withdrawal. J Cell Biol 124:537–546.

Edwards SN, Buckmaster AE, Tolkovsky AM (1991): The death programme in sympathetic neurons can be suppressed at the posttranslational level by NGF, cyclic AMP and depolarization. J Neurochem 57:2140–2143.

Eischen CM, Dick CJ, Leibson PJ (1994): Tyrosine kinase activation provide an early and requisite signal for Fas-induced apoptosis. J Immunol 153:1947–1954.

Ellis RE, Yuan JY, Horvitz HR (1991): Mechanisms and functions of cell-death. Annu Rev Cell Biol 7: 663–698.

Fisher DE (1994): Apoptosis in cancer therapy—crossing the threshold. Cell 78:539–542.

Franklin JL, Johnson EM (1992): Suppression of programmed neuronal death by sustained elevation of cytoplasmic calcium. Trends Neurosci 15:501–508.

Friedmann T (1994): Gene-therapy for neurological disorders. Trends Genet 10:210–214.

Garcia I, Martinou I, Tsujimoto Y, Mardnou J-C (1992): Prevention of programmed cell death of sympathetic neurons by the bcl-2 protooncogene. Science 258:302–304.

Harrington EB, Bennet MR, Fanidi A, Evan GI (1994): c-Myc–induced apoptosis in fibroblasts is inhibited by specific cytokines. EMBO J 13:3286–3295.

Hendry IA, Campbell J (1976): Morphometric analysis of rat superior cervical ganglion after axotomy and NGF treatment. J Neurocytol 5:351–360.

Hengartner MO, Ellis RE, Horvitz HR (1992): Caenorhabditis elegans gene ced-9 protects cells from programmed cell-death. Nature 356:494–499.

Hengartner MO, Horvitz HR (1994) C elegans cell-survival gene ced-9 encodes a functional homolog of the mammalian protooncogene bcl-2. Cell 76:665–676.

Heumann R (1994): Neurotrophin signalling. Curr Opin Neurobiol 4:668–679.

Heusel JW, Wessleschmidt RL, Shresta S, Russell JH, Ley TJ (1994): Cytotoxic lymphocytes require granzyme B for the rapid induction of DNA fragmentation and apoptosis in allogeneic target cells. Cell 76:977–987

Ip NY, Nye SH, Boulton TG, Davis S, Taga T, Yanping L, Birren SJ, Yasukawa K, Kishimoto T, Anderson DJ, Stahl N, Yancopoulos GD (1992): CNTF and LIF act on neuronal cells via shared signalling pathways that involve the IL-6 signal transducing receptor component gp 130. Cell 69:1121–1132.

Jacobson MD, Evan GI (1994): Breaking the ICE. Curr Biol 4:337–340.

Johnson PA, Miyanohara A, Levine F, Cahill T, Freidmann T (1992): Cytotoxicity of a replication-defective mutant of herpes simplex type I. J Virol 66:2952–2965.

Kaufmann SH, Desnoyers S, Ottaviano Y, Davidson NE, Poirier GG (1993): Specific proteolytic cleavage of poly(ADP-ribose) polymerase—An early marker of chemotherapy-induced apoptosis. Cancer Res 53:3976–3985.

Kerr JFR, Wyllie AH, Currie AR (1972): Apoptosis: A basic biological phenomenon with wide ranging implications in tissue kinetics. Br J Cancer 26:239–257.

Kerr JFR, Winterford CM, Harmon BV (1994): Apoptosis—its significance in cancer and cancer therapy. Cancer 73:2013–2026.

Kessler JA, Ludlam WH, Freidin MM, Hall DH, Michaelson MD, Batter DK (1993): Cytokine-induced programmed death of cultured sympathetic neurons. Neuron 11:1123–1132.

Koike T, Martin DP, Johnson EM (1989): Role of calcium channels in the ability of membrane depolarization to prevent neuronal death induced by trophic factor deprivation: Evidence that levels of internal calcium determine NGF dependence of sympathetic ganglion cells. Proc Natl Acad Sci USA 86:6421–6425.

Korsmeyer SJ, Shutter JR, Veis DJ, Merry DE, Oltvai ZN (1993): Bcl-2/bax—a rheostat that regulates an antioxidant pathway and cell death. Semin Cancer Biol 4:327–332.

Kumar S, Kinoshita M, Noda M Copeland NG, Jenkins

NA (1994): Induction of apoptosis by the mouse nedd2 gene, which encodes a protein similar to the product of the *Caenorhabdids elegans* cell death gene *ced-3* and the mammalian Il-l-beta–converting enzyme. Genes Dev 8:1613–1626.

Lazebnik YA, Kaufmann SH, Desnoyers S, Poirier GG, Earnshaw WC (1994): Cleavage of poly(ADP-ribose) polymerase by a proteinase with properties like ICE. Nature 371:346–347.

Levi-Montalcini R (1965): The nerve growth factor: Its mode of action on sensory and sympathetic nerve cells. Harvey Lectures 60:217–259.

Maller JL (1993): On the importance of protein-phosphorylation in cell-cycle control. Mol Cell Biochem 128:267–281.

Marshall CJ (1994): MAP kinase kinase kinase, MAP kinase kinase and MAP kinase. Curr Opin Genet Dev 4:82–89.

Martin DP, Schmidt RE, DiStephano PS, Lowry OH, Carter JG, Johnson EM Jr (1988): Inhibitors of protein synthesis and RNA synthesis prevent neuronal cell death caused by nerve growth factor deprivation. J Cell Biol 106:829–844.

Martin DP, Ito A, Horigome K, Lampe PA, Johnson EM. (1992): Biochemical characterization of programmed cell-death in NGF-deprived sympathetic neurons. J Neurobiol 23:1205–1220.

Minshull J (1993): Cyclin synthesis—who needs it? Bioessays 15:149–155.

Miura M, Zhu H, Rotello R, Hartwieg EA, Yuan JY (1993): Induction of apoptosis in fibroblasts by Il- l-beta—converting enzyme, a mammalian homolog of the *C. elegans* cell-death gene *ced-3*. Cell 75: 653–660.

Mcgahon A, Bissonnette R, Schmitt M, Cotter KM, Green DR, Cotter TG (1994): bcr-abl maintains resistance of chronic myelogenous leukemia cells to apoptotic cell-death. Blood 83:1179–1187.

Ng NFL, Shooter EM (1993): Activation of p21(ras) by nerve growth-factor in embryonic sensory neurons and PC12 cells. J Biol Chem 268:25329–25333.

Nobes CD, Buckmaster AE Tolkovsky AM (1991): Mediators of nerve growth factor action in cultured rat superior cervical ganglion neurons. J Autonom Nerv Syst 33:213–214.

Nobes CD, Tolkovsky AM (1995): Neutralizing anti-p21^{ras} Fabs suppress rat sympathetic neuron survival induced by NGF, LIF, CNTF and cyclic AMP. Eur J Neurosci 7:344–350.

Nunez G, London L, Hockenberry D, Alexander M, McKearn J, Korsmeyer S (1990): Deregulated bcl-2 gene expression selectively prolongs the survival of growth factor deprived haemopoietic cell lines. J Immunol 144:3602–3610.

Oltvai ZN, Milliman CL, Korsmeyer SJ (1993): Bcl-2 heterodimerizes *in vivo* with a conserved homolog, bax, that accelerates programmed cell-death. Cell 74:609–619.

Oppenheim RW (1991): Cell death during development of the nervous system. Annu Rev Neurosci 14:453–501.

Puil L, Liu JX, Gish G, Mbamalu G, Bowtell D, Pelicci PG, Arlinghaus R, Pawson T (1994): Bcr-Abl oncoproteins bind directly to activators of the ras signaling pathway. EMBO J 13:764–773.

Rabizadeh S, OH J, Zhong LT, Yang J, Bitler CM, Butcher LL, Bredesen DE (1993): Induction of apoptosis by the low-affinity NGF receptor. Science 261:345–348.

Raff MC (1992): Social controls on cell survival and cell death. Nature 356:397–400.

Ratan RR, Murphy TH, Baraban JM (1994): Macromolecular synthesis inhibitors prevent oxidative stress-induced apoptosis in embryonic cortical-neurons by shunting cysteine from protein synthesis to glutathione. J Neurosci 14:4385–4392.

Reed JC (1994): Bcl-2 and the regulation of programmed cell-death. J Cell Biol 124:1–6.

Ruckenstein A, Rydel RE, Greene LA (1991): Multiple agents rescue PC12 cells from serum-free cell death by translation- and transcription-independent mechanisms. J Neurosci 11:2552–2563.

Rydel RE, Greene LA (1988): cAMP analogs promote survival and neurite outgrowth in cultures of rat sympathetic and sensory neurons independently of nerve growth factor. Proc Natl Acad Sci USA 85:1257–1261.

Savill J, Fadok V, Henson P, Haslett C (1993): Phagocyte recognition of cells undergoing apoptosis. Immunol Today 14:131–136.

Shi LF, Kam CM, Powers JC, Aebersold R, Greenberg AH (1992): Purification of 3 cytotoxic lymphocyte granule serine proteases that induce apoptosis through distinct substrate and target–cell interactions. J Exp Med 176:1521–1529.

Smeyne RJ, Klein R, Schnapp A, Long LK, Bryant S, Lewin A, Lira SA, Barbacid M (1994): Severe sensory and sympathetic neuropathies in mice carrying a disrupted TRK/NGF receptor gene. Nature 368:246–249.

Smith RL, Pizer LI, Johnson EM, Wilcox CL (1992) Activation of 2nd-messenger pathways reactivates latent herpes-simplex virus in neuronal cultures. Virology 188:311–318.

Snider WD (1994): Functions of the neurotrophins during nervous-system development—what the knockouts are teaching us. Cell 77:627–638.

Spaete RR, Frenkel N (1982): The herpes simplex virus amplicon: A new eucaryotic defective virus cloning-amplifying vector. Cell 30:295–304.

Tolkovsky AM, Buckmaster EA (1989): Deprivation of nerve growth factor rapidly increases purine efflux from cultured sympathetic neurons. FEBS Lett 255:315–320.

Tomkins CE, Edwards SN, Tolkovsky AM (1994): Apoptosis is induced in postmitotic neurons by arabinosides and topoisomerase II inhibitors in the presence of NGF. J Cell Sci 107:1499–1507.

Vaux D, Cory S, Adams J (1988): Bcl2 gene promotes haemapoietic cell survival and cooperates with c myc to immortalize pre B cells. Nature 335:440–442.

Vaux DL, Haecker G, Strasser A (1994) An evolutionary perspective on apoptosis. Cell 76:777–779.

Vaux DL, Weissman IL (1993): Neither macromolecular synthesis not Myc is required for cell death via the mechanism that can be controlled by Bcl2. Mol Cell Biol 13:7000–7005.

Virdee K, Tolkovsky AM (1995): The activation of p42 and p44 MAP kinases is not essential for the survival of rat sympathetic neurons. Eur J Neurosci 7 (in press).

Wallace TL, Johnson EM Jr (1989): Cytosine arabinoside kills postmitotic neurons: Evidence that deoxycytosine may have a role in neuronal survival that is independent of DNA synthesis. J Neurosci 9:115–124.

Wright LL, Cunningham TJ, Smolen AJ (1983): Developmental neuron death in the rat superior cervical ganglion: Cell counts and ultrastructure. J Neurocytol 12:727–738.

Wong GHW, Goeddel DV (1994): Fas antigen and p55 TNF receptor signal apoptosis through distinct pathways. J Immunol 152:1751–1755.

Wood MJA, Byrnes AP, Pfaff DW, Rabkin SD, Charlton HM (1994): Inflammatory effects of gene transfer into the CNS with defective HSV-1 vectors. Gene Ther 1:283–291.

Wyllie AH, Kerr JBS, Currie AR (1980): Cell death: The significance of apoptosis. Int Rev Cytol 68:251–306.

Yuan JY, Horvitz HR (1992): The *caenorhabditis elegans* cell-death gene ced-4 encodes a novel protein and is expressed during the period of extensive programmed cell-death. Development 116:309–320.

Yuan JY, Shaham S, Ledoux S, Ellis HM, Horvitz HR (1993): The *C. elegans* cell-death gene *ced-3* encodes a protein similar to mammalian interleukin-1-beta–converging enzyme. Cell 75:641–652.

ABOUT THE AUTHORS

AVIVA TOLKOVSKY is a University Lecturer at the Department of Biochemistry at the University of Cambridge, UK, where she teaches Biochemistry and Neurobiology to undergraduates and graduates. Her university education was pursued at the Hebrew University, Jerusalem, Israel, where she gained her Ph.D. in Biophysics in 1975. During her postdoctoral work with Alex Levitzki she combined experimental work with the development of mathematical models to distinguish between mechanisms by which β-adrenergic receptors couple to adenylate cydase, demonstrating one of the first examples of "collision coupling" whereby one receptor activates many effector molecules. Her interest in the SCG system developed while holding an Amersham Senior Research Fellowship at Cambridge, returning there recently from a combined lectureship and tutorial fellowship at the University of Oxford and Somerville College. Her main interests lie in the study of signal transduction related to death and survival in primary neurons, for which she tries to combine a variety of biochemical techniques with the predictive powers of mathematical modeling.

E. ANNE BUCKMASTER completed her B.Sc. in Biology at Sussex University and Ph.D. at the University of Cambridge in the Department of Pathology working on herpes simplex virus with Tony Minson. She continued her postdoctoral research in the same department and then at the AFRC Institute for Animal Research at Houghton before moving to Aviva Tolkovsky's group in Oxford, where she worked on the use of herpes simplex virus as a vector for introducing and expressing genes in cultured neurons. She is now working on the mechanisms of neuronal degeneration with Michael Brown and Hugh Perry in the University Laboratory of Physiology at Oxford.

SUSAN EDWARDS completed her B.Sc. and M.A. in Natural Sciences at the University of Cambridge before moving to A. Tolkovsky's group at the University of Oxford, where she earned her Ph.D. for work on the mechanism of neuronal cell death in sympathetic neurons. She is currently working on the role of syndecan-1 proteoglycan in the control of cell behavior in Markku Jalkanen's group, Centre for Biotechnology, University of Turku, Finland.

CATHERINE D. NOBES completed her B.Sc. in Biochemistry at University College, London, and her Ph.D. at University of Cambridge, where she worked with Martin Brand on the regulation of respiration in hepatocytes. She moved to Oxford to work with Helen Saibil on rhodopsin-linked G-proteins, moving to A. Tolkovsky's laboratories to work on the role of p21 ras in neuronal survival. She is currently a postdoctoral research fellow working in Alan Hall's laboratory at the MRC Laboratory of Molecular Cell Biology at University College, London, on the small GTP-binding proteins Rac and Rho.

KANWAR VIRDEE is a postdoctoral research associate in Aviva Tolkovsky's group studying the intracellular signaling mechanisms by which NGF and other neurotrophic factors prevent apoptotic death of sympathetic neurons. He received his B.Sc. from the University of Dundee, Scotland, and went on to earn a Ph.D. for work on the intracellular signaling mechanisms of fibroblast growth factor at the University of Sheffield with Barry Brown. Since 1989 Dr. Virdee has held postdoctoral fellowships at INSERM U211, Nantes, France, working on HILDA/LIF signaling with Yannick Jacques, and at the Medical College of Georgia, working on signaling in iris-sphincter smooth muscle with Ata Abdel-Latif.

Cellular Aging and Cell Death: 283–302
© 1996 Wiley-Liss, Inc.

Neuronal Loss in Aging and Disease

Benjamin Wolozin, Yongquan Luo, and Katherine Wood

I. INTRODUCTION

In this chapter we discuss the mechanisms of cell loss occurring in some of the major neurodegenerative disorders, including Parkinson's disease (PD), amyotrophic lateral sclerosis (ALS), Alzheimer's disease (AD), and Huntington's chorea. The etiology of these illnesses is unknown for most people, except for rare families in which the diseases are transmitted through hereditary mechanisms. However, valuable model systems have been developed for each of these illnesses that suggest potential pathways for the cell loss.

These illnesses share common themes in both the development of the model systems and the resulting mechanism suggested for cell loss. For instance, the models that have been developed draw greatly from the study of small subsets of patients whose illnesses result from readily defined causes, such as genetic mutations or environmental toxins. Another general theme is the importance of calcium and free radicals in the mechanism of cell death. In each of these illnesses, influx of calcium into the cytoplasm and the generation of free radicals appear to be important to the pathophysiology of the illness, either as a primary cause of the illness, as in familial ALS, or as secondary factors in the illness, as in PD and AD. Taken as a whole, the study of the pathophysiology of these illnesses points to the important roles of calcium and free radicals in the mechanism for cell loss of these accelerated forms of neurodegeneration.

During development and throughout life in the immune system calcium and free radicals also function as important mediators of pro-grammed cell death, termed *apoptosis*. The involvement of similar mediators in neurodegenerative disorders suggests that apoptosis may also be important in neurodegenerative diseases. Although it remains unclear whether the mechanism of cell death in neurodegenerative disorders involves apoptosis, there are many similarities between the two types of processes, and the study of apoptosis has led to novel approaches in the study of neurodegenerative illnesses. We will also examine the relationship between the pathophysiology of the neurodegenerative disorders and mechanisms important to apoptosis.

Finally, the study of neurodegenerative illnesses may have important clues about the mechanisms of normal aging on the brain. Many of the processes that occur in the neurodegenerative illnesses may be accelerated forms of normal physiological changes that occur with age. By identifying physiological changes that occur with disease, we may, therefore, gain further insight into which age-related changes are most harmful to the central nervous system and what interventions may best protect against these changes. For instance, the neurofibrillary pathology that occurs in AD can also be found in brains from nondemented elderly subjects at lower abundance. This suggests that the processes occurring in AD also occur in normal aging, but to a lesser degree. Thus, the changes that occur in the Alzheimer brain may provide valuable information to enhance our understanding of some of the changes that occur with normal aging.

Of the illnesses that result from primary disorders of neuronal metabolism, AD is the most common cause of age-related cognitive deficits, while PD, Huntington's disease, and ALS account for the majority of motor difficulties. Of course, the most common sources of neurologic difficulty are stroke and multi-infarct dementia; however, because these two disorders are based on age-related vascular changes we do not discuss them in this chapter. The impact of the primary neurodegenerative illnesses on our society is increasing because of a dual trend. One trend is demographic changes in our population, namely, the baby boomers are getting older and as they age the proportion of elderly people increases. Approximately 15% of the population over 65 years are afflicted with Alzheimer's disease, which by the year 2000 will affect more than 8 million individuals in the United States alone. In addition, other individuals are affected by PD, ALS, and other neurodegenerative disorders. Compounding this trend is an increase in the longevity due to medical advances that allow more people to live longer. The total cost of these illnesses in both health care dollars and nursing home expenses is immense and, unfortunately, growing.

II. PARKINSON'S DISEASE

PD presents with a progressive loss in motor ability. Patients have difficulty using their limbs, experience a resistance to movement classically defined by the term *cogwheel rigidity*. In addition, PD patients are increasingly unable to initiate motor action and, ultimately, are unable to move at all. A subset of PD patients also show a dementia, termed *subcortical dementia*. The brains of PD patients show a loss of dopaminergic neurons, as well as the development of neurofibrillary structures, termed *Lewy bodies*. The PD patients with dementia also show cortical pathology in the form of neuritic plaques.

PD is one of the first diseases for which a model system was developed. The origins of this model began on the streets of Los Angeles over a decade ago. To avoid government regulations and jail, drug dealers began modifying drugs so that they no longer were the same as the drugs that had been declared illegal by the FDA. These drugs were then sold on the street and essentially tested on human subjects without any prior pharmacological characterization on either people or animals. Perhaps not surprisingly, some of these drugs had toxic side effects. The first cases were identified by Langston and colleagues [1983]. One of the first interesting cases occurred when a young individual came to a California teaching hospital. This young man was only about 30 years old, but had developed acute PD with all of the classic signs, including cogwheel rigidity and lack of motor initiation. Careful analytic detective work identified one chemical, 1-methyl-4-phenyl-1,2,3,6-tetrahydropyridine (MPTP), as the toxin that was common to all of these patients, which was reported by Langston and colleagues in 1983.

After a decade of work the mechanism of MPTP-induced neurotoxicity has been largely elucidated (Fig. 1) [Snyder and D'Amato, 1986; Tipton and Singer, 1993]. After ingestion, MPTP readily crosses the blood–brain barrier, where it comes into contact with neurons and glia. MPTP itself is largely harmless. However, MPTP is metabolized by the glial enzyme monoamine oxidase B (MAO-B) first into 1-methyl-4-phenyl-1,2-dihydropyridinium ion (MPDP$^+$) and then into the toxic compound 1-methyl-4-phenyl-1,2-phenylpyridinium ion (MPP$^+$) [Tipton et al., 1986; McCrodden et al., 1990]. Interestingly, although ingestion of MPTP ultimately kills dopaminergic neurons, the conversion of MPTP to the toxic MPP$^+$ does not occur at dopaminergic nerve terminals. Rather, MAO-B is located in glia at sites separate from the dopaminergic nerve terminals [O'Carroll et al., 1987]. The importance of MAO-B as a bioactivator of MPTP may also account for the relative insensitivity of rodents to MPTP toxicity, because mice and rats have lower levels of glial MAO-B than do primates but higher levels of hepatic and vascular MAO-B; thus, a greater proportion of MPTP is converted to MPP$^+$ outside the central nervous system where it can be cleared without harm [Tipton, 1986; Riachi et al., 1988]. Once con-

Pathway for MPTP Toxicity

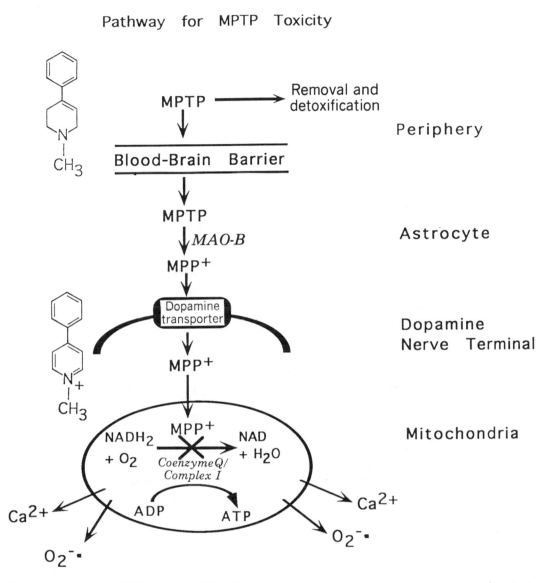

Fig. 1. *Pathway for MPTP toxicity. MPTP diffuses across the blood–brain barrier, where it is converted to MPP+ by the glial enzyme MAO-B. MPP+ is then concentrated in dopaminergic neurons by the dopamine transporter. Once inside the neuron, it blocks the action of coenzyme Q–complex I in the mitochondrial membrane, which inhibits redox activity and ATP production. Free radical production and calcium efflux follow.*

verted to MPP+, the selective toxicity arises from its specific interaction with the dopamine transporter. MPP+ binds to the dopamine transporter with a K_i of 13 nM. In contrast, its affinity for the noradrenergic and serotonergic transporters is much less, exhibiting K_i's in the high micromolar range [Javitch, 1985]. As a result of uptake by the dopamine transporter, MPP+ is selectively concentrated in dopaminergic neurons to levels high enough to be toxic.

The mechanism of toxicity of MPP+ arises from its accumulation in the mitochondrial matrix and the subsequent loss of energy metabolism. Following the uptake, MPP^+ is concentrated over 100-fold in the mitochondrial matrix [Ramsay and Singer, 1986]. As the concentration increases, MPP^+ inhibits the enzyme NADH dehydrogenase, which prevents reducing equivalents of NADH from reaching coenzyme Q [Ramsay et al., 1991a,b]. MPP^+ is equally toxic to all mitochondria, so at high doses it will kill all types of cells, regardless of whether they have selective transporters or not. As a result of this blockade of NADH dehydrogenase, mitochondrial respiration, oxidative phosphorylation, and ATP production are all blocked.

While the mechanism for selective toxicity of MPTP is highly specific to this compound, the research done on the consequences of mitochondrial blockade has important implications for understanding mechanisms of neuronal death in general. Blockade of mitochondrial respiration appears to have two consequences, production of free radicals and increases in cytoplasmic calcium levels. Both of these events appear to be toxic to neurons, and both may contribute to the mechanism of cell death. Inhibition of NADH dehydrogenase by a variety of compounds, including MPP^+ and rotenone, leads to production of the free radical superoxide (Fig. 1) [Hasegawa et al., 1990; Cleeter et al., 1992]. A significant amount of experimental evidence indicates that this production of superoxide is toxic to neurons. Cadet and colleagues have used a transgenic mouse having up to three times the activity of Cu/Zn-superoxide dismutase (Cu/Zn-SOD) to study MPTP toxicity [Przedborski, et al., 1992]. This enzyme reduces superoxide levels by converting superoxide to hydrogen peroxide, where it can be converted to water by the enzyme catalase. Having increased protection against the superoxide radical, the Cu/Zn-SOD transgenic mice show no toxicity in response to intraperitoneal injection of MPTP, while the nontransgenic littermates show marked dopaminergic cell loss in response to MPTP administration [Przedborski et al., 1992]. Further evidence supporting the role of the super-

oxide ion in MPTP toxicity comes from studies with L-deprenyl, which show that it protects against MPTP toxicity in both rodents and humans. Interestingly, the dose at which L-deprenyl is neuroprotective is below the dose required for MPTP inhibition. Chiueh and colleagues [1993] have shown that L-deprenyl is a free radical scavenger, and they suggest that this mechanism may account for the protection against MPTP toxicity. These separate lines of evidence all indicate that reduction of superoxide formation in response to MPTP administration protects against neurotoxicity, which emphasizes the importance of free radicals in the mechanism of MPTP toxicity.

Similar types of protection experiments also suggest a role for calcium in the mechanism of MPTP-induced cell death. Pretreatment of rats with the NMDA antagonists AP7, CPP, or MK-801 all provide protection against MPP^+ toxicity [Turski et al., 1991]. The NMDA receptor mediates glutamate- and aspartate-induced excitation, which is coupled to calcium influxes. Excitotoxicity has been implicated in a number of other conditions, including Huntington's disease and stroke [Beal, 1992]. The mechanism for the excitotoxicity in response to MPTP is proposed to result from a loss of membrane potential due to decreased ATP availability and the subsequent relief of the voltage-dependent block of the NMDA receptor by Mg^{2+} [Turski et al., 1991]. Such changes may be exacerbated by an efflux of calcium from mitochondrial stores in response to the blockade of mitochondrial respiration (Fig. 1). Thus, both calcium and superoxide radicals may contribute to the mechanism of cell death in MPTP-induced neurotoxicity.

An important question is how the MPTP model relates to PD in general. Research on the pathology and epidemiology of PD, combined with the therapeutic benefits of L-deprenyl, lend credence to the relevance of MPTP to PD. Measurements of NADH-coenzyme Q reductase activity in idiopathic PD show disease-related decreases in activity in substantia nigra, caudate, putamen, and platelets of patients [Mizuno et al., 1989, 1990; Parker et al., 1989; Schapira et al., 1989, 1990]. Such changes are consistent with decreased mitochondrial function, as would be expected if environmental toxins, some of which are similar to MPTP, were relevant to PD [Snyder and D'Amato, 1986]. Further support comes from epidemiological studies that suggest that pesticides may be a source of potential toxins [Tanner, 1989]. In developed countries, the incidence of PD is greater in rural communities and appears to be associated with areas where vegetable farming or wood pulp mills predominate [Tanner, 1989]. One could imagine that such environmental toxins might exacerbate the general loss of dopaminergic neurons that occurs with aging and contribute to PD.

Recent clinical experience with L-deprenyl lends support for the role of mitochondrial toxins and free radicals in PD. L-deprenyl is an inhibitor of MAO-B, which converts MPTP to the active toxin MPP$^+$. Tetrud and Langston [1989] reported that PD patients treated with L-deprenyl plus L-DOPA live longer than patients receiving only L-DOPA. Subsequent studies did not repeat the findings of increased longevity, but they did find that use of L-deprenyl delayed the requirement for L-DOPA 1 year [Group, 1989; Tetrud and Langston, 1989]. Knoll [1988] has also reported increased longevity in rats treated with L-deprenyl, but other investigators have failed to repeat these results [Ingram et al., 1993]. While L-deprenyl is best known as an MAO-B inhibitor, the doses at which it exerts neuroprotective action in PD and in MPTP toxicity are both below the K_i for the enzyme. Thus, blockade of MAO-B action (which could prevent activation of putative toxins) does not explain the neuroprotective action of L-deprenyl. Rather, the proposed mechanism for the neuroprotection is through scavenging of free radicals or prevention free radical formation [Chiueh et al., 1993]. As a result of these two lines of evidence, clinicians have been supplementing regimens of PD patients with both L-deprenyl and free radical scavengers such as vitamin E. However, to date

no benefits have been seen from the addition of such supplements to the regimens of PD patients [Group, 1993]. Although the final conclusions have yet to come in, the information obtained from the studies of PD suggest that the MPTP model is an excellent model for the illness. The rapid pace of research using the MPTP system provides support for its power, and the information obtained from the system has increased our knowledge greatly, implicating environmental toxins in the pathogenesis of PD and identifying the roles of calcium and free radicals in the pathway of cell death.

III. AMYOTROPHIC LATERAL SCLEROSIS

As with PD, where a small number of patients has led to the development of a model of the illness, the study of ALS has also moved ahead due to the identification of the genetic defect of a small number of patients who have a familial form of ALS. ALS, also known as Lou Gehrig's disease, is an uncommon illness in which patients gradually lose motor function, beginning with the distal limbs and proceeding proximally. The pathological changes characteristic of this illness include loss of nerve cells in the anterior horns of the spinal cord and motor nuclei of the lower brainstem, as well as sclerosing of the lateral motor columns, hence the name ALS. On a microscopic level, the dying neurons show neurofibrillary changes. In approximately 10% of cases of ALS, the disease is transmitted as an autosomal dominant disorder. A subset of ALS families, about 20% of the familial ALS cases, show localization of the disorder to chromosome 21q22.1 [Deng et al., 1993]. Subsequent analysis identified the affected gene as the Cu/Zn-superoxide dismutase (SOD) [Deng et al., 1993; Rosen et al., 1993]. While, the type of mutation differs from family to family, the mutations are all missense changes in either exon 1, 2, 4, or 5 of SOD. Interestingly, there are no mutations at the active site in exon 3. Biochemical studies suggest that the mutations decrease the stability of the enzyme, so that there is less SOD present in the affected cells

[Gurney et al., 1994]. Interestingly, while most loss-of-function mutations are recessive, the SOD mutations are dominant. SOD is a dimer; thus loss of function in one of the subunits decreases the stability of the entire protein, even when there is a normal subunit present [Deng et al., 1993]. This hypothesis is supported by findings that familial ALS patients heterozygous for the SOD mutations have only 50%–60% of the normal level of SOD activity in their red blood cells and brains [Deng et al., 1993].

Loss of SOD function is deleterious presumably because of the concomitant increase in free radical damage. The Cu/Zn-SOD catalyzes the conversion of $O_2^{\cdot-}$ to H_2O_2 and thereby reduces the amount of $O_2^{\cdot-}$, which also reduces peroxynitrite anion and hydroxyl radical formation and protects against free radicals generated metabolically or by environmental toxins [Beckman et al., 1990]. Thus, cells with decreased SOD activity are subject to increased free radical–mediated damage. While it is unclear why loss of SOD would specifically harm the neurons of the anterior horn, transgenic studies have elegantly confirmed that such mutations can lead to conditions identical to ALS. Transgenic mice expressing a human SOD gene containing a glycine to alanine substitution at position 93, which is a mutation that has been documented in familial ALS, develop a motor illness very similar to ALS [Gurney et al., 1994]. The animals become paralyzed in their distal limbs and die at 5–6 months [Gurney et al., 1994]. The pathology of these animals is also similar to ALS, showing loss of motor neurons in the ventral spinal cord and loss of myelinated axons from the ventral motor roots, while the dorsal sensory roots were relatively spared [Gurney et al., 1994]. Thus, decreases in antioxidant defenses, such as those seen with dominant negative mutations to SOD, can lead to progressive neurodegeneration of specific populations of nerve cells. Put differently, the studies of familial ALS provide strong evidence that free radical–mediated damage can lead to late onset neurodegeneration.

Interestingly, the etiologies of sporadic and familial ALS appear to differ, although both appear to involve agents implicated in programmed cell death. Studies of the etiology of sporadic ALS implicate the excitatory amino acids glutamate and aspartate rather than free radicals [Patten, 1994]. Glutamate and aspartate levels are elevated by up to 200% in patients with ALS [Patten et al., 1878; Rothstein et al., 1990]. The elevated levels of these amino acids appear to result from decreases in their transport or uptake, which may increase the levels of glutamate and aspartate in the synaptic cleft [Rothstein et al., 1992]. Consistent with this is the finding that levels of glutamate and aspartate in brain matter are actually reduced by up to 40% in ALS [Plaitakis et al., 1988]. Thus, a picture of altered excitatory amino acid metabolism is beginning to appear in sporadic ALS. Decreased uptake or transport results in increased levels of excitatory amino acids in the synaptic cleft and cerebral spinal fluid with correspondingly decreased levels of excitatory amino acids in the neuropil. The increased levels of glutamate and aspartate result in increased calcium influx. As discussed later in this chapter, the increased levels of calcium may stimulate a form of apoptosis.

Excitatory amino acids also appear to be important in a variant of ALS, Guamian ALS–Parkinson dementia complex. This illness, which occurs only on the island of Guam, occurs probably due to the presence of high levels of the excitatory amino acid β-N-methylamino-L-alanine [Spencer et al., 1987]. Guamian natives use the cycad nut, which contains high levels of β-N-methylamino-L-alanine, as a staple part of their diet leading to high levels of this excitotoxic amino acid [Spencer et al., 1987]. The resulting toxicity is likely similar to that for sporadic ALS.

Thus, in familial, sporadic, and Guamian ALS both calcium and free radicals appear to be important: calcium for sporadic and Guamian ALS and free radicals for familial ALS. Taken together, these findings emphasize the importance of these two agents in the pathway leading to cell death in neurodegenerative illnesses.

IV. ALZHEIMER'S DISEASE

Our knowledge of the processes contributing to AD has largely evolved out of studies of the neuropathology of the illness, though, as with ALS, information from the study of the molecular genetics in specific families has propelled forward our understanding of the illness. The pathophysiology of AD appears to derive largely from the action of the β-amyloid (aβ) peptide [Selkoe, 1994]. As in PD and ALS, the mechanism of aβ toxicity appears to involve calcium ions and free radicals. Although the pathway of discovery has been different for each of these illnesses, the pathway leading to neuronal death shows common themes.

AD is a relentless, progressive dementia characterized by a progressive loss of cognitive function and affects approximately 15% of individuals over 65 years of age. At autopsy, AD brains show severe cortical atrophy, which on the microscopic level is characterized by extensive neuronal loss, gliosis, and an abundance of neuritic plaques and neurofibrillary tangles. As a result of comprehensive research, the components of the neuritic plaques and neurofibrillary tangles are now known. β-Amyloid, also termed aβ, is a 4 kD peptide that is the principal constituent of neuritic plaques and appears to be central to the pathology of AD. In 1984, Glenner and colleagues sequenced this peptide, and, shortly thereafter, the parent gene coding for the peptide was cloned and sequenced [Glenner and Wong, 1984; Kang et al., 1987, This parent gene, termed *amyloid precursor protein* (APP), encodes a ubiquitously expressed 116 kD membrane-bound protein that is rapidly turned over, with a half-life of only hours, and is essential for neurite extension [Saitoh et al., 1989; Weidermann et al., 1989]. Depending on the cell type, APP can be either internalized or secreted, where it functions in part as a protease inhibitor [Haass et al., 1992a]. Regardless of the type of processing, APP is processed to

generate aβ, which is constitutively secreted by all cell types as both a 3 and a 4 kDa form [Haass et al., 1992b]. Neurons tend to secrete a predominance of 4 kD aβ, while peripheral cells secrete predominantly 3 kD aβ. The discovery that aβ is constitutively secreted was surprising, because most researchers assumed that neuritic plaques must accumulate due to new synthesis of aβ.

As with ALS, the discovery of a small number of families with mutations in a critical gene, APP, propelled our mechanistic understanding of the disease forward. In about 5% of the cases of AD, the illness is transmitted as an aggressive, early onset, autosomal dominant form of the disorder usually striking individuals when they are in their 50s. A few of these patients have mutations in the APP gene, which is located on chromosome 21.q22, that are associated with a progressive early onset form of AD, termed *familial AD* [Hardy, 1992]. Three types of mutations occur. Two mutations are at positions flanking the aβ domain, one being a Val to Ile change at residue 717 and the other being a Glu to Gln at residue 693 of APP_{751} [Levy et al., 1990; Chartier-Harlin et al., 1991; Goate et al., 1991]. The Glu to Gln mutation actually causes a disease that is distinct from Alzheimer's disease, hereditary cerebral hemorrhage, which probably relates to an important role that APP plays in the regulation of blood clotting [Oltersdorf et al., 1989; Van Nostrand et al., 1989]. A third mutation, found in a Swedish family, is actually a double mutation substituting a Lys-Met at position 695–6 to Asn-Leu [Mullan et al., 1992]. The Swedish mutation results in a six- to eightfold increase in the secretion of the 4 kD form of aβ from cells [Citron et al., 1992]. The mutation at 717, on the other hand, acts in a more subtle manner by increasing the length of secreted aβ from 40 to 42 residues, which increases the tendency of aβ to aggregate into a toxic, polymeric form of the molecule [Suzuki et al., 1994]. The mutations at 717 and 693, which were identified first, were important for showing that changes in the biology of APP could cause AD. The Swedish mutation, though, was particularly important because it showed that increased production of aβ was sufficient to cause AD. Thus, work in molecular genetics, combined with the elegant cell biology that followed, conclusively identified APP and aβ as central elements in AD.

Evidence implicating aβ in the pathophysiology of AD also comes from studies of the actions of aβ in experimental systems. Yankner, Neve, and colleagues began studying the action of APP on cells grown in culture. They found that a carboxy-terminal fragment of APP, which contained the aβ domain, was toxic to PC12 cells after the cells were differentiated by addition of nerve growth factor [Yankner et al., 1989]. Yankner et al. [1990] later extended these studies by showing that the aβ peptide itself was the toxic agent. Interestingly, their system using primary cultures of hippocampal neurons, the toxicity of aβ was apparent only after the neurons had matured for 3–7 days [Yankner et al., 1990]. Prior to this, aβ was actually trophic and increased neurite extension from the neurons. In fact, a peptide consisting of $a\beta_{1-28}$ is only trophic; the toxicity resides in 10 residues located at $a\beta_{25-35}$ [Whitson et al., 1989; Yankner et al., 1990]. Another compelling observation about the action of aβ was that not only did it cause neuronal death, but the dying neurons express markers for neurofibrillary pathology in the process [Yankner et al., 1990]. Application of subnanomolar doses of aβ elicit reactivity over 24 hours of incubation with a sensitive marker of neurofibrillary tangle formation, *termed Alz-50* [Wolozin et al., 1986]. The discovery that aβ elicits Alz-50 reactivity as part of the sequence of events leading to cell death advanced our understanding of the pathophysiology of AD greatly by suggesting that the neurofibrillary tangles that accumulate in the AD brain are a result of the action of aβ. This work provides added support for the hypothesis that aβ toxicity is central to the pathophysiology of AD. Neurofibrillary tangles are intracellular structures that are composed of bundles of highly phosphorylated forms of microtubule-associated Tau protein. Tau protein regulates

the assembly of tubulin, its binding to intermediate filaments such as neurofilament and the outgrowth of neurites from developing neurons [Cleveland et al., 1977; Caceres and Kosik, 1990; Chen et al., 1992]. The activity of Tau is regulated by phosphorylation, and Tau can be phosphorylated at up to 14 different sites by a plethora of kinases, including protein kinase C, casein kinase, protein kinase A, tau kinase, calcium/calmodulin-dependent protein kinase, p^{cdc34}, and MAP kinase [Baumann et al., 1993]. However, the phosphorylation events critical for the conversion of tau to paired helical filaments are located at specific sites on Tau. Careful work has shown that of the kinases listed above, only Map kinase, p^{cdc34}, and casein kinase are capable of converting Tau to paired helical filaments [Baumann et al., 1993]. Selective antibodies, such as the antibody Alz-50 and PHF1, recognize phosphorylation of these particular sites and thus are early markers of neurofibrillary tangle formation both in cell culture and in postmortem brain tissue [Wolozin et al., 1986]. Following phosphorylation of Tau at the Alz-50 epitope other events, such as ubiquitinylation, may also occur. The full sequence of biochemical events that lead to neurofibrillary tangle formation have yet to be worked out. Nonetheless, antibodies such as Alz-50 are useful reagents for identifying early steps in tangle formation.

The pathway by which aβ kills neurons is beginning to be understood (Fig. 2). The aβ peptide has a large number of potential sites of action, because it is a hydrophobic peptide with a great avidity for membranes and therefore interacts with many proteins. One possible mechanism of aβ toxicity is via calcium. In the micromolar range, aβ has many affects on calcium metabolism. *In vitro*, micromolar levels of aβ polymerize to form membrane channels that allow the influx of calcium [Arispe et al., 1993a,b]. This polymerization is increased by oxidation of the aβ peptide, which is consistent with cell culture studies showing that oxidized aβ is more toxic than monomeric aβ [Pike et al., 1991; Dyrks et al., 1992]. Studies

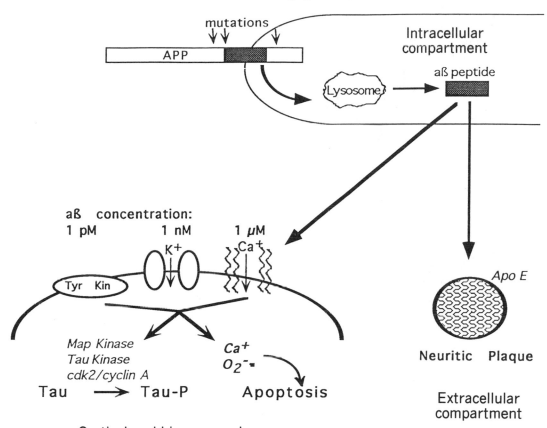

Fig. 2. *Mechanism of action of aβ. The aβ peptide is derived from the parent molecule, amyloid precursor protein. Following secretion, aβ can oxidize, which results in precipitation and aggregation to form neuritic plaques. Apolipoprotein E may enhance the formation of neuritic plaques. aβ also has many concentration-dependent cellular actions. It can aggregate to form calcium channels, it can also block potassium channels, and it can activate tyrosine phosphorylation. Activation of these processes then leads to phosphorylation of Tau protein (which reduces microtubule formation) increases cytosolic calcium levels and production of free radicals. Calcium and free radicals can both stimulate apoptosis.*

using hippocampal neurons grown in culture lend support for a calcium channel hypothesis by showing that increased levels of cytosolic calcium can be detected in the neurons 72 hours after addition of aβ [Mattson et al., 1992]. As mentioned above, Tau phosphorylation is sensitive to calcium levels, and, in fact, Alz-50 reactivity is induced by application of calcium ionophores, such as A23187, to cultures of hippocampal neurons [Mattson, 1990]. Increasing calcium could also activate apoptosis, and aβ application has been shown to elicit apoptosis in cultures of hippocampal neurons [Loo

et al., 1993]. Thus, aβ-induced elevation of intracellular calcium levels provides a plausible mechanism to account for neurofibrillary tangle formation and neuronal death. Interestingly, the lipid composition of the membranes greatly influences the rate of aβ channel formation, so it is possible that the changes in membrane composition of the AD brain that have been documented could exacerbate the influence of aβ on calcium.

There are some problems with the aβ/calcium hypothesis, though. The requirement of micromolar levels of aβ in order to activate calcium influxes conflicts with estimates of aβ levels in the brain and body fluids that put aβ levels in the picomolar range rather than the micromolar range [Seubert et al., 1992]. Thus, the concentration of aβ in our bodies appears to be too low to activate calcium influxes. The calcium influxes elicited by aβ could result from the artificially high levels of aβ achievable with *in vitro* systems. Being a hydrophobic molecule, aβ could easily cause many membrane changes using the artificially high concentrations achievable in the laboratory, but this does not mean that this is the process that is actually occurring in the body. The importance of apoptosis to aβ action and AD pathology is also unclear. Although cultured hippocampal neurons show apoptosis in response to aβ, PC12 cells, which are also very sensitive to aβ, do not show apoptosis; they die by necrosis without the rapid DNA fragmentation or membrane blebbing characteristic of apoptosis. Also, evidence of apoptosis in the AD brain is lacking, although there are many technical problems making analysis of apoptosis in post mortem brains difficult. Apoptotic neurons might be phagocytosed, which would preclude detection. Conversely, postmortem decay might lead to artifactual DNA fragmentation and false positives. Nonetheless, given the massive amount of cell death in the AD brain, it seems likely that evidence of apoptosis should be present. Thus, while the aβ/calcium hypothesis proposes a clear, logical pathway for AD pathology, significant lines of evidence argue against this hypothesis.

Recent work in our laboratory suggests a mechanism whereby aβ may affect calcium levels at physiological concentrations. We have found evidence that aβ activates tyrosine phosphorylation at picomolar levels of aβ [Luo et al., 1994]. The phosphorylation is rapid, detectable within 1 minute of addition of aβ, and is maintained for up to 24 hours (Fig. 3). This phosphorylation is not seen with a peptide of similar composition but reversed sequence. The site of action of aβ may be a growth factor receptor, because many growth factor receptors are tyrosine kinases. This activation of

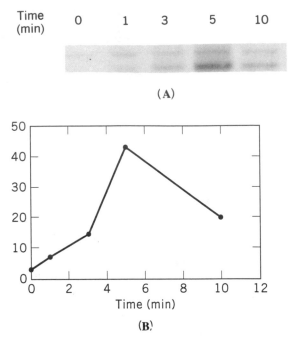

Fig. 3. *aβ increases tyrosine phosphorylation. (A): Application of 100 pM aβ₁₋₄₀ increases the level of tyrosine phosphorylation within 1 minute in PC12 cells. PC12 cells were incubated in serum-free medium overnight, and 100 pM aβ₁₋₄₀ was added for the amount of time shown. The cells were then harvested and immunoblotted with the antiphosphotyrosine antibody PY20 (Transduction Labs, Kentucky). (B): The immunoreactivity increases 10-fold within 5 minutes, as shown by videodensitometry.*

tyrosine phosphorylation may be an important key toward elucidating the mechanism of action of aβ. As mentioned, application of aβ to cells causes little immediate change in calcium metabolism of neurons under basal conditions [Mattson et al., 1992]. However, under conditions where tyrosine phosphorylation is potentiated, such as during inhibition of tyrosine phosphatases with sodium orthovanadate, 100 pM aβ can in fact stimulate an immediate influx of calcium. In this context, it is interesting to note that many tyrosine kinase–linked receptors can also activate Map kinase and pCDC34, which are the kinases involved in the conversion of Tau to paired helical filaments. Thus, at low levels aβ may be acting via a tyrosine kinase-linked receptor to stimulate calcium influx and activate kinases that phosphorylate Tau protein.

Two alternative pathways for aβ toxicity have been proposed. Recently, aβ has been shown to block K$^+$ channels at nanomolar levels of aβ [Etcheberrigaray et al., 1994]. Because K$^+$ channels are important for memory processes, such a blockade could contribute to the memory dysfunction in AD. The relationship between K$^+$ channel blockade and AD pathology, though, is unclear. An alternative hypothesis that is quite intriguing points at a hydrogen peroxide–mediated mechanism that presumably involves a free radical mechanism similar to that seen in ALS [Behl et al., 1994]. A sensitive indicator of aβ toxicity turns out to be an assay for cellular redox activity based on enzymatic reduction of 3-[4,5-dimethylthiazol-2-yl]-2,5-diphenyltetrazolium bromide (MTT), termed the MTT reduction assay [Mosmann, 1983; Shearman et al., 1994]. This assay measures the transfer of hydride ion from NADH or NADPH, which measures the activity of the mitochondrial reduction chain, and also turns out to be a sensitive indicator of aβ toxicity. The MTT assay shows decreases in activity parallel to, but preceding, evidence of cell death by more classic measures of cytotoxicity such as lactate dehydrogenase release or trypan blue exclusion assays.

Sensitivity of this redox assay to aβ suggests that free radicals are involved in the process of aβ toxicity. In support of this proposal are data showing that free radical scavengers, such as tocopherol, reduce aβ toxicity in PC12 cells [Behl et al., 1992]. Nitric oxide, an important free radical generated in the brain, has been excluded as the mediator of aβ toxicity in the PC12 system because nitric oxide does not potentiate aβ toxicity, and the nitric oxide synthase inhibitor NNL-arginine does not protect cells against aβ toxicity [Behl et al., 1994]. On the other hand, hydrogen peroxide, which is linked to the superoxide and hydroxyl radicals, does appear to mediate aβ toxicity in PC12 cells. Hydrogen peroxide production is detected in PC12 cells after addition of aβ and inhibitors of hydrogen peroxide production block aβ toxicity [Behl et al., 1994]. Hydrogen peroxide can be generated by a number of enzymes, including the enzyme complex NADPH oxidase. The NADPH oxidase inhibitors diphenyl iodonium and neopterin both block the toxicity of aβ in PC12 cells and primary cultures of hippocampal neurons [Behl et al., 1994]. Thus, aβ appears to induce hydrogen peroxide formation, and, since hydrogen peroxide is a precursor to the superoxide and hydroxyl free radicals, aβ therefore appears to induce free radical formation. Free radicals could also impact on neurofibrillary tangle formation by oxidizing Tau and increasing the tendency of phosphorylated Tau to aggregate and/or precipitate. The involvement of free radicals in the pathophysiology of AD is also appealing because it puts the pathophysiology of this illness into a similar mechanistic framework as PD and ALS.

Thus, a picture is beginning to emerge as to a general mechanism for AD (Fig. 2). As with PD and ALS, both calcium and free radicals are implicated in the pathogenesis of the illness. Some patients produce more aβ and develop familial AD. Other patients do not produce more aβ, but may have an increased tendency to form polymerized aβ. For instance, patients with the apolipoprotein E4, which is associated with an elevated risk of AD and increased plaque size, may have an increased tendency to develop polymerized aβ [Corder et al., 1993]. The increased plaque size may result from increased association between apolipoprotein E4 and aβ or, alternatively, it may increase the secretion of aβ resulting in a greater accumulation of aβ [Strittmatter et al., 1993]. As aβ accumulates due to oxidation and precipitation the levels of aβ in the neuropil get high enough to activate receptors, which results in either calcium influxes or the production of hydrogen peroxide and the hydroxyl free radicals. This then induces neurofibrillary tangle formation and causes lipid peroxidation, protein oxidation, DNA damage, and, ultimately, cell death. Such a mechanism also points to many potential avenues for therapy

of AD. Agents that interfere with the generation of aβ, the polymerization of aβ, the binding of aβ to its membrane receptor or the
post-membrane actions of aβ could all be thera-
peutically beneficial in AD.

V. NEURODEGENERATIVE DISEASES: DIFFERENT ENTITIES, SIMILAR PATHWAYS

Although the causal mechanisms from disease to disease differ, many of the resulting
processes are similar. Many of the neurodegenerative disorders show features of pathology in common [Trojanowski et al., 1993]. For
instance, PD, ALS, and AD all show varying
forms of neurofibrillary pathology. PD patients
develop Lewy bodies, AD patients develop
neurofibrillary tangles, and ALS patients develop variants of Lewy bodies as well as structures similar to neurofibrillary tangles, termed
spheroids and *hyaline bodies* [Hirano, 1991;
Leigh and Swash, 1991; Trojanowski et al.,
1993]. In addition, they also have other features of pathology in common, such as
granulovacuolar degeneration [Goldman and
Yen, 1986]. PD, ALS, and AD are among the
most common illnesses among a larger family
of neurodegenerative disorders that also show
neurofibrillary pathology, such as supranuclear
palsy, Pick's disease, diffuse Lewy body disease, and Gerstmann-Straussler syndrome
[Goldman and Yen, 1986; Prusiner, 1991;
Trojanowski et al., 1993]. Because of the similarities in the pathologies, it is reasonable to
speculate that these illnesses also share similarities in their mechanism of neurodegeneration. In
fact, the discussion of the three illnesses presented in detail above shows how these illnesses do share common features in their
mechanism of neurodegeneration. In each of
these three illnesses, damage from free radicals appears to play an important role causing
cumulative oxidation that interferes with cellular function and ultimately kills the neurons.
In addition, calcium also appears to play an
important role, perhaps stimulating harmful
proteases and DNases.

Many investigators have speculated that neurons are particularly prone to neurodegeneration
because of their requirement for many mitochondria to support their large energy needs.
As we can see with the MPTP model, processes
that interfere with mitochondrial function, for
instance, by inhibiting the complexes involved
in oxidative phosphorylation, increase free
radical formation. In addition, because mitochondria have large supplies of calcium, loss
of mitochondrial function also causes release
of calcium into the cytoplasm, which can
stimulate proteases as well as the DNases involved in apoptosis. Exclusive of the accumulation of any toxin, such as MPTP, the process
of aging is known to reduce mitochondrial
function, and such changes have been shown
to increase the susceptibility of animals to
MPTP-related toxicity [Sershen et al., 1985].
The increased susceptibility could be due to
the affects of aging on cytochrome I, which
is the target of MPP⁺. The activity of cytochrome I decreases up to 22% over normal
aging [Bowling et al., 1993]. The result of this
is that mitochondria grow increasingly weak
with age and become increasingly vulnerable
to other stressors.

With the neurodegenerative diseases, we
must then add onto this background of age-
related loss of mitochondrial function potential damage from the action of the harmful
process that is characteristic of each illness,
such as the accumulation of MPTP or aβ. The
combination of these processes results in an
increased burden on the neuron. Damaged mitochondria release calcium and free radicals.
In some cases the damage is exacerbated by
toxins like MPTP, which results in increased
release of calcium and increased production of
free radicals. In other cases, as with aβ, age-
related damage to mitochondria may contribute calcium and free radicals to a cellular
cytoplasm that already has increased levels of
these species due to the action of aβ. Together,
aβ and mitochondrial damage may produce an
additive effect that increases calcium levels or
free radical production to a point where they
are harmful. Thus, while the differing pre-

sentations of the various neurodegenerative illnesses probably result from differences in etiologies, there also appears to be common pathways through which the neurodegeneration proceeds.

VI. PROGRAMMED CELL DEATH: THE PATHWAY FOR CELL DEATH?

Our understanding of the pathways by which calcium in particular and free radicals lead to cell death has been greatly facilitated by the discovery of programmed cell death, which occurs during development. Neuronal loss is an integral part in the normal development of a functional integrated nervous system, and 50%, or more, of all neurons die before adulthood [Oppenheim, 1991]. Competition for neurotrophic factors produced by target cells and input from afferent neurons regulates neuron number [Oppenheim, 1991; Raff, 1993]. In simplistic terms each surviving neuron requires appropriate input from afferent neurons, correct target matching to efferent neurons, and the necessary supplementary factors from other surrounding cells to survive. Thus, during development an initial excess of neurons is produced, and competition within the neuronal population leads to survival of only those neurons that are functionally, temporally, and spatially correct [Cowan et al., 1984].

The neurons that do not survive the competition die as a result of an intrinsic cell suicide program, termed *apoptosis*. The process of apoptosis involves the induction of specific proteins, such as p53, and activation of degradative enzymes. What follows is a characteristic pattern that involves rapid nuclear collapse and DNA destruction, as is discussed in previous chapters of this book. The mechanism for apoptosis appears to proceed via entry into an abortive cell cycle. Thus, the same machinery that contributes to the generation of cells through cell division also can contribute to the death of cells through apoptosis when cells receive conflicting signals relating to cell division. Apoptosis occurs in adult life as well as during development. For instance, apopto-

sis is a common response in immune cells. In the majority of organs and in the immune system the controlled loss of cells due to the induction of apoptosis is not detrimental, as the lost cell numbers can be replenished through new cell production. However, neurons cannot replenish their numbers through cell division. Loss of neurons via apoptosis during development is beneficial, but apoptosis that occurs in the mature brain, as may be occurring in neurodegenerative illnesses, is harmful.

Interestingly, the signals that initiate apoptosis appear to be similar to signals that occur in neurodegenerative disorders, as well as in other situations, such as chemotherapy. In many cell types, cell suicide is an apparent controlled physiological response to insults that include generation of free radicals, x-ray irradiation, calcium influx, viral infection and direct DNA damage [Wyllie and Kerr, 1980; Meyaard, et al., 1992]. Calcium and free radicals, in particular, are able to induce proteins that are involved in apoptotic pathways, such as p53. Other proteins, such as Bcl-2, inhibit apoptosis. Thus, cells normally exist perched in an equilibrium between forces that induce apoptosis and forces that prevent apoptosis.

The role of oxidants as inducers of apoptosis has been highlighted by the discovery of the oncogene bcl-2, which may act in an antioxidant pathway to protect cells against free radical damage [Hockenberry et al., 1993]. Bcl-2 is a membrane-bound protein concentrated at intracellular locations in the cell that are all sites of free radical production, namely, the mitochondrial, nuclear, and endoplasmic reticular membranes [Krajewski et al., 1993]. Bcl-2 can protect against apoptosis caused by a wide variety of inducers, including hydrogen peroxide and other agents that cause free radical damage, such as radiation. Interestingly, bcl-2 has been demonstrated to prevent necrotic cell death in cells depleted in glutathione [Kane et al., 1993]. Transgenic mice lacking functional Bcl-2 protein develop normally, but as they age they exhibit characteristics that mimic free radical damage, including kidney and immune system dysfunction and hypopigmentation of hair [Veis et al., 1993]. Bcl-2 prevents apoptosis caused by withdrawal of interleukin-3 from an IL-3–dependent hematopoietic cell line. It has recently been demonstrated that the repartitioning of calcium that occurs on IL-3 withdrawal is blocked by bcl-2 [Baffy et al., 1993]. On IL-3 withdrawal, cytosolic levels of free calcium and nonmitochondrial calcium concentrations fall, whereas mitochondrial calcium levels are increased. Bcl-2 was shown to block the redistribution of calcium and prevent apoptosis. It has also been shown that bcl-2 can block the induction of apoptosis by a specific inhibitor of the endoplasmic reticulum–associated calcium pump [Lam et al., 1994].

Although much of the work relating to Bcl-2 function has concentrated on cells of the immune system, there is accumulating evidence that this protein can also function in neurons to prevent apoptosis [Garcia et al., 1992; Allsopp et al., 1993; Zhong et al., 1993]. We have studied the induction of apoptosis in precursor cerebellar granule cells *in vivo* by ionizing radiation [Wood et al., 1993]. Radiolytic attack on water molecules leading to free radical formation is postulated to be the initiating damage leading to apoptosis in response to radiation. We have shown that radiation-induced apoptosis and, by inference, free radical-induced apoptosis, are dependent upon expression of functional p53 protein. However, the physiological developmental apoptosis occurs normally. This suggests that apoptosis can be induced by more than one pathway.

Because the signals for apoptosis are the same molecules that are produced during oxidative stress, it is increasingly evident that oxidative stress is a common inducer of apoptosis. Many of the agents that can induce apoptosis are oxidants or stimulate the production of free radicals through cellular metabolism. DNA damage itself may also be a signal for apoptosis. Conversely, inhibitors of apoptosis such as Bcl-2 exhibit antioxidant properties. Neurons may be particularly susceptible to inappropriate activation of apoptosis because their metabolic rates are high and free radicals

are produced as a normal part of cell metabolism. As the processes of aging or disease impair mitochondrial activity, modify DNA and induce protein modifications, the equilibrium between survival and apoptosis shifts toward apoptosis. For instance, it has been estimated that many thousands of DNA bases are modified each day due to free radical action and it seems probable that some of this damage escapes repair [Richter et al., 1988]. Such changes are exacerbated by decreasing efficiency of the repair mechanisms that occurs with increasing age. Damage to the cell with age combined with less than 100% efficiency of repair of damage may ultimately lead to an imbalance in the cell, thereby triggering the apoptotic mechanisms that have been in place but inactive since neuron genesis.

Neurodegenerative diseases may mimic an accelerated aging process by increasing the levels of free radicals or disruption of calcium homeostasis. For instance, the dysfunctional copper/zinc superoxide dismutase gene found in familial ALS probably increases free radical levels, which results in increased DNA damage, activation of apoptosis, and neurodegeneration. Similarly, MPTP disrupts mitochondrial functioning, which results in the generation of free radicals and extrusion of calcium into the cytoplasm; this also may activate apoptosis. β-Amyloid appears to stimulate calcium release directly and hydrogen peroxide/superoxide production, which, again, could lead to apoptosis. Production of free radicals and influx of calcium also occur in stroke; presumably, other neurodegenerative processes follow similar pathways. Thus, each of the neurodegenerative disorders discussed in this chapter involves processes that are important to the triggering of apoptosis. Perhaps, in the future, we may see inhibitors of apoptosis become useful therapies for AD, PD, and ALS.

VII. NEURODEGENERATIVE DISORDERS AS MODELS FOR NORMAL AGING

A final point worth addressing is what the study of neurodegenerative disorders tells us about normal aging. AD, PD, and other neuro-degenerative diseases may, in a sense, be models for normal aging. Each disorder may represent an example of a system where, out of the many changes that occur with aging, one particular age-related pathway is accelerated. For instance, let us assume that one of the important features in AD is that a small fraction of aβ gets oxidized, precipitates, and, over time, accumulates in the brain. This process probably occurs in everyone, but in Alzheimer patients the rate of accumulation appears to be increased. However, with time, plaques would be expected to accumulate in most people, whether or not they have AD. Bradley Hyman has suggested that anyone who lives long enough may develop the neuritic and neurofibrillary changes seen with AD, and, in fact, neuritic plaques and neurofibrillary tangles are commonly seen in the brains of neurologically normal individuals, but at a lower abundance than in the Alzheimer brain [Trojanowski et al., 1993]. Thus, while neuritic and neurofibrillary pathology normally accumulate with age, in AD we can see the effects of the accelerated accumulation of aβ and associated neuritic and neurofibrillary pathology.

Similar analogies can be made for free radical damage, as shown by ALS, and mitochondrial damage, as shown by MPTP-induced PD. In ALS, deficits in free radical defenses lead to increased levels of free radicals and increased oxidative damage to target cells, which in this case are the motor neurons of the anterior horns of the spinal tract. In the MPTP model of PD we see the effect of mitochondrial impairment, which increases the production of free radicals and may increase cytosolic calcium levels. Due to selective accumulation of MPP^+ in dopaminergic neurons, these cells are particularly vulnerable to the affects of MPP^+. With age, all of these processes may occur and contribute to the process of aging, but in the neurodegenerative disorders we see the effects when one particular process takes precedence. Thus, the neurodegenerative disorders can help us to identify processes that occur with aging that may be particularly harmful to neurons.

As treatments are developed for these disorders, these treatments might impact on normal aging or other age-related illnesses, in addition to being therapeutic for a particular illness such as PD or AD. One interesting example of this phenomenon was found in studies on L-deprenyl, which, although controversial, are intriguing. L-deprenyl has been found to be a useful treatment for PD; however, its value may well extend beyond PD. Knoll and colleagues suggest that L-deprenyl actually increases longevity in all people. Knoll [1993] has found that rats treated with L-deprenyl live longer than untreated rats. Unfortunately, other laboratories have had mixed results. Ingram and colleagues [1993], for instance, did not find that L-deprenyl increased longevity in their studies. A second example is the use of free radical scavengers. As mentioned, multiple studies implicate free radicals in neurodegenerative disorders. Epidemiological studies suggest that diets with increased amounts of free radical scavengers reduce the occurrence of cataracts, cardiovascular ailments, and many cancers, all of which can be thought of as manifestations of age-related damage [Ames et al., 1993]. Decreasing free radicals may also impact on fundamental rates of aging as well. Studies of transgenic *Drosophila* show that increases in catalase and SOD, two important antioxidant defense enzymes, can increase the longevity of the flies up to 30% [Orr and Sohal, 1994]. Thus, treatments useful in one age-related illness may impact on a number of age-related illnesses and perhaps on the process of aging itself, because many of these illnesses derive from the accelerated forms of normal aging.

And so, the study of neurodegenerative diseases tells us much about aging itself. However, some realism is also worthwhile, and so let us end with a brief anecdote. One of the authors (B.W.) was traveling with his uncle, a physicist named Edwin. Edwin works extensively with computers, and, like computers, he is a binary sort of fellow—zero or one, he is either off or on. He never goes through a relaxed, chatty phase, he either says nothing or he directly cuts to the essence of the matter and says something very poignant. While the two were traveling together alone in Edwin's truck, they got onto the subject of eating habits, or, more to the point, B.W. got into a long monologue about his eating habits because Edwin was in "off" mode—he was silent and not responding. For fifteen minutes B.W. explained in multiple ways how he tried to eat many fruits and vegetables and little meat and fat in order to maximize his health. At first it was enjoyable to explain his well thought out (if not somewhat neurotic) eating habits, but as Edwin's silence and B.W.'s monologue dragged on the monologue became increasingly uncomfortable and B.W. was running out of things to say. Thankfully, they began to approach town. Seeing that the end of the car ride was coming, Edwin switched into "on" mode. Edwin looked over at B.W. and said one thing and only one thing. "Ben," he said "you are going to die."

The pathway leading to cell death may not be as mysterious from a mechanistic standpoint as it is from a global, philosophic standpoint. There appear to be specific cascades of events that actually stimulate the cell to self-destruct. In the neurodegenerative disorders, specific defects accelerate these processes in vulnerable cells. As our understanding of the mechanism of these processes increases, our ability to intervene also should increase. The cells in our bodies are perched in a homeostatic balance between apoptosis, basal cellular functioning, and mitosis. In cancer, the altered signals lead to mitosis. Therapy for cancer aims to inhibit mitosis (which, ironically, can lead to apoptosis), but the regenerating nature of cancer cells makes it difficult to eradicate the illness. In neurodegenerative disorders, altered signals may lead to the other end of the homeostatic spectrum, apoptosis. Like mitosis, apoptosis has a definable pathway of events. Intervention may be able to inhibit this process, and, because neurons are nondividing, the cells are unlikely to mutate and escape from therapy, as occurs with cancer. Thus, our increased understanding of the mechanism of cell death in neurodegenerative disorders, although occurring later than our understanding of cancer, may ultimately lead to therapies that are more efficacious than with cancer.

REFERENCES

Allsopp TE, Wyatt S, et al. (1993): The proto-oncogene bcl-2 can selectively rescue neurotrophic factor-dependent neurons from apoptosis. Cell 73:295–307.

Ames B, Shigenaga M, et al. (1993): Oxidants, antioxidants, and the degenerative diseases of aging. Proc Natl Acad Sci USA 90:7915–7922.

Arispe N, Pollard H, et al. (1993a): Giant multilevel cation channels formed by Alzheimer disease β-protein [AβP-(1-40)] in bilayer membranes. Proc Natl Acad Sci USA 90:10573–10577.

Arispe N, Rojas E, et al. (1993b): Alzheimer disease amyloid β protein forms calcium channels in bilayer membranes: Blockade by tromethamine and aluminum. Proc Natl Acad Sci. USA 90:567–571.

Baffy G, Miyashita T, et al. (1993): Apoptosis induced by withdrawal of interleukin-3 (IL-3) from an IL-3-dependent hematopoietic cell line is associated with repartitioning of intracellular calcium and is blocked by enforced bcl-2 oncoprotein production. J Biol Chem 268:6511–6519.

Baumann K, Mandelkow E, et al. (1993): Abnormal Alzheimer-like phosphorylation of tau-protein by cyclin-dependent kinases cdk2 and cdk5. FEBS Lett 336:417–424.

Beal M (1992): Does impairment of energy metabolism result in excitotoxic neuronal death in neurodegenerative illnesses? Ann Neurol 31:119–130.

Beckman J, Beckman T, et al. (1990): Apparent hydroxyl radical production by peroxynitrite: Implications for endothelial injury from nitric oxide and superoxide. Proc Natl Acad Sci USA 87:1620–1624.

Behl C, Davis J, et al. (1992): Vitamin E protects nerve cells from amyloid β protein toxicity. Biochem Biophys Res Commun 186:944–950.

Behl C, Davis J, et al. (1994): Hydrogen peroxide mediates amyloid β protein toxicity. Cell 77:1–20.

Bowling A, Mutisya E, et al. (1993): Age-dependent impairment of mitochondrial function in primate brain. J Neurochem 60:1964–1967.

Caceres A, Kosik K (1990): Inhibition of neurite polarity by tau antisense oligonucleotides in primary cerebellar neurons. Nature 343:461–463.

Chartier-Harlin M, Crawford F, et al. (1991): Early-onset Alzheimer's disease caused by mutations at codon 717 of the β–amyloid precursor protein gene. Nature 353:844–846.

Chen J, Cowan N, et al. (1992): Projection domains of MAP2 and tau determine spacings between microtubules in dendrites and axons. Nature 360:674–676.

Chiueh C, Miyake H, et al. (1993): Role of dopamine autoxidation, hydroxyl radical generation, and calcium overload in underlying mechanisms involved in MPTP-induced parkinsonism. Adv Neurol 60:251–258.

Citron M, Oltersdorf T, et al. (1992): Mutation of the β-amyloid precursor protein in familial Alzheimer's disease increases β-protein production. Nature 360:672–674.

Cleeter M, Cooper J, et al. (1992): Irreversible inhibition of mitochondrial complex I by 1-methyl-4-phenylpyridinium (MPP$^+$): Evidence for free radical involvement. J Neurochem 58:786–789.

Cleveland DW, Hwo SY, et al. (1977): Purification of tau, a microtubule associated protein that induces the assembly of microtubules from purified tubulin. J Mol Biol 116:207–225.

Corder E, Saunders A, et al. (1993): Gene dose of apolipoprotein E type 4 allele and the risk of Alzheimer's disease in late onset families. Science 261:921–923.

Cowan WM, Fawcett JW, et al. (1984): Regressive events in neurogenesis. Science 225:1258–1265.

Deng H, Hentati H, et al. (1993): Amyotrophic lateral sclerosis and structural defects in Cu, Zn superoxide dismutase. Science 261:1047–1051.

Dyrks T, Dyrks E, et al. (1992): Amyloidogenicity of βA4 and βA4-bearing amyloid protein precursor fragments by metal-catalyzed oxidation. J Biol Chem 267:18210–18217.

Etcheberrigaray R, Ito E, et al. (1994): Soluble β-amyloid induction of Alzheimer's phenotype for human fibroblast K$^+$ channels. Science 264:276–279.

Garcia I, Matinou I, et al. (1992): Prevention of programmed cell death of sympathetic neurons by the bcl-2 proto-oncogene. Science 258:302–304.

Glenner GG, Wong CW (1984): Initial report of the purification and characterization of a novel cerebrovascular amyloid protein. Biophys Biochem Res Commun 120:885–890.

Goate A, Chartier-Harlin MC, et al. (1991): Segregation of a missence mutation in the amyloid precursor protein gene with familial Alzheimer's disease. Nature 349:704–706.

Goldman JE, Yen SH (1986): Cytoskeletal protein abnormalities in neurodegenerative diseases. Ann Neurol 19:209–223.

Group PS (1989): Effect of deprenyl on the progression of disability in early Parkinson's disease. N Engl J Med 321:1364–1371.

Group PS (1993): Effects of tocopherol and deprenyl on the progression of disability in early Parkinson's disease. N Engl J Med 328:176–183.

Gurney M, Pu H, et al. (1994): Motor neuron degeneration in mice that express a human Cu/Zn superoxide dismutase mutation. Science 264:1772–1775.

Haass C, Koo E, et al. (1992a): Targeting of cell-surface β-amyloid precursor protein to lysosomes: Alterna-tive processing into amyloid-bearing fragments. Nature 357:500–503.

Haass C, Schlossmacher M, et al. (1992b): Amyloid β-peptide is produced by cultured cells during normal metabolism. Nature 359:322–324.

Hardy J (1992): Framing β-amyloid. Nature Gen 1:233–234.

Hasegawa E, Takeshiga K, et al. (1990): 1-Methyl-4-phenylpyridinium (MPP$^+$) induces NADH-dependent superoxide formation and enhances NADH-dependent lipid peroxidation in bovine heart submitochondrial particles. Biochem Biophys Res Commun 170:1049–1055.

Hirano A (1991): Cytopathology of amyotrophic lateral sclerosis. Adv Neurol 56:91–101.

Hockenberry DM, Oltvai ZN, et al. (1993): Bcl-2 functions in an antioxidant pathway to prevent apoptosis. Cell 1993:241–251.

Ingram D, Wiener H, et al. (1993): Chronic treatment of aged mice with L-deprenyl produces marked striatal MAO-B inhibition but no beneficial effects on survival, motor performance, or nigral lipofuscin accumulation. Neurobiol Aging 14:431–440.

Javitch J, D'Amato R, et al. (1985): Parkinsonism-inducing neurotoxin N-methyl-4-phenyl-1,2,3,6-tetrahydropyridine: Uptake of the metabolite N-methyl-4-phenylpyridine by dopamine neurons explains selective toxicity. Proc Natl Acad Sci USA 82:2173–2177.

Kane DJ, Sarafian TA, et al. (1993): Bcl-2 inhibition of neural cell death: Decreased generation of reactive oxygen species. Science 262:1274–1277.

Kang J, Lamaire HG, et al. (1987): The precursor of Alzheimer's disease amyloid A4 protein resembles a cell-surface receptor. Nature 325:733–736.

Knoll J (1988): The striatal dopamine dependency of life span in male rats. Longevity study with (–)-deprenyl. Mech Ageing 46:237–262.

Krajewski S, Tanaka S, et al. (1993): Investigation of the subcellular distribution of bcl-2 oncoprotein: Residence in the nuclear envelope, endoplasmic reticulum, and outer mitochondrial membranes. Cancer Res 53:4701–4714.

Lam M, Dubyak G, et al. (1994): Evidence that bcl-2 represses apoptosis by regulating endoplasmic reticulum-associated calcium fluxes. Proc Natl Acad Sci USA 91:6569–6573.

Langston J, Ballard P, et al. (1983): Chronic Parkinsonism in humans due to a product of meperidine-analog synthesis. Science 219:979–980.

Leigh P, Swash M (1991): Cytoskeletal pathology in motor neuron diseases. Adv Neurol 56:115–124.

Levy E, Carman M, et al. (1990): Mutation of the Alzheimer's disease amyloid gene in hereditary cerebral hemorrhage, Dutch-type. Science 248:1124–1126.

Loo D, Copani A, et al. (1993): Apoptosis is induced by β-amyloid in cultured central nervous system neurons. Proc Natl Acad Sci USA 90:7951–7955.

Luo Y, Li, et al Y. (1994): Activation of tyrosine phosphorylation by β-amyloid. J Neurochem (in press).

Mattson M, Cheng B, et al. (1992): β-Amyloid peptides destabilize calcium homeostasis and render human cortical neurons vulnerable to excitotoxicity. J Neurosci 12:376–389.

Mattson MP (1990): Antigenic changes similar to those seen in neurofibrillary tangles are elicited by glutamate and Ca^{2+} in cultured hippocampal neurons. Neuron 2:105–117.

McCrodden J, Tipton K, et al. (1990): The neurotoxicity of MPTP and the relevance to Parkinson's disease. Pharmacol Toxicol 67:8–13.

Meyaard L, Otto SA, et al. (1992): Programmed death of T cells in HIV-1 infection. Science 257:217–219.

Mizuno Y, Ohta S, et al. (1989): Deficiencies of complex I of the respiratory chain in Parkinson's disease. Biochem Biophys Res Commun 163:1450–1455.

Mizuno Y, Suzuki K, et al. (1990): Postmortem changes in mitochondrial respiratory enzymes in brain and preliminary observations in Parkinson's disease. J Neurol Sci 96:49–57.

Mosmann T (1983): Rapid colorimetric assay for cellular growth and survival: Application to proliferation and cytotoxicity assays. J Immun Methods 65:55–63.

Mullan M, Houlden H, et al. (1992): A locus for familial early-onset Alzheimer's disease on the long arm of chromosome 14, proximal to the alpha 1-antichymotrypsin. Nature Gen 2:340–342.

O'Carroll A, Tipton K, et al. (1987): Intra- and extra-synaptosomal deamination of dopamine and noradrenaline by the two forms of human brain monoamine oxidase. Implications for the neurotoxicity of N-methyl-4-phenyl-1,2,3,6-tetradydropyridine in man. Biogenic Amines 4:165–178.

Oltersdorf T, Fritz LC, et al. (1989): The secreted form of the Alzheimer's amyloid precursor protein with the Kunitz domain is protease nexin-II. Nature 341:144–147.

Oppenheim R (1991): Cell death during development. Annu Rev Neurosci 14:453–501.

Orr W, Sohal R (1994): Extension of life-span by over-expression of superoxide dismutase and catalase in *Drosophila melanogaster*. Science 263:1128–1130.

Parker W, Boyson S, et al. (1989): Abnormalities of the electron transport chain in idiopathic Parkinson's disease. Ann Neurol 26:719–733.

Patten B (1994). Excitatory Amino Acid Transmitters. London: Chapman and Hall Medical.

Patten B, Harati V, et al. (1878): Free amino acid levels in amyotrophic lateral sclerosis. Ann Neurol 3:305–309.

Pike C, Walencewicz A, et al. (1991): *In vitro* aging of β-amyloid protein causes peptide aggregation and neurotoxicity. Brain Res 563:311–314.

Plaitakis A, Constantakakis E, et al. (1988): The neuroexcitotoxic amino acids glutamate and aspartate are altered in the spinal cord and brain in amyotrophic lateral sclerosis. Ann Neurol 24:446–449.

Prusiner S (1991): Molecular biology and transgenetics of prion diseases. Crit Rev Biochem Mol Biol 26:397–438.

Przedborski S, Kostic V, et al. (1992): Transgenic mice with increased Cu/Zn-superoxide dismutase activity are resistant to N-methyl-4-phenyl-1,2,3,6-tetrahydropyridine-induced neurotoxicity. J Neurosci 12:1658–1667.

Raff M (1993): Programmed cell death and the control of cell survival: Lessons from the nervous system. Science 262:695–700.

Ramsay R, Krueger M, et al. (1991): Interaction of 1-methyl-4-phenylpyridinium (MPP^+) and its analogs with the rotenone/piericidin binding site of NADH dehydrogenase. J Neurochem 56:1184–1190.

Ramsay R, Krueger M, et al. (1991): Evidence for the inhibition of sites of neurotoxic amine 1-methyl-4-phenylpyridinium (MPP^+) with mitochondria. Biochem J 273:481–484.

Ramsay R, Singer T (1986): Energy-dependent uptake of N-methyl-4-phenyl-pyridinium, the neurotoxic metabolite of 1-methyl-4-phenyl-1,2,3,6-tetrahydropyridine, by mitochondria. J Biol Chem 261: 7585–7587.

Riachi N, Harik S, et al. (1988): On the mechanisms underlying 1-methyl-4-phenyl-1,2,3,6-tetrahydropyridine neurotoxicity. II. Susceptibility among mammalian species correlates with the toxin's metabolic patterns in brain microvessels and liver. J Pharmacol Exp Ther 244:443–448.

Richter C, Park J, et al. (1988): Normal oxidative damage to mitochondrial and nuclear DNA is extensive. Proc Natl Acad Sci USA 85:6465–6467.

Rosen D, Siddique T, et al. (1993): Mutations in Cu/Zn superoxide dismutase gene are associated with familial amyotrophic lateral sclerosis. Nature 362: 59–62.

Rothstein J, Martin L, et al. (1992): Decreased glutamate transport by the brain and spinal cord in amyotrophic lateral sclerosis. New Engl J Med 326:1464–1468.

Rothstein J, Tsai G, et al. (1990): Abnormal excitatory amino acid metabolism in amyotrophic lateral sclerosis. Ann Neurol 28:18–25.

Saitoh T, Sundsmo M, et al. (1989): Secreted form of amyloid β protein precursor is involved in the growth regulation of fibroblasts. Cell 58:615–22.

Schapira A, Cooper J, et al. (1989): Mitochondrial complex I deficiency in Parkinson's disease. J Neurochem 54:823–827.

Schapira A, Mann V, et al. (1990): Anatomic and disease specificity of NADH-CoQ$_1$ reductase (complex I) deficiency in Parkinson's disease. J Neurochem 55:2142–2145.

Selkoe D (1994): Normal and abnormal biology of the β-amyloid precursor protein. Annu Rev Neurosci 17:489–517.

Sershen H, Mason M, et al. (1985): Effect of N-methyl-4-phenyl-1,2,3,6-tetrahydropyridine (MPTP) on age-related changes in dopamine turnover and transporter function in the mouse striatum. Eur J Pharm 113:135–136.

Seubert P, Vigo-Pelfrey C, et al. (1992): Isolation and quantification of soluble Alzheimer's β-peptide from biological fluids. Nature 359:325–327.

Shearman M, Ragan C, et al. (1994): Inhibition of PC12 cell redox activity is a specific, early indicator of β-amyloid-mediated cell death. Proc Natl Acad Sci USA 91:1470–1474.

Snyder S, D'Amato R (1986): MPTP: A neurotoxin relevant to the pathophysiology of Parkinson's disease. Neurology 36:250–258.

Spencer PS, Nunn PB, et al. (1987): Guam amyotrophic lateral sclerosis–parkinsonism–dementia linked to a plant excitant neurotoxin. Science 237:517–522.

Strittmatter W, Saunders A, et al. (1993): Apolipoprotein E: High-avidity binding to β-amyloid and increased frequency of type 4 allele in late-onset familial Alzheimer disease. Proc Natl Acad Sci USA 90:1977–1981.

Suzuki N, Cheung T, et al. (1994): An increased percentage of long amyloid β protein secreted by familial amyloid β protein precursor (βAPP$_{717}$) mutants. Science 264:1336–1340.

Tanner C (1989): The role of environmental toxins in the etiology of Parkinson's disease. TINS 12:49–54.

Tetrud J, Langston J (1989): The effect of deprenyl (selegiline) on the natural history of Parkinson's disease. Science 245:519–522.

Tipton K (1986): Oxidation and enzyme-activated irreversible inhibition of rat liver monoamine oxidase by 1-methyl-4-phenyl-1,2,3,6-tetrahydropyridine. Biochem J 240:379–383.

Tipton K, McCrodden J, et al. (1986): Oxidation and enzyme-activated irreversible inhibition of rat liver monoamine oxidase-B by 1-methyl-4-phenyl-1,2,3,6-tetrahydropyridine (MPTP). Biochem J 240:379–383.

Tipton K, Singer T (1993): Advances in our understanding of the mechanisms of the neurotoxicity of MPTP and related compounds. J Neurochem 61:1191–1206.

Trojanowski J, Schmidt M, et al. (1993): Altered tau and neurofilament proteins in neurodegenerative diseases: Diagnostic implications for Alzheimer's disease and Lewy body dementias. Brain Pathol 3:45–54.

Turski L, Bressler K, et al. (1991): Protection of substantia nigra from MPP$^+$ neurotoxicity by N-methyl-D-aspartate antagonists. Nature 349:414–418.

Van Nostrand WE, Wagner SL, et al. (1989): Protease Nexin II, a potent antichymotrypsin, shows identity to amyloid β-protein precursor. Nature 341:546–549.

Veis DJ, Sorenson CM, et al. (1993): Bcl-2–deficient mice demonstrate fulminant lymphoid apoptosis, polycystic kidneys, and hypopigmented hair. Cell 1993:229–240.

Weidermann A, Konig G, et al. (1989): Identification, biogenesis and localization of the precursors of Alzheimer's disease A4 amyloid protein. Cell 571:115–126.

Whitson JS, Selkoe DJ, et al. (1989): Amyloid β protein enhances the survival of hippocampal neurons in vitro. Science 243:1488–1490.

Wolozin BL, Pruchnicki A, et al. (1986): A neuronal antigen in the brains of Alzheimer patients. Science 232:648–650.

Wood K, Dipasquale B, et al. (1993): In situ labeling of granule cells for apoptosis-associated DNA fragmentation reveals different mechanisms of cell loss in developing cerebellum. Neuron 11:621–632.

Wyllie A, Kerr J (1980): Cell death: The significance of apoptosis. Int Rev Cytol 68:251–306.

Yankner B, Dawes L, et al. (1989): Neurotoxicity of a fragment of the amyloid precursor associated with Alzheimer's disease. Science 245:417–420.

Yankner BA, Duffy LK, et al. (1990): Neurotrophic and neurotoxic effects of amyloid β protein: Reversal by tachykinin neuropeptides. Science 250:279–282.

Zhong L-T, Sarafian T, et al. (1993): Bcl-2 inhibits death of central neural cells induced by multiple agents. Proc Natl Acad Sci USA 90:4533–4537.

ABOUT THE AUTHORS

BENJAMIN WOLOZIN is a senior staff fellow in the Section of Geriatric Psychiatry at the National Institute of Mental Health. After receiving a B.A. *magna cum laude* with honors in Chemistry from Wesleyan University in Middletown, CT, he entered the Medical Scientist Training Program at the Albert Einstein College of Medicine in the Bronx, NY, where he earned an M.D., Ph.D. in 1988. His Ph.D. thesis was done in the laboratory of Peter Davies and focused on the cell biology of Alzheimer's disease. After spending a year at Mt. Sinai Medical Center studying Molecular Biology, Dr. Wolozin joined the Section of Geriatric Psychiatry at NIMH in 1989. He is currently investigating the cell biology of Alzheimer's disease using cell culture, immunochemical, and molecular methods. Dr. Wolozin was awarded the Bennet Prize in 1993 for the best paper in Biological Psychiatry by a young investigator, the Lindsley Prize in 1989 for the best thesis in Neuroscience, and the Hawk Prize in 1980 for excellence in biochemical research.

YONGQUAN LUO is a fellow in the Section of Geriatric Psychiatry at the National Institute of Mental Health. After receiving a B.S. from Shanghai Medical University in Shanghai, China, in 1987, he came to the United States to pursue a Ph.D. at the SUNY Health Science Center in Brooklyn, NY, where he received a Ph.D. in 1989 for work on the chemoattractant signal transduction pathway in the vomeronasal epithelium of garter snakes. He is currently doing a postdoctoral fellowship with Dr. Wolozin, studying signal transduction pathways of β-amyloid.

KATHERINE WOOD is a visiting fellow in the Surgical Neurology Branch, NINDS. After practicing as a registered nurse, she obtained a B.S.C. from the University of Warwick, Coventry, U.K., with joint honors in Microbiology and Virology in 1988. She continued at the University of Warwick, where she received a Ph.D. in 1991 for her study of plant toxins with Prof. J. M. Lord. Dr. Wood joined Richard Youle for postdoctoral fellowship in 1991, where she has developed *in situ* methods for detection of apoptosis.

Index